Lupins as Crop Plants

Biology, Production and Utilization

Lupins as Crop Plants
Biology, Production and Utilization

Edited by

J.S. Gladstones

Formerly of the Department of Agriculture,
Western Australia

C.A. Atkins and **J. Hamblin**

Centre for Legumes in Mediterranean Agriculture,
University of Western Australia

CAB INTERNATIONAL

CABI is a trading name of CAB International

CABI Head Office
Nosworthy Way
Wallingford
Oxfordshire OX10 8DE
UK

Tel: +44 (0)1491 832111
Fax: +44 (0)1491 833508
Email: cabi@cabi.org
Web site: www.cabi.org

CABI North American Office
875 Massachusetts Avenue
7th Floor
Cambridge, MA 02139
USA

Tel: +1 617 395 4056
Fax: +1 617 354 6875
Email: cabi-nao@cabi.org

A catalogue record for this book is available from the British Library, London, UK.

Library of Congress Cataloging-in-Publication Data
Lupins as crop plants: biology, production and utilization/edited by J.S. Gladstones, C.A. Atkins and J.Hamblin
p. cm.
Includes index.
ISBN 0-85199-224-2 (alk.paper)
1. Lupins. I. Gladstones, J.S. II. Atkins, C.A. III. Hamblin, J.
SE177.L85L86 1998
633.3'67---dc21 97-35255
CIP

ISBN-13: 978-0-85199-224-2
ISBN-10: 0-85199-224-2

First published 1998
Transferred to print on demand 2007

Printed and bound in the UK by CPI Antony Rowe, Eastbourne.

Contents

Contributors

J.G. Allen, *Agriculture Western Australia, 3 Baron-Hay Court, South Perth, Western Australia 6151, Australia*

C.A. Atkins, *Centre for Legumes in Mediterranean Agriculture (CLIMA), University of Western Australia, Nedlands, Western Australia 6907, Australia*

R.J. van Barneveld, *South Australian Research and Development Institute – Pig and Poultry Production Institute, Nutrition Research Laboratory, The University of Adelaide, Roseworthy, South Australia 5371, Australia*

R. Brennan, *Agriculture Western Australia, 444 Albany Highway, Albany, Western Australia 6330, Australia*

A.G.P. Brown, *Agriculture Western Australia, 3 Baron-Hay Court, South Perth, Western Australia 6151, Australia*

B.J. Buirchell, *Agriculture Western Australia, 3 Baron-Hay Court, South Perth, Western Australia 6151, Australia and Centre for Legumes in Mediterranean Agriculture, University of Western Australia, Nedlands, Western Australia 6907, Australia*

P.D.S. Caligari, *Department of Agricultural Botany, Plant Science Laboratories, University of Reading, Whiteknights, Reading RG6 6AS, UK*

W.A. Cowling, *Agriculture Western Australia, 3 Baron-Hay Court, South Perth, Western Australia 6151, Australia and Centre for Legumes in Mediterranean Agriculture, University of Western Australia, Nedlands, Western Australia 6907, Australia*

B. Cox, *Grain Pool of Western Australia, 172 St Georges Terrace, Perth, Western Australia 6000, Australia. Present address: SGB Aust. Pty. Ltd, Level 11, 256 Adelaide Terrace, Perth, Western Australia 6000, Australia*

M.J. Dilworth, *Centre for Legumes in Mediterranean Agriculture (CLIMA), University of Western Australia, Nedlands, Western Australia 6907, Australia and Murdoch University, Murdoch, Western Australia 6150, Australia*

M. Dracup, *Agriculture Western Australia, 3 Baron-Hay Court, South Perth, Western Australia 6151, Australia and Centre for Legumes in*

Mediterranean Agriculture (CLIMA), University of Western Australia, Nedlands, Western Australia 6907, Australia

A.C. Edwards, *ACE Livestock Consulting Pty Ltd, PO Box 108, Cockatoo Valley, South Australia 5351, Australia*

R.J.N. Emery, *Centre for Legumes in Mediterranean Agriculture (CLIMA), University of Western Australia, Nedlands, Western Australia 6907, Australia*

I.R.P. Fillery, *Centre for Legumes in Mediterranean Agriculture (CLIMA), University of Western Australia, Nedlands, Western Australia 6907, Australia and CSIRO Private Bag, Wembley, Western Australia 6019, Australia*

R.J. French, *Agriculture Western Australia, 3 Baron-Hay Court, South Perth, Western Australia 6151, Australia and Centre for Legumes in Mediterranean Agriculture (CLIMA), University of Western Australia, Nedlands, Western Australia 6907, Australia*

J.S. Gladstones, *27 Pandora Drive, City Beach, Western Australia 6015, Australia. Formerly Department of Agriculture, Western Australia, Australia*

S. Gupta, *Faculty of Agriculture, University of Western Australia, Nedlands, Western Australia 6907, Australia*

J. Hamblin, Director, *Centre for Legumes in Mediterranean Agriculture (CLIMA), University of Western Australia, Nedlands, Western Australia 6907, Australia*

J.G. Howieson, *Centre for Legumes in Mediterranean Agriculture (CLIMA), University of Western Australia, Nedlands, Western Australia 6907, Australia and Agriculture Western Australia, 3 Baron-Hay Court, South Perth, Western Australia 6151, Australia Present address: Centre for Rhizobium Studies, Murdoch University, South Street, Murdoch, Western Australia 6150, Australia*

C. Huyghe, *Station d'Amélioration des Plantes Fourragères, Institut National de la Recherche Agronomique (INRA), 86600 Lusignan, France*

R. Jarvis, *Agriculture Western Australia, 3 Baron-Hay Court, South Perth, Western Australia 6151, Australia*

M.G.K. Jones, *School of Biological and Environmental Sciences, Murdoch University, Murdoch, Western Australia 6150, Australia*

R.A.C. Jones, *Agriculture Western Australia, 3 Baron-Hay Court, South Perth, Western Australia 6151, Australia and Centre for Legumes in Mediterranean Agriculture (CLIMA), University of Western Australia, Nedlands, Western Australia 6907, Australia*

A.B. Legocki, *Institute of Bioorganic Chemistry, Polish Academy of Sciences, Noskowskiego 12/14, 61-704 Poznań, Poland*

N. Longnecker, *Centre for Legumes in Mediterranean Agriculture (CLIMA), and Soil Science, Plant Nutrition and Faculty of Agriculture, University of Western Australia, Nedlands, Western Australia 6907, Australia*

F.R. Minchin, *Institute of Grassland and Environmental Research, Plas Gogerddan, Aberystwyth, Dyfed SY23 3EB, UK*

P. Nelson, *Agriculture Western Australia, 3 Baron-Hay Court, South Perth, Western Australia 6151, Australia and Grain Pool of Western Australia, 172 St Georges Terrace, Perth, Western Australia 6000, Australia*

J. Palta, *Centre for Legumes in Mediterranean Agriculture (CLIMA), University of Western Australia, Nedlands, Western Australia 6907, Australia and CSIRO Plant Industry, Private Bag, Wembley, Western Australia 6019, Australia*

D.J. Pannell, *Centre for Legumes in Mediterranean Agriculture (CLIMA), University of Western Australia, Nedlands, Western Australia 6907, Australia*

J.S. Pate, *Centre for Legumes in Mediterranean Agriculture (CLIMA), University of Western Australia, Nedlands, Western Australia 6907, Australia*

M.W. Perry, *Centre for Legumes in Mediterranean Agriculture (CLIMA), University of Western Australia, Nedlands, Western Australia 6907, Australia and Agriculture Western Australia, 3 Baron-Hay Court, South Perth, Western Australia 6151, Australia*

D.S. Petterson, *Agriculture Western Australia, 3 Baron-Hay Court, South Perth, Western Australia 6151, Australia*

M. Reader, *Centre for Legumes in Mediterranean Agriculture (CLIMA), University of Western Australia, Nedlands, Western Australia 6907, Australia and Agriculture Western Australia, 3 Baron-Hay Court, South Perth, Western Australia 6151, Australia*

A. Robson, *Centre for Legumes in Mediterranean Agriculture (CLIMA), and Soil Science, Plant Nutrition and Faculty of Agriculture, University of Western Australia, Nedlands, Western Australia 6907, Australia*

I. Rowland, *Agriculture Western Australia, 3 Baron-Hay Court, South Perth, Western Australia 6151, Australia*

M.M. Sikorski, *Institute of Bioorganic Chemistry, Polish Academy of Sciences, Noskowskiego, 12/14, 61-704 Poznań, Poland*

P.M.C. Smith, *Centre for Legumes in Mediterranean Agriculture (CLIMA), University of Western Australia, Nedlands, Western Australia 6907, Australia*

T. Stepkowski, *Institute of Bioorganic Chemistry, Polish Academy of Sciences, Noskowskiego 12/14, 61-704 Poznań, Poland*

M.W. Sweetingham, *Agriculture Western Australia, 3 Baron-Hay Court, South Perth, Western Australia 6151, Australia and Centre for Legumes in Mediterranean Agriculture (CLIMA), University of Western Australia, Nedlands, Western Australia 6907, Australia*

W. Swiecicki, *Poznań Plant Breeders, 63-004 Tulce, Kasztanowa 5, Poland*

Caixan Tang, *Centre for Legumes in Mediterranean Agriculture (CLIMA), University of Western Australia, Nedlands, Western Australia 6907, Australia*

N.C. Turner, *Centre for Legumes in Mediterranean Agriculture (CLIMA), University of Western Australia, Nedlands, Western Australia 6907, Australia and CSIRO Plant Industry, Centre for Mediterranean Agricultural Research, Private Bag, PO Wembley, Western Australia 6014, Australia*

Preface

More than 2000 years ago Virgil (70–19 BC) noted the benefits of growing lupins in rotation with wheat: 'Or, changing the season, you will sow there yellow wheat, whence before you have taken up the joyful pulse, with rustling pods, or the vetch's slender offspring and the bitter lupin's brittle stalks and rustling grove.'

Despite this recognition of their value in rotations, it was not until the early parts of this century that the bitter lupin of Virgil was modified to turn the species from a wild or semi-domesticated species into a crop plant. This was first done in Germany for white (*Lupinus albus*) and yellow lupin (*L. luteus*) and was in response to the pressures of World War I and the need for high protein crops, adapted to temperate conditions. It was much later, during the period 1954–1967, that the narrow-leafed lupin (*L. angustifolius*) was fully domesticated by John Gladstones in Western Australia. In these three cases genes for low alkaloids (sweetness) were combined with non-shattering pods and soft seeds. This combination of characteristics meant that all three species could be planted and even germination would then occur; later in the cycle, as the crops ripened, they retained their pods and so could be harvested; and the produce was edible by stock and humans without complex processing to remove the alkaloids.

This story of the domestication of three new species from one genus in such a short period of time is probably unique in the history of agriculture. What is even more remarkable is the fact that a further four species from the genus have since been domesticated or at least have all the domestication genes available and will be fully domesticated before the end of the millennium. The species are *L. mutabilis*, *L. cosentinii*, *L. atlanticus* and *L. pilosus*. Of these seven species, six are of Old World origin and only one, *L. mutabilis*, is of New World origin.

Not only has domestication been rapid in this genus, but also the first genetically engineered cultivars generated using recombinant DNA technology are currently being grown in the field for both *L. angustifolius* and *L. luteus*, and it is likely that we will be able to transform *L. albus* in the near future. Whether this will occur for the newer, domesticated species will depend on their potential in the market place and their ability to attract scarce resources for this activity.

The success of recent domestication efforts in this genus has been due to an understanding of the factors required for domestication, an ability to search for or create new variation and the development of rapid screening techniques for alkaloids.

Once all the domestication genes were accumulated in a species there was considerable local interest in growing them. For *L. albus* and *L. luteus* the region of major interest initially was in Germany, Poland and the former USSR. The key uses were for animal feed and green manure. With the advent of high yielding cereals in that region of North Eastern Europe, the appeal of lupins has fallen, particularly as it is difficult to produce high quality seed due to the onset of autumn, with cool temperatures and high humidities, at harvest time.

The major interest in *L. angustifolius* was in Western Australia where John Gladstones developed the first true crop types of this species. In the 30 years since the release of the first variety (Uniwhite) in 1967, production has grown from zero to more than 1.4 million tonnes in 1995. This species is the only one to enter international trade on any scale and exports from Australia have exceeded 800,000 tonnes per annum.

In tandem with the domestication of this genus, there has been a major effort to improve the crop agronomy and physiology, assess the role of lupins in rotations, manage the pest and disease threat, examine the use of lupins in animal feed rations and develop markets.

This book brings together the current state of knowledge on this genus for agriculture. As the area of lupins has increased, so has the agronomic and physiological knowledge increased, as well as the range of pests and diseases that affect production. The outlook of the book is international, but because of the success of lupins in Western Australia much of the information is based on that experience. At the same time the range of environments in which lupins can be grown successfully has increased. It is likely that the area of production will increase further in Australia for several reasons. There is an expanding suite of improved varieties being produced, as well as more species becoming available adapted to a wider range of soil types. At the same time the yield potential in farmers' fields has been increasing by 2% per annum. There is a growing knowledge of the crop and its role in animal feeding and in human food. This book is an up-to-date statement of the development of the genus and its impact in agricultural production systems. The potential of this genus to contribute to improved profitability of farming on very poor sandy and acidic soils has had and will continue to have a major impact on the economic viability of farming in Australia and beyond.

Besides lupin specialists, this book will appeal to anyone interested in crop domestication and development and it provides an excellent case study for students on how an intensive research effort by many different specialists can develop a new industry. It is also a monument to the pioneers of lupin research who had the vision to persist with a crop that has been on the fringes of agriculture for millennia, but only in the last 70 years has gone from being a species of minor interest to being an important component of farming systems. Their story is documented here by many of the people who were actively involved in making it a success.

John Hamblin
Director
Centre for Legumes
in Mediterranean Agriculture
University of Western Australia
Nedlands
Western Australia

1 Distribution, Origin, Taxonomy, History and Importance

J.S. Gladstones

27 Pandora Drive, City Beach, Western Australia 6015, Australia; formerly Department of Agriculture, Western Australia, Australia

Present Natural Distribution and Habitats of the Genus *Lupinus*

Lupinus Tourn. (family *Leguminosae*, subfamily *Papilionoideae*) is a genus of herbaceous annuals and herbaceous to shrubby perennials. Greatest species numbers are in the coastal and mountain regions of western North America, from Alaska to the Mexican border; in Mexico; in the Andean Highlands of Peru and neighbouring regions; and in Brazil, Uruguay and Argentina. A few extend through southern and east coastal USA. Dunn (1984) described tentative groupings of the 500-odd New World taxa which he recognized as warranting description at the species or subspecies level. The best-studied are those of North and Central America. Planchuelo (1984, 1998) and Planchuelo and Dunn (1984) described the beginnings of work on the hitherto poorly documented lupins of South America.

With few exceptions the species of the Pacific seaboard and adjacent mountains are taxonomically ill-defined, with a predominant chromosome number of $2n = 48$ and much interspecific fertility. Their formal classification poses difficulties which may never be fully resolved. The smaller number of American Atlantic seaboard species appear to fall into a different group or groups, with closer links to Old World species as discussed in the Origin and Evolution of *Lupinus* section.

The Old World has only 12 recognized lupin species, ranging around the Mediterranean and into east Africa (Gladstones, 1970, 1974; Plitmann and Heyn, 1984). Unlike the bulk of the New World taxa, the Old World species fall into distinct cyto-taxonomic groups, several of which are monospecific.

Zhukovsky (1929) surveyed the distributions and habitats of both Old and New World lupin species, as known following the explorations of N.I. Vavilov and colleagues. Relevant later references are noted above, but none deals with the genus worldwide.

Wild lupins cover a wide climatic range: from the sub-arctic climate of Alaska and the High Rockies, through mediterranean and semi-desert climates, to the highlands of east Africa, Mexico and the Andes, and finally the subtropical

 Lupins as Crop Plants: Biology, Production and Utilization (eds J.S. Gladstones, C.A. Atkins and J. Hamblin)

lowlands of eastern South America and south-eastern USA. Growing seasons are typically cool to mild, those of the annual species in subtropical lowlands being winter–spring, despite uniform or predominantly summer rainfall.

Photoperiodic responses presumably vary accordingly. The Mexican and Andean highland populations grow through the short, summer rainy seasons of those areas. They therefore reproduce under declining day-lengths, unlike most other members of the genus.

One universal characteristic of lupins is that they are plants of open and sunny habitats, intolerant of much shading. In wooded areas they grow in clearings and on open banks, especially as pioneer plants on recently disturbed soils. The arctic and high-alpine species are typically low shrubs of mountain prairies or meadows. Those of Mexico, for instance, grow on volcanic mountains and, together with grasses, form a substantial part of the prairie vegetation above 2500 m. Similarly they are prominent in the shrubby prairie and meadow populations of Bolivia and Peru at 2000 to 4500 m. The lupins of California, on the other hand, range from coastal sandhills, through lowland grassy and oak woodland formations, to pine forest at above 2000 m. Those of the North American Atlantic littoral grow among oaks, pines and *Magnolia* species.

The soil preferences of lupins have much in common throughout the genus. Typically soils are light-textured and acid to neutral, with a few ranging up to moderately alkaline. Those with heavier textures are usually stony or rubbly, as on mountain scree slopes. Normally they are well drained and fairly deep. Always they are low in nitrogen. These soil characteristics accord with the deep-penetrating and sparse root systems of lupins, which enable them to scavenge deeply and widely in soils of low water and nutrient-holding capacity. They represent a co-adaptation with the 'slow-growing', nitrogen-fixing root symbiont *Bradyrhizobium* sp. (*Lupinus*), which appears to have coevolved from a tropical/subtropical background as postulated by Norris (1956).

Origin and Evolution of *Lupinus*

A common assumption is that *Lupinus* evolved out of the tropical and subtropical *Sophoreae*, the primitive generalized tribe of the subfamily *Papilionoideae*. The *Sophoreae* are thought in turn to have arisen from the still more primitive legume subfamily *Caesalpinioideae*, perhaps in the Palaeocene or Eocene epochs of the lower Tertiary some 40–70 million years BP (Polhill *et al.*, 1981; Raven and Polhill, 1981). On the subsequent appearance of *Lupinus*, presumably some time in the mid to late Tertiary, two schools of thought have contended.

The Northern Hemisphere *Genisteae* pathway

This mainstream school places *Lupinus* in the tribe *Genisteae* (*sensu stricto* of Polhill, 1976). Bisby (1981) went further and proposed that, as a distinctive group within the *Genisteae*, *Lupinus* should be elevated to become the monogeneric subtribe *Lupininae*. The remaining genera of the tribe, comprising the *Cytisus–Genista* complex, would make up the subtribe *Genistinae*.

Polhill (1976) and Turner (1981) suggested that the *Genisteae* tribe evolved out of the African *Sophoreae* via the *Thermopsideae*, a tribe of temperate papilionoid shrubs and perennial herbs found only in the Northern Hemisphere. Following that, the annual habit of the Old World lupins, together with their limited surviving species numbers and the wide genetic gaps among them, could be ascribed to increasing aridity from the Miocene (10–25 million years BP) onwards.

The presence of *Lupinus* on both sides of the Atlantic pointed to early disjunction between the respective populations, but how early remained unclear. Raven and Polhill (1981) held that Eurasia and North America remained in effective contact right up to the Miocene, via both the North Atlantic and the Bering Straits. Legume seeds can be transported over quite long ocean distances by migrating birds etc., due to their impermeability to water and digestion.

Plitmann (1981) postulated that *Lupinus* evolved in North America from an earlier-migrated Old World thermopsoid ancestor, in the same way as the solely North American genus *Baptisia*; and that early species in turn spread to the Mediterranean and South America by such long-distance transport. The latter route would have been supplemented by direct land migration once the Americas joined up about 5.7 million years BP (Raven and Axelrod, 1974; Raven and Polhill, 1981).

Much evidence has since accumulated, however, which seems to rule out a single North American origin and subsequent radiation. Nowacki (1960) showed marked dissimilarities between Old and New World lupins in their alkaloid compositions. Later studies using serology (Nowacki and Jaworski, 1978; Cristofolini, 1989), isoenzymes (Wolko and Weeden, 1990a,b) and most recently DNA structure (Käss and Wink, 1994, 1998; Ainouche and Bayer, 1998) have confirmed the depth (but partial inconsistency) of the main transatlantic disjunction. Only the recent DNA analyses have included an adequate sampling of the American species. They reveal a contrast between those of the west and east coasts, the eastern (Atlantic) species grouping more with some from the Old World than with the main body of American species from the west coast and mountains.

Cristofolini (1989) showed *Lupinus* to be serologically much closer to the remaining *Genisteae*, all with exclusively Old World distributions, than to the American *Sophoreae* and *Thermopsideae*. The DNA evidence (Käss and Wink, 1994, 1995, 1998) likewise places *Lupinus* clearly among the *Genisteae*. Further evidence linking *Lupinus* to the remaining *Genisteae* lies in their evident polyploidy (Bisby, 1981; Goldblatt, 1981).

The well-established diploid genetic behaviour of the Old World lupin species shows that polyploidy among these, at least, must be fairly ancient. Polyploidy was thus probably fundamental to the evolution of the genus, or of its broad ancestral group. The DNA studies of Käss and Wink (1998) suggest, on the other hand, that differentiation within the genus itself took place fairly recently, within the last 12–14 million years. If this is anywhere near correct, it rules out a genus origin before the separation of the Americas from Eurasia and Africa by continental drift. Species or proto-species dispersion must therefore have taken place in part by long-distance seed transport across oceans. The commonest explicit or implicit assumption (e.g. Nowacki and Jaworski, 1978; Raven and Polhill, 1981; Wolko and Weeden, 1990a) is that this took place westwards to the Americas from Europe or Africa.

Against that, Cristofolini (1989) argued for an original evolution from Asian *Sophoreae*, along with other *Genisteae*, followed by independent migration of the ancestral American lupins eastwards across the Bering Straits or a preceding land bridge, and of the ancestors of the Old World lupins westwards to the Mediterranean and Africa. Both of the present distribution ranges, according to this idea, are refuges from climatic deterioration in Asia itself, and have been centres of independent evolution from early in the genus's existence.

The Southern Hemisphere pathway

Dunn (1984) and Gross (1986) presented a radically different hypothesis: that *Lupinus* originated in South America, from *Crotalaria*.

Once included as a subtribe under the *Genisteae* (*sensu lato* of Bentham, 1865), the *Crotalarieae* are now regarded as a separate tribe, having a tropical or subtropical distribution centred largely on southern Africa, but with extensions to neighbouring continents (Polhill, 1976, 1981b). *Lupinus*, then, would have evolved from such a background, probably in South America following the Atlantic disjunction, and radiated secondarily from there.

This hypothesis sees the *Crotalaria*-like perennial, simple-leaved lupins of subtropical Brazil as the ancestral form. It sees compound leaves and the annual habit as evolving later, perhaps as an opportunistic adaptation to seasonal shortness or aridity in neighbouring regions such as the Andean Highlands. Migration to North America could have taken place both early, by long-distance transport, and later, directly by land. Evidence for the former is seen in the presence of two simple-leaved species in south-eastern USA.

Dunn's (1984) notion that the Mediterranean lupins were introduced from South America by humans must surely be discounted, despite certain species similarities (e.g. of the South American annuals *L. linearis* Desr. and the *L. gibertianus* C.P. Smith complex to *L. angustifolius*). It is, of course, possible that the opposite occurred, i.e. that modern Mediterranean forms were introduced by early settlers and some became naturalized in South America, as they did in Australia (see Botany of the Individual Agriculturally Significant Species section). Either way, Zhukovsky (1929) considered *L. linearis* to be so like *L. angustifolius* as to be probably identical with it. More detailed evidence is needed to settle this question.

Descent of *Lupinus* from *Crotalaria* would require large changes in chromosome number. Goldblatt (1981) lists *Crotalaria* as having $2n = 14$–16, and the *Crotalarieae* generally, $2n = 14$–18. Little is known of the cytology of Atlantic South American lupin species, but known chromosome numbers for the North American and Andean species range from $2n = 36$ to $2n = 96$, with $2n = 48$ by far the commonest. Dunn (1984) and Gross (1986) postulate the necessary changes by aneuploidy in the base number from 7 to 6, followed by repeated chromosome doubling associated with the evolution of annual and compound-leaved forms and adaptation to cooler or more arid climates.

In this context it is interesting that of the two genera of the present *Crotalarieae* having strictly South American distributions, the only two species (of *Anarthrophyllum*) with recorded chromosome counts are anomalous for the tribe in having $2n = 24$. Polhill (1981b) and Goldblatt (1981) argue from this and a highly modified morphology that they might be better transferred to the tribe *Genisteae*. The simple-leaved lupins, then, are not the only South American group which seems borderline between the two tribes.

Lupin evolution: a final synthesis?

Combined with the earlier work, recent evidence from DNA analysis (Käss and Wink, 1994, 1998; Ainouche and Bayer, 1998) points finally to what I believe could be the true history of lupin evolution. It is broadly consistent with the traditional belief of a Northern Hemisphere *Genisteae* pathway, but incorporates part of the *Crotalaria* hypothesis and also has novel elements.

The DNA evidence shows *Lupinus* not as descended directly from *Sophoreae*, *Thermopsideae* or *Crotalarieae*, but as an advanced and highly evolved end-product of a stream of evolution from which these tribes were successively diverging branches. The analysis supports subdivision of the *Genisteae* into the subtribes *Lupininae* (= *Lupinus*) and *Genistinae*, and also shows the *Crotalarieae* and *Genisteae* as quite recently diverged sister tribes.

Within *Lupinus* there is progressive development and branching off, first of the simple-leaved species of the South American east coast; then of the rough-seeded group of North Africa and adjacent areas (see the Quaternary Evolution of *Lupinus* section); and thirdly of the pan-Mediterranean smooth-seeded species. The North and South American west coastal and mountain species are an end-point, the most highly evolved group of all. Käss and Wink (1998) estimated from their molecular data that divergence of *Lupinus* from its *Genisteae* background was as recent as 12–14 million years BP, and the final divergence of the western American smooth-seeded lupins from their Mediterranean counterparts only 3–4 million years BP.

Such an evolutionary tree agrees well with the present distributions and phenologies of the respective *Lupinus* groups. A somewhat older evolution than the Käss and Wink (1998) estimate would not materially alter its hypothesized nature.

Thus, the perennial, largely simple-leaved lupins of east coastal South America are the most primitive, both in their morphology and in their still subtropical warm-temperate adaptation. Although clearly enough within the evolutionary pathway of the *Genisteae* and *Lupininae*, their similarities to *Crotalaria* suggest that branching from that pathway took place before the tribal differentiation was complete. Polhill (1981b) and Van Wyk and Schutte (1995) describe other such groups in tropical/subtropical Africa, which was almost certainly the focus of the tribal differentiation (Raven and Polhill, 1981). Early transatlantic transport of these proto-lupins to the climatically comparable east of South America would have allowed their persistence there, largely unchanged. A small presence of similar species in subtropical south-eastern USA might have resulted directly from independent introduction, or indirectly by later long-distance transport from eastern South America, as Dunn (1984) has suggested.

The Old World rough-seeded lupins appear to be relict descendants of an early northwards evolution out of the subtropics, with specialized adaptation to fluctuating semi-desert and warm Mediterranean conditions.

The Old World smooth-seeded lupins are a further step away from the tropics and deserts, finding their present home mainly in the cooler and wetter northern Mediterranean climates.

Finally, the west coastal and mountain species of North, Central and South America appear to have evolved from, or from the same stock as, the Old World smooth-seeded lupins. This at first sight suggests westward long-distance transport across the Atlantic, perhaps a number of times, followed by local evolution and perhaps secondary long-distance or direct land spread to South America.

More likely is migration via northern Asia and the Bering Strait, as suggested by Cristofolini (1989). But rather than Cristofolini's hypothesized origin of *Lupinus* (and other *Genisteae*) in the Sino-Himalayan region of Asia, followed by spread westwards as well as eastwards, I see a subtropical African origin followed by land spread northwards and thence eastwards, with progressive waves of adaptation to colder climates.

The most likely region of origin for the immediate ancestors of today's smooth-seeded Mediterranean and western American lupins is temperate western Asia, e.g. Turkey, Syria and surrounding areas. Locally surviving species would then have spread westwards independently along both shores of the Mediterranean. This might help to explain, for instance, the mild genetic barrier now present within *L. angustifolius*, between accessions from Morocco and cultivated material of probably Iberian derivation (personal observation). A parallel genetic barrier, together with a clear morphological distinction, exists between Moroccan and other populations of *Trifolium subterraneum* (personal observation).

Further cold adaptation and migration eastwards across northern Asia could then readily lead to lupins crossing the present Bering Strait into North America. The cold climate adaptation retained by some North American species, e.g. *L. arcticus*, lends credibility to this idea. However, glaciation during the Quaternary would have obliterated any remnants in north Asia, so that continued evolution of the branch was necessarily based on genotypes already in North America.

If this hypothesis is correct, final evolution in the Americas must have included reverse adaptation towards warmer climates and to the annual habit found in some of the central and southern species. The latter may well be descended from some of the first immigrants, judging by the apparently early branching and further DNA evolution of South American species such as *L. mutabilis* (Käss and Wink, 1994, 1998).

Quarternary evolution of *Lupinus*

Like that of most plants, the history of *Lupinus* in the Northern Hemisphere during the Quaternary period (from 2 million years BP) was dominated by the coming and going of the ice ages. Particularly in North America the advances and retreats of the ice sheets created rapid changes of environment and a continuing state of evolution. Lupins and their nodule bacteria were especially effective colonizers of the virtually nitrogen and organic matter-free 'soils' that were exposed by the retreating ice sheets. One result was strong selection to develop or maintain cold tolerance.

By definition, all the present northerly and high-altitude populations must have immigrated and/or evolved *in situ* since the end of the last ice age, about 12,000 years BP. This explains, as discussed more fully by Dunn (1984), their continued interspecific fertility and poor taxonomic definition.

Main exceptions to the common chromosome number of $2n = 48$ are $2n = 96$ in some species towards the northern and high-altitude limits, and $2n = 36$ in the two Texan species for which counts are recorded, *L. subcarnosus* Hook. and *L. texensis* Hook. (Turner, 1959). The pattern conforms to the thesis of greatest polyploidy following active recent evolution and colonization (Stebbins, 1985). The $2n = 36$ Texan species are possibly *in situ* survivors from early immigration, whether via the Bering Strait or even (perhaps) direct across the Atlantic. Their further study in this last respect would be of interest.

The Quarternary advances and retreats of the glaciers in Europe may have been less dramatic than in North America, but ecologically they were complicated by shifting water and desert barriers which at times closed the routes for ready species migration and survival. Nor were there such universal contrasts of altitude as those that allowed North American species to adapt locally by migrating up and down mountainsides. These factors appear to have added to the exigencies of late Tertiary times in eliminating many species, and creating wide genetic gaps among those that survived.

Among the Old World lupins, the rough-seeded group, or section *Scabrispermae* of Plitmann and Heyn (1984), occupy a clearly special evolutionary position. Morphological, genetic, serological and other biochemical evidence (Gladstones, 1974; Plitmann, 1981; Plitmann and Heyn, 1984; Cristofolini, 1989; Wolko and Weeden, 1990a; Carstairs *et al.,* 1992) show them to be an unusually homogeneous group, other than in their diversity of chromosome number. They are at least as distinct from the smooth-seeded Old World species as either group is from the main New World lupins.

I speculate that the present rough-seeded lupins proliferated from a narrow genetic base surviving in late Tertiary times. Pluvial periods associated with repeated Quaternary glaciation further north probably supported flourishing populations across the present Saharo-Sinai region. Lupins would have been efficient colonizers of the sandy soils, low in nitrogen and organic matter, left by former deserts of the interglacials. Returning aridity would then have restricted survival to isolated, better-watered highlands or to northern escapes such as the Mediterranean's eastern shore; or alternatively, would have enforced adaptation as a desert ephemeral, as in the case of *L. digitatus*. Present species distributions (next section) reflect this.

Consistent with the North American distribution of chromosome numbers, we find the lowest numbers in *L. cosentinii* ($2n = 32$) and *L. digitatus* ($2n = 36$), species adapted to the warmest part of the range. *L. atlanticus* ($2n = 38$) and *L. princei* ($2n = 38$) occupy highland refuges in north-western and eastern Africa, respectively; while *L. palaestinus* and *L. pilosus*, at the east/north-eastern limit, both have $2n = 42$. The last two show evidence of continuing evolution.

Crossing among species within the rough-seeded group (Roy and Gladstones, 1985, 1988; Carstairs *et al.*, 1992; Gupta *et al.*, 1994, 1996) shows substantial mutual compatibility despite the diverse chromosome numbers. This seems consistent with recent increases in chromosome number, which must have been by individual chromosome duplication or subdivision rather than full polyploidy. Presumably the advantages for evolution in changeable environments are comparable.

The smooth-seeded Old World lupin species conform to the same pattern, having mostly northern distributions and still higher chromosome numbers. These range from $2n = 40$ in *L. angustifolius* to $2n = 50$ in *L. albus*, and $2n = 52$ in *L. micranthus*, *L. luteus* and *L. hispanicus*. The last two species form a close complex which appears still to be evolving at the cool limit of lupins in Europe. As previously noted by Gladstones (1974), the species with highest chromosome numbers and current or recent active evolution appear also to have the highest incidence of natural outcrossing.

The smooth-seeded Old World lupins comprise four distinct groups, all monospecific apart from the *L. luteus/L. hispanicus* complex. Genetic, biochemical and serological studies all show them to be fairly widely separated (Nowacki, 1960; Nowacki and Prus-Glowacki, 1971; Nowacki and Jaworski, 1978; Cristofolini, 1989; Wolko and Weeden, 1990b). However, DNA analysis (Käss and Wink, 1994, 1998; Ainouche and Bayer, 1998) suggests a weak grouping of *L. luteus/L. hispanicus* with *L. angustifolius*, and of *L. micranthus* with *L. albus*. This and the earlier serological (Cristofolini, 1989) and isoenzyme (Wolko and Weeden, 1990a) evidence suggest further that the latter two species are genetically closer than the others to the rough-seeded group.

Botany of the Individual Agriculturally Significant Species

The lupin species of present or potential significance come almost exclusively from Africa and the Mediterranean, and indeed include all known species from that region. By contrast only one of the many New World species, *L. mutabilis*, has large seeds and an established history of human use or evident potential for it.

Table 1.1 lists chromosome numbers, main synonyms and distributions in brief for all the Mediterranean/African species and *L. mutabilis*. Appendix 1.1 gives a short key to the former group; for full taxonomic treatment and references, see Gladstones (1974). Buirchell and Cowling (Chapter 2, this volume) detail species habitats. The remainder of this section treats the botany of the individual species.

L. mutabilis Sweet

Under the common names tarwi and chocho (see Buirchell and Cowling, Chapter 2, this volume), *L. mutabilis* has been grown in the Andes Highlands of Bolivia, Peru and Equador since pre-Columbian times (Brücher, 1968; Antunez de Mayola, 1982). Crop types have indehiscent pods and large, permeable, mostly white

Table 1.1. Species, main synonymy, chromosome numbers, distributions and common names of lupins with present and potential agricultural use.

Species	Synonyms	Chromosome no. (2*n*)	Distribution and common names
New World			
L. mutabilis Sweet		48	S. America: Andean Highlands, native and cult. 'Tarwi'; 'pearl lupin'
Old World, smooth-seeded			
1. *L. albus* L.			
var. *albus*	*L. termis* Forsk.	50	Mediterranean etc., cult. 'European white lupin', 'albus lupin'
var. *graecus* (Boiss. and Sprun.) Gladst.	*L. graecus* Boiss. and Sprun. *L. jugoslavicus* Kazim. and Now. *L. vavilovi* Atab. and Maiss.	50	NE Mediterranean, native
2. *L. angustifolius* L.	*L. varius* L., *sensu* Sp. Pl. edn 1 (1753) *L. linifolius* Roth *L. reticulatus* Desv. *L. opsianthus* Atab. and Maiss.	40	Pan-Mediterranean, native and semi-cult.; Australia, naturalized and cult.; elsewhere, cult. 'Narrow-leafed lupin'; 'blue lupine' (USA)
3. *L. micranthus* Guss.	*L. hirsutus* L., *sensu* Sp. Pl. edn 2 (1763)	52	Pan-Mediterranean, native. 'Lesser hairy blue lupin'
4. *L. luteus* L.		52	W Iberia, scattered pan-Mediterranean, native or semi-cult.; elsewhere, cult. 'Yellow lupin'
5. *L. hispanicus* Boiss. and Reut.			
subsp. *hispanicus*		52	C and S Spain; ? Algeria, Greece, Turkey; native
subsp. *bicolor* (Merino) Gladst.	*L. rothmaleri* Klink.	52	C and NW Spain, Portugal, ? Greece; native
Old World, rough-seeded			
1. *L. pilosus* Murr.	*L. hirsutus* L., *sensu* Sp. Pl. edn 1 (1753) ? *L. anatolicus* Swiec.	42	NE Mediterrranean, native and ? semi-cult. 'Greater hairy blue lupin'
2. *L. cosentinii* Guss.		32	W Mediterranean, Morocco, native; Australia, naturalized. 'Sandplain lupin'
3. *L. digitatus* Forsk.	*L. semiverticillatus* Desr. *L. tassilicus* Maire	36	Sahara, native
4. *L. atlanticus* Gladst.		38	S Morocco, native. 'Moroccan lupin'
5. *L. princei* Harms		38	E Africa, native
6. *L. somaliensis* Baker		?	E Africa, native

seeds (Blanco, 1982), but retain the wild characteristic of high alkaloid content. They are also cultivated in other temperate climates as ornamentals, for their conspicuous multi-coloured flowers and attractive scent.

Fully wild ancestors grow within the region of traditional *L. mutabilis* cropping. Their leaves are smaller, with narrow leaflets; the pods smaller and fully dehiscent (shattering) at maturity; and the seeds much smaller, black or marbled, and impermeable to water (Blanco, 1982). Natural cross-pollination is very common and to some degree obligate (Pakendorf, 1970), giving rise to the specific epithet meaning 'changeable'.

L. albus L. (Fig. 1.1)

The Linnean Type of the species is of the large-seeded, cultivated form which has been long grown around the Mediterranean and in the Nile Valley. This has variably bluish-white flowers and, like cultivated *L. mutabilis*, now has indehiscent pods and permeable seeds, but high alkaloid content. The Nile Valley populations, extending to Ethiopia, were once incorrectly thought to be a separate species *L. termis* Forskål. Simpson (1986a,b), Neves-Martins (1994) and Buirchell and Cowling (Chapter 2, this volume) describe geographical and cultivar races.

The wild ancestors of *L. albus* differ markedly in superficial characteristics, having blue flowers, shattering pods and smaller, dark-coloured, impermeable seeds. These again have been described as separate species, under names such as *L. graecus* Boissier and Spruner, *L. jugoslavicus* Kazimierski and Nowacki, and *L. vavilovi* Atabekova and Maissurian. But as reviewed elsewhere (Gladstones, 1974), all freely interbreed with cultivated *L. albus*, differing from it only in simple domestication genes and some physiological characters selected under cultivation. Intermediate plant types occur wherever cultivated and wild populations

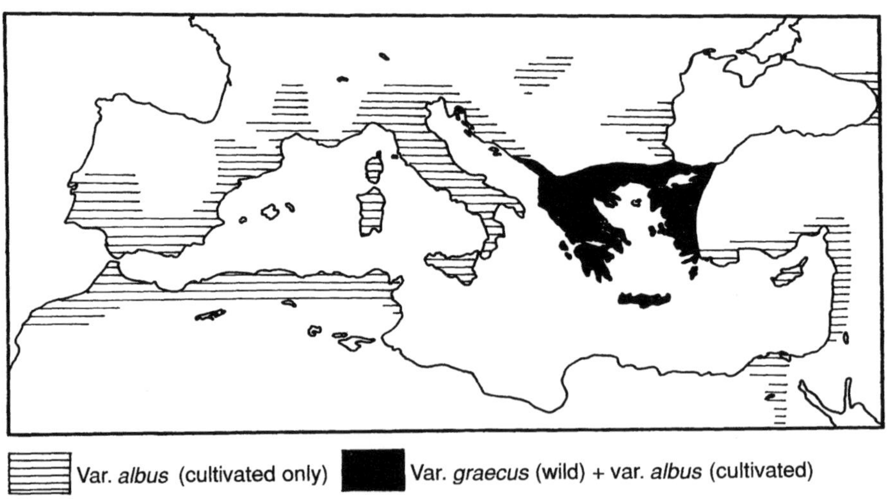

Fig. 1.1. Distribution of wild (var. *graecus*) and cultivated (var. *albus*) *Lupinus albus* L. in the Mediterranean region.

overlap, such as in southern Greece and Crete (Simpson, 1986b). Certainly there is no evidence at this stage of speciation in its true botanical sense, or even of subspeciation. I have therefore preferred to maintain the wild forms merely as a botanical variety, var. *graecus*.

Neither the wild nor the cultivated form of *L. albus* appears to have become naturalized anywhere else in the world.

L. angustifolius L. (Fig. 1.2)

Phenotypic contrasts between semi-cultivated and wild forms of *L. angustifolius*, although less extreme than in *L. albus*, led to similar confusions in the early years of Linnaean taxonomy. The Linnaean Type of the species was of a relatively coarse-leafed plant, such as was cultivated in south-west France and elsewhere for green manure and possibly cattle fodder at the time. The simultaneously described *L. varius* L. of *Species Plantarum* Edn 1 (1753) proved, later, merely to describe a wild form of the species, with smaller leaves and seeds (Gladstones, 1974; Lee and Gladstones, 1979). The sinking of *L. varius* L. under synonymy with *L. angustifolius* L. is now formally recognized (Anon., 1983). Later synonyms, as listed in Table 1.1 and by Gladstones (1974), also mostly describe small-seeded wild types.

Confusion over the identity of *L. varius* L. was compounded by Linnaeus's own later misapplication, in *Species Plantarum* Edn 2 (1763), of the same epithet to what we now know as *L. pilosus* Murr. Later botanists were in turn to misapply it to *L. cosentinii* Guss. (q.v.) and *L. digitatus* Forsk.

Despite the wide geographical range and phenotypic diversity of *L. angustifolius*, no convincing evidence has been adduced for its taxonomic subdivision. The phenotypic variation is largely continuous, apart from very recently selected major domestication characters (see Cowling *et al.*, Chapter 4, this volume). Figure 1.3 illustrates the modern cultivar Merrit. 'Wild' types from

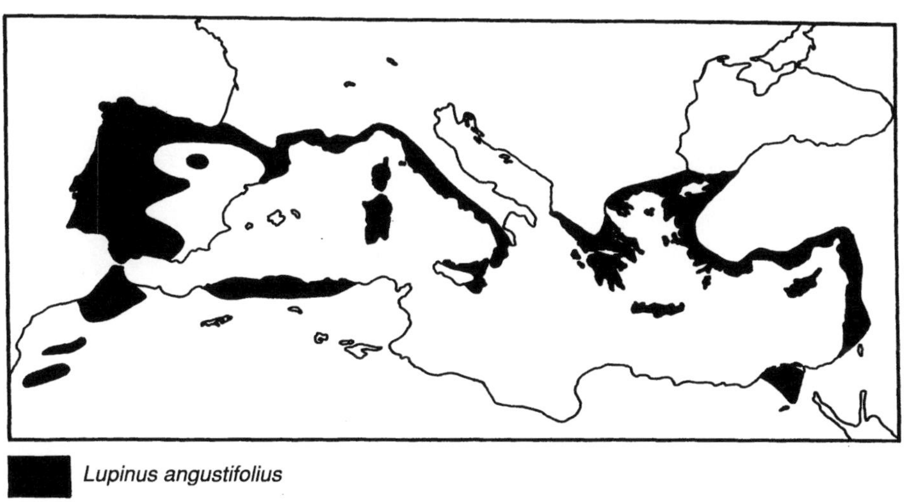

Fig. 1.2. Distribution of *Lupinus angustifolius* L. in the Mediterranean region.

North Africa and Iberia do, however, show a bimodal distribution between small-seeded truly wild and larger-seeded semi-cultivated forms (Gladstones and Crosbie, 1979). The former come disproportionately from natural or near-natural habitats, such as beach sands and forest margins, and the latter mainly from inland more fertile soils where they are often associated with present or putative past

Fig. 1.3. *Lupinus angustifolius* L. cv. Merrit. Drawn by Miss Ellen Hickman, Albany, Western Australia. Reproduced with approval from the owners P. and C. Atkins.

olive groves, vineyards etc. Their location and large seed size are obviously related to past cultivation (probably for green manure) and human selection. Mild fertility disturbances in crosses between modern cultivars (of probably Iberian background) and wild types from Morocco and the eastern Mediterranean are readily eliminated in normal breeding (personal observation). They do not seem a sufficient basis for postulating subspeciation.

Figure 1.2 shows one change from the distribution indicated by Figure 2 of Gladstones (1974), which depicted the natural distribution of *L. angustifolius* as including south-west France, north to the Loire. The specimens and references for this appear now to have been based on cultivated and perhaps related volunteer populations, which a subsequent personal search suggests are extinct. Probably the species' true range on the Atlantic seaboard does not extend north of the Pyrenees.

At least one more or less truly naturalized population of *L. angustifolius* exists outside the Mediterranean region; in the vicinity of the Swan Valley and at Gingin, Western Australia, where it has been known since the 1880s (Gladstones, 1958b). This type is hard-seeded and moderately late flowering, with medium-sized and distinctively round seeds. It resembles some of the recent accessions from southern Portugal and Spain, and could conceivably have been introduced direct from there in the mid 19th century (see section for *L. cosentinii* Guss. below). The later-used 'New Zealand blue lupin' was a soft-seeded green-manuring type of undoubtedly northern European commercial origin, first imported from New Zealand in the late 1920s (Gardner and Elliot, 1929). It volunteers commonly in the high-rainfall areas of south-western Australia, but is only doubtfully naturalized.

L. micranthus Guss. (Fig. 1.4)

The taxonomic history of *L. micranthus* has been greatly confused by its conflation with *L. pilosus* Murr. in much of the early (and some later) botanical literature, many considering the two to be conspecific under the Linnaean epithet *hirsutus* of *Species Plantarum* Edn 1 (1753). The diagnosis and references there for *L. hirsutus* clearly enough described what we now know as *L. pilosus*, but no corresponding specimen labelled *L. hirsutus* survives. The surviving specimens so labelled are of *L. micranthus*, and were probably received by Linnaeus between 1755 and 1757, i.e. between the two editions of *Species Plantarum* (Gladstones, 1974). Changes in the diagnosis for *L. hirsutus* in Edn 2 (1763) reflect this. The resulting incongruity among specimens, original diagnosis and subsequent botanical usage led Gladstones (1974) and Lee and Gladstones (1979) to propose rejection of *L. hirsutus* as a *nomen confusum*, a proposal later approved formally under Article 69 of the International Code of Botanical Nomenclature (Anon., 1983). That left the way open for adoption of the first unequivocal names *L. pilosus*, of Murray in Linnaeus (1774), and *L. micranthus*, of Gussone (1828).

L. micranthus is widespread throughout the Mediterranean region, but not especially common. The literature on it is obscure, most references to '*L. hirsutus*' probably being not to it, but to *L. pilosus*. I have found no published evidence of it having been cultivated. Nevertheless the individual seed size of around 100 mg (Gladstones and Crosbie, 1979; Clements and Cowling, 1991) suggests past and possible future use.

Early literature reports of chromosome number for *L. micranthus*, cited by Gladstones (1970, 1974), indicated $2n = 50$. This has now been amended. The correct number is $2n = 52$ (Pazy *et al.*, 1977).

L. luteus L. (Fig. 1.5)

The taxonomic identity of *L. luteus* has never been in dispute, but we do not know how far it is truly native, and how far it has become naturalized following

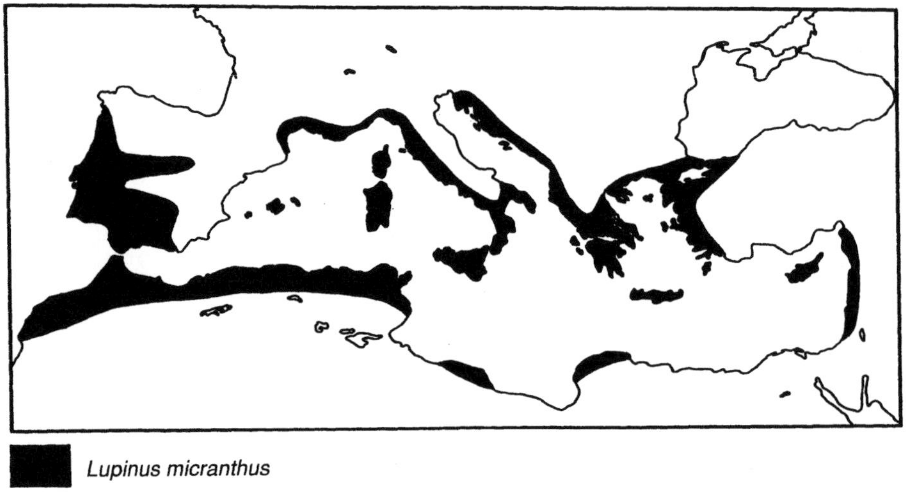

Fig. 1.4. Distribution of *Lupinus micranthus* Guss. in the Mediterranean region.

Fig. 1.5. Distribution of *Lupinus luteus* L. in the Mediterranean region.

human distribution and cultivation. The species' attractive golden yellow flowers and sweet scent have long assured wide cultivation as an ornamental, while its outstanding production of high-nitrogen green and dry matter on infertile acid soils has led to a traditional western Mediterranean use for green manuring and forage.

Portuguese populations of bitter *L. luteus* include both relatively large-seeded and early-flowering cultivated types, and some which are smaller seeded and later flowering – the latter presumably being ancestral. Many have a greenish brown seed colour, with a speckling of paler background showing through including a distinct pale band around the hilum, as illustrated by Klinkowski (1938), Kazimierski and Kazimierska (1975) and Almeida and Maltez (1977). Some Portuguese and nearly all internationally cultivated bitter strains, on the other hand, have a sparser dark speckling on an off-white background, and a more or less colourless band around the hilum. Yellow lupins I have collected or seen in southern Italy (Calabria) and Morocco have been of this type, as are scattered semi-naturalized populations which have persisted following cultivation as an ornamental or for green manuring in high-rainfall parts of south-western Australia. In contrast Klinkowski (1938) and Kazimierski and Kazimierska (1975) note that Sicilian forms have a distinctive, much darker and thicker arc around the hilum, a pattern which Klinkowski says also typifies Greek forms. These are presumably true natives.

The collections from Morocco have been characteristically early flowering and large seeded (Clements and Cowling, 1991). Israeli strains in the Australian collection have been early flowering and still larger seeded (Gladstones and Crosbie, 1979). Likewise Klinkowski (1938) described both North African and Israeli *L. luteus* as early flowering.

Kazimierski and Kazimierska (1975), on the other hand, maintained that Israeli *L. luteus* strains are later-flowering and have smaller seeds than those from elsewhere, as well as having other morphological differences and at least one cytological difference leading to partial F1 sterility in crosses. On this basis they proposed that Israeli *L. luteus* should be classified as belonging to a distinct subspecies, *orientalis*.

This seems insufficient evidence to support taxonomic subdivision; on the other hand it does seem clear that some *L. luteus* outside the Iberian Peninsula is truly indigenous. Native presence at both ends of the Mediterranean in turn lends weight to the possibility of an eastern origin (see section on Origin and Evolution, above), and to a greater presence throughout at some time in the past. Given the specific adaptation (at least in the Western Mediterranean) of *L. luteus* to areas of high rainfall, this would presumably have coincided with pluvial conditions across the region.

A related question is how far native *L. luteus* still exists in the Aegean region. Sibthorp and Smith (1830) mention and illustrate *L. luteus* in the Flora of Greece, but we currently have no specimen from there in the Australian collection. Two lines recently collected in northern Greece, initially thought to be *L. luteus*, were later reassigned to *L. hispanicus* (Clements and Cowling, 1990). Both Boissier (1872) and Chamberlain, in Davis (1970), list *L. hispanicus* in western Turkey, but not *L. luteus*. Against that Klinkowski (1938) speaks of apparently native

Greek *L. luteus*, while Kazimierski and Kazimierska (1975) and Kurlovich (1994) list both Greek and Anatolian (Turkish) ecotypes. The Kazimierski seed illustration and plant description are definitely of *L. luteus*, though of a type apparently resembling western cultivated bitter forms. Their genetic data also show full inter-fertility with western, rather than Israeli, genotypes.

L. hispanicus **Boiss. and Reuter** (Fig. 1.6)

a) subsp. *hispanicus*
b) subsp. *bicolor* (Merino) Gladst.

The original diagnosis of Boissier and Reuter (1842) was for the form here described as subsp. *hispanicus*, which is common in the mainly schistose and other metamorphic rock hills of central and south-west Spain. It resembles *L. luteus* and has the same chromosome number, but the flowers are pale violet and unscented, and the seeds are slightly rough and plain olive-brown or white. While always recognized as closely related to *L. luteus*, its status as a separate species has not been questioned. Kazimierski (1982) found complete F1 sterility in crosses with *L. luteus*.

The status of subsp. *bicolor* has been more controversial. It was originally described by Merino in 1905 as variety *bicolor*, alternatively of *L. hispanicus* or *L. luteus* (for references see Gladstones, 1974). Rothmaler (1935) raised it to species status. However *bicolor* as a species epithet proved to be illegitimate, having already been used by Lindley for an American *Lupinus* species. Klinkowski (1938) accordingly renamed it *L. rothmaleri* Klink.

This form is common in central and northern Portugal and central to north-west Spain, under conditions of generally higher altitude and/or rainfall, and poorer drainage than subsp. *hispanicus*, most typically on granite soils (Glad-

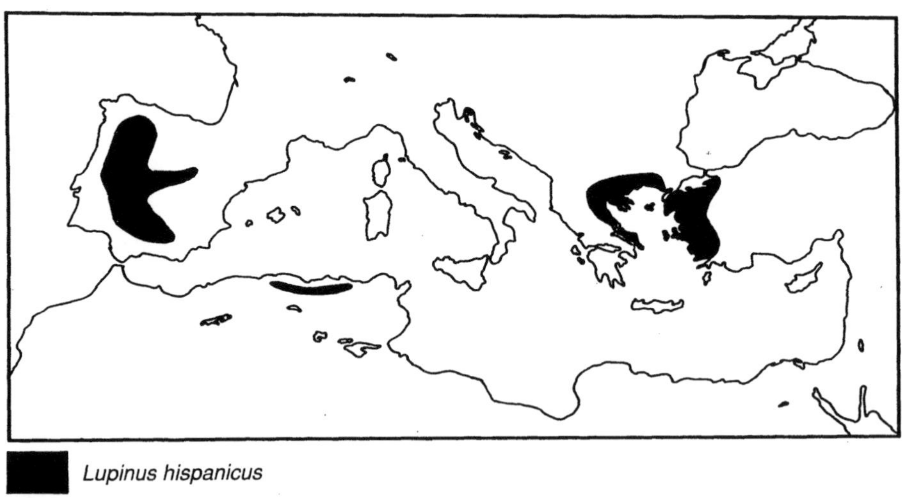

Fig. 1.6. Distribution of *Lupinus hispanicus* Boiss. and Reuter in the Mediterranean region.

stones, 1976; Gladstones and Crosbie, 1979). Morphologically it is in some ways intermediate between subsp. *hispanicus* and *L. luteus*, having flowers which start pale yellow but turn lilac after anthesis (hence '*bicolor*'), and smallish, smooth seeds with *luteus*-like markings.

In general the two subspecies of *L. hispanicus* occupy different habitats, but some overlap occurs in central Spain. Where it does, intermediate types are common (Gladstones, 1974, 1976). The combination of largely distinct morphologies and habitats with evident inter-fertility, in contrast to the known near-complete sterility between subsp. *bicolor* and *L. luteus* (Lamberts, 1958; Kazimierski and Kazimierska, 1965, 1970) led me to propose a new status of both as subspecies of *L. hispanicus* (Gladstones, 1974). This also allowed reinstatement of Merino's original epithet *bicolor*, it being no longer used at species level. Kazimierski (1982) disagreed and maintained that the two should be retained as separate species, but offered only minor and indirect evidence in support. Subsequent crossings in all combinations among four accessions of each subspecies, selected from all parts of their ranges, showed virtually no genetic barriers (Gladstones, 1984). Cordero *et al.* (1988) likewise recorded apparent natural hybrids between the two forms, and supported their status as subspecies rather than species. Consistent with this Wolko and Swiecicki (1988) found electrophoretic differences between the two, but these were much smaller than between *L. hispanicus* and *L. luteus*.

The recorded presence of *L. hispanicus* in Greece and Turkey (Boissier, 1872; Zhukovsky, 1929; Chamberlain, in Davis, 1970; Clements and Cowling, 1990) adds a hitherto unexplored dimension to its botany, including its relationship with *L. luteus*. Boissier described Turkish *L. hispanicus* as resembling its Spanish violet-flowered and brown-seeded (subsp. *hispanicus*) form, and said that in addition to Spain it grows wild in Algeria, near Constantine; on sandy alluvia of the Meander River, in western Turkey; and on the Aegean island of Naxos. By contrast, Clements and Cowling (1990) describe their two northern Greek accessions as having pale cream flowers and white seeds: characters which seem to straddle the two subspecies. Do we have here a further distinct form? If so, it could lend support to the idea that this is a common centre of origin for all the Old World smooth-seeded lupins, including *L. hispanicus* and *L. luteus*, and that most later migrated westwards. Clearly a detailed search for, and study of, related forms in the region is well warranted. It will need to be done soon, because genetic erosion of lupins there is already acute (P.H. Davis, University of Edinburgh, 1976, personal communication; Cowling, 1984; Clements and Cowling, 1990).

L. pilosus Murr. (Fig. 1.7)

This was the first to be described of the Old World group of rough-seeded lupins as defined by Gladstones (1974), and when Plitmann and Heyn (1984) formalized the group as the subgeneric Section *Scabrispermae* they nominated *L. pilosus* as the Type Species. Buirchell and Cowling (1992) give good illustrations of its flowers, leaves and seeds, together with those of other members of the section apart from *L. somaliensis*.

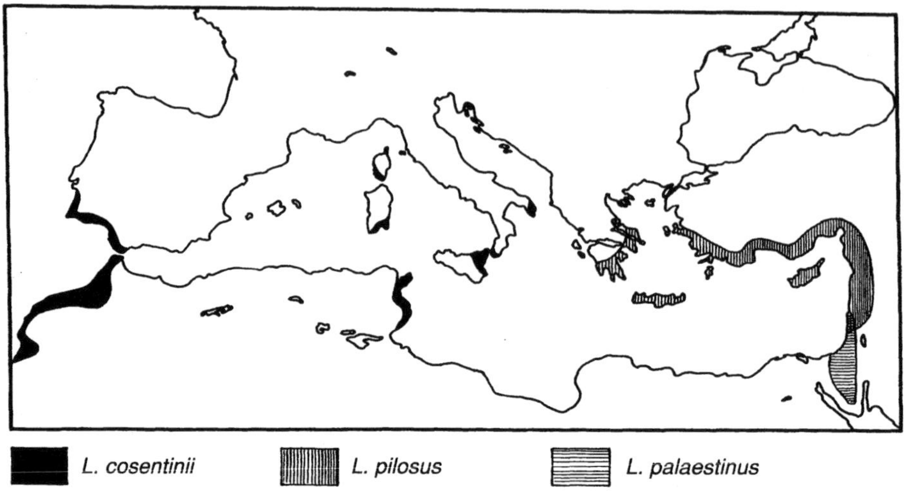

Fig. 1.7. Distribution of the Mediterranean rough-seeded lupins (*Lupinus pilosus* Murr., *L. palaestinus* Boiss. and *L. cosentinii* Guss.).

The confused taxonomy of *L. pilosus* is treated in part above under *L. angustifolius* and *L. micranthus.* The one surviving Linnaean specimen of it (898.3, a pink-flowered form) is labelled *L. varius*, but disagrees with Linnaeus's own *L. varius* diagnosis, which has since been shown to describe a form of *L. angustifolius.* Linnaeus's diagnosis and references for '*L. hirsutus*' in *Species Plantarum* Edn 1 do describe *L. pilosus*, but no concordantly labelled Linnaean specimen survives. Surviving Linnaean specimens labelled *L. hirsutus* are clearly of *L. micranthus.*

Most subsequent botanical literature refers to *L. pilosus* as either '*L. hirsutus*' or '*L. varius*'. Franco and da Silva, in *Flora Europaea* (1968), mistakenly unite it further with *L. digitatus* (see below), under '*L. varius* subsp. *orientalis*'. However, with the recent official rejection of the epithet *hirsutus* and submergence of *varius* under synonymy with *angustifolius,* Murray's unequivocal epithet *pilosus* in Linnaeus (1774) now takes precedence.

Seventeenth century pre-Linnaean references (Gladstones, 1974) show that *L. pilosus* had been introduced into Western Europe by the beginning of that century. Several describe or imply use for human consumption in its eastern Mediterranean home, a practice confirmed in more recent publications (Chamberlain, in Davis, 1970; Kalcas, 1974; Plitmann *et al.*, 1980; Plitmann, 1981). Kalcas (1974) describes two forms of wild lupin commonly growing around the edges of cultivated fields in Aegean Turkey, of which one was almost certainly *L. pilosus.* The seeds 'are eaten dry and raw for stomach disorders, but when powdered they are also used for food by the poor . . . just as they were in classical times'. On the other, Hanelt (1960) writes of *L. pilosus* being cultivated in Turkey and Crete for green manuring only.

The natural variability of *L. pilosus* is only now starting to be documented. Herbarium and botanic garden specimens often have pink or (occasionally) purple or white flowers and nearly all are early flowering, whereas virtually all field collections have dark blue flowers and vary widely among accessions in flowering time. However, all have large to very large seeds (Clements and Cowling, 1991), consistent with past selection and distribution by humans (e.g. Plitmann *et al.*, 1980; Plitmann, 1981). The presence in the Kew collection of specimens from as far south as Ethiopia (Gladstones, 1974) likewise points to past distribution and use by humans.

Cornut (1635) described *L. pilosus* (under the pre-Linnaean binomial of *Lupinus indicus*, or 'Indian lupin') in his *Canadensium Plantarum Historia*, suggesting that early French settlers took the species with them to Quebec. Chevalier and Trochain (1937) deny this, saying that Cornut included it only as being grown in the Paris Botanic Garden. There is, however, other evidence for its carriage by settlers to new countries. In 1962 I found pink-flowered *L. pilosus* persisting naturally in a back paddock of the Benedictine Monastery at New Norcia, Western Australia (founded by Spanish monks in 1847). While not proving early introduction, the finding in a remote paddock does point to use other than as an ornamental. Historical records (see section for *L. cosentinii*, below) suggest that it could have been introduced from Spain by the first monks.

Whether *L. pilosus* was ever truly or widely naturalized in Western Australia remains a mystery. Gardner and Elliot (1929) and Gardner and Bennetts (1956) wrote of '*L. hirsutus*' as a second type, along with what we now know as *L. cosentinii*, under the common name 'Western Australian blue lupin' and naturalized in the Geraldton–Northampton region. This may have been a misconception based on taxonomic confusion, but the possibility cannot be ruled out. Whatever the history, no evident trace exists now.

The botany of *L. pilosus* in its native region also raises unanswered questions. Swiecicki *et al.* (1994) describe a putative new species *L. anatolicus*, collected in south-western Turkey, having characteristics mostly intermediate between those of *L. pilosus* and *L. micranthus* but with smooth seeds like the latter. Against that, Clements *et al.* (1996) found smooth-seeded individuals associated with typical *L. pilosus* in Syria, which appeared identical with the Swiecicki form, but which in chromosome number, crossing behaviour and other aspects of phenotype seemed merely to be a variant of *L. pilosus*. They concluded that elevation of the smooth-seeded form to species status is premature. See Buirchell and Cowling (Chapter 2, this volume) for further discussion.

The existence of an apparently smooth-seeded *L. pilosus* is itself of evolutionary interest, coming as it does from near the poleward limit of the Section *Scabrispermae*, in the region where I suggest the smooth-seeded Old World species first evolved.

Finally, certain collections from Wadi Milk in Israel, CSIRO (Australia) CPI numbers 62284–62287, have been classified as *L. pilosus* but show individual characteristics closer to *L. palaestinus* (below), including *Bradyrhizobium* compatibility and flower form (Gladstones, 1974; Gladstones and Crosbie, 1979; Clements *et al.*, 1996). The last authors also observed that the flowers need tripping for efficient self-pollination, like *L. palaestinus* but unlike *L. pilosus*. The existence and nature of such apparent intermediates needs further study. Plitmann *et al.* (1980) and Pazy *et al.* (1981) have reported natural hybrids between *L. pilosus* and *L. palaestinus* where grown together in artificial culture, but doubt that this would occur now in nature because the species are never found in the same localities.

L. palaestinus Boiss. (Fig. 1.7)

A common chromosome number of $2n = 42$ and capacity for intercrossing (Kazimierski, 1961; Plitmann *et al.*, 1980; Pazy *et al.*, 1981) point to a close relationship between *L. palaestinus* and *L. pilosus*. Nevertheless the experimental crossings revealed a fairly substantial genetic barrier, while the morphological and ecological distinctions described are, for the most part, clear and consistent. Nobody has suggested combining the two species.

Whereas *L. pilosus* grows widely on heavy soils and in hilly country, and has undoubtedly spread under (or in association with) cultivation, the *L. palaestinus* populations appear more truly native and in Israel are confined to light soils of the coastal strip (Plitmann *et al.*, 1980). However, as these authors discuss, the seeds still seem large enough to suggest an influence of human selection at some past time.

L. cosentinii Guss. (Fig. 1.7)

L. cosentinii is a species of western Mediterranean coastal sandy soils. It is also widely naturalized on the southern west coast of Western Australia, where it is locally known as the sandplain (or Western Australian, or Geraldton) blue lupin.

First critically described by Joanne Gussone in her *Flora of Sicily* (1828), the sandplain lupin's true identity was long unrecognized in Australia. Australian botanists (Black, 1924; Gardner and Elliot, 1929) identified it first with *L. pilosus* Murr., and later (e.g. Gardner and Bennetts, 1959) with *L. varius* L. The latter was the officially recognized name in Western Australia through the 1940s, 1950s and (in the State Department of Agriculture) most of the 1960s. Franco and da Silva (1968) in *Flora Europaea* called it *L. varius* subsp. *varius*, as opposed to subsp. *orientalis* which supposedly included *L. pilosus* and *L. digitatus*.

My initial investigations (Gladstones, 1954) suggested the sandplain lupin to be *L. cosentinii* (or more strictly, *L. cosentini* as it appeared in the early literature). Later (Gladstones, 1958a,b) I recognized precedence of the earlier *L. digitatus* of Forskål (1775), which the best information at the time suggested was the same species. That is how the sandplain lupin came to be known as *L. digitatus* in much of the 1960s literature (i.e. other than by the Western Australian Department of Agriculture, which continued to prefer *L. varius*).

Final resolution came in 1968 (Gladstones, 1970, 1974) when I was able to study the Kew, British Museum and Linnaean Society herbarium collections. This work confirmed identity with *L. cosentinii*, but showed *L. digitatus* to be clearly distinct – a conclusion confirmed later by the first chromosome counts for *L. digitatus*. The sandplain lupin was also unquestionably distinct from *L. pilosus* and from other species previously confounded with it. The correct naming (modified to the conventional genitive form *cosentinii*) has been recognized more or less universally since 1970.

L. cosentinii appears genetically most closely related to *L. digitatus* and *L. atlanticus* (see below), with which crossing and limited gene exchange have been possible despite differing chromosome numbers (Roy and Gladstones, 1985, 1988; Carstairs *et al.*, 1992; Buirchell, 1994, 1998; Gupta *et al.*, 1994, 1996). A reported more or less fully fertile cross between '*L. varius*' and '*L. digitatus*' (Kazimierska and Kazimierski, 1995) was in fact between two accessions of *L. cosentinii* (B.J. Buirchell, Western Australia, 1996, personal communication).

It is of some interest to speculate on the timing and history of *L. cosentinii* naturalization in Australia. The major wild populations of Western Australia extend along 800 km of mainly coastal sandy soils where both climate and soils closely match those of the species' western Mediterranean home. It is common on disturbed sites. Rarer occurrences in South Australia may, in odd cases, be similarly naturalized from an early date, but that is uncertain. Any in western Victoria and coastal New South Wales are almost certainly descended from later plantings for forage or dune stabilization.

The Western Australian populations north of Perth are brown-seeded, and it was this type which early in the present century was taken into cultivation for soil improvement and summer grazing, and later named cv. Chapman. Those from south of Perth are mainly grey-seeded and a little later flowering (personal observations, 1954–1970). The two types overlap around Perth. Presumably they originated from separate introductions. Neither has the brown pod seams typical of North African *L. cosentinii*, so they must have come from the northern Mediterranean shore.

Press correspondence and other records from the 1920s together with personal reminiscences collected in the mid 1950s (Gladstones 1958b, 1969, 1972) show that both populations were present by the 1870s or 1880s, i.e. 50 years after first European settlement. A 1922 report by Mr A.W. Patrick referred to the existence of a mill, said to have been used to grind lupins for human consumption, at the convict settlement at Lynton, north of Geraldton. If true, this would suggest the presence of *L. cosentinii* there by the late 1860s, since the convict system ended in 1868.

Direct introduction by British settlers, whether deliberate or accidental, seems unlikely before 1870. The species would have been unknown to them, while prior to the opening of the Suez Canal (1869) shipping from Britain bypassed Mediterranean and (normally) Atlantic seaboard ports. There was, however, one historically recorded early voyage from southern Spain which, from its timing and circumstances, might explain how such a species could have reached Australia.

The memoirs of Bishop Rosendo Salvado (Stormon, 1977) record in detail the first years of the Spanish Benedictine monastery and mission at New Norcia, 130 km north of Perth, Western Australia. The small group who founded the mission in 1847 had travelled via England, but in 1849 a larger party of 39 new recruits sailed direct from Cadiz, in southern Spain.

This is within the very limited area of Europe where *L. cosentinii* is native (Fig. 1.7). Morison (1680) described what was almost certainly *L. cosentinii* as *Lupinus minoris caerulei Gadensis* (lesser blue lupin of Cadiz). Puerto de Santa Maria, just across Cadiz Bay from the city, is where Bourgeau collected some of his *L. cosentinii* specimens in the mid 19th century. The voyage being a special one (at royal expense) to a new agricultural mission, there can be no doubt that the ship carried a comprehensive set of agricultural equipment and seeds. Whether *L. pilosus* was among these remains speculative, but it was known and cultivated in western Europe at the time, and was certainly present at the mission later. *L. angustifolius* and *L. luteus* were established green manuring plants in Spain and Portugal, while some *L. albus* was also grown. A similar local plant around Cadiz could therefore hardly have escaped the attention of the agriculturally experienced members of the party, especially considering that an unexpected month's delay in sailing (Stormon, 1977, pp.102–104) would have given them ample time to explore the district and seek out local seeds which might be useful. The fact that the wild *L. angustifolius* naturalized not far from the mission at Gingin is of a type found characteristically in southern Portugal and Spain further suggests the possibility that both species were collected and introduced at the same time.

The location of New Norcia on the main land route from Perth to Geraldton, then being settled, would in turn have given ready passage for any such plant to the latter district. Once there the ideal climate and red sandy soils, combined with the presence of *Bradyrhizobum* strains from native legumes which were fully compatible with *L. cosentinii* (Lange, 1960), ensured rapid natural spread of that species throughout the district. By contrast the greater soil fertility requirement of *L. angustifolius*, its relative inability to nodulate with bradyrhizobia available directly from native legumes and the late maturity of the strain introduced would have severely limited any potential for widespread early naturalization.

L. digitatus Forsk. (Fig. 1.8)

The true *L. digitatus* is a pan-Saharan species, growing as an ephemeral on the sandy bottoms of occasionally flooded watercourses. When originally described by Forskål (1775), it was a common volunteer in fields of the Egyptian Nile Valley and Delta; but with changes in agriculture and irrigation practice following building of the Aswan dam, it appears more or less to have disappeared from there (A. Bari, Syria, 1997, personal communication).

Chevalier and Trochain (1937) give an extensive account of it as *L. tassilicus* Maire, in the western Saharan region. It is of special interest that the recorded western distribution extends south to the Senegal Valley at latitude 16°N, where the short rainy season is in summer. The lupins grow in winter, most likely as a volunteer among irrigated winter cereals.

Chevalier and Trochain acknowledged similarity of *L. tassilicus* to the earlier named *L. digitatus* of Egypt, but wrongly maintained the latter to be synonymous with *L. pilosus*. This mistake probably traces back to the incorrect description of *L. digitatus* in Muschler's *Flora of Egypt* (1912), and reappears in several subsequent regional floras which conflate *L. digitatus* with *L. pilosus*. Franco and da Silva, in *Flora Europaea* (1968) compound the error by including *L. digitatus* with *L. pilosus* under '*L. varius* subsp. *orientalis*'. However the subsequent chromosome counts of $2n = 36$ for authentic Egyptian *L. digitatus* (Plitmann and Heyn, 1984; Carstairs *et al.*, 1992), following an identical count for *L. tassilicus* by Eichorn (1949), supports the morphological evidence of Gladstones (1974), Lee and Gladstones (1979) and Plitmann and Heyn (1984) that *L. digitatus* and

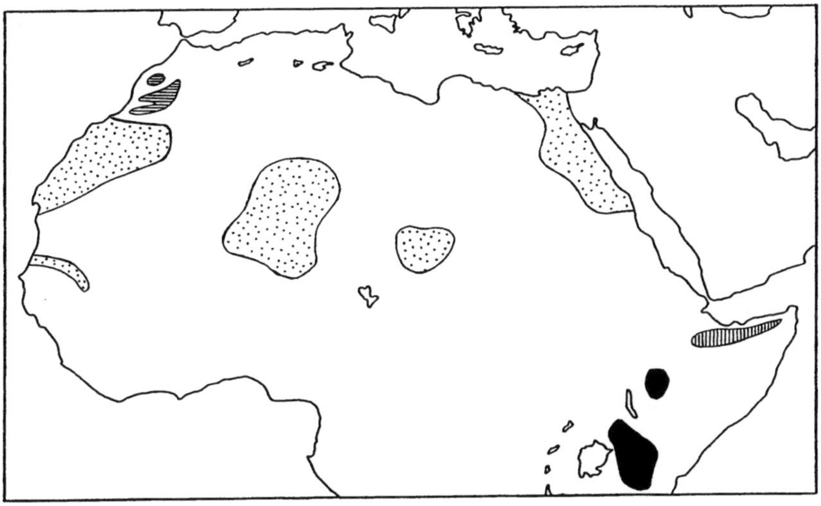

Fig. 1.8. Distribution of the African rough-seeded lupins (*Lupinus digitatus* Forsk., *L. atlanticus* Gladst., *L. princei* Harms and *L. somaliensis* Baker).

L. tassilicus are the same species. It is unquestionably distinct from *L. cosentinii* ($2n = 32$), *L. pilosus* ($2n = 42$) and all the other Old World rough-seeded lupins.

L. atlanticus Gladst. (Fig. 1.8)

Good specimens of this species were in the Kew herbarium from the time of Hooker and Ball's expedition to Morocco in 1871, but were labelled *L. pilosus* or *L. varius*. My 1968 examination showed it to be a distinct species of rough-seeded lupin, apparently peculiar to the Atlas foothills and Anti Atlas of southern Morocco. Subsequent seed collecting (Gladstones, 1976; Gladstones and Crosbie, 1979; Buirchell, 1991) showed that it grows mainly on rubbly slopes and as a volunteer in crops or fallow fields, at altitudes from about 500 to 1600 m. The formal diagnosis appears in Gladstones (1974), with the 1936 Kew specimen 2846 of E.K. Balls nominated as Type of the species.

A chromosome number of $2n = 38$ (Pazy, cited by Plitmann *et al.*, 1980; Carstairs *et al.*, 1992) distinguishes *L. atlanticus* from all other Old World rough-seeded lupins except *L. princei* (see below). Crossing behaviour (Roy and Gladstones, 1985, 1988; Carstairs *et al.*, 1992) has suggested closest genetic affinity to *L. cosentinii* ($2n = 32$) and *L. digitatus* ($2n = 36$). More recent studies of crossing and chromosome behaviour (Gupta *et al.*, 1994, 1996; Chapter 3, this volume), together with some morphological features have further suggested an affinity to *L. pilosus* ($2n = 42$).

The large seeds of *L. atlanticus* and its usual association with cultivated land (Gladstones and Crosbie, 1979; Buirchell, 1991) indicate very likely past and perhaps present use for human consumption.

L. princei Harms (Fig. 1.8)

This species is recorded for the highlands of Kenya, Tanzania and southern Ethiopa, at 1700–3000 m. The Gillett specimen 14199 (Kew), noted as being abundant in the Mega district of southern Ethiopia, is atypical for a wild lupin in having white flowers. This, and the presence of both *L. pilosus* and *L. albus* in Ethiopia, again suggests that such large-seeded lupins have all at some stage been cultivated or semi-cultivated there.

Although morphologically a typical member of the rough-seeded group of Old World lupins, and having the same chromosome number as *L. atlanticus* ($2n = 38$), *L. princei* has always failed to produce viable seed when crossed with others of the group (Carstairs *et al.*, 1992; Gupta *et al.*, 1994, 1996). It also fails to nodulate with standard lupin bradyrhizobia (personal observations, 1966–1970; Buirchell and Cowling, 1992). Against that, the DNA analyses of Käss and Wink (1998) show *L. princei* to be fully part of the group. Growing in an isolated and adaptively different habitat in the equatorial highlands presumably explains its physiological and genetic specialization.

L. somaliensis Baker (Fig. 1.8)

To the best of our knowledge this is known only from its Type Specimen, collected some time before 1895 at 1500 m in the Golis Range of Somalia. The specimen has no mature pods or seeds, but in other respects is a typical enough Old World rough-seeded lupin. Gladstones (1974) gives as full a description as the specimen allows.

Plitmann and Heyn (1984) question whether *L. somaliensis* is distinct from *L. princei*, pointing out that subsequent specimens supposedly of the former often turn out to be *L. princei*. Nevertheless the characteristics of the Type do appear genuinely unique, particularly in the hairless upper surface of the leaflets. As compared with *L. princei* the stem and petiole hairs are also longer (about 2 mm compared with at most 1 mm in *L. princei*) and the leaflets appear a little broader and the pedicels longer. The flowers are longer relative to their height; their central pale area is a clearly defined white spot as opposed to the pale yellowish band with only pale blue upper margin of *L. princei*; and the keel tip is more or less pointed whereas that of *L. princei* is fairly blunt.

I remain certain that *L. somaliensis* is a distinct species within the rough-seeded group. Whether it still exists in living form is uncertain. Its original description as a perennial is also open to doubt.

Pre-history and History of Lupin Use

Pre-agricultural use

The literature gives many hints that other species besides *L. albus* and *L. mutabilis* have been used in historical and earlier times for human consumption, although in recent times mainly by the poorer classes or in times of famine. Evidence on some of the individual rough-seeded Old World lupins has been noted above. Common names such as 'lupino salvatico' for Italian wild *L. angustifolius* (Bauhin *et al.*, 1651; Savi, 1798) attest to use of this species as well. Earlier writers considered wild lupins more effective for medicinal purposes than the cultivated *L. albus* (Parkinson, 1640).

As discussed in greater detail elsewhere (Gladstones, 1970, 1974; Plitmann *et al.*, 1980; Plitmann, 1981), use of the rough-seeded lupins for human consumption is probably very ancient and has arguably been central to their final evolution as large-seeded species. I speculate that they have been a significant food source from early in human evolution, and that they have been so throughout the common distribution area of lupins and early humans in north and east Africa and the eastern Mediterranean. Very extensive lupin populations could well have existed on the sands of the Sahara during pluvial periods associated with the ice ages. In that context they would have been subject to intermittent human influences over hundreds of thousands, possibly millions of years.

Rough-seeded lupins are peculiarly suited to a role ancillary to human evolution. The growing plants are bitter and resist predators, but whether naturally or from long selection, the seeds have only moderate alkaloid contents (Beck *et al.*, 1979; Allen *et al.*, 1990; Buirchell *et al.*, 1994), and are not especially bitter (e.g. Cornut, 1635; Morison, 1680; personal observation). Some direct use of untreated seeds is therefore possible. Technology to crack them and leach out much of the alkaloid is in any case very simple, while the natural tendency of lupins to grow in dense patches, which ripen synchronously, makes seed collection easy for observant hunter-gatherers despite pod shattering. Reports by A. Bari (Syria, 1997, personal communication) give an example of present opportunistic collection from a remote stand of *L. digitatus* in Egypt.

The readiness with which wild lupins volunteer on disturbed sites means that they commonly establish around human encampments, there to grow protected by their bitterness while green. Sources favoured and harvested for their seed size or lack of seed bitterness would thus be favoured for reproduction. It is easy to envisage eventual encouragement in this by seed scattering and deliberate soil disturbance, in a primordial form of agriculture.

Only hardy and self-sufficient plants such as lupins could allow such first steps towards agriculture. Limited technology for land clearing and cultivation must also have confined these activities to lightly vegetated sites and lighter soil types (Cole, 1967). Lupins might thrive, but few other potential crops could grow under the conditions.

All this changed with the coming of the neolithic revolution, with its shift to permanent settlement on the fertile soils of the river valleys, and unprecedented technology for the cultivation, irrigation and manuring of the newly domesticated crops adapted to them. Lupins would appear then to have fallen into neglect, though still used as needed where they grew naturally or as a weed.

It seems likely that there were attempts to domesticate lupins, but that these were unsuccessful except partly in the case of *L. albus* in the Mediterranean and *L. mutabilis* in South America. The higher soil fertility of the new agriculture held no advantage for lupins, while their continuing ability to grow unattended on waste sites, or as weeds, maintained a powerful selective advantage for wild characteristics such as bitterness, shattering pods and hard seeds in the absence of strong and continuous artificial selection to the contrary.The Old World lupins have thus remained for the most part stubbornly wild, albeit probably larger seeded, and in some cases perhaps less bitter seeded, than their pre-human ancestors.

Part-domestication and early cultivation

The part-domestication and early historical cultivation of *L. albus* is extensively described in a mostly German literature (Knapp, 1931; Hanelt, 1960; Hondelmann, 1984, 1996).

Both linguistic evidence and the known distribution of ancestral var. *graecus* (Fig. 1.1) show that domestication started in the Aegean region, whence the Greek name *thermos* (meaning 'hot', presumably in reference to bitterness) was carried in a range of variants throughout the classical eastern Mediterranean world and as far as India. Buirchell and Cowling (Chapter 2, this volume) trace common namings through various other regions and languages.

The famous physician Hippocrates of Cos (400–356 BC), cited by Hanelt (1960), mentioned nutritional, medicinal and cosmetic uses for the seeds. Theophrastus (372–288 BC), a pupil of Aristotle, also wrote extensively about *L. albus* in his *Enquiry into Plants*, which remained a leading text through Roman times and well into the Middle Ages. Interestingly he mentioned that harvesting is best done after rain, to prevent the seeds jumping from the pods. The Roman writer Pliny the Elder (AD 23–79) repeated this, and also noted that the plant seeds down and volunteers naturally. We may conclude that *L. albus* was not yet fully non-shattering, although possibly soft-seeded.

Roman agriculture depended greatly on legume crops and green manures to maintain soil fertility. The writings of Columella (*fl.* 1st century AD) and Pliny show that *L. albus* was prominent among them, and that it was favoured for both green manuring and seed harvest. The Latin *Lupinus* corresponds to 'wolf bean', a reference perhaps to the plant's ability to grow in rough and wild places, where soils were poor and its bitterness was protection enough against insect and animal predators.

Uses for the seeds included both animal and human consumption. The seeds, then as now around the Mediterranean, were boiled and then leached for long periods to remove most of their bitter but water-soluble alkaloids. One method was to hang the boiled seeds in bags for a few days in the ocean or running streams. Many farms had a special basin in the farmyard for leaching lupins, called a *labrum lupiniarum* (Hanelt, 1960). Pliny (Book 17, 26) recommended pouring the water in which lupins had been boiled or leached around fruit trees. The variously cooked, salted, or otherwise seasoned seeds were commonly sold in the streets and markets, as a snack food.

The pharmacological text *De Materia Medica* of Dioscorides (*fl.* 1st century AD) mentioned numerous medical uses, including lupin meal mixed with honey or vinegar as a cure for worms, and infusions or poultices of lupin for various skin complaints and boils. Meal cooked in rainwater to a slimy texture was widely used as a skin lotion.

Lupins had social roles as well. Guillaume (1923) recounts that generals who had been accorded a triumph or citizens who aspired to public office distributed lupins among the Roman population – hence the cry: 'Lupins for the people!'

Seeds of *L. albus* also represented money in plays and games. *Nummus lupinus* (lupin money), like *aurum canicum* (dog's gold), came to describe fictitious or worthless currency.

Few accounts exist of *L. albus* cultivation after the fall of Rome, but some use as a minor subsistence crop has persisted to the present throughout the Mediterranean area, the Black Sea surrounds and the Nile Valley south to Ethiopia.

One story appears in the 15th century Persian–Arabian *Book of the Thousand Nights and One Night*. On the 947th night, in the Mardrus and Mathers translation:

> As the girls turned away, an old woman entered the Hammām, bearing a bowl on her head, filled with fried earth nuts and roast lupins. Some of the young women bought her goods for a penny, a half-penny, or two pence; and Dalāl, wishing to forget her troubles, called to her saying: 'Give me a pennyworth of lupins, good aunt.' The old woman sat down by the bench and filled a horn measure with lupins . . .

The parallel history of *L. mutabilis* in the Andean highlands of South America is less well documented. Hondelmann (1984) cited archaeological evidence of its cultivation in the 6th or 7th century BC, while in the Nazca (AD 100–800) and subsequent civilizations it formed a regular part of the crop rotation. Brücher (1968) says that the mountain Indians debittered the seeds by washing them in bags in mountain streams for some weeks. Alkaloid removal was not complete, and poisoning sometimes still occurred. Seeds were also steeped in containers and the leachings used as an insecticide.

Lopez-Bellido and Fuentes (1986) mention reputed fertility-enhancing properties of lupins, and their use in Andean religious rites and festivals; also curative roles of the alkaloids in cardiac diseases, rheumatism, malaria and internal parasitic infections, similar to those attributed to *L. albus* in Rome.

The arrival of the Spanish conquistadors in the 16th century started a progressive decline in native Andean agriculture. The new dominant society disdained lupins and their cultivation became restricted to marginal subsistence agriculture in a few very high-altitude environments. Interest in lupins has only started to re-emerge in the last decade or two.

Bitter lupin cultivation in northern Europe

The agricultural history of bitter lupins in northern Europe can be said to have started in 1781. First impetus came from no less a person than King Frederick II (Frederick the Great) of Prussia, who personally sent for seed of *L. albus* from Italy with a view to its use for improving the poor soils of northern Germany. Hondelmann (1984, 1996) gives full details.

This and later northern European efforts with *L. albus* largely failed, because the soils were too poor for it and the available genotypes too late maturing in the region's short summer growing season. Then in 1841 a farmer named Borchard tried growing the garden yellow lupin, *L. luteus*. This was much more successful – so successful, in fact, that despite continuing official support for *L. albus* the cultivation of *L. luteus* had by the 1860s become an essential part of agriculture right across the acid, sandy soils of the Baltic coastal plain. Farms in Brandenburg and Pomerania had as much as 50% of their area under yellow lupins, either for forage or for green manuring (Hackbarth and Husfeld, 1939). Lupins became a staple of the Saxony merino wool industry.

The appearance of lupinosis and some heavy sheep losses after 1875 reduced lupin growing in the following decades. Nevertheless their role for green manuring remained important, thanks in part to energetic advocacy by writers such as Dr Albert Schultz of Lupitz (known as Schutz-Lupitz) and later Professor Julius Kühn. At the turn of the century lupins were still sown on some 500,000 ha annually in Germany: 400,000 ha for green manure or forage, and 100,000 ha for seed.

L. angustifolius also became established in northern Europe during this period, but its history remains obscure. The genetic stocks are generally agreed to have originated in Iberia. Hanelt (1960) and Hondelmann (1996) note early 19th century cultivation in south-west France for cattle fodder, and at least some use in Germany for green manuring. Farmers in Suffolk, England, were using *L. angustifolius* for sheep folding and improvement of light soils at least as early as 1859 (Oldershaw, 1920, 1925). A particular virtue of *L. angustifolius* in northern latitudes was its tolerance of frosts, and ability to keep growing well into the autumn when sown after cereal harvest.

The distribution maps of Fischer and Sengbusch (1935, reproduced by Hondelmann, 1996) show bitter *L. luteus* and *L. angustifolius* cultivation extending eastwards well into Russia and the northern Ukraine, on the region's acid podzolic soils. On the other hand cultivation of *L. albus* in areas of summer growing season remained limited, confined to the warmer climates and more fertile soils of eastern central Europe and the south-west of the Ukraine.

Lupin cultivation fell further in the advanced economies after about 1900. The Saxony wool industry declined because of competition from imports, while ever cheaper nitrogen fertilizers and rising land values favoured more intensive and continuous production of non-leguminous cash crops, at the expense of leguminous fodders and green manures.

World War I, and the accompanying blockade of Germany, brought a renewed interest in lupins as a source of desperately needed protein. Much effort went into industrial processes for de-bittering the seeds; but although technically successful, these could not compete commercially once the emergency had passed. Researchers did, however, show the feasibility of using lupins for many food and other purposes, as demonstrated by the famous 1918 dinner of the Association of Applied Botany in Hamburg. Petterson (Chapter 13, this volume) gives details.

Roasted *L. angustifolius* seeds, along with chicory, were in any case much used in wartime Germany for making *ersatz* coffee, a use which Guillaume (1923) states had long been traditional in France, and had given rise to common names for the species such as *lupin à café*. Nineteenth century French botanical references, cited by Gladstones (1974), confirm this. Cowling (1984) likewise recorded the traditional name of 'wild coffee' for *L. angustifolius* in north-eastern Greece. But as Hanelt (1960) observed darkly: 'On the qualities of the resulting beverage, the literature remains silent.'

A result of the wartime interest was a growth in German private breeding of *L. luteus* and, especially, *L. angustifolius* during the 1920s. New bitter varieties of the latter gave advances in maturity and (apparently) soft-seededness, developments which underlie the growth soon after of bitter *L. angustifolius* cultivation for forage and soil improvement in the south-eastern USA, in the Cape Province of South Africa (until devastated by mildew) and on the Canterbury Plains of southern New Zealand, whence seed was exported to Western Australia from 1929 onwards. During this period a strong tradition of lupin research started in Poland, under the leadership of Professor Stefan Barbacki.

Lupins still lacked some of the essential attributes of a crop plant, however. Only *L. albus* and *L. mutabilis* had pods which did not immediately shatter at ripeness. Only in those species and *L. angustifolius* had soft-seededness been attained, allowing reliable germination. Above all, every species except, to some extent, the rough-seeded group had high seed alkaloid contents, precluding food use except after troublesome processing.

The idea of breeding cultivars free from alkaloids emerged early in the present century. Von Rümker (1913, cited by Hondelmann, 1996) appears to have been the first to suggest it as a theoretical possibility. A later advocate was Professor Erwin Baur (1875–1933), who in the 1920s was Director of the Kaiser Wilhelm Institute for Genetic Research in Berlin. Baur cited Vavilov's Law of Parallel Variation as reason to think that natural mutants for alkaloid freedom could be found if sought hard enough, but that they would be rare, because susceptibility to insect and animal predators must in normal circumstances lead to their selective elimination.

One listener to Baur's 1927 lectures was Reinhold von Sengbusch (1898–1986), then a guest researcher at the Institute. After studying the earlier published work of the Russian researcher D.N. Pryanishnikov on techniques for lupin alkaloid analysis, von Sengbusch took on the challenge of devising a quicker and simpler method, for use in screening large populations. To that end he was appointed on 1 October, 1927, to the newly founded Kaiser Wilhelm Institute for Plant Breeding Research at Müncheberg, just outside Berlin, of which Baur was Director until his death in 1933. The success of von Sengbusch over the next few years, until politically related disagreements with the Institute's new National Socialist-aligned administration forced his departure in 1937, was to become the foundation of modern lupin breeding. Chapter 4 (Cowling *et al.*, this volume) takes up that story.

Importance and Agricultural Future of Lupins

Lupins have a unique place among crop legumes. From a speculated pre-agricultural and proto-agricultural role as a human food source, they have been largely neglected since the neolithic revolution. Their weedy adaptation to poor soils and rough husbandry outside settled arable agriculture (previous section) militated against successful domestication. This will remain a practical problem wherever wild lupins are present, due to the difficulty of keeping crop varieties free from contamination by admixture or, where the species are the same, by out-crossing. It is no accident that the most successful example of domestication and commercial cultivation, that of *L. angustifolius* in Western Australia, has been in areas totally devoid of that species in its less developed forms.

A second reservation is that because this success has been on virgin lupin soils, it may have been to some extent because diseases or pests specific to lupins are not yet present. This is but a temporary advantage, albeit one to be prolonged as far as possible by careful quarantine and seed hygiene.

Against these factors must be considered the changing nature of agriculture at its frontiers. The world's growing human population has meant bringing into cultivation soils that previously were thought marginal, or worse. Many of these are well suited (in appropriate climates) to lupins but to few other crops, and certainly to few, if any, other temperate crop legumes.

But it is precisely on such soils that legumes have their greatest agronomic and economic advantage. They are low in nitrogen, allowing legumes to compete best against non-leguminous weeds. Nitrogen application to ill-adapted other crops tends to be inefficient, and leads readily to leached nitrate accession into the groundwater. By contrast the nitrogen from breaking down of legume roots, nodules and above-ground residues becomes available to alternating rotation crops at steady rates, depending largely on temperature and moisture, and is therefore more or less matched to the potential growth rate and nutrient needs of the crop. The early sowing of all crops into stubble that this system allows also maximizes usable growing season length, and therefore yield potential, and minimizes the risk of erosion on otherwise often very erodible soils. In these ways a legume/non-legume crop rotation can be agronomically sound and profitable on soils too poor for almost any other farming system, as has been well demonstrated on the Western Australian sandplains.

Lupins have the special advantage for mediterranean climates that they can be sown with the very first seasonal rains, or even dry before them. This fact, combined with their deep root systems, means that they both dovetail well with other cropping operations and use a very high proportion of total seasonal rainfall. Groundwater drainage and leaching of soluble salts from the uplands and slopes, to create rising water tables and salting problems in the valleys, is thereby minimized.

All these factors mean that lupins can be expected to fill a growing role in agriculture wherever they are adapted but that the individual lupin species will continue to be constrained by their fairly specific climate and soil ranges, in common with most other crop legumes. Future lupin breeding will therefore need to broaden these ranges as far as possible, as well as seeking to anticipate diseases and pests which may not yet have appeared or become serious. Part of the gain will come from domesticating further large-seeded lupin species, perhaps especially rough-seeded species adapted to marginal alkaline soils.

One problem to be faced is that having plants such as lupins can tempt *Homo sapiens* into clearing land which perhaps should never be cleared, either for native flora and fauna conservation, or because it is too ecologically fragile for sustainable agriculture of any sort. These are moral questions which will become central in the decades to come. So far lupins have served mainly to underpin ecologically improved farming systems on already cleared infertile or degraded land; but perhaps the stage has already been reached where some of this should be allowed to revert – as far as can be achieved – to its native state, or used for more benign non-agricultural purposes.

For all these reasons, the future breeding of lupins will take place in a wider philosophical context than in the past. That said, their potential to contribute remains vast. Unlike nearly all other food crops, lupins are still only at the beginning of their genetic improvement for cropping. Conventional breeding still has much to give, while the emerging complementary technologies of genetic engineering (Atkins *et al.*, Chapter 3, this volume) will extend the possibilities much further. There will be accompanying advances in field and food-processing technology, enabling lupins to be exploited more fully, yet retaining their unique hardiness and tolerance of poor soils, born of long evolution in the wild and on the historical and geographical fringes of agriculture. Lupins are destined to come in from the cold and become one of the great food crops of the future.

References

Ainouche, A. and Bayer, R.J. (1998) Phylogenetic relationships among the Old World and New World *Lupinus* species (*Fabaceae*) based on internal transcribed spacers of nuclear ribosomal DNA. *Proceedings of the 8th International Lupin Conference, Pacific Grove, USA* (in press).

Allen, D.G., Greirson, B.N., Harris, D.J. and Spadek, Z.E. (1990) Alkaloid profiles of various lupin species. Unpublished report, Chemistry Centre, Western Australian Department of Mines, Perth, p. 13.

Almeida, J.R. Marques de and Maltez, A.C. (1977) Prospecção de possiveis ecotipos Portugueses de tremocilha (*Lupinus luteus* L.). *Agronomia Lusitana* 38, 137–161.

Anon. (1983) Proposals 474 and 475. *Lupinus varius* L. and *Lupinus hirsutus* L. (*Leguminosae*). *Taxon* 32, 624.

Antunez de Mayolo, S. (1982) Tarwi in ancient Peru. In: Gross, R. and Bunting, E.S. (eds) *Agricultural and Nutritional Aspects of Lupines: Proceedings of the First International Lupin Workshop, Lima-Cuzco, Peru, 1980.* Deutsche Gesellschaft für Technische Zusammenarbeit (GTZ), Eschborn, pp. 1–11.

Bauhin, J., Cherlero, J.H., Chabraeus, D. and Agraffenried, Fr L. (1651) *Historiae Plantarum Universalis.* Ebroduni.

Beck, A.B., Goldspink, B.H. and Knox, J.R. (1979) A re-examination of the alkaloids of *Lupinus cosentinii. Journal of Natural Products* 42, 385–398.

Bentham, G. (1865) In: Bentham, G. and Hooker, J.D., *Genera Plantarum*, Vol. 1. Reeve, London.

Bisby, F.A. (1981) Genisteae. In: Polhill, R.M. and Raven, P.H. (eds) *Advances in Legume Systematics*, Part 1. Royal Botanic Gardens, Kew, pp. 409–425.

Black, J.M. (1924) *Flora of South Australia.* Government Printer, Adelaide.

Blanco, O. (1982) Genetic variability of Tarwi (*Lupinus mutabilis* Sweet). In: Gross, R. and Bunting, E.S. (eds) *Agricultural and Nutritional Aspects of Lupines: Proceedings of the First International Lupine Workshop, Lima-Cuzco, Peru, 1980.* Deutsche Gesellschaft für Technische Zusammenarbeit (GTZ), Eschborn, pp. 33–49.

Boissier, E. (1872) *Flora Orientalis.* H. Georg, Geneva and Basle.

Boissier, E. and Reuter, G. (1842) *Diagnoses Plantarum Novarum Hispanicarum.* F. Ramboz, Geneva.

Brücher, H. (1968) Die genetischen Reserven Südamerikas für die Kulturpflanzenzüchtung. *Theoretical and Applied Genetics* 38, 9–22.

Buirchell, B.J. (1991) *Collection of Wild* Lupinus *spp. in Morocco and the Canary Islands.* Department of Agriculture, Western Australia, Perth, p. 14.

Buirchell, B.J. (1994) Domestication and potential of the rough-seeded lupins. In: Neves Martins, J.M. and Beirão da Costa, L. (eds) *Advances in Lupin Research: Proceedings of the 7th International Lupin Conference, Evora, Portugal, 1993,* pp. 19–24.

Buirchell, B.J. (1998) Genetic diversity in rough-seeded lupins. *Proceedings of the Eighth International Lupin Conference, Pacific Grove, California* (in press).

Buirchell, B.J. and Cowling, W.A. (1992) Domestication of rough-seeded lupins. *Journal of Agriculture, Western Australia* 33(4), 131–137.

Buirchell, B.J., Cowling, W.A., Allen, D., Greirson, B. and Harris, D. (1994) Seed alkaloids of the rough-seeded lupins. In: Dracup, M. and Palta, J. (eds) *Proceedings of the First Australian Lupin Technical Symposium.* Department of Agriculture, Western Australia, Perth, p. 255.

Carstairs, S.A., Buirchell, B.J. and Cowling, W.A. (1992) Chromosome number, size and interspecific crossing ability of three Old World lupins, *Lupinus princei* Harms, *L. atlanticus* Gladstones and *L. digitatus* Forskål, and implications for cyto-systematic relationships among the rough-seeded lupins. *Journal of the Royal Society of Western Australia* 75, 83–88.

Chevalier, A. and Trochain, J. (1937) Histoire de trois lupins. *Revue de Botanique Appliquée et d'Agriculture Tropicale* 17, 85–97.

Clements, J.C. and Cowling, W.A. (1990) Summary of wild and semi-domesticated lupin accessions introduced to Australia from 1974 to 1989. *Australian Plant Introduction Review* 21(2), 1–14.

Clements, J.C. and Cowling, W.A. (1991) *1991 Catalogue of the Australian Lupin Collection.* Department of Agriculture, Western Australia, Perth, p. 186.

Clements, J.C., Buirchell, B.J. and Cowling, W.A. (1996) Relationship between morphological variation and geographical origin or selection history in *Lupinus pilosus. Plant Breeding* 115, 16–22.

Cole, S. (1967) *The Neolithic Revolution,* 4th edn. British Museum (Natural History), London, p. 68.

Columella, Lucius Junius Moderatus (1st century AD) *De Re Rustica,* tr. E.S. Forster and E. Heffner. Heinemann, London, revised edn 1968, 3 vols.

Cordero, S.A., Arrieta, V. and Muzquiz, M. (1988) *Lupinus hispanicus* breeding. *Proceedings of the 5th International Lupin Conference, Poznań, Poland.* PWRil, Poznań, pp. 156–169.

Cornut, J. (1635) *Canadensium Plantarum Historia.* Simonem le Moyne, Paris.

Cowling, W.A. (1984) *Report of a Mission to Collect Wild* Lupinus *in Greece.* Department of Agriculture, Western Australia, Perth, p. 14.

Cristofolini, G. (1989) A serological contribution to the systematics of the genus *Lupinus (Fabeceae). Plant Systematics and Evolution* 166, 265–278.

Davis, P.H. (1970) *Flora of Turkey,* Vol.3. University Press, Edinburgh.

Dioscorides, Pedanios (1st century AD) *De Materia Medica,* tr. J.Berendes. Stuttgart, 1902. (Cited by Hanelt, 1960.)

Dunn, D.B. (1984) Genetic resources, cytotaxonomy and distribution of New World lupin species. *Proceedings of the 3rd International Lupin Conference, La Rochelle, France,* pp. 68–85.

Eichorn, A. (1949) A propos de la carologie du *Lupinus tassilicus* Maire. *Revue de Cytologie et de Biologie Végétales* 11, 333–350.

Fischer, A. and Sengsbusch, R. von (1935) Geschichte des Lupinenanbaus und die Verbreitung der Lupinen in Deutschland, sowie die Möglichkeiten der Erweitung des Lupinenanbaus. *Züchter* 7, 182–187, 206–207.

Forskål, P. (1775) *Flora Aegyptico-Arabica*, ed. C.Niebuhr.
Franco, J. do A. and da Silva, A.R.P. (1968) In: Tutin, T.G., Heywood, V.H., Burges, N.A., Moore, D.M., Valentine, D.H., Walters, S.M. and Webb, D.A. (eds) *Flora Europae*, 2, 105–106. Cambridge University Press.
Gardner, C.A. and Bennetts, H.W. (1956) *The Toxic Plants of Western Australia.* West Australian Newspapers, Perth, p. 288.
Gardner, C.A. and Elliot, H.G. (1929) The common blue lupin. *Journal of Agriculture, Western Australia* 6(2), 414–420.
Gladstones, J.S. (1954) A study of five *Lupinus* species. BSc Agric (Hons) thesis, University of Western Australia, Perth.
Gladstones, J.S. (1958a) The naturalized and cultivated species of *Lupinus* (*Leguminosae*) recorded for Western Australia. *Journal of the Royal Society of Western Australia* 41(2), 29–33.
Gladstones, J.S. (1958b) A study of the genetics and agronomy of the genus *Lupinus* with particular reference to varieties of potential value in Western Australia. PhD thesis, University of Western Australia, Perth.
Gladstones, J.S. (1969) Lupins in Western Australia. 1. Species and varieties. *Journal of Agriculture, Western Australia* 10(4), 318–324.
Gladstones, J.S. (1970) Lupins as crop plants. *I.A.B. Field Crop Abstracts* 23, 123–148.
Gladstones, J.S. (1972) *Lupins in Western Australia.* Bulletin 3834, Department of Agriculture, Western Australia, Perth, p. 37.
Gladstones, J.S. (1974) *Lupins of the Mediterranean Region and Africa.* Technical Bulletin 26, Department of Agriculture, Western Australia, Perth, p. 48.
Gladstones, J.S. (1976) Observations on the distribution and ecology in Iberia and North Africa of some annual legumes adapted to neutral and acid soils. *Plant Introduction Review, Division of Plant Industry, CSIRO (Australia)* 11, 9–23.
Gladstones, J.S. (1984) Present situation and potential of Mediterranean/African lupins for crop production. *Proceedings of the 3rd International Lupin Conference, La Rochelle, France,* pp. 18–37.
Gladstones, J.S. and Crosbie, G.B. (1979) *Wild Lupin Types Introduced into Western Australia to 1973: Collection Site Data, Preliminary Ratings of Field Characteristics and Disease Reactions, and Measurements of Seed Protein and Oil Contents.* Technical Bulletin 43, Department of Agriculture, Western Australia, Perth, p. 44.
Goldblatt, P. (1981) Cytology and phylogeny of *Leguminoseae.* In: Polhill, R.M. and Raven, P.H. (eds) *Advances in Legume Systematics, Part 2.* Royal Botanic Gardens, Kew, pp. 427–463.
Gross, R. (1986) Lupins of the old and new world – a biological cultural coevolution. First Reinhold von Sengbusch Memorial Lecture. *Proceedings of the 4th International Lupin Conference, Geraldton, Western Australia.* Department of Agriculture, Western Autralia, Perth, pp. 244–277.
Guillaume, M. (1923) Les lupins horticoles et de grande culture. Leurs emplois. *Revue de Botanique Appliquée et d'Agriculture Coloniale* 3, 758–770.
Gupta, S., Buirchell, B.J. and Cowling, W.A. (1994) Compatability/incompatibility of the rough-seeded *Lupinus* species in interspecific hybridization. In: Dracup, M. and Palta, J. (eds) *Proceedings of the First Australian Lupin Technical Symposium, Perth, Western Australia, 1994.* Department of Agriculture, Western Australia, South Perth, p. 267.
Gupta, S., Buirchell, B.J. and Cowling, W.A. (1996) Interspecific reproductive barriers and genomic similarity among the rough-seeded *Lupinus* species. *Plant Breeding* 115(2), 123–127.
Gussone, J. (1828) *Flora Siculae Prodromus.* Naples.

Hackbarth, J. and Husfeld, B. (1939) *Die Süsslupine*. Paul Parey, Berlin, p. 89.

Hanelt, P. (1960) *Die Lupinen*. A. Ziemsen, Wittenberg, p. 104.

Hondelmann, W. (1984) The lupin – ancient and modern crop plant. *Theoretical and Applied Genetics* 68, 1–9.

Hondelmann, W. (1996) *Die Lupine: Geschichte und Evolution einer Kulturpflanze*. Bundesforschungsanstalt für Landwirtschaft Braunschweig-Völkenrode (FAL), Landbauforschung Völkenrode, Sonderheft 162, p. 247.

Kalcas, E.L. (1974) *Food from the Fields: Edible Wild Plants of Aegean Turkey*. Birlik Matbaasi, Ismir.

Käss, E. and Wink, M. (1994) Molecular phylogeny of lupins. In: Neves Martins, J.M. and Beirão da Costa, L. (eds) *Advances in Lupin Research: Proceedings of the 7th International Lupin Conference, Evora, Portugal, 1993*, pp. 267–270.

Käss, E. and Wink, M. (1995) Molecular phylogeny of the *Papilionoideae* (Family *Leguminosae*): *Rbc*L gene sequences versus chemical taxonomy. *Botanica Acta* 108, 163–168.

Käss, E. and Wink, M. (1998) Molecular phylogeny and phylogeography of *Lupinus* (*Leguminosae*) inferred from nucleotide sequences of the *rbc*L gene and ITS 1+2 regions of rDNA. *Plant Systematics and Evolution* (in press).

Kazimierska, E.M. and Kazimierski, T. (1995) Cytoembryology of infertile segregants of the hybrid *Lupinus varius* L. and *L. digitatus* Forsk. *Journal of Applied Genetics* 36, 215–227.

Kazimierski, T. (1961) Interspecific hybridization of *Lupinus*. *Genetica Polonica* 2, 97–102.

Kazimierski, T. (1982) Cytogenetics of species and hybrids in the *lutei* section the genus *Lupinus*. In: Gross, R. and Bunting, E.S. (eds) *Agricultural and Nutritional Aspects of Lupines: Proceedings of the First International Lupine Workshop, Lima-Cuzco, Peru, 1980*. Deutsche Gesellschaft für Technische Zusammenarbeit (GTZ), Eschborn, pp. 51–68.

Kazimierski, T. (1988) An attempt to present lupin evolution of the Old World: materials and impressions. In: Twardowski, T. (ed.) *Proceedings of the 5th International Lupin Conference, Poznań, Poland*, PWRiL, Poznań, pp. 103–109.

Kazimierski, T. and Kazimierska, E.M. (1965) Studies on a species hybrid *Lupinus rothmaleri* Klink. × *Lupinus luteus* L. *Genetica Polonica* 6, 125–140.

Kazimierski, T. and Kazimierska, E.M. (1970) Further studies on the hybrid *Lupinus rothmaleri* Klink. × *Lupinus luteus* L. *Genetica Polonica* 11, 187–205.

Kazimierski, T. and Kazimierska, E.M. (1975) Morphological and cytological differences within the species *Lupinus luteus* L. *Acta Societatis Botanicorum Poloniae* 44, 265–275.

Klinkowski, M. (1938) Das iberisch – nordafrikanisch Heimatgebiet von *Lupinus luteus* L., *Lupinus angustifolius* L., *Lupinus hirsutus* L., *Lupinus hispanicus* Boiss. et Reut. und *Lupinus rothmaleri* Klink. *Züchter* 10, 113–126.

Knapp, O. (1931) *Lupinus albus*. Eine historische sowie botanische-variations-statistische Studie. *Zeitschrift für Pflanzenzüchtung* 16, 279–335.

Kurlovich, B.S. (1994) Ecogeographic classification of *Lupinus albus*, *L. angustifolius* and *L. luteus*. In: Neves Martins, J.M. and Beirão da Costa, L. (eds) *Advances in Lupin Research: Proceedings of the 7th International Lupin Conference, Evora, Portugal, 1993*. ISA Press, Lisbon, pp. 59–62.

Lamberts, H. (1958) Ein neuer Typus mit abbrechenden Hülsen und ein neuer Artbastard bei Lupinen. *Züchter* 28, 15–17.

Lange, R.T. (1960) The occurrence, characteristics and classification of *Rhizobium* associated with indigenous *Leguminosae* in south-western Australia. PhD thesis, University of Western Australia, Perth.

Lee, A.T. and Gladstones, J.S. (1979) Proposals to reject *Lupinus varius* and *Lupinus hirsutus* of Europe (*Fabaceae*). *Taxon* 28, 616–621.

Linnaeus, C. (1753) *Species Plantarum*, 1st edn. Laurentii Salvii, Stockholm.

Linnaeus, C. (1763) *Species Plantarum*, 2nd edn. Laurentii Salvii, Stockholm.

Linnaeus, C. (1774) *Systema Vegetabilum*, 13th edn, ed. J.A. Murray. G.C. Dieterich, Göttingen.

Lopez-Bellido, L. and Fuentes, M. (1986) Lupin crop as an alternative source of protein. *Advances in Agronomy* 40, 239–295.

Mardrus, J.C. and Mathers, P. (tr.) *The Book of the Thousand Nights and One Night*, in 4 vols, 7th impression 1949. Routledge and Kegan Paul, London.

Morison, R. (1680) *Plantarum Historiae Universalis Oxonensis*. Sheldon, Oxford.

Muschler, R. (1912) *Manual of Flora of Egypt*. Berlin.

Neves Martins, J.M. (1994) Numerical taxonomy on the study of *Lupinus albus* accessions. In: Neves Martins, J.M., and Beirão da Costa, L. (eds) *Advances in Lupin Research: Proceedings of the 7th International Lupin Conference, Evora, Portugal, 1993*. ISA Press, Lisbon, pp. 84–89.

Norris, D.O. (1956) Legumes and the *Rhizobium* symbiosis. *Empire Journal of Experimental Agriculture* 24, 247–270.

Nowacki, E. (1960) Systematics of *Genisteae* in the light of chemical analyses. *Genetica Polonica* 1, 119–143.

Nowacki, E. and Jaworski, A. (1978) A study on the evolution of the genus *Lupinus* by serological methods. *Genetica Polonica* 19, 153–163.

Nowacki, E. and Prus-Glowacki, W. (1971) Differentiation of protein fractions in species and varieties of the genus *Lupinus* with the use of serological methods. *Genetica Polonica* 12, 245–260.

Oldershaw, A.W. (1920) The value of lupins on poor light land. *Agriculture, London* 26, 982–991.

Oldershaw, A.W. (1925) Lupins and light land. *Agriculture, London* 32, 316–325.

Pakendorf, K.K. (1970) Male sterility in *Lupinus mutabilis* Sweet. *Zeitschrift für Pflanzenzüchtung* 63, 227–236.

Parkinson, J. (1640) *Theatricum Botanicum, or An Universall and Compleate Herball*. Tho. Cotes, London.

Pazy, B., Heyn, C.C., Herrnstadt, I. and Plitmann U. (1977) Studies in populations of the Old World *Lupinus* species. 1. Chromosomes of the east-Mediterranean lupines. *Israel Journal of Botany* 26, 115–127.

Pazy, B., Plitmann, U. and Heyn, C.C. (1981) Genetic relationships between *Lupinus pilosus* and *L. palaestinus* (*Fabaceae*). *Plant Systematics and Evolution* 137, 39–44.

Planchuelo, A.M. (1984) Taxonomic studies of *Lupinus* in South America. *Proceedings of the 3rd International Lupin Conference, La Rochelle, France,* pp. 40–54.

Planchuelo, A.M. (1998) Biodiversity of lupins in South America. *Proceedings of the 8th International Lupin Conference, Pacific Grove, USA* (in press).

Planchuelo, A.M. and Dunn, D.B. (1984) The simple leaved lupines and their relatives in Argentina. *Annals of the Missouri Botanical Garden* 71, 92–103.

Pliny the Elder (1st century AD) *Naturalis Historia,* tr. H. Rackham. Heinemann, London, 1950, vol. 5 of 10 vols.

Plitmann, U. (1981) Evolutionary history of the Old World lupines. *Taxon* 30, 430–437.

Plitmann, U. and Heyn, C.C. (1984) Old World lupins: taxonomy, evolutionary relationships, and links with New World species. *Proceedings of the 3rd International Lupin Conference, La Rochelle, France,* pp. 56–66.

Plitmann, U., Heyn, C.C. and Pazy, B. (1980) Biological Flora of Israel. 7. *Lupinus palaestinus* Boiss. and *L. pilosus* Murr. *Israel Journal of Botany* 28, 108–130.

Polhill, R.M. (1976) *Genisteae* (Adans.) Benth. and related tribes (*Leguminosae*). *Botanical Systematics* 1, 143–368.

Polhill, R.M. (1981a) *Papilionoideae.* In: Polhill, R.M. and Raven, P.H. (eds) *Advances in Legume Systematics, Part 1.* Royal Botanic Gardens, Kew, pp. 191–208.

Polhill, R.M. (1981b) *Crotalarieae.* In: Polhill, R.M. and Raven, P.H. (eds) *Advances in Legume Systematics, Part 1.* Royal Botanic Gardens, Kew, pp. 399–402.

Polhill, R.M., Raven, P.H. and Stirton, C.H. (1981) Evolution and systematics of the *Leguminosae.* In: Polhill, R.M. and Raven, P.H. (eds) *Advances in Legume Systematics, Part 1.* Royal Botanic Gardens, Kew, pp. 1–26.

Raven, P.H. and Axelrod, D.I. (1974) Angiosperm biogeography and past continental movements. *Annals of the Missouri Botanical Garden* 61, 539–673.

Raven, P.H. and Polhill, R.M. (1981) Biogeography of the *Leguminosae.* In: Polhill, R.M. and Raven, P.H. (eds) *Advances in Legume Systematics, Part 1.* Royal Botanic Gardens, Kew, pp. 27–54.

Rothmaler, W. (1935) Plantae novae vel criticae Peninsulae Ibericae. *Cavanillesia* 7, 111–121.

Roy, N.N. and Gladstones, J.S. (1985) Prospects for interspecific hybridization of *Lupinus atlanticus* Gladst. with *L. cosentinii* Guss. *Theoretical and Applied Genetics* 71, 238–241.

Roy, N.N. and Gladstones, J.S. (1988) Further studies with interspecific hybridization among Mediterranean/African lupin species. *Theoretical and Applied Genetics* 75, 606–609.

Savi, G. (1798) *Flora Pisana.* P. Giacomelli, Pisa.

Sibthorp, J. and Smith, J.E. (1830) *Flora Graeca.* Richard Taylor, London.

Simpson, M.J.A. (1986a) Geographical variation in *Lupinus albus* L. I. Iberia. *Zeitschrift für Pflanzenzüchtung* 96, 232–240.

Simpson, M.J.A. (1986b) Geographical variation in *Lupinus albus* L. II. Northwest Spain, the Nile Valley, the Balkans and Turkey. *Zeitschrift für Pflanzenzüchtung* 96, 241–251.

Stebbins, G.L. (1985) Polyploidy, hybridization, and the invasion of new habitats. *Annals of the Missouri Botanical Garden* 72, 824–832.

Stormon, E.J. (ed. and tr.) (1977) *The Salvado Memoirs.* University of Western Australia Press, Perth, p. 300.

Swiecicki, W., Wolko, B. and Jach, K. (1994) *Lupinus anatolicus* – an unknown species of the Old World. In: Neves Martins, J.M. and Bierão da Costa, L. (eds) *Advances in Lupin Research: Proceedings of the 7th International Lupin Conference, Evora, Portugal, 1993.* ISA Press, Lisbon, pp. 63–64.

Theophrastus (4th century BC) *Enquiry into Plants,* tr. A. Hort. Heinemann, London, 1948.

Turner, B.L. (1959) *The Legumes of Texas.* University of Texas, Austin.

Turner, B.L. (1981) *Thermopsideae.* In: Polhill, R.M. and Raven, P.H. (eds) *Advances in Legume Systematics, Part 1.* Royal Botanic Gardens, Kew, pp. 403–407.

Van Wyk, B.-E. and Schutte, A.L. (1995) Phylogenetic relationships in the tribes *Podalyrieae, Liparieae* and *Crotalarieae.* In: Crisp, M. and Doyle, J.J. (eds) *Advances in Legume Systematics 7: Phylogeny.* Royal Botanic Gardens, Kew, pp. 283–308.

Wolko, B. and Swiecicki, W.K. (1988) The application of electrophoretic separation of isozymes for genetical characterization of genus *Lupinus.* In: Twardowski, T. (ed.)

Proceedings of the 5th International Lupin Conference, Poznań, Poland. PWRiL, Poznań, pp. 103–109.

Wolko, B. and Weeden, N.F. (1990a) Isozyme number as an indicator of phylogeny in *Lupinus. Genetica Polonica* 31, 179–187.

Wolko, B. and Weeden, N.F. (1990b) Relationships among lupin species as reflected by isozyme phenotype. *Genetica Polonica* 31, 189–197.

Zhukovsky, P.M. (1929) (The genus *Lupinus* Tourn.) *Bulletin of Applied Botany, Genetics and Plant Breeding, Leningrad* 21, 241–292 (in Russian). (CSIRO, Australia, Translation 2839.)

Appendix 1.1 Key to the Mediterranean and African Species of *Lupinus* (after Gladstones, 1974)

1. Seeds ± smooth.
   2. Lower flowers of inflorescences alternate.
      3. Upper lip of calyx entire; seeds ± square, compressed; flowers white to violet-blue.

         1. *L. albus* L.

      3. Upper lip of calyx deeply 2-partite; flowers normally blue.
         4. Leaflets linear; lower lip of calyx entire to slightly 3-toothed; seeds round; plant sparsely sericeous.

            2. *L. angustifolius* L.

         4. Leaflets short and broad; lower lip of calyx deeply 3-toothed; seeds lenticular or compressed, brownish; plant coarsely hirsute.

            3. *L. micranthus* Guss.

   2. Flowers fully verticillate.
      5. Flowers golden yellow, scented; upper surface of leaflets villous.

         4. *L. luteus* L.

      5. Flowers cream to mauve, not scented; upper surface of leaflets ± glabrous.

         5. *L. hispanicus* Boiss. and Reut.

1. Seeds rough; flowers subverticillate to verticillate, usually blue.
   6. Hairs on stems ≤ 1 mm; pedicels ≤ ⅓ length of calyx, rigid; both surfaces of leaflets sericeous.
      7. Lower lip of calyx shallowly 3-toothed; tip of keep blunt, pigmented blue; seeds about 8 × 6 × 3 mm, brown or greyish with blackish markings.

         6. *L. cosentinii* Guss.

      7. Lower lip of calyx ± entire; tip of keel pointed, not pigmented blue.
         8. Pods 9–12 mm broad; seeds about 7 × 6 × 3 mm, mottled reddish brown with a smooth hump over the hilum.

            7. *L. digitatus* Forsk.

         8. Pods 16–18 mm broad; seeds about 10 × 8 × 4 mm, mottled reddish brown on pale yellow.

            8. *L. princei* Harms

   6. Hairs on stem > 1 mm; lower lip of calyx entire.
      9. Hairs on stem 3–4 mm; pedicel ≥ ½ length of calyx, lax.

10. Flowers very large, deep blue (rarely pink) with a white band up the centre of the standard; leaflets softly villous; seeds 10–14 × 9–12 × 6–8 mm, mottled brownish red; plants erect.

9. *L. pilosus* Murr.

10. Flowers pale yellow to pale pink, tinged with blue; leaflets sericeous; seeds slightly smaller than 9, mottled brown; plants rosetted.

10. *L. palaestinus* Boiss.

9. Hairs on stem 1–2 mm; pedicel ≤ ½ length of calyx, rigid.

11. Leaflets sericeous; pedicel ≤ ⅓ length of calyx; standard with a central white band, broadening towards the upper margin; seeds about 8–11 × 6–8 × 5 mm, mottled brown to pinkish brown with a smooth hump over the hilum.

11. *L. atlanticus* Gladst.

11. Upper surface of leaflets glabrous; pedicel about ½ length of calyx; rounded white spot on standard, not approaching the upper margin.

12. *L. somaliensis* Baker

2 Genetic Resources in Lupins

BEVAN J. BUIRCHELL AND WALLACE A. COWLING

Agriculture Western Australia, 3 Baron-Hay Court, South Perth, Western Australia 6151 and the Centre for Legumes in Mediterranean Agriculture, University of Western Australia, Nedlands, Western Australia 6907

Introduction

Genetic resources have been important in the development of many crop species in the past 100 years of 'modern' plant breeding (Shands and Weisner, 1991, 1992), although they are often underutilized. In the case of lupins they are critical.

The natural genetic diversity of a species is normally taken to include wild or weedy types, landraces or 'primitive cultivars', and breeding lines or cultivars developed in the past 100 years (Frankel, 1977). This chapter will also include genetic stocks such as mutants, selections and transgenic or interspecific types, which are becoming increasingly important for lupin domestication and improvement.

Sweet cultivated lupins have been fully domesticated in only the past 60 years and are genetically very close to their wild and landrace parents. Primary crosses between domesticated types and wild or landrace types continue to play an extremely important role. For example, *L. angustifolius* cultivars Gungurru and Merrit have considerable advantages over previous varieties and together occupy 89% of the area of lupins in Western Australia (Garlinge, 1996). Both are derived from the same primary cross between a cultivated and wild type of *L. angustifolius* (Gladstones, 1994). In addition, new species are currently undergoing domestication directly from wild forms. Several examples will be referred to in this chapter.

The importance of collection, documentation, evaluation, conservation and utilization (in breeding programmes) of wild and landrace lupin species was recognized by early breeders in Germany, Poland and Holland, several of whom collected lupins in Mediterranean countries in the 1930s and 1940s (Hackbarth and Troll, 1956). Early maturity in *L. albus* and *L. luteus* was found in landrace types from Palestine and Egypt, and resistance to powdery mildew and *Fusarium* wilt in *L. luteus* from Portugal (Hackbarth and Troll, 1956).

Much loss of lupin genetic resources occurred in Germany during World War II (Hackbarth and Troll, 1956), but further collections were made from both east

 Lupins as Crop Plants: Biology, Production and Utilization (eds J.S. Gladstones, C.A. Atkins and J. Hamblin)

and west into the Mediterranean region in the late 1960s and 1970s (Gladstones, 1973, 1976; Lehman and Hammer, 1983) and 1980s (Mota *et al.*, 1981, 1982; Simpson and McGibbon, 1982a,b; Cowling, 1986; Papineau and Huyghe, 1989; Buirchell, 1992; Hammer and Perrino, 1995). This work added many new introductions to the Australian Lupin Collection between 1968 and 1989 (Gladstones and Crosbie, 1978; Clements and Cowling, 1990a).

Lupin genetic resources have been exchanged and shared freely among researchers internationally, and often collaborative projects occur with the host country – for example, the fruitful collaboration between Morocco and Australia on wild lupins in Morocco (Buirchell, 1992; Bounejmate *et al.*, 1994). Lupin production in the 1990s in Australia, Russia, Ukraine, Belarus, Poland, Chile, South Africa and the USA is based on improved varieties from crosses with this shared genetic resource. There has been close collaboration between the major breeding programmes, and researchers and breeders continue to exchange both wild and domesticated material. The International Lupin Association played a large role in developing this collaboration in the 1980s through conferences, an international newsletter and a survey on lupin genetic resources that was prepared by the Association for publication by the International Board of Plant Genetic Resources (Bennencourt *et al.*, 1989).

In this chapter we have largely restricted our discussion to the annual Mediterranean and North African lupin species, of which the distributions are shown in Chapter 1. They include all the domesticated lupin species grown as commercial crops (*L. angustifolius*, *L. albus*, *L. luteus*); recently domesticated species not in commercial production (*L. cosentinii* and *L. atlanticus*); lupin species that are currently undergoing domestication (*L. pilosus*); and species that are completely wild and as yet have no economic value (*L. micranthus*, *L. hispanicus*, *L. princei*, *L. palaestinus*, *L. digitatus* and *L. somaliensis*).

Apart from *L. mutabilis*, we shall not attempt to detail the immense genetic resources of wild lupins in the New World, which have been poorly studied, documented and collected for conservation in gene banks. This does not imply that they have no potential value. *L. mutabilis* has a long history of cultivation in the Andes of South America which parallels the domestication of *L. albus* by early farmers in the Mediterranean region (Hondelmann, 1984). More than 83 species of the genus *Lupinus* have been identified in the Andean region of South America (Hernández Bermejo and León, 1994). The ornamental Russell lupin is derived from interspecific crosses among North American lupin species and is of international importance to horticulture. Several perennial species such as *L. arboreus* or Russell lupins have been spread intentionally around the world for erosion control or soil improvement. Many of the New World species are perennial and genetic conservation is not normally possible in countries with Mediterranean climates.

Genetic Resource Collections

Major collections of lupin genetic resources (> 200 accessions) exist in several countries (Table 2.1). The dominant species in collections usually reflects the major breeding objectives of that country. For instance, Australia has a large collection of *L. angustifolius* and has a good representation of the rough-seeded lupins; Poland and other eastern European countries have substantial collections of *L. luteus*; and western Europe has substantial collections of *L. albus*. South American institutions specialize in collections of *L. mutabilis*. The proportion of wild and landrace types in international collections (in contrast to mutants, breeding lines or cultivars) is above 80% in most lupin species (Table 2.1) which reflects the importance of these types in breeding and its short history. Although the four main agricultural lupin species (*L. angustifolius, L. albus, L. luteus* and *L. mutabilis*) appear to be well represented in collections (Table 2.1), there is a high level of duplication of accessions due to exchange of seed among institutions.

Inter- and intraspecies variation in the Australian Lupin Collection

The Australian Lupin Collection has grown from its origins as a small breeder's collection in Dr J.S. Gladstones' lupin breeding programme at the University of Western Australia in the 1950s and 1960s (Gladstones, 1994) to a fully documented and well conserved entity with short-term storage and evaluation at Agriculture Western Australia in South Perth (known as the Agriculture WA International Lupin Collection) and long-term storage in the Australian Temperate Field Crops Collection in Horsham, Victoria. Australia now has the world's largest and most comprehensive collection of Old World lupins (Table 2.1). All Old World lupin species are represented except for *L. somaliensis*, which is described only by a single botanical specimen (Gladstones, 1974; Chapter 1, this volume) and is possibly extinct. Only a few accessions of *L. digitatus* exist in the collection and these originated from one collection site in Egypt. Collection site and evaluation data on the collection was first published by Gladstones and Crosbie (1978) and updated by Clements and Cowling (1990b, 1991). Work is underway to make the database available for international access on the Internet.

A major focus of this chapter is to summarize and discuss collection site and evaluation data of species in the Australian Lupin Collection.

Collection site pH

L. angustifolius, L. luteus and *L. hispanicus* are generally found on acidic soils (Table 2.2) though some accessions of *L. angustifolius* and *L. luteus* have been found on alkaline soils that do not contain free lime. *L. albus* has been found on a range of soils from pH 5 through to pH 9.5. The *L. albus* collection is biased by the large number of Egyptian accessions from high pH soils: the soils in the Nile Valley are very alkaline and yet *L. albus* grows well in this region, and is also found on limestone-derived soils in Greece (Simpson and McGibbon, 1983). *L. micranthus* is found across a range of pH from 5.5 to 7.5.

L. pilosus, L. atlanticus and *L. cosentinii* are found on soils ranging from slightly acid (pH 5.7) to highly alkaline (pH 9.5). The only accession of *L. princei* which has a soil pH recorded came from soils of pH 5.7 in the highlands of Kenya. No soil pH data are available for *L. digitatus* although it has been reported to come from calcaric fluvisol type soils. The soils of the Nile Valley, where *L. digitatus* has been reported to grow in the past (Gladstones, 1974), are highly

Table 2.1. Summary of lupin genetic resources held in major national collections worldwide (includes collections with more than 200 lupin accessions identifed to species level). The total number of accessions for each species includes wild/weedy (W) and landrace (L) types, mutants, breeding lines, cultivars and other genetic stocks.

		Lupin species											
		L. albus[2]		*L. angustifolius*		*L.luteus*		*L.hispanicus*		*L. micranthus*		*L. cosentinii*	
Country	Source[1]	W + L[3]	Total	W + L	Total	W + L	Total	W + L	Total	W + L	Total	W + L	Total
Australia	a	637	775	1229	1756	90	207	74	74	43	43	103	239
Bolivia	c												
Canada	b	n.a.	42	n.a.	34		58						
France	b	n.a.	1398	n.a.	168	n.a.	148						
Germany-A	b	20	115	291	373	71	169	97	97	3	3	33	33
Germany-B	b	n.a.	190	n.a.	465	n.a.	792	n.a.	75	n.a.	20	n.a.	6
Peru-A	b			60	60	50	50	30	30				
Peru-B	c												
Poland	b	127	250	119	248	144	421	61	61	2	2	7	7
Portugal-A	b	270	270	240	240	210	210	108	108	6	6	11	11
Portugal-B	b	377	377	37	37	78	78					5	5
Russia	b	n.a.	423	n.a.	687	n.a.	774	n.a.	17	n.a.	7	n.a.	32
Spain-A	b	n.a.	567	537	537	n.a.	240	177	177			13	13
Spain-B	b	600	656	500	519	300	335	150	150				
Spain-C	b	0	6	330	410	10	80	253	253				
United Kingdom	b	n.a.	1100										
USA	b	n.a.	157	n.a.	150	n.a.	61	n.a.	34	n.a.	1		
% W + L (where available)		83		80		61		100		100		56	
TOTAL			6326		5684		3623		1076		82		346

Table 2.1. *(continued)*

	Lupin species															
	L. pilosus		*L.atlanticus*		*L.palaestinus*		*L. digitatus*		*L. princei*		*L. mutabilis*		Other[4]		Total all species	
Country	W + L	Total	W + L	Total	W + L	Total	W + L	Total	W + L	Total	W + L	Total	W + L	Total	W + L	Total
Australia	140	146	62	71	10	10	3	3	6	6	43	123	59	63	2499	3516
Bolivia											220	220			220	220
Canada											n.a.	2	n.a.	78	n.a.	214
France											n.a.	332			n.a.	2046
Germany-A	6	6	18	18	4	4					993	998	52	57	1588	1873
Germany-B	n.a.	11									n.a.	25	n.a.	90	n.a.	1674
Peru-A											1800	1800			1940	1940
Peru-B											1725	1725			1725	1725
Poland	7	7	9	9	4	4	1	1			14	14	20	25	515	1049
Portugal-A													10	10	855	855
Portugal-B	10	10									149	149	5	5	661	661
Russia	n.a.	20	n.a.	8							n.a.	132	n.a.	256	n.a.	2356
Spain-A											20	20			n.a.	1554
Spain-B											1238	1238			2788	2898
Spain-C															593	749
United Kingdom											n.a.	200			n.a.	1300
USA	n.a.	4	n.a.	2							n.a.	68	n.a.	264	n.a.	741
% W + L (where available)	96		91		100		100		100		99		91		86	
TOTAL		204		108		18		4		6		7046		848		25371

Table 2.1. *(continued)*

[1]Reproduced with permission of the International Plant Genetic Resources Institute (IPGRI), from *Promoting the Conservation and Use of Underutilized and Neglected Crops Lupin* – Lupinus *spp.* by W.A. Cowling, B.J. Buirchell and M.E. Tapia, IPK, Gatersleben/IPGRI, Rome, 1998. [2]*L. albus* includes *L. albus* var. *albus, L. albus* var. *graecus, L. graecus* and *L. termis.* [3]W + L represents the sum of Wild and Landrace types in the collection. [4]'Other' species are mostly New World species from North and South America.

Countries/Institutions:

Australia:	International Lupin Collection, Agriculture WA, 3 Baron-Hay Court, South Perth, Western Australia 6151, Australia (Curator – Dr W.A. Cowling), with copies in long-term storage at the Australian Temperate Field Crops Collection, Victorian Institute for Dryland Agriculture, Private Bag 260, Horsham, Victoria 3400, Australia (Curator – A. McIntyre)
Bolivia:	Centro Fitotechnico y Ecogenetico de Pairumani, km 20 Carr. Cochabamba-Sipe Sipe, Cochabamba, Bolivia
France:	INIA Station d'Amélioration des Plantes Fourragères, 86600 Lusignan, France (Lupin breeder – Dr C. Huyghe)
Germany-A:	Institut für Pflanzenbau und Pflanzenzüchtung (FAL), Bundesallee 50, 38116 Braunschweig, Germany
Germany-B:	Genebank, Institute of Plant Genetics and Crop Plant Research (IPK), Correnstrasse 3, 06466 Gatersleben, Germany (Database – Dr H. Knüppfer)
Peru-A:	Centro de Investigatión en Cultivos Andinos, Universidad del Cusco, Apartado 921, Cusco, Peru
Peru-B:	INIA Santa Ana Research Station, Huancayo, Peru
Poland:	Plant Experiment Station, Wiatrowo, 62-100 Wagrowiec, Poland (Curator – Prof. W.K. Swiecicki)
Portugal-A:	Germplasm Bank – Genetics, National Agricultural Research Station, Quinto do Marquês, 2780 Oeiras, Portugal (Curator – E. Bettencourt)
Portugal-B:	Instituto Superior de Agronomia, Universidade Técnica de Lisboa, Tapada da Ajuda, 1399 Lisboa Codex, Portugal (Curator – Dr J.M. Neves Martins)
Russia:	N I Vavilov Scientific Research Institute of Plant Industry, 44 Herzen Street, St Petersburg 190000 Russia
Spain-A:	Centro de Recursos Fitogeneticos, Institute Nacional de Investigaciones Agrarias, Autovía de Aragón km 36, Finca La Canaleja, 28800 Alcalá de Henares (Madrid), Spain (Curator – C. de la Cuadra)
Spain-B:	Servicio de Investigación y Desarrollo Tecnológico, Consejería de Agricultura y Comercio, Junta de Extremadura, Apartado 22, 06080 Badajoz, Spain (Head of Genebank – Dr F. González)
Spain-C:	Departamento de Pastos y Forrajes, Servicio de Investigación Agraria, Cordel de Medina s/n, Apartado Oficial, 37008 Salamanca, Spain (Lupin breeder – Dr C. Crespo)
UK:	Agricultural Botany, Plant Science Laboratories, University of Reading, PO Box 221, Whiteknights, Reading RG6 2AS, UK
USA:	Western Region Plant Introduction Station, USDA Agricultural Research Service, 59 Johnson Hall, Washington State University, Pullman, Washington 99164 6402, USA (Curator Legumes – Dr C. Simon)

alkaline (pH 9.0). *L. palaestinus* comes from the alkaline soils of coastal Israel. It is clear that the rough-seeded lupins are generally more adapted to alkaline soils than the smooth-seeded types.

Collection site altitude

L. angustifolius and *L. albus* grow from sea level near the Mediterranean coast into nearby and inland mountain ranges (Table 2.3). Landraces of *L. albus* were more widely selected by early farmers in extremely diverse environments of the Mediterranean region, Atlantic Ocean islands (such as the Canary and Azores Islands) and north and east Africa. The highest elevation of *L. albus* landraces (2460 m) was recorded in the highlands of Kenya. *L. luteus* is restricted to altitudes below 1000 m, while *L. hispanicus* and *L. micranthus* extend into the mountains to 1350 m.

L. cosentinii is predominantly found in the lower altitudes along the coastal plains (Table 2.3). There have only been a few collections from higher altitudes in Morocco, and these may be valuable as a source of frost tolerance in this species. *L. pilosus* is found from the low coastal altitudes to the higher altitudes (1100 m) in the mountains, especially in Crete. *L. atlanticus* only grows in the

Table 2.2. Soil pH of collection sites for wild and landrace *Lupinus* spp. in the Australian Lupin Collection.

Lupin species	Min.	Mean	Max.	SD	*n*
L. albus	5.0	8.25	9.50	1.24	172
L. angustifolius	4.20	6.73	9.00	0.72	445
L. atlanticus	6.50	7.82	9.50	0.78	35
L. cosentinii	6.20	7.64	9.00	0.70	37
L. hispanicus	4.50	5.90	7.00	0.49	26
L. luteus	5.50	6.46	8.5	0.85	22
L. micranthus	5.50	6.31	7.50	0.55	27
L. mutabilis	6.50	6.75	7.00	0.35	2
L. pilosus	5.50	7.60	9.50	1.01	46
L. princei	–	5.7	–	–	1

Table 2.3. Altitude of collection sites for wild and landrace *Lupinus* spp. in the Australian Lupin Collection.

Lupin species	Min.	Mean	Max.	SD	*n*
L. albus	1	480	2460	518	94
L. angustifolius	1	427	1800	323	610
L. atlanticus	460	985	1630	248	37
L. cosentinii	10	138	1015	206	45
L. hispanicus	149	745	1250	313	47
L. luteus	10	166	500	163	25
L. micranthus	10	499	1350	474	16
L. mutabilis	2200	2807	3500	399	7
L. pilosus	3	231	1100	258	44
L. princei	2100	2280	2460	255	2

mountainous areas of Morocco between 450 and 1630 m. *L. princei* is restricted to the eastern highlands of Africa at high altitudes – higher altitudes on average than any other wild lupins in the Old World. It is found on Mt Kenya at 2200 to 2600 m. *L. digitatus* is reported to occur in the mountainous regions of the southern Sahara and the valleys of the Nile and Senegal Rivers over a range of altitudes (Gladstones, 1974). *L. palaestinus* grows in coastal plains and into the mountains of Israel at altitudes of probably less than 1000 m.

Collection site rainfall

Most lupin species grow in environments with at least 300 mm average annual rainfall, with the exception of *L. angustifolius, L. cosentinii* and *L. atlanticus*, which extend to lower rainfall (Table 2.4). The latter two species, along with *L. palaestinus* and presumably *L. digitatus*, occur mainly where rainfall is less than 600 mm. The remaining species are normally found in rainfall regions above 600 mm, with some collections of *L. pilosus* at sites with up to 2000 mm annual rainfall.

Flowering time

The flowering time of accessions is measured over winter–spring in the mild mediterranean climate of Perth, Western Australia, at approximately 32°S latitude (Table 2.5). The earliest of most species flower in 70–80 days (indicating very little vernalization requirement), with one very early flowering accession of *L. albus* (59 days) from Syria. *L. micranthus* and especially *L. hispanicus* and *L. princei* tend to flower later and have long rosette stages with a strong vernalization requirement for initiation. These species tend to be found at higher elevations. Days to flowering within *L. angustifolius* were negatively correlated with winter temperature at collection sites, but positively with collection site rainfall, and weakly so with altitude (Clements and Cowling, 1994).

Of the rough-seeded lupins, *L. pilosus* has the widest range of flowering times with some accessions flowering as early as 71 days and others as late as 134 days. *L. atlanticus* and *L. cosentinii* also have a large range of flowering times among

Table 2.4. Rainfall of collection sites for wild and landrace *Lupinus* spp. in the Australian Lupin Collection (from Cowling, 1994).

Lupin species	Min.	Mean	Max.	SD	*n*
L. albus	350	562	1200	189	36
L. angustifolius	200	643	1500	222	351
L. atlanticus	200	421	650	116	35
L. cosentinii	100	412	700	135	36
L. hispanicus	450	766	1500	215	44
L. luteus	300	645	1100	229	18
L. micranthus	525	701	1000	181	11
L. mutabilis	500	1150	1700	493	4
L. palaestinus	500	566	600	58	3
L. pilosus	350	675	2000	331	33
L. princei	584	792	1000	294	2

Table 2.5. Days to flowering of *Lupinus* spp. evaluated in the field in Western Australia.

Lupin species	Min.	Mean	Max.	SD	*n*
L. albus	59	92	137	19	562
L. angustifolius	75	103	144	9	1127
L. atlanticus	82	102	126	12	53
L. cosentinii	76	103	134	10	50
L. digitatus	84	93	102	9	3
L. hispanicus	118	143	157	9	63
L. luteus	88	111	150	14	75
L. micranthus	92	114	144	11	33
L. mutabilis	69	89	113	12	36
L. palaestinus	81	91	103	7	9
L. pilosus	71	92	134	16	121
L. princei	136	145	150	8	3

accessions. Days to flowering in *L. pilosus* were not correlated with altitude of collection site, but were strongly so with rainfall (Clements *et al.*, 1996). Several ornamental selections of *L. pilosus* from European botanical gardens were early flowering with little vernalization requirement, as a result of selection for thermoneutrality by early horticulturalists (Clements *et al.*, 1996). In *L. atlanticus* the northern population in Morocco, which is late flowering, comes from intermediate altitudes. *L. princei* has the longest time to flowering of all the rough-seeded lupin species. This delay in flowering time makes it unsuitable at present for cultivation in a mediterranean climate. *L. palaestinus* has a narrow range of flowering times, possibly reflecting the narrowness of its habitat.

Seed weight

L. albus has the widest range of seed weights, which reflects selection among landraces for this character by ancient farmers. Selection appears to have occurred towards opposite extremes, with the smallest types in Ethiopia (16 g/100 seeds) and the largest in Italy (77.0 g/100 seeds). *L. angustifolius* has not been selected for human consumption as in *L. albus*. Nevertheless, forage and green manuring accessions tend to have large and in some cases permeable seeds compared with the smaller non-permeable wild types. On average *L. angustifolius* weighs 11.3 g/100 seeds with some accessions as small as 2.9 and others as large as 24.4 g/100 seeds. *L. luteus* has a narrower range of seed weights and on average has smaller seeds than *L. angustifolius*.

L. pilosus covers a range of seed sizes from 21.6 to 73.1 g/100 seeds (Table 2.6), which is nearly as wide as the range for *L. albus*. Within *L. pilosus* there is also considerable variation in seed coat colour and markings and a small number of accessions with relatively smooth seed (Clements *et al.*, 1996). *L. atlanticus* seed size varies from small types in southern Morocco, weighing 19.9 g/100 seeds, to the larger seeded types in the north at 47 g/100 seeds. *L. cosentinii* has the smallest seeds of the rough-seeded lupins, except for *L. digitatus*, with some accessions as small as 5.8 g/100 seeds. *L. palaestinus* and *L. princei* range from

20 to 40 g/100 seeds. The seed size for the sole collection of *L. digitatus* is 12 g/100 seeds.

Height at maturity

L. micranthus, *L. hispanicus* and *L. digitatus* are the shortest of the Old World lupins (Table 2.7). *L. micranthus* is easily distinguished from the other species by its very low stature. Some *L. albus* accessions grow to 2 m high and these have very thick stems. *L. mutabilis* is another lupin species that is very tall at maturity. Some of the species or genotypes within them, especially those requiring vernalization, form rosettes in the early stages of growth and it is only in the warmer part of spring that the main stem elongates.

Pods on main stem

Pod set on the main stem is quite variable among and within species. *L. albus* and *L. angustifolius* on average set 6 to 10 pods on the main stem but some

Table 2.6. Weight (g) of one hundred seeds of the different *Lupinus* spp. evaluated in the field in Western Australia.

Lupin species	Min.	Mean	Max.	SD	*n*
L. albus	16.0	31.7	77.0	8.4	536
L. angustifolius	2.9	11.3	24.4	3.9	586
L. atlanticus	19.9	31.6	47.0	6.8	37
L. cosentinii	5.8	16.3	22.7	3.8	39
L. digitatus	–	12.0	–	–	1
L. hispanicus	3.8	6.4	13.4	2.3	56
L. luteus	5.3	9.7	15.3	2.1	66
L. micranthus	5.7	8.7	13.9	2.4	21
L. mutabilis	1.7	19.0	25.4	5.1	26
L. palaestinus	19.0	21.9	26.5	4.1	3
L. pilosus	21.6	43.7	73.1	10.6	62
L. princei	–	33.1	–	–	1

Table 2.7. Height at maturity (cm) of *Lupinus* spp. evaluated in the field in Western Australia.

Lupin species	Min.	Mean	Max.	SD	*n*
L. albus	40	124	200	33.5	379
L. angustifolius	10	85	170	22.3	1117
L. atlanticus	25	67	126	17	45
L. cosentinii	30	96	160	25.4	29
L. digitatus	35	49	62	19.1	2
L. hispanicus	15	48	73	17.3	14
L. luteus	30	96	150	26.4	53
L. micranthus	17	35	55	9.1	28
L. mutabilis	82	132	180	23.4	32
L. palaestinus	30	61	80	14.1	10
L. pilosus	35	82	140	18.8	114
L. princei	74	82	88	7.4	3

Table 2.8. Number of pods on the main stem of *Lupinus* spp. evaluated in the field in Western Australia.

Lupin species	Min.	Mean	Max.	SD	*n*
L. albus	0	6	19	4.5	148
L. angustifolius	1	10	27	5.7	122
L. atlanticus	1	11	20	5.6	22
L. cosentinii	1	6	32	8.8	12
L. hispanicus	13	20	26	4.8	6
L. luteus	1	17	30	7.1	19
L. micranthus	1	11	16	3.4	16
L. mutabilis	0	5	12	4.1	12
L. pilosus	1	8	18	3.0	39
L. princei	–	12	–	–	1

accessions have set up to 27 pods. *L. luteus* normally sets more pods on the main stem than the previous two species, and the maximum observed so far is 30 pods (Table 2.8). *L. hispanicus* seems to be the most consistent at setting pods on the main stem with a minimum of 13 and a maximum of 26 in different accessions. The results for *L. palaestinus* are not presented because this species requires insect pollination for reasonable pod set and these experiments were conducted in an insect-proof screenhouse.

Genetic diversity in lupin species

Lupinus albus

There are two varieties of albus lupins, *L. albus* var. *graecus* (the wild ancestor) and *L. albus* var. *albus* (bitter cultivated landraces) (Gladstones, 1974; Chapter 1, this volume).

L. ALBUS VAR. *GRAECUS.* Simpson and McGibbon (1983) collected var. *graecus* on mainland Greece as well as from the Greek islands of Kea and Andros. They also observed natural hybrids between the two varieties on mainland Greece. There have been reports of non-permeable white seeded accessions from the island of Tinos that may also represent hybrids between var. *albus* and var. *graecus*. Cowling (1986) collected var. *graecus* on the island of Andros, and later received from the University of Thessaloniki a donation of var. *graecus* from mainland northern Greece. *L. albus* var. *graecus* was found by Perrino *et al.* (1984) in Calabria in southern Italy in 1983, perhaps taken there by the early Greek settlers of the area. This is the first report of wild *L. albus* beyond the Balkans.

L. albus var. *graecus* is a late-flowering annual with small, non-permeable seeds coloured red or mottled red. The pods shatter freely and seeds are bitter. The mainstem pod set is substantial and this character has been used by breeders to increase pod set in domesticated forms of *L. albus* var. *albus*.

L. ALBUS VAR. *ALBUS.* *L. albus* var. *albus* has been found in all countries bordering the Mediterranean sea and northern and eastern African countries such as Sudan, Ethiopa and Kenya. The islands in the Atlantic Ocean such as the Azores, Madeira and Canary Islands are also sources of *L. albus* landraces.

There is strong evidence from ancient writings that *L. albus* var. *albus* was used by ancient farmers in Greece and became well established in the Roman Empire (Gladstones, 1976; Chapter 1, this volume). Many common names exist for it in Mediterranean countries derived from the Latin *lupus* (wolf) or the Greek *thermos* (hot). One can trace the movement of lupins around the Mediterranean region by examining the present use of the Latin derivative in northern and western Mediterranean countries (words such as lupin, lupina, lupine, lupino, lubinu) and the Greek derivative in eastern Mediterranean and north African countries (words such as termis, turmas, turmus, tarmus, turmusa, furmesa) (Neves Martins, 1991). Local dialects in Portugal and Spain appear to have a mixture of the two derivatives, probably as a result of the Moorish invasions of this region from the south (words such as altramuz, tremoços, tramoços, entramoços, chochos). In Turkey the common name is 'aci bakla', which invokes a hypothesis that there were different origins for the Turkish landraces (Neves Martins, 1991). Many Mediterranean landraces have been carried to other parts of the world with emigrants from Italy, Spain and Portugal. *L. albus* var. *albus* is probably the most widely grown lupin in the world.

Simpson and Neves Martin (1984) and Simpson (1986a,b) studied the geographical variation among *L. albus* collected from the Iberian Peninsula, the Balkans, the Nile Valley and Turkey (including material collected by M. Lunoble and J. Papineau of INRA, Lusignan, France). Using principal component analysis and non-metric multidimensional scaling they were able to characterize four distinct geographical races among the wild and cultivated *L. albus.*

1. The Iberian race had a wide variation in morphological features. According to Simpson (1986a) there was a cline running from south Portugal through south-west Spain to north Portugal and north-west Spain. Time to flowering, leaf size and leaf number increased from south to north, while pod and seed size tended to decrease. Within the Iberian populations they recognized two distinct local population types. The first type from the Azores Isands, Catalonia, Cadiz and central Portugal was characterized by large leaves, pods and seeds. The second local ecotype that was commonly cultivated in the northern provinces of Palencia and Leon was distinguished by small leaves, pods and seeds as well as a relatively short time to flowering. Simpson (1986b) concluded that the Iberian race was intermediate for most characters and lacked features to distinguish it from other races.

2. The Nile Valley race had strong affinities with southern Iberian populations but tended to have smaller and fewer main-stem leaves, as well as shorter rosette stages (Simpson, 1986b). It has been referred to as *L. termis* Forskål in the past, but the name is considered synonymous with *L. albus* (Gladstones, 1974; Chapter 1, this volume) and the Linnaean name takes precedence.

3. The Turkish race was a quite distinct and homogeneous group characterized by unusually small pods, seeds and leaves, few main-stem leaves, few primary pods, more secondary pods and numerous tertiary and higher order level pods (Simpson, 1986b). The earliest flowering genotypes are from this region and they also lack a vernalization response (Simpson, 1984). In addition, there were populations that contained pink flowers, a feature not found in any other group. (As stated above, the different common name of lupins in Turkey invokes a hypothesis of a different origin of this group.)

4. The Balkan race consisted of both wild and cultivated populations. This race was characterized by short stature and abundant primary and secondary pods.

The wild types (var. *graceus*) were distinguished by their shorter stems, deep blue flowers, shattering pods, and coloured seed with impermeable testa.

The conclusions drawn by Simpson and Neves Martin (1984) and Simpson (1986a,b) allow breeders to make use of the genetic resources from these countries to improve specific characteristics such as early flowering, early maturity, pod setting, and seed size.

Recent collections from the Azores Islands by Papineau and Huyghe (1989) have been analysed for genetic variability (Huyghe *et al.*, 1990). Flowering time ranged from late spring types to very late winter types. A large range of variation was also observed for vegetative development, with interesting short late genotypes, and for the yield components of pod setting and seed size. In Western Australia, the Azores material is late flowering with very strong, thick stems and large leaves. Huyghe *et al.* (1990) also suggested that the Azorean lines may be sources of resistance for some lupin diseases such as rust (*Uromyces lupinicolus*), *Pleiochaeta setosa* and anthracnose (*Colletotrichum gloeosporioides*) since these diseases were prominent on the Islands. Yang *et al.* (1996) showed that the Azorean material as well as lines from Crete were resistant to *Pleiochaeta setosa* root rot.

The Egyptian and Sudanese types that have been collected and studied in Western Australia show very early rapid growth and early flowering. They seem to have been selected for rapid spring growth and early harvest to fit the tight rotations under flood irrigation in the Nile Valley. Further analysis of Egyptian material by Hoballah *et al.* (1990) showed that there was a broad range of variation for yield, yield components and the chemical composition of the seed. The Ethiopian lines are interesting in that there seem to be two distinct populations. Firstly, the normal large-seeded types resemble those found in Sudan and Egypt. The second comprise a small-seeded type (< 20 g/100 seeds) which also has small leaves and does not readily form hybrids with other *albus* lines. The great diversity in *L. albus* var. *albus* results from farmer selection over the ages. Landraces were most likely taken with invading armies to the extremes of its climatic tolerance in diverse regions of the Mediterranean and northern Africa during the past few millennia. Breeding programmes in Australia, France, Britain, Ukraine, Russia, Poland, Germany, USA, Chile, South Africa, Egypt, Italy, Spain and Portugal are exploiting this genetic variation to breed and adapt the species to their target environments. The addition of mutant lines with low alkaloids, determinancy, early flowering, dwarfism, self-finishing and so on have assisted in the spread of cultivated sweet albus lupins to a diverse array of climates around the world including the continental climates in northern Europe (Cowling, Chapter 4, this volume).

L. angustifolius

The Australian Lupin Collection has the largest and most diverse collection of *L. angustifolius* in the world (Table 2.1). Many accessions have been contributed by other international genetic resource centres but the majority have been collected from the wild by Australian agricultural researchers interested in Mediterranean legumes (Gladstones, 1973, 1976; Clements and Cowling, 1990a). Before the Gladstones collecting missions of 1968 and 1973 there were few wild *L. angustifolius* in international genebanks outside of eastern Europe.

Gladstones' collections of *L. angustifolius* in Spain, Portugal, Morocco, Tunisia, Italy and Sicily and earlier collections from Israel were grown out in Western Australia in 1975 and evaluated for agronomic traits, disease susceptibility and chemical composition. This information (Gladstones and Crosbie, 1978) represents a substantial view of the genetic diversity in *L. angustifolius.*

The time to flowering varied from 90 to 120 days. A distinct early-flowering group (90–95 days) was observed in accessions collected at low altitudes of coastal Israel and Morocco, and from the Anti Atlas mountains of southern Morocco. Tallness was inversely related to branching. The taller types tended to be less branching while the short early rosetted types were strongly branching. The latter were usually found at high altitude and were fine textured and small seeded. Another distinctive group consisted of the short, profusely branching, fine, small-seeded yet early-flowering types from the sandy Moroccan lowlands and the Anti Atlas mountains, along with the later-flowering but otherwise similar lines from two sandy locations in the extreme coastal area of south west Morocco.

Seed number per pod varied from four to six or seven, with larger numbers being associated with small seeds. Gladstones and Crosbie (1978) observed that individual seed weight in *L. angustifolius* showed a bimodal distribution with genotypes about equally divided between a small-seeded group (mostly 50–100 mg) and a medium to large-seeded group (mostly 120–160 mg, with a skewed tail up to 200 mg). True large-seeded types were found in Iberia and Morocco. The large-seeded Iberian types were found below 900 m, and one confined to inland regions, while the Moroccan large-seeded types were found exclusively in northern Morocco between 300 m and 1000 m. There is some evidence that the large-seeded types may have been used in cultivation. The true small-seeded types are found throughout the species' natural range and extend from sea level to 1300 m. The smallest-seeded types were from the central Spanish highlands, the sandy coastal lowlands of Morocco and southern Spain, and the Anti Atlas mountains of Morocco.

Clements and Cowling (1994) more recently analysed the diversity in *L. angustifolius* collected from the Aegean region using the material collected earlier by Cowling (1986) and other collectors in the Aegean (Clements and Cowling, 1990a). They observed that 'the unique geographical and climatic features of the Aegean region, with its many isolated islands and variable microclimates, have assisted the generation of an unusually high degree of genetic variation in wild *L. angustifolius.*' Using hierarchical cluster analysis on

19 morphological traits from 157 accessions they were able to identify 13 groups that accounted for 81% of genotype and 41% of genotype × trait sums of squares. There were two groups originating from the Dhodhekanisos Islands that had very desirable agronomic characteristics such as large seeds and many pods on the main stem. Accessions from the Kikladhes Islands were extremely variable and belonged to a large number of the groups. The accessions from northern Greece were later-flowering, shorter and smaller-seeded types and grouped with some accessions from the southern Greek islands. The earliest-flowering (73 days) wild *L. angustifolius* was collected from Aegina Island south of Athens. This study was valuable in identifying geographical areas that warranted further collecting due to the occurrence of accessions with useful agronomic traits.

L. atlanticus

L. atlanticus was named by Gladstones following a study in 1968 of the Kew Herbarium collection and a collecting mission to Morocco in 1973 (Gladstones, 1974). Buirchell (1992) has made further collections. This distinctive species is restricted to the Atlas and Anti Atlas mountains of Morocco at altitudes less than 1700 m but greater than 460 m. The northern extent of its habitat is around Beni Mellal and it has only been found on the western slopes of the Atlas mountains. Some plants have been collected in the deeper valleys of the Atlas mountains but the main area seems to be the foothills of the western slopes. In the Anti Atlas mountains it is found in the valleys of the central mountains around Tanalt, in the valleys around Tafraoute and on the Kerdous Plateau. The southern boundary is unknown but it is probably restricted to above 250 mm annual rainfall.

PHENOLOGICAL DATA. Upward of 20 different phenotypic traits were measured on 54 accessions of *L. atlanticus* collected from the Atlas and Anti Atlas mountains of Morocco, when grown in the field in Perth, Western Australia. Based on the four geographical regions nominated above, principal coordinate analysis (PCA) of these data resulted in the northern group being easily distinguishable from the other three groups (Fig. 2.1). Because of the large variation among accessions within the other groups they could not be separated from each other and overlapped considerably on the PCA graph.

Accessions from the Northern region had early rapid growth with maximum values for height, were late flowering and had exceptionally large seed (Table 2.9). The accessions from the Valley region were slow growing (minimum values for height), were late flowering and had the most pods on the main stem. Accessions from the Southern and Central regions had similar growth characteristics, flowered earlier than those from the other two regions and had larger leaves.

ALKALOIDS. The ratio of epilupinine and multiflorine varied considerably among the accessions of *L. atlanticus*, ranging from 0.48 to 2.55 (Table 2.10). At the northern extremity (average 0.73) multiflorine is the dominant alkaloid, while in the south epilupinine (average 0.19) is dominant (Table 2.10). Total seed alkaloid content varied from 0.29% to 0.50% of dry weight.

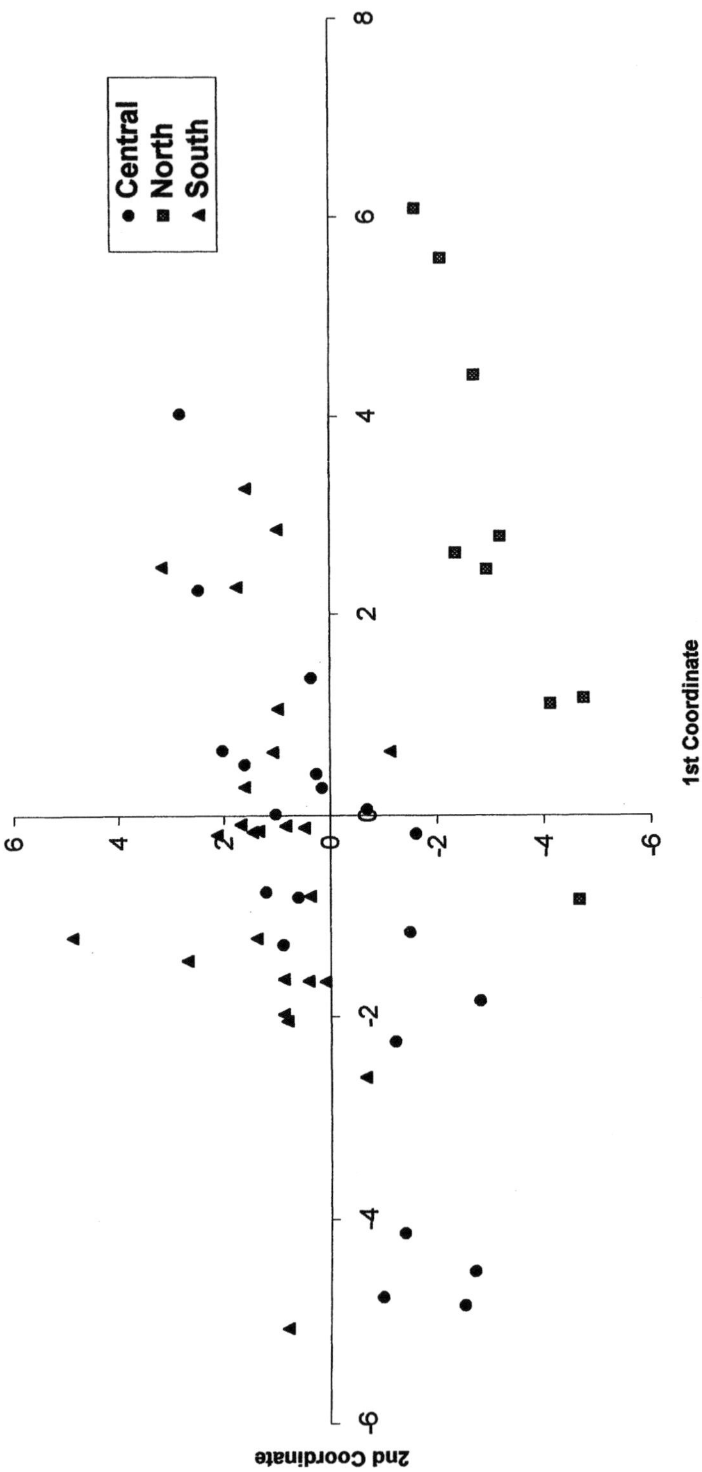

Fig. 2.1. Principal coordinate analysis of *L. atlanticus* collection from Morocco.

Table 2.9. Means for morphological traits (maxima in bold and minima in italics).

Region	ht5	ht9	ht13	ht17	fht	htpd	dtf	upbr	lwbr	pdms	ydms	yd/pt	lfw	lfl	dymsrip	sdwt
North	**11.9**	**18.8**	**26.5**	**79.0**	**88.4**	**62.5**	123.4	**4.9**	1.6	8.8	13.2	**28.5**	*1.5*	*6.2*	59.6	45.6
Valley	*6.8*	*11.9*	*17.6*	*61.2*	*75.1*	53.2	**125.6**	*3.6*	0.8	**13.4**	**13.4**	*18.9*	1.7	6.9	*58.8*	28.0
Central	8.1	14.8	22.8	76.1	79.9	*49.2*	112.2	3.9	*0.7*	11.8	12.1	25.5	1.8	7.2	**68.3**	27.5
South	8.9	16.1	23.8	76.4	76.7	50.3	*110.0*	3.8	0.9	*7.8*	*8.2*	22.8	**2.0**	**7.4**	66.4	26.5

ht5, plant height at 5 weeks; ht9, plant height at 9 weeks; ht13, plant height at 13 weeks; ht17, plant height at 17 weeks; fht, final plant height; htpd, height to first main stem pod; dtf, days to flowering; upbr, number of upper branches; lwbr, number of lower branches; pdms, number of pods per main stem; yd/pt, yield per plant; lfw, leaflet width; lfl, leaflet length; dymsrip, days to ripening of main stem pods; sdwt, weight of 100 seeds (g).

Table 2.10. The seed alkaloid content variation in genotypes of *L. atlanticus* from Morocco.

	Region		
Alkaloid	North	Central	South
No. accessions	10	15	21
Epilupinine %	0.16	0.15	0.19
Multiflorine %	0.22	0.14	0.16
Total %	0.38	0.29	0.35
Ratio E/M	0.73	1.07	1.19

L. cosentinii

L. cosentinii is mostly restricted to the coastal lowlands of western Morocco, the extreme south-west of Spain and Portugal, eastern Tunisia, Sicily, Corsica and Sardinia. Until 1990, it was thought to be restricted to within 40 km of the coast and at lower altitudes (Gladstones, 1976). However, seeds collected in 1990 from the region between the Atlas and Anti Atlas mountains in Morocco were from over 50 km inland and at altitudes up to 800 m (Buirchell, 1992).

In Morocco, *L. cosentinii* is found mainly on neutral red sands of varying depth over limestone to the north of Rabat and on black, moderately alkaline sands and loamy sands over shallow limestone to the south of Rabat. South of Agadir and to the east they were found on the slightly alkaline sands usually associated with limestone (Gladstones, 1976).

This species was probably introduced to Western Australia over 100 years ago (Gladstones, 1994; Chapter 1, this volume) and became naturalized on the coastal sand plains where it is used successfully as a natural legume pasture for rotation with cereal crops and for sheep feed. Its natural distribution and diversity is poorly documented.

L. digitatus

L. digitatus is known from botanical collections in a number of regions in the southern Sahara, mostly associated with mountainous areas. It has been found on the border of Chad and Libya, Niger and Algeria, in the Senegal Valley, the North West Sahara and in the Nile Valley (Gladstones, 1974). The only seed in international seed banks is from a single collection from the Wadi Karit in south-east Egypt. (This collection was selected in the field in Western Australia, resulting in two distinct selections with different seed colours; therefore three accessions, including the original seed, are listed in Table 2.1 for the Australian Lupin Collection.) The species was noted as a common winter volunteer in cultivated fields in the Nile Valley but has disappeared as a consequence of the damming of the Nile at Aswan. The current range of *L. digitatus* is likely to be more restricted than indicated in Gladstones (1974), due to desertification across much of the range. It is clearly a species for which there is an urgent need for collection and conservation. Our knowledge of the genetic potential of this species is based on the single Egyptian collection.

L. hispanicus

L. hispanicus is confined mainly to inland Spain and Portugal, on acid soils up to 1500 m (Gladstones, 1976). Gladstones (1974; Chapter 1, this volume) described two distinct subspecies, *L. hispanicus* subsp. *hispanicus* and *L. hispanicus* subsp. *bicolor*. The former grows in south and central Spain, mainly in mountainous areas. It has violet flowers and nearly glabrous leaves. Seeds are similar in size to those of *L. luteus*. The latter subspecies has a cream corolla becoming lilac, smaller seeds and hairier leaves, and extends into the high altitudes of central and north-western Spain and central and northern Portugal on mainly flat and sometimes poorly drained granitic soils. Both species are very late flowering but subsp. *bicolor*, with its northerly distribution and high altitude habitat, is later than subsp. *hispanicus* (Gladstones and Crosbie, 1978).

The extent and nature of *L. hispanicus* growing outside Spain and Portugal is much less certain (Gladstones, Chapter 1, this volume). Cowling (1986) made two collections in northern Greece of a short wild lupin with smooth leaves, pale yellow flowers and pure white seeds. Although originally nominated as *L. luteus*, the accession is very late flowering and probably a form of *L. hispanicus*. We have at present no other forms in the Australian collection.

L. luteus

The natural habitat of *L. luteus* in the Iberian Peninsula and coastal Israel and Lebanon is confined more or less to neutral to acid sandy soils. As discussed in Chapter 1, the status of other populations scattered around the Mediterranean is less certain, some perhaps being merely naturalized. Semi-naturalized populations have also been reported in Madeira (Lowe, 1868) and the Azores Islands (Papineau and Hugyhe, 1989), as well as the South of France (Thellung, 1912). Semi-domesticated varieties and wild types have escaped from field plantings in South Africa and Australia and have persisted for many years. Throughout, the adaptation is to acid and sandy soils and mainly to low altitudes.

Gladstones and Crosbie (1978) looked at genetic variation within 11 genotypes of *L. luteus*. The eight accessions collected from the high rainfall area of northern Portugal and north-west Spain were late maturing and relatively small seeded. They suggested that these accessions approached the true 'wild type' while cultivated lines in the same regions of Spain and Portugal (not collected) were larger seeded and earlier maturing (Gladstones and Crosbie, 1978). The data presented by Clements and Cowling (1991) tend to support this and indicate that accessions of *L. luteus* from Morocco were even earlier flowering than most other wild or semi-domesticated types. These and other accessions from Israel appear particularly promising for breeding purposes. The authors have noted that the Polish cultivar Teo is very early flowering and lines from the Azores Islands are very late flowering. There is a large variation in seed coat colour in *L. luteus* from pure white or slightly speckled to heavily speckled and nearly totally black seed coats.

L. micranthus

L. micranthus has been little collected. This species is wild, weedy, late flowering and very short and has hairy leaves and pods. It is an unattractive species from the domestication point of view, and this may explain why it is poorly represented in gene banks. From botanical records this species is widely distributed throughout the Mediterranean basin. A diversity in collection sites has been recorded by Australian researchers collecting in the Mediterranean region. Gladstones (1976) collected one accession in Morocco from a gritty, slightly alkaline clay at 600 m altitude just south of Rabat. Buirchell (1992) also collected *L. micranthus* in Morocco near Khenifra and has observed the species in the Atlas mountains east of Marrakech. Cowling (1986) found several accessions on Greek islands on poor acidic soils. A more recent collection by C. Francis (personal communication) was made in Crete from an alkaline clay environment. *L. micranthus* from southern Spain collected by M. Judges (personal communication) had a very prostrate habit, and inside these stands of *L. micranthus* were occasional wild *L. luteus* plants that mimicked the prostrate habit of *L. micranthus*.

Our observations on *L. micranthus* are that it is a small prostrate plant, very hairy and of no potential for domestication as a crop plant. However, it may be useful as a natural regenerating pasture on alkaline fine-textured soils if the alkaloid levels in the seed could be reduced to acceptable levels for grazing by ruminants.

L. palaestinus

L. palaestinus is a low-growing species, highly branched, with long inflorescences. Its natural range extends from the central and southern plains of Sharon and Philistia, down the coast of Israel. A pocket of the species is also found on the Sinai Peninsula in the Jebel El Tih. In the Sinai this species probably takes advantage of moist conditions in the bottom of wadis. *L. palaestinus* seems to require insect pollination. It does make an excellent garden plant but has little potential as a crop species because of its short and spreading stature.

L. pilosus

L. pilosus has a diverse distribution from the mountainous regions of Crete and Turkey to the Greek Islands and around the coast to Israel and Syria. Most reports indicate that it prefers sandy soils to loamy clays with limestone present. Most reports indicate that it prefers sandy soils of neutral to alkaline pH. The species seems well adapted to a diverse range of environments and there is considerable variation within this species.

There is great diversity in flowering time. The early-flowering types usually have early rapid growth and require no vernalization for flowering. The late-flowering types require vernalization and tend to have a long period in a rosette state before rapid growth near flowering. The seed size varies from intermediate (42 g/100 seeds) to very large (72 g/100 seeds).

PHENOLOGICAL GROUPS. Clements *et al.* (1996) examined 71 accessions of *L. pilosus* for variation in morphological characters. Using hierarchical cluster analysis they were able to choose ten groups of accessions that accounted for 79% of genotype sums of squares and 69% of genotype × character sums of squares.

These ten groups fell into four different categories:

1. The first category consisted of Group 132, which was distinguished by poor pod set, few seeds and small leaves. The plants were poorly nodulated and had pale yellow leaves. This group was easily distinguishable from the other accessions of *L. pilosus.* Pod set was improved with artificial tripping of flowers. Interestingly, the receptacles on stigmas were very small and the peristigmatic collar had long epidermal hairs. This is a similar characteristic to *L. palaestinus* and this group may either represent a distinct variation of *L. pilosus* or natural hybrids of *L. pilosus* and *L. palaestinus.* However, natural hybrids are made less likely by the geographical separation of *L. palaestinus* on the coastal areas of Israel and Group 132 in northern and eastern Israel.

2. The second category was formed from Groups 126, 128 and 130. These groups had short early growth, late flowering and short main-stem inflorescences. Members of this category were also very fertile with many pods on the main stem and many seeds per pod. Group 128 from Crete was late flowering and highly branched. Group 130 had large leaves and seed and originated at moderate to high altitudes in Greece, Syria and Israel.

3. The third category of Groups (125, 127, 129 and 131) included accessions that had early rapid growth and early flowering. Group 127 included most of the botanical garden selections from Europe and Australia, which were early flowering and lacked a vernalization requirement. Groups 125, 129 and 131 included the accessions from Greece, Syria and Israel that had rapid early growth and early flowering. Group 129 had accessions with the largest leaves and seeds.

4. The fourth category included Groups 107 and 111 that had long main-stem inflorescences. These two groups were dissimilar in other characters. Group 107 was closer aligned to Group 132 with poor pod set and low yield. Group 111 was late flowering and very rosetted in the early stages of growth. The accessions in Group 111 were collected in Turkey; they had the smallest seeds of all the accessions and came from high altitudes.

SMOOTH-SEEDED *L. PILOSUS*. Within *L. pilosus* we have found three accessions with smooth seeds. One of the three accessions, when collected, consisted of a mixture of smooth and rough seeds. All three accessions were collected in the Tartus and Homs regions of southern Syria.

Swiecicki *et al.* (1996) have also collected a smooth-seeded lupin from Turkey which in the field looked similar to *L. pilosus*, but they proposed that it should be given status as a new lupin species *L. anatolicus*. We have grown this Turkish accession alongside other smooth-seeded and rough-seeded *L. pilosus* accessions from the eastern Mediterranean, and the Turkish accession fits within the range of variation within *L. pilosus* reported by Clements *et al.* (1996). The two types are similar, based on taxonomic descriptors and crossing ability. The smooth-seeded Turkish accession shares the same chromosome number as *L. pilosus* ($2n = 42$) and we obtained fertile F1 and F2 plants from crosses of the accession labelled *L. anatolicus* and other *L. pilosus* accessions. We conclude that the smooth seed coat is only a variation within *L. pilosus*. One of us (B.J.B.) has recently selected a smooth-seeded artificial mutant of *L. pilosus*. The gene (or genes) for smooth seed coats has potential value for the domestication of this species.

ALKALOID CONTENT. Alkaloid levels in *L. pilosus* ranged from 0.13 to 0.85% with an average of 0.55% (Table 2.11). Multiflorine was the dominant alkaloid in *L. pilosus* and the epilupanine/multiflorine ratio varied from 0.10 to 0.67. The Crete populations had the highest ratios of epilupinine/multiflorine and the Syrian–Turkish populations the lowest. The populations from the Kikladhes Islands of the Mediterranean Sea and Israel had intermediate levels.

YIELD DATA. A range of accessions of wild *L. pilosus* were evaluated for yield in 1992 and 1993 at Mt Barker, a long-season site in Western Australia (Table 2.12). Yields varied from two to four tonnes per ha and accessions in Group 130 (Clements *et al.*, 1996) were the highest yielding. This was predictable as Group 130 had the highest main-stem yield, many pods on the main stem, large seed and large leaves. Group 131 was the next highest yielding group. Groups 131 and 130 both had early rapid growth with early or mid flowering times, which are needed to exploit the long-season mediterranean environment at Mt Barker. Group 132 was not included in the experiment, due to lack of nodulation. Accessions from Groups 129, 130 and 131 will be most useful for a breeding programme.

Table 2.11. Average of seed alkaloids (% dry weight) in *L. pilosus* from the eastern Mediterranean (from Buirchell *et al.*, 1993). Ratio is that of epilupinine/multiflorine.

Region	Number of accessions	Epilupinine	Multiflorine	Total	Ratio
Syria/Turkey	7	0.08	0.32	0.41	0.25
Israel	12	0.14	0.42	0.56	0.30
Kikladhes Is.	7	0.11	0.44	0.55	0.25
Andros	5	0.16	0.40	0.57	0.40
Crete	4	0.19	0.47	0.68	0.45

Table 2.12. Variation in yield for *L. pilosus* in 1992 and 1993 at Mt Barker, Western Australia.

Morphological group (Clements *et al.*, 1996)	Characteristics	Yield (t ha^{-1})
126	Less branching, short	2.21
128	Branching, tall	2.18
125	Short, small seeds, small leaves	2.45
127	Tall, low seeds/pod, early flowering	2.48
129	Tall, large leaves, large seeds	3.34
131	High yield, high seeds/pod, large leaves, large seeds	3.57
130	High yield, high number pods, large leaves, large seeds	4.03

L. princei

L. princei is an erect lupin species found in the highlands of Kenya, southern Ethiopia and northern Tanganyika at high altitudes in open scrub or woodlands (Gladstones, 1974). This species has not been extensively collected and is represented by only six accessions in the Australian Lupin Collection (Table 2.1), all from the highlands of Kenya. The extent of its natural habitat in Ethiopia is unknown.

There is limited genetic variation in *L. princei* among the accessions in the Australian Lupin Collection. One accession has white flowers while the others are blue flowered. *L. princei* does not nodulate with the standard Group G *Bradyrhizobium* strain used in lupin culture in Australia but it readily nodulates with another *Bradyrhizobium* isolated from *Ornithopis* sp. (J. Howieson, 1992, Perth, personal communication). *L. princei* flowers very late compared with all other lupin species and it is not unusual for it to take more than 150 days to flower in the mildly vernalizing winter growing season of Perth, Western Australia. The seeds are large and have characteristic markings that distinguish them from seed of other lupin species.

L. somaliensis

This species is known from a single herbarium specimen (Gladstones, Chapter 1, this volume). Nothing is known of its genetic resources, or, indeed, whether it still exists in the wild.

L. mutabilis

The cultivated Andean lupin, *L. mutabilis*, has an ancient history of domestication and utilization parallel to that of *L. albus* in the Mediterranean region. However, there is one critical difference between *L. albus* and *L. mutabilis*: whereas cultivated *L. albus* var. *albus* has a wild relative in existence in the Mediterranean region (*L. albus* var. *graecus*) (Gladstones, 1974), there are no known wild relatives of *L. mutabilis* in South America. Some evidence suggests an ancient history of domestication in South America of at least 600 or 700 BC (Hondelmann, 1984), and *L. mutabilis* was drawn on ceramic vessels of the pre-Inca era (Tapia *et al.*, 1988). According to Tapia and Vargas (1982), *L. mutabilis* is closely related to *L. praestabilis* in southern Peru; based on speculation of Kazimierski and

Novacki (1961) it may be the result of interspecific hybridization among two North American species.

L. mutabilis exists today in the high Andes (> 2000 m) of Peru, Colombia, Ecuador, Bolivia and probably Chile. It has been largely displaced in the post-Hispanic era by European crops (Hernández Bermejo and León, 1994), and is now grown by village people on a very small scale – fewer than 2000 ha per year in Peru in the 1970s (Tapia *et al.*, 1988).

L. mutabilis has various types: 'chocho' is a tall annual or biennial plant, adapted to the long growing seasons of the Andes in the north of Peru and in Colombia and Ecuador (growing season 240–300 days), which has long branches; and 'tarwi' is an annual type from high altitudes in the central Andes of Peru and Bolivia, with a shorter form known as 'tauri' at 3800 m around Lake Titicaca (borders of Peru and Bolivia), which has a dominant main stem and few branches and a shorter growing season of about 155 days (Tapia, 1990).

L. mutabilis flowers in both short tropical days and longer summer days in the temperate zone, but maturity is delayed in long summer days of temperate latitudes (40–60°S or N). The slow and uneven ripening is a major constraint to more mechanical forms of cultivation. The seed is highly nutritious in both oil and protein (Hernández Bermejo and León, 1994), but requires a labour-intensive alkaloid extraction process for human consumption. One sweet mutant, *L. mutabilis* cv. Inti, has been released in Chile. In our experience, this cultivar has about 0.06% alkaloids, or about three times the limit set by health authorities in Australia and Britain for human consumption. The disease anthracnose (*Colletotrichum* sp.) is also considered a major constraint to production, and seed disinfection against this disease is necessary (Hernández Bermejo and León, 1994).

Collection of *L. mutabilis* was supported by IBPGR and GTZ in Germany in the 1970s and 1980s. *L. mutabilis* seed is conserved in working collections at several experimental stations and universities in the Andean region in Peru (Cuzco, Santa Ana, Illpa, Camacani, Baños del Inca and Canáan), Bolivia (Patacamaya, Toralapa and Pairumani), and Ecuador (Santa Catalina) (Hernández Bermejo and León, 1994). However, storage conditions are not optimal and there is a deficiency of collection site and evaluation data. Mujica *et al.* (1991) have published a catalogue for the collection at the University of Puno in Peru. In some cases, duplicate collections exist in Europe or other countries and evaluation work has also taken place there.

References

Bennencourt, E., Konopka, J. and Damania, A.B. (1989) *Directory of Crop Germplasm Collections.* 1. I. Food legumes. *Arachis, Cajanus, Cicer, Lens, Lupinus, Phaseolus, Pisum, Psophocarpus, Vicia and Vigna.* International Board for Plant Genetic Resources (IBPGR), Rome.

Bounejmate, M., Buirchell, B., Birouk, A., Bouizgaren, A. and Saidi, N. (1994) Distribution naturelle au Maroc de trois espèces de lupin en relation avec certains facteurs du milieu. *Al Awamia* 84, 29–42.

Buirchell, B.J. (1992) Collecting wild *Lupinus* spp. in Morocco. *FAO/IBPGR Plant Genetic Resources Newsletter* 90, 36–38.

Buirchell, B.J., Cowling, W.A., Allen, D., Greirson, B. and Harris, D. (1993) Seed alkaloids of the rough-seeded lupins. *7th International Lupin Conference, Evora, Portugal,* Abst. 2–3.

Chevalier, A. and Trochan, J. (1937) Histoire de trois Lupins. *Revue de Botanique Appliquée et d'Agriculture Tropicale* 17, 85–99.

Clements, J.C. and Cowling, W.A. (1990a) Summary of wild and semi-domesticated lupin accessions introduced to Australia from 1974 to 1989. *Australian Plant Introduction Review* 21, 1–14.

Clements, J.C. and Cowling, W.A. (1990b) *The Australia Lupin Collection. Passport Data for Wild and Semi-domesticated Accessions Introduced to Australia to 1990.* Research Report 1/90, Department of Agriculture Western Australia, South Perth, 149 pp.

Clements, J.C. and Cowling, W.A. (1991) *Catalogue of the Australian Lupin Collection. Including Field Evaluation Data for Wild, Semi-domesticated and Fully Domesticated Accessions.* Research Report 3/91, Department of Agriculture Western Australia, South Perth, 186 pp.

Clements, J.C. and Cowling, W.A. (1994) Patterns of morphological diversity in relation to geographical origins of wild *Lupinus angustifolius* from the Aegean region. *Genetic Resources and Crop Evolution* 41, 109–122

Clements, J.C., Buirchell, B.J. and Cowling, W.A. (1996) Relationship between morphological variation and geographical origin or selection history in *Lupinus pilosus. Plant Breeding* 115, 16–22.

Cowling, W.A. (1986) Collection of *Lupinus* in Greece. *FAO/IBPGR Plant Genetic Resources Newsletter* 65, 20–22.

Cowling, W.A. (1994) Use of lupin genetic resources in Australia. In: Neves Martins, J.M., and Beirão da Costa, L. (eds) *Advances in Lupin Research.* ISA Press, Lisbon. pp. 9–18.

Cowling, W.A., Buirchell, B.J. and Tapia, M.E. (1998) *Promoting the Conservation and Use of Underutilized and Neglected Crops Lupins* – Lupinus *spp.* IPK, Gatersleben/IPGRI, Rome.

Frankel, O.H. (1977) Genetic resources. *Annals of the New York Academy of Science* 287, 332–344.

Garlinge, J. (ed.) (1996) *1997 Crop Variety Sowing Guide for Western Australia.* Bulletin No. 4330, Department of Agriculture Western Australia, South Perth, 324 pp.

Gladstones, J.S. (1973) Observations on the environment and ecology of some annual legumes in southern Italy. *Australian Plant Introduction Review* 9, 11–29.

Gladstones, J.S. (1974) *Lupins of the Mediterranean Region and Africa.* Technical Bulletin No. 26, Department of Agriculture of Western Australia, South Perth, 48 pp.

Gladstones, J.S. (1976) Observations on the distribution and ecology in Iberia and North Africa of some annual legumes adapted to neutral and acid soils. *Australian Plant Introduction Review* 11, 9–23.

Gladstones, J.S. (1994) An historical view of lupins in Australia. In: Dracup, M. and Palta, J. (eds) *Proceedings of the First Australian Lupin Technical Symposium.* Department of Agriculture Western Australia, South Perth, pp. 1–38.

Gladstones, J.S. and Crosbie, G.B. (1978) *Lupin Wild Types Introduced into Western Australia to 1973.* Technical Bulletin No. 43, Department of Agriculture of Western Australia, South Perth, 44 pp.

Goodman, M.M. (1990) Genetic and germ plasm stocks worth conserving. *Journal of Heredity* 81, 11–16.

Hackbarth, J. and Troll, H.-J. (1956) Lupins as grain legumes and fodder plants. In: Kappert, H. and Rudorf, W. (eds) *Handbook of Plant Breeding*, Part IV, 2nd edn. Verlag Paul Parey, Berlin and Hamburg, pp. 1–51 (in German).

Hammer, K. and Perrino, P. (1995) Plant genetic resources in South Italy and Sicily – studies toward *in situ* and on-farm conservation. *FAO/IBPGR Plant Genetic Resources Newsletter* 103, 19–23.

Hernández Bermejo, J.E. and León, J. (1994) *Neglected Crops: 1492 From a Different Perspective.* FAO Plant Production and Protection Series No. 26, Food and Agriculture Organization of the United Nations, Rome.

Hoballah, A.A., Ibrahim, A.F. and Jahn-Deesbach, W. (1990) Evaluation of yield components and seed quality in white lupins (*Lupinus albus*). *Proceedings 6th International Lupin Conference, Temuco, Chile*, Abst. A5, p. 12.

Hondelmann, W. (1984) The lupin – ancient and modern crop plant. *Theoretical and Applied Genetics* 68, 1–9.

Huyghe, C., Papineau, J. and Vergnaulte, E. (1990) Genetic variability among the Azoren populations: potential uses in breeding. *Proceedings 6th International Lupin Conference, Temuco, Chile*, Abst. H17, p. 150.

Kazimierski, T. and Kazimierska, E.M. (1975) Morphological and cytological differences within the species *Lupinus luteus* L. *Acta Bot. Soc. Polon.* 44, 265–275.

Kazimierski, T. and Novacki, E. (1961) Indigenous species of lupine regarded as initial forms of the cultivated species. *Lupinus albus* L. and *Lupinus mutabilis* Sweet. *Flora* 151, 202–209.

Lehman C.O. and Hammer, K. (1983) Botanical results of a mission to Spain 1978 for the collection of cultivated and wild legumes. *Kulturpflanze* 31, 185–206 (in German).

Lowe, R.T. (1868) *A Manual Flora of Madeira*. John van Voorst, London.

Mota, M., Gusmão, L. and Bettencourt, E. (1981) *Lupinus* and *Secale* collection in Portugal. *FAO/IBPGR Plant Genetic Resources Newsletter* 47, 26–27.

Mota, M., Gusmão, L. and Bettencourt, E. (1982) *Lupinus* and *Secale* collection in Portugal II. *FAO/IBPGR Plant Genetic Resources Newsletter* 50, 22–23.

Mujica, A., Huapaya, F., Rodriguez, M., Pino, S. and Apaza, V. (1991) *Catálogo de Cultivos Andinos. Tarwi.* Universidad Nacional del Altiplano, Escuela de Postgrado. Puno IBPGR. Puno, Perú.

Neves Martins, J.M. (1991) Caracterização e avaliação de germoplasma de *Lupinus albus* L. PhD thesis, Universidade Técnica de Lisboa, Instituto Superior de Agronomia, Lisbon, Portugal (in Portuguese).

Papineau, J. and Huyghe, C. (1989) Collecting white lupin in the Azores. *FAO/IBPGR Plant Genetic Resources Newsletter*, 88/89, 77–78.

Perrino, P., Hammer, J. and Hanelt, P. (1984) Collection of land-races of cultivated plants in southern Italy 1983. *Kulturpflanze* 32, 207–216.

Pinto, P.A. (1994) The role of *Lupinus luteus* in a Portuguese environment friendly agroforestry system. In: Neves Martins, J.M. and Beirão da Costa, L. (eds) *Advances in Lupin Research*. ISA Press, Lisbon, pp. 290–295.

Plitmann, U. and Heyn, C.C. (1984) Old World *Lupinus*: taxonomy, evolutionary relationships, and links with New World species. In: *Proceedings of the Third International Lupin Conference, La Rochelle, France*, pp. 56–66.

Shands, H.L. and Weisner, L.E. (1991) *Use of Plant Introductions in Cultivar Development. Part 1.* CSSA Special Publication Number 17, Crop Science Society of America, Madison, Wisconsin.

Shands, H.L. and Weisner, L.E. (1992) *Use of Plant Introductions in Cultivar Development. Part 2.* CSSA Special Publication Number 20, Crop Science Society of America, Madison, Wisconsin.

Simpson, M.J.A. (1984) Geographical variation in *Lupinus albus* L. PhD thesis, University of Reading, UK.

Simpson, M.J.A. (1986a) Geographical variation in *Lupinus albus* L. I. Iberia. *Plant Breeding* 96, 232–240.

Simpson, M.J.A. (1986b) Geographical variation in *Lupinus albus* L. II. Northwest Spain, the Nile Valley, the Balkans and Turkey. *Plant Breeding* 96, 241–251.

Simpson, M.J.A. and McGibbon, R. (1982) Collection of *Lupinus* in the Iberian Peninsula. *FAO/IBPGR Plant Genetic Resources Newsletter* 50, 14–19.

Simpson, M.J.A. and McGibbon, R. (1983) *Lupinus* collection in Greece and Yugoslavia. *FAO/IBPGR Plant Genetic Resources Newsletter* 52, 28–30.

Simpson, M.J.A. and Neves Martin, J.M. (1984) Distribution of plant types in *Lupinus albus* L. In: *Proceedings of the Third International Lupin Conference, La Rochelle, France*, pp. 88–101.

Swiecicki, W., Swiecicki, W.K. and Wolko, B. (1996) *Lupinus anatolicus* – a new lupin species of the Old World. *Genetic Resources and Crop Evolution* 43, 109–117.

Tapia, M. (1990) *Cultivos Andinos Subexplotados y su Aporte a la Alimentación*. FAO, Santiago de Chile.

Tapia, M. and Vargas, C. (1982) Wild lupine of the Andes of Southern Perú. In: *Agricultural and Nutritional Aspects of Lupines. Proceedings of the First International Lupine Workshop. Lima, Cuzco, Perú*, 23–31.

Tapia, M.E., Riva, E.A. and Hernández-Bravo, G. (1988) Cool season food legumes in South and Central America. In: Summerfield, R.J. (ed.) *World Crops: Cool Season Food Legumes*. Kluwer Academic, Dordrecht, The Netherlands, pp. 1051–1057.

Thellung, A. (1912) *La Flore Adventice de Montpellier*. Le Maout, Cherbourg.

Vavilov, N.I. (1951) The origin, variation, immunity and breeding of cultivated plants. Selected writings of N.I. Vavilov translated from the Russian by K. Starr Chester. *Chronica Botanica* 13, 1–366.

Yang, H.A., Sweetingham, M.W. and Cowling, W.A. (1996) The leaf infection process and resistance to *Pleiochaeta setosa* in three lupin species. *Australian Journal of Agricultural Research* 47, 787–799.

3 Genetics, Cytology and Biotechnology

Craig A. Atkins[1], Penelope M.C. Smith[1], Sanjiv Gupta[2], Michael G.K. Jones[3] and Peter D.S. Caligari[4]

[1]Centre for Legumes in Mediterranean Agriculture, University of Western Australia, Nedlands, Western Australia 6907; [2]Faculty of Agriculture, University of Western Australia, Nedlands, Western Australia 6907; [3]School of Biological and Environmental Sciences, Murdoch University, Murdoch, Western Australia 6150; [4]Department of Agricultural Botany, Plant Science Laboratories, University of Reading, Whiteknights, Reading RG6 6AS, UK

Introduction

The tools of molecular biology and recombinant DNA technology combined with new procedures for cell and tissue culture have opened the way for direct genetic engineering of plants. Among crop species, members of the *Solanaceae* and *Brassicaceae* have proved to be the most easily engineered but recently a number of cereals and leguminous species have also been transformed to express novel genes. Of the pulses, most attention has been given to soybean (*Glycine max*) and to a lesser extent to pea (*Pisum sativum*) and *Vigna* species (Christou, 1994). Of the pasture legumes, a number of clovers and medics, including alfalfa (*Medicago sativa*), have been successfully transformed (Christou, 1994). In many cases the genes transferred have simply conferred expression of a selectable marker, such as resistance to an antibiotic, or of an enzyme which is easily assayed *in vivo*, such as the blue coloration due to β-glucuronidase (the *GUS* gene). However, genes are now being transferred which introduce novel traits with agronomic utility or which increase the value of harvested grain. These include: resistance to broad spectrum herbicides or to insect pests and diseases; altered oil, carbohydrate or protein content and quality; plant architecture and features of reproductive development; and even promotion of the synthesis of a particular organic molecule, effectively using the plant as a 'bioreactor' (Poirier *et al.*, 1995; Van Der Leij and Witholt, 1995).

Routine procedures for transforming narrow-leafed lupin (*Lupinus angustifolius*), based on *Agrobacterium tumefaciens*-mediated gene transfer, have also been developed (Pigeaire *et al.*, 1997). Initial transformations to confer herbicide tolerance have been successful and transgenic lines progressed to the field trial stage. A range of other genetic engineering strategies have been developed to alter the species in other ways (Atkins *et al.*, 1994). These include

 Lupins as Crop Plants: Biology, Production and Utilization (eds J.S. Gladstones, C.A. Atkins and J. Hamblin)

viral resistance, enhanced harvest index and modifications to seed composition (Molvig *et al.*, 1994). Many of these alterations may also apply to other lupins and effective methods for their transformation and regeneration are being sought.

Among the diversity of lupin species a number of desirable traits have been identified which would enhance the value and agronomic perfomance of other members of the genus if they could be transferred through interspecific hybridization. These include resistance to viral and other diseases, oil and protein content and quality of harvested grain, tolerance of environmental stress, including waterlogging, and adaptation to a wider range of climate and soil types. Consequently there has been renewed interest in the cytogenetics of *Lupinus* and the application of cell and protoplast fusion techniques, as well as those of more conventional crossbreeding, to generate viable interspecific hybrids.

This chapter provides an overview of the nature, number and behaviour of the chomosomes of many lupin species and identifies cytogenetic restrictions and outcomes from studies aimed at their hybridization. Newer approaches to introduce novel traits through this means are indicated. Current protocols for genetic engineering of *L. angustifolius* and strategies aimed at exploiting transformation to improve yield potential through the introduction of genes for herbicide tolerance or resistance to viral disease or to alter source/sink relations in the species are outlined.

Cytogenetics of *Lupinus* Species

Cytogenetic studies provide a basis for developing an evolutionary relationship between the members of a genus as well as contributing to a breeding programme by identifying some of the problems associated with interspecific hybridization. Such studies also help to develop cytogenetic stocks which may serve as a valuable resource for applying molecular approaches to identify genes and their location on chromosomes.

Methodology

It is worth outlining successful methods because of the difficulty frequently experienced in preparing high quality chromosome spreads for lupins. Counting can be done at either mitosis or meiosis, but counts made at both stages of the cell cycle are preferable. Metaphase and anaphase of mitosis and diakinesis and metaphase I of meiosis are ideal stages for counting. Anaphase I of meiosis is seldom used for counts because sometimes the univalent chromosomes divide at the equatorial plate during cell division and lead to miscounting. Mitotic metaphases and anaphases may be examined in dividing root tip cells. Lupin seeds are scarified (in the case of rough-seeded lupins) and germinated on filter paper in Petri dishes at 22–27°C for a few days. Root tips (2–3 cm long) are cold treated with ice water, with or without 0.1% colchicine, at 2–4°C for 2–4 h, then fixed in ethanol : glacial acetic acid (3 : 1) for 5 min at 50°C and transferred to new fixative at 4°C. After 18–48 h the root tips are treated with 0.25N HCl at 45–55°C for approximately 2 min before being squashed in either lacto-aceto orcein (Darlington and La Cour, 1969) or 4% acetocarmine after heating the root tips gently for 1–2 min. Metaphase chromosome lengths may be determined from photographic enlargements. Chromosomes of the same size are then matched to prepare the karyotype. Meiosis in *Lupinus* species seems to be a rapid process. Small flower buds are fixed in ethanol : glacial acetic acid (3 : 1) at 50°C for 1 h and held overnight at room temperature in the same solution before transfer to 70% ethanol for preservation. Pollen mother cells (PMCs) may be studied either by bulk staining in alcoholic hydrochloric acid carmine (Snow, 1963) for at least five days before analysis, or directly by squashing young anthers on a glass slide with gentle heating in 4% acetocarmine solution.

Chromosome number and size

Studies to determine the chromosome number of lupin species have been reported periodically over almost 60 years (Tuschnjakowa, 1935; Malheiros, 1942; Eichorn, 1949; Gladstones, 1958, 1974, 1984; Pazy *et al.,* 1977; Plitmann and Heyn, 1984; Plitmann and Pazy, 1984; Carstairs *et al.*, 1992). The chromosomes of lupin species are relatively small in size (about 1–4 μm) but large in number ($2n = 32–52$) and, as a consequence, considerable difficulty has been experienced in defining the number for all species. Gladstones (1984) indicated the exact chromosome number of at least nine species from the cytotaxonomic studies published by Kazimierski (1960, 1961, 1964). Pazy *et al.* (1977) confirmed chromosome numbers for at least seven of these species and Carstairs *et al.* (1992) counted chromosome numbers of *L. princei*, *L. atlanticus* and *L. digitatus* and prepared the karyotype of *L. princei* (S. Carstairs, Perth, 1992, personal communication). Table 1.1 is a compilation of confirmed chromosome numbers for the genus.

The arm lengths of chromosomes in *Lupinus* species are indistinguishable except in case of *L. princei* and a few of the chromosomes of *L. cosentinii*, *L. digitatus* and *L. atlanticus* (Carstairs *et al.*, 1992). The chromosomes of *L. princei* have been found to be the largest among the species (1.77–4.40 μm), being 1.5–1.65 times the size of those of *L. atlanticus* and *L. digitatus*. Pazy *et al.*

Table 3.1. Reported work on interspecific hybridization in lupins in the recent literature.

Species	Reference
L. angustifolius × *L. luteus*	Nam and Zayakin, 1998
L. angustifolius × *L. albus*	Nam and Zayakin, 1998
L. atlanticus × *L. cosentinii*	Gupta and Buirchell, 1998
L. atlanticus × *L. digitatus*	Buirchell, 1994
L. luteus × *L. hispanicus*	Wolko and Weeden, 1994
L. mutabilis × *L. polyphyllus*	Römer *et al.*, 1998
L. mutabilis × *L. albus*	Sawicka-Sienkiewicz, 1998; Siamasonta and Caligari, 1998
L. mutabilis × *L. graecus*	Sawicka-Sienkiewicz, 1998
L. mutabilis × *L. elegans*	Sawicka-Sienkiewicz, 1998
L. mutabilis × *L. pubescens*	Sawicka-Sienkiewicz, 1998
L. mutabilis × *L. hartwegii*	Sawicka-Sienkiewicz, 1998
L. mutabilis × *L. nanus*	Sawicka-Sienkiewicz, 1998
L. mutabilis × *L. hartwegii*	Wolko and Weeden, 1994

(1977) determined *L. cosentinii* chromosomes to be about 3 μm in length and comparable with those of *L. atlanticus* and *L. digitatus*. Centromeres are located in the median and submedian regions of the mitotic chromosomes in *L. princei*, *L. atlanticus* and *L. digitatus*, the large pair of which also contain secondary constrictions (Carstairs *et al.*, 1992). Identification of centromeres in the smaller sized chromosomes of other species has proved difficult.

Comparison of genomes

On the basis of cytogenetic studies of meiosis, the genome of *L. atlanticus* is apparently more similar to that of *L. digitatus* than to that of *L. cosentinii* and that of *L. pilosus*, which is in turn more similar to the genome of *L. atlanticus* than to that of *L. cosentinii*. These species have at least two genomes partially in common, but include inversions and translocations. *L. princei* appears to have an isolated genome within this group. Some of the lupin species have identical chromosome numbers (Table 1.1) but number alone does not necessarily indicate relatedness, and vice versa. For example, *L. luteus* and its nearest relative *L. hispanicus* have the same number of chromosomes ($2n = 52$) but are only partially related, while *L. micranthus*, with an equal chromosome number, belongs to a different group. Similarly *L. princei* has the same chromosome number as *L. atlanticus* but the two species appear to be genomically isolated.

The partial similarity of chromosomes and consequently of their genomes indicates that the *Lupinus* species are evolutionary polyploids (Pazy *et al.*, 1977; Plitmann and Pazy, 1984; Gupta *et al.*, 1996). The basic chromosome number (x) is invariably 6 in the American lupins (Dunn and Gillett, 1966). Pazy *et al.* (1977) have suggested a series of basic chromosome numbers of 5, 6 (or 9), 7, 8, and 13. Gupta *et al.* (1996) further suggested a basic chromosome number of 6 on the basis of 12 chromosome bivalents found consistently in the rough-seeded lupin group. These species, with partial genomic similarity and chromosomal abnormalities (inversions and translocations), might have evolved through allopolyploidization. On the other hand species like *L. princei*, which has very large chromosomes, might have evolved through autopolyploidization, and have remained genomically isolated.

Future studies

Cytological studies in *Lupinus* species are not well advanced, compared with those of, for example, wheat, maize, tomato, etc. Efforts have not been made to develop cytogenetic aneuploid stocks like monosomic series ($2n - 1$), trisomic ($2n + 1$), addition ($2n + 1$) or substitution ($2n - 1 + 1$) lines which can be used as tools to locate useful genes on chromosomes. Fertility of interspecific hybrids between *L. cosentinii*, *L. digitatus* and *L. atlanticus* obtained by Roy and Gladstones (1985, 1988) and Gupta *et al.* (1996) and survival of these interspecific hybrids with different chromosome numbers of $2n = 34$, 36 and 38 (S. Gupta, Perth, 1994, personal communication) in the successive generations provide enough evidence that rough-seeded lupins, at least, can withstand loss or additions of chromosomes. Embryo rescue can further help to recover interspecific hybrids with embryo–endosperm incompatibility (see below). Studies on these hybrids can provide vital information on the chromosome similarity, genomic constitution and its evolution.

Proximity of genomes involved in *Lupinus* species lacks complete understanding. Furthermore there is ample scope to resolve and construct genomic constitution. To unravel genomic relatedness, techniques like the use of pachytene karyotypes are suitable for species with small chromosomes (Singh and Hymowitz, 1988), and molecular cytogenetic techniques, like chromosomal banding, *in situ* hybridization, or chromosomal painting (Young, 1994) to probe and image interchanges should be considered for a species like *L. princei* which has large chromosomes (Gill *et al.*, 1991). Molecular mapping of lupin using AFLP has been initiated (M.G.K. Jones, Perth, 1996, personal communication) and eventually this will also contribute to a comparison of genomes.

Interspecific Hybridization

A brief search through any of the current literature database systems quickly reveals a paucity of references to interspecific crosses or the production of hybrids between *Lupinus* spp (Table 3.1). This is despite their obvious interest and potential value in breeding lupins; the combination of traits in such crosses would have clear merit and could greatly enhance the use of the species under domestication. Why should such a paucity exist? There are, as ever, several reasons. Firstly, the number of researchers and breeders working on this genus is restricted. Secondly, the use of a range of 'domesticated' species, rather than just one, vitiates against a single focus for commercially attractive interspecific crosses. Thirdly, the practical difficulties with many aspects of controlled reproduction (few seeds per pollination, high levels of flower drop, poor pod setting, etc.), even in intraspecific crosses, makes obvious practical difficulties. Fourthly, although *L. angustifolius*, and perhaps *L. luteus*, is amenable to the newer techniques of biotechnology and tissue culture (see Genetic Engineering section, below), lupins were never destined to be 'model' species, which benefit from a wider breadth of work that 'models' have focused on them. Fifthly, and contributed to by the above, the basic information which would allow a more informed approach to be taken to attempts aimed at obtaining successful interspecific hybridization has been lacking.

That is not to say that interspecific hybridization has not been attempted and has not met with some successes; indeed there are scattered reports of work on interspecific hybridization throughout the literature, and recent references are listed in Table 3.1. Some of the early work was reported by Fruwith (1910) while later Gollmick (1937) tried a more detailed investigation of the underlying causes of failure – a theme returned to by later workers. An example of success is perhaps most clearly given by the exploitation of crosses between *L. atlanticus* and *L. cosentinii* in Australia (Roy and Gladstones, 1985; Buirchell, 1994) and serves as a useful signal for what might be possible and the advantages to be gained by further efforts in this area. In addition Table 3.2 indicates that some cytogenetic analysis has been achieved for hybrids among members of the rough-seeded lupins and for a cross of the two related species, *L. luteus* and *L. hispanicus*.

Table 3.2. Average chromosome configurations in fertile interspecific hybrids of *Lupinus* species at meiosis (compiled from Kazimierski, 1980; Plitmann *et al*., 1980; Gupta *et al*., 1996).

Cross	Chromosome number (2*n*)	Diakinesis/Metaphase I			
		I	II	III	IV+
L. digitatus × *L. cosentinii* and its reciprocal	34	6.33	13.09	0.24	0.18
L. atlanticus × *L. cosentinii* and its reciprocal	35	5.65	12.53	0.61	0.48
L. atlanticus × *L. digitatus*	37	1.17	16.65	0.36	0.59
L. pilosus × *L. cosentinii*	37	4.61	12.22	0.61	1.17
L. pilosus × *L. atlanticus* and its reciprocal	40	1.45	14.94	0.68	1.45
L. palaestinus × *L. pilosus*	42	0.00	21.00	0.00	0.00
L. luteus × *L. hispanicus*	52	13.00	18.6	0.00	0.45

Causes of failure of interspecific crosses

The simple observation of a failure to produce plants after pollination with another species hides a multitude of possible causes. A crude list of possibilities would include the following.

- Simply that the pollination failed for 'environmental' reasons.
- Failure of pollen to germinate.
- Failure of pollen to develop pollen tubes that penetrate the stigma.
- Failure of pollen to grow, or grow far enough, down style.
- Failure of pollen to find ovary and penetrate ovule.
- Failure of ovary stimulation and early embryo development.
- Failure of later embryo development.
- Inviability of developed embryo.

Such a list inevitably could be made more refined with aspects within any of these 'stages' being more clearly defined, possibly in biologically meaningful ways. However, this list serves to emphasize the array of possible 'failures' and a framework for reviewing evidence for any of these being predominant in lupins.

'Environmental' reasons for pollination failure

A whole range of possibilities can be raised under this heading. It is well accepted that controlled hand pollination within lupins is not always successful and factors such as temperature and humidity are likely influencing variables. Another consideration is that when pollinating flowers such as lupins, one pollination produces one seed (as opposed to species such as *Nicotiana* where one pollination gives the potential for several hundred seeds); thus minor reductions in potential seed production will often lead to no seed at all.

Pollen germination

Pollen needs not only to alight on the stigma but also to initiate the processes that give rise to germination. Therefore it needs to imbibe water and begin the complex of reactions that lead to the development of a pollen tube. Despite the possibilities of a failure at this stage, there is little evidence in the literature of problems with pollen germination. In recent experiments no suggestion of a lack of germination was observed with a range of crosses between species (Siamasonta, 1996).

Pollen tube development and growth in the stigma

Penetration by pollen tubes, their subsequent success in penetrating the stigmatic surface and their growth down the style are the activities that would be affected by the 'established' incompatibility (gametophytic or sporophytic) systems in plants. They would thus seem to be a likely point for problems in lupins in relation to pollen from species interacting with another. In reality, however, there seems little evidence for incompatibility at this level, as was deduced by Faluyi (1980) and Williams *et al.* (1980). There are reports in the literature of unilateral incompatibility, where the pollen grows when the cross is made in one direction but where failure is observed when the pollen and female parent are reversed (Williams *et al.*, 1980; Schäfer-Menuhr *et al.*, 1988; Busmann-Loock *et al.*, 1992). The validity of this phenomenon is probably in need of further investigation since recent work with fluorescence microscopy has shown pollen tube growth in reciprocal crosses of *L. albus* and *L. mutabilis*, one of the combinations reported to show such unilateral responses (Siamasonta, 1996). This may not be surprising since most workers have found (and some reported) genotypic effects for components affecting interspecific hybridization in lupins and so choice of genotypes within a species may be crucial (Roy and Gladstones, 1985; Kasten *et al.*, 1991).

Pollen tube growth in the style

There are reports of somewhat reduced growth rates for pollen tubes with certain species combinations (Jaranowski, 1962; Williams *et al.*, 1980) but there is no clear evidence of problems which would be decisive in determining the outcome of the pollination.

Penetration of ovule, ovary stimulation and early embryo development

There is little evidence in the literature to suggest that, given suitable conditions, this will not occur (Jaranowski, 1962; Williams *et al.*, 1980). However, these events are difficult to observe directly and so caution must be exercised in assuming that they occur.

Later embryo development

This seems to be the stage where most researchers highlight the problems they have in actually achieving successful hybrids. One must always be careful about reducing the possibilities of earlier stages being insignificant, since completely negative results are always more difficult to report and have published, and are naturally discarded as programmes progress. Nevertheless, most workers report observations of ovule/embryo development. However, there are no reports of detailed knowledge of the genetic composition of the 'developing embryo' (i.e. most studies have not been able to test positively for the presence of both maternal and paternal DNA); they simply observe some development at this stage, often followed by degeneration of the embryo. It is following these observations that embryo rescue techniques have been attempted, whereby the developing embryo is dissected out and placed on defined media in an attempt to encourage further embryo development, which is assumed to be aborted *in vivo* because of endosperm failure (Kasten *et al.*, 1991; Schäfer-Menuhr *et al.*, 1988; Kasten and Kunert, 1991; Siamasonta and Caligari, 1998). It is also the case that in the development of the embryo at these later stages 'its own' genotype will start being expressed and so genomic problems might start to become apparent. Clearly an investigation attempting to characterize the genotype of the very early developing ovaries/embryos to confirm their genetic composition would be very informative and indicate the direction of future effort.

Inviability of plantlets/plants

The possibilities of inviability, and certainly sterility, are common in any interspecific crossing programmes. Factors such as the potential for chromosomal imbalance or lack of coevolved gene function are obvious and so the establishment of mature plants is always difficult and is usually expressed under conditions where the researcher has very few plantlets (often only one), so there is little scope to optimize the conditions needed. Micropropagation can help where protocols are in existence. Thus many plantlets are lost at this stage and they are often sterile if they reach maturity. Chromosome doubling (particularly in the case of odd-numbered polyploids) can sometimes be employed to restore fertility by a doubling of the ploidy level. Even having managed to get an embryo to develop does not guarantee that a usable mature, whole, green plant will follow.

Cytological studies of interspecific hybrids

Because of the incompatibility barriers noted above there have been relatively few cytological studies of meiosis in successful interspecific lupin hybrids (Kazimierski, 1980; Plitmann *et al.*, 1980; Gupta *et al.*, 1996). Such studies depend largely on the formation of bivalents and multivalents between species. Table 3.2 is a compilation of data for the meiotic behaviour of chromosomes in the limited number of interspecific lupin hybrids that are available (Kazimierski, 1961; Williams *et al.*, 1980; Pazy *et al.*, 1981; Roy and Gladstones, 1988; Busmann-Loock *et al.*, 1992; Gupta *et al.*, 1996). Detailed studies on the similarity of chromosomes among the rough-seeded lupin group have been conducted for the first time by Gupta *et al.* (1996). This group consists of six species: *L. cosentinii* (n = 16), *L. digitatus* (n = 18), *L. atlanticus* (n = 19), *L. princei* (n = 19), *L. pilosus* (n = 21) and *L. palaestinus* (n = 21). The seventh species, *L. somaliensis* (n = ?), appears to be extinct. Meiotic behaviour of chromosomes was reported on eight successful interspecific F1 hybrids.

The behaviour of chromosomes in *L. cosentinii* × *L. digitatus* and its reciprocal cross indicates that on an average 12–14 of the possible 17 pairs of chromosomes form bivalents in every cell and are homologous or homoeologous. A few chromosomes appear to have structural rearrangements (inversions and

Table 3.3. Time course and frequency of recovery of stable transformants following transformation and regeneration of *Lupinus angustifolius* L. cv. Merrit with a *bar*-CaMV 35S-*nos* construct to confer tolerance to the herbicide Basta (glufosinate).

Step		Number	Months
1.	Embryonic explants infected with *Agrobacterium* (AgL0/pCGP963) and the primary Basta selection applied at the sites of infection (2 mg glufosinate ml^{-1})	2463	0
2.	Shoots into culture for micropropogation with 0.2 mg glufosinate ml^{-1} in the medium		
3.	T_0-herbicide resistant shoots	9	5
4.	Grafts made to root stocks		
5.	T_0-seed	57	8–10
6.	T_1-seedlings	28	
7.	T_1-resistant plants selected with a leaf spray of 0.1 mg glufosinate ml^{-1}	15	14
8.	T_1-seed	8[2]	
9.	T_2-seedlings	39	
10.	T_2-resistant plants selected with a leaf spray of 0.1 mg glufosinate ml^{-1}	33	
11.	T_2-seed	2522	20
12.	T_3-seed from outside screenhouse trials[1]	*c.* 0.5×10^6	26
13.	T_4-seed from field trial plots[1]	*c.* 40×10^6	38

[1]Screenhouse and field trials approved by the Genetic Manipulation Advisory Committee (GMAC) of the Commonwealth of Australia.
[2]Eight plants from only one accession, three other accessions held in reserve.

translocations) as they form multivalents in low frequency. The rest of the chromosomes are non-homologous. The average number of multivalents is slightly higher in the crosses between *L. cosentinii* × *L. atlanticus* and its reciprocal but meiosis is similar to *L. cosentinii* × *L. digitatus* and its reciprocal. The higher number of bivalents, high chiasmata frequency and lower number of univalents in the cross between *L. atlanticus* × *L. digitatus* indicate that there is greater homology of chromosomes between these two species than *L. atlanticus* or *L. digitatus* with *L. cosentinii*. The interspecific crosses of *L. pilosus* with *L. atlanticus* and *L. cosentinii* indicate that *L. pilosus* is more homologous to *L. atlanticus* than to *L. cosentinii* on the basis of frequency of chiasmata in bivalents, univalents and multivalent formation. The chromosomes of *L. pilosus* may be completely homologous with those of *L. palaestinus* (Kazimierski, 1961; Pazy *et al.*, 1981) but further studies to test this idea have been precluded by unsuccessful attempts at hybridization (Gupta *et al.*, 1996). The chromosomes of *L. princei* seem to be isolated or are non-homologous with those of any of the other rough-seeded lupin chromosomes, as there are no reports of interspecific hybrids. Moreover the large chromosomes of this species make it cytogenetically distinct compared with other lupin species.

Kazimierski (1980) studied meiotic behaviour of chromosomes in the *lutei* section by crossing *L. luteus* and *L. hispanicus*. Similar to the rough-seeded group, chromosome configurations which include bivalents, univalents and tetravalents have been found. In spite of the two species having the same chromosome number, there is a difference of one translocation and Kazimierski (1980) concluded that there is partial homology of chromosomes in this section of the genus.

More than 50% of chromosomes of the species involved in these hybrids are homo(oe)logous (Table 3.2). To define further the prevalence of chromosomes that are in common among these species, more work is required on the application of banding techniques of chromosomes. Such studies would permit distinction between heterochromatic and euchromatic regions and lead to individual identification of metaphase chromosomes from the same or different species. Similarly, the application of *in situ* hybridization which permits molecular analysis of chromosome structure and physical localization of specific DNA sequences in individual chromosomes can provide vital information on the evolution of lupin chromosomes.

Overcoming the interspecific barriers?

An approach that attempts to circumvent many of the above problems is the use of protoplast fusion. Using this technique, cells have their walls removed (the cells are then termed protoplasts) in culture and are subsequently stimulated (by chemicals or electric pulses) to fuse. If mixed cultures contain protoplasts of two species, some of the fusions will be interspecific. The need is then to be able to select the interspecific products from the intraspecific ones and from unfused protoplasts. The most suitable systems have the potential to apply a selective differential which allows only the interspecific hybrids to develop – for example, if one species carried a herbicide resistance gene and the other an antibiotic resistance factor, then on medium containing both herbicide and antibiotic only the hybrids would develop. At present such methods are still being developed in lupins (Wetten, 1994; Wetten *et al.*, 1998) but whole plants have been regenerated from protoplasts without fusion (Schäfer-Menuhr, 1989). The use of genetic transformation procedures potentially provides a much more realistic method of introducing such selectable markers into lines that it is desired to fuse, but this is not yet routine with a range of lupin species.

Interspecific hybrids are a valuable route in transfering useful variation into commercially exploitable breeding lines and hence allow such variation to be incorporated into a desired background, following rounds of backcrossing. They are also potentially a source of completely new species that might be exploitable, as was probably the case with the Russell (garden) lupin (Gorer, 1988). Their potential value has to be seen against the existence of variation within a species which can be readily exploited, the potential to induce new variation via mutagenesis and the new possibilities of introducing genes via direct genetic transformation (see Genetic Engineering, below). As always, different species, different characters, different availability of facilities and different situations will all contribute to the decision about whether to invest time and energy in trying to optimize successful protocols to obtain hybrids. At present the inputs have been modest (but see Table 3.2) while the possible rewards are substantial.

Genetic Engineering

A number of protocols for organogenesis and regeneration of whole plants have been described for a number of *Lupinus* species. The initial studies by Lee (1955a,b) showed that explants from mature embryos of *L. hartwegii* could be induced to callus or regenerate rooted plants; Sator (1985) extended the protocols to include three other lupin species (*L. angustifolius*, *L. luteus* and *L. polyphyllus*). Sroga (1987) established callus cultures for *L. angustifolius*, *L. albus*, *L. luteus*, *L. hartwegii* and *L. polyphyllus* from embryogenic tissue as well as from mature vegetative and cotyledonary tissues. High frequencies of multiple shoot induction were demonstrated for calluses from embryonic axes and hypocotyls for all species except *L. albus* and *L. luteus*, with lower frequencies for those from leaves and cotyledons and none for calluses from roots. A high proportion (96%) of shoots formed roots in culture and 33–65% formed viable plants. Survival of regenerated plants was 26% higher for *L. angustifolius* compared with *L. polyphyllus*. More recently Daza and Chamber (1993) reported regeneration of viable *L. luteus* plants following shoot induction from hypocotyl explants in culture, and Rahim and Caligari (1998) have described an efficient *in vitro* shoot regeneration protocol for *L. mutabilis* based on explants from developing seeds. Protoplasts have been prepared from leaf tissue of *L. angustifolius*, *L. albus*, *L. luteus*, *L. mutabilis*, *L. hispanicus*, *L. pilosus*, *L. polyphyllus* and *L. palaestinus* using conventional enzymic digestion of cell walls (Wetten *et al.*, 1994). The yields were suitable for interspecific crosses using dielectrophoresis (Wetten *et al.*, 1998) and potentially could also be used for direct transformation using electroporation.

These cell and tissue culture methods have not been extended and effectively exploited for regeneration of genetically transformed plants. However, recently two Australian groups have independently described *Agrobacterium*-mediated transformation and regeneration of *L. angustifolius*. Molvig (1992) and Molvig *et al.* (1993, 1994) used slices of the cotyledonary node of young seedlings or of the developing embryonic axis. The technique was essentially the same as had proved successful for transformation and regeneration of *Pisum sativum* (Schroeder *et al.*, 1993) but, while transgenic peas were recovered with a frequency of 2–4%, a much lower frequency (< 0.01%) was obtained for *L. angustifolius* cv. Warrah (Molvig *et al.*, 1993). The method developed by Pigeaire *et al.* (1994) used explants from germinating seeds and relied on gene transfer to the shoot apex with subsequent development of shoots from transformed axillary buds. The frequency of recovery of transgenic plants ranged from 1.5 to 6%. Both of these protocols have led to the generation of transgenic lines but because of the much greater utility of that developed by Pigeaire *et al.* (1994, 1995, 1997) the details of their methods are summarized below. The method described is reproducible and is in routine use in genetically engineering lines with a range of novel traits, including alterations to growth regulator synthesis and sensitivity, changes to the nutritional quality of the grain, tolerance to herbicides and resistance to viruses.

Preliminary screening of lupin cultivar and *Agrobacterium* strain

Transient *GUS* expression in 19 different lupin cultivars was compared using four different strains of *Agrobacterium tumefaciens* (AgL0, LBA4404, K61, EHA101), each carrying the same binary vector, which contained a *gus*-CaMV*35S* construct (pCGP83; Janssen and Gardner, 1989). *A. tumefaciens* strain AgL0 (Lazo *et al.*, 1991) gave the highest expression when used to infect cv. Unicrop and so the transformation protocol was developed with this cultivar. The methods have been tested successfully with a wide range of cultivars including Merrit, advanced breeding lines and lines with restricted branching habit. Strain AgL0 has also proved superior for transformation of other lupin species (see section on Transformation and regeneration of other lupin species, below).

The constructs used in development of the transformation protocol were pCGP963 or pCGP1258 in the *Agrobacterium* strain AgL0. The plasmid pCGP963 was derived from pGA492 (An, 1986) and contained the *bar* gene controlled by the CaMV-35S promoter and an *ocs* (octopine synthase) terminator. The plasmid pCGP1258 is the same as pCGP963 except that it contains a *gus* gene from pKIWI101 (Janssen and Gardner, 1989) under control of the 35S promoter.

Explants and inoculation

The explant for transformation is the shoot apex of the embryonic axis excised from a germinated seed. To prepare explants, surface sterilized seed was germinated overnight at 25°C or until the root was 2–10 mm long. The seed coat, the cotyledons and the two pairs of leaves were then excised under a dissecting microscope to leave the plumule with an apical dome and the third pair of leaf primordia. To facilitate infection by *Agrobacterium* the apical dome was stabbed with a 30G needle before being plated upright (i.e. with the root in solid medium) on a co-cultivation medium containing MS salts (Murashige and Skoog, 1962) with B_5 vitamins, 10 mg BAP and 1 mg NAA per litre, adjusted to pH 5.8 and solidified with 0.25% (w/v) gelrite (Sigma). Subsequently it has been found that addition of glucose (10 mM) and acetosyringone (20 μM) to this medium increases the frequency of transformation (P.M.C. Smith, N. Fletcher and C.A. Atkins, Perth, 1996, unpublished).

A drop (≈5 μl) of *Agrobacterium* culture (resuspended in MS salts at 5×10^8 cells per ml) was placed on top of the wounded apex. The explants were co-cultivated with the bacterium for 2 days in low light at 25°C, after which they were transferred to regeneration medium (MS medium supplemented with one-tenth the level of growth regulators and 150 mg l^{-1} timentin to inhibit growth of the *Agrobacterium*). Transformed cells were selected for by applying a solution containing glufosinate (2 mg ml^{-1}) and 0.1% Tween 20 to the apex. The explants were grown in the same culture medium under low light at 25°C with an 8 h photoperiod.

Selection and regeneration of transgenic axillary shoots

Almost 100% of explants co-cultivated with pCGP963/AgL0 or pCGP1258/AgL0 regenerated shoots (10–15 mm long), but there was no regeneration where explants were co-cultivated with the host strain AgL0 without the *bar* construct. Two weeks after selection the axillary shoots were excised and placed on a 'micropropagation' medium (as above but with a further one-tenth dilution of the growth regulators, 150 mg timentin, 20 mg glufosinate per litre and solidified with 0.9% agar) which encouraged the development of transformed axillary buds. Green shoots were transferred to fresh medium every 2 weeks until they began to produce axillary shoots. Unlike control shoots, which usually produce axillary shoots on four sides (when grown on media without glufosinate), the transformed shoots often produced a shoot on only one side (secondary shoot). This indicates the chimeric nature of the primary transformants produced using this method and highlights the need for rigorous and repeated selection.

The secondary shoots were excised and cultured in the presence of glufosinate and in turn produced axillary shoots (usually on two sides) which were excised and cultured. In many cases the primary and secondary shoots did not survive the selection over a long period (3 months) and so it was almost invariably the tertiary and quaternary shoots which were retained for eventual transfer to the glasshouse.

Grafting transgenic shoots

In a series of preliminary experiments transformed lupin shoots did not root, even with changes in hormone levels to promote rooting, and so a grafting protocol was developed. More recently much higher frequencies of root initiation and adequate root development have been obtained (see Root development, below).

Several shoots (clones) representing each transgenic event were grafted on to the cotyledonary node of 10–14-day-old seedlings of their corresponding cultivar growing in sand with a complete nutrient solution (equivalent to Hoagland's solution with 5 mM KNO_3), or potting mix supplemented with nutrients. The tissue above the cotyledonary node was excised and the remaining 5 mm long piece of stem was split longitudinally so that the small cultured shoot could be inserted and held securely. The strength of the stem was sufficient to hold the grafted shoot in place without the need for tape or support. The grafts with their root stock were covered in a plastic vial, itself wrapped in aluminium foil for 3–4 days, after which the foil was removed. Over this period the graft union formed and new shoot tissue was evident. After a further 1–2 days the vial was removed and the plants were grown on under normal glasshouse conditions to obtain viable seed. The extent to which the transgenic shoots grew depended on the time to flowering. If shoots were held in tissue culture for extended periods (more than 4 months) before grafting, they flowered too soon and seed was not set. This could be overcome to some extent if the floral organs on the main axis were removed once the graft was established, leaving the axillary shoots to flower later when sufficient vegetation had been formed to support seed set and filling.

Root development on lupin shoots

To induce root formation, the base of a shoot was first dipped in 1 mg ml^{-1} indolebutyric acid (IBA) and the shoot was then grown on micropropagation media containing 3 mg IBA l^{-1} but no other hormones (P.M.C. Smith, N. Fletcher and C.A. Atkins, Perth, 1996, unpublished). Roots developed after approximately 2 weeks in culture and the plantlets could be transferred to the glasshouse after 3 weeks.

Assays for genetic transformation

Transformed axillary shoots which developed on the glufosinate medium (20 mg ml^{-1}) were tolerant of a 2 mg ml^{-1} glufosinate solution (in 0.1% Tween 20) painted on excised leaflets. Grafted shoots and foliage which expressed the *bar* gene in subsequent generations were tolerant to application of 0.1 mg glufosinate ml^{-1} in a leaf paint assay and, for a whole plant spray treatment, Basta at 0.1 mg glufosinate ml^{-1}. Resistance of leaflets or whole plants was assessed after 6–14 days. Stable integration of the *bar* gene was also confirmed by Southern analysis of genomic DNA and by northern analysis of RNA isolated from leaf tissue.

Inheritance patterns in the first generation of transgenic seed (T_1) equated roughly to a 3 : 1 Mendelian ratio, i.e. a segregation expected for a dominant gene at a single locus in a self-pollinating species. Genetic analysis of inheritance and expression of the *bar* gene in subsequent generations of transgenic lines is in progress.

Transformation frequency and time required to make a new genotype

Table 3.3 indicates the outcome and time required for successful transformation of cv. Merrit, a current commercial lupin cultivar, to express tolerance to Basta at the field trial stage of development. Overall the transformation frequency was about 0.4%, but this varies considerably with cultivar and frequencies of 2.8% have been found for an older variety, cv. Unicrop. Confirmation of the initial transformation required about 5 months, at the end of which a series of transformed shoots were available for grafting. The T_1 seed was available after a further 5 months and sufficient seed was generated in the T_3 generation for field and screen house trials. These trials have been approved by the regulatory authority of the Commonwealth Government of Australia (Genetic Manipulation Advisory Committee). From the initial explant to the point where sufficient transgenic seed had been generated for controlled field site trials at three locations required just over three years. Although the full regulatory process has not yet been applied to the transgenic lines of lupin which have been produced and Basta is not yet approved for broad-scale use on lupin crops in Western Australia, potentially the new technology of a herbicide resistant lupin could be available to the farmer some six years after the initial transformations were made.

Since these early transformations with the *bar* gene the transformation and regeneration system has been improved to give average frequencies of 2.5% for cv. Merrit and 6% or more for Unicrop (P.M.C. Smith, N. Fletcher, R. Walker and C.A. Atkins, Perth, 1996, unpublished).

Transformation and regeneration of other lupin species

There have been no published reports of routine protocols for the stable transformation of any other species of lupin. Mugnier (1988) demonstrated transformation of *L. albus* and *L. polyphyllus* with *A. rhizogenes* and established a hairy root culture for each species but did not regenerate transformed plants. Infection of embryonic axis explants of *L. albus* with AgL0/pCGP1258 (the vector that proved successful with *L. angustifolius*) resulted in low levels of transient GUS expression, but no Basta-tolerant shoots were generated and attempts to graft control shoots were not successful (P.M.C. Smith, N. Fletcher and C.A. Atkins, Perth, 1996, unpublished). The main difficulty with grafts was microbial contamination of the union as a result of prolonged exudation of phloem (containing 10–20% sucrose and amino acids) at the wound.

Some progress has been made with *L. luteus* (Somsap *et al.*, 1993, 1994) and a binary plasmid system in *A. tumefaciens* AGL1 carrying a *GUS–bar* fusion (similar to that used above for *L. angustifolius*). Recently transgenic plants of genotypes WDT 6172, Popiel and Teo carrying constructs to confer viral resistance (see next section) have been regenerated (M.G.K. Jones, Perth, 1996, personal communication). The transformation and regeneration protocols for *L. luteus* are similar to those outlined above for narrow-leafed lupin except that an *in vitro* grafting technique has been used to increase the frequency of successful grafts in the species, reducing possible losses in the glasshouse due to either desiccation or disease. The *Agrobacterium* strain used was critical to the frequency of transformation, AgL0 proving superior to AgL1. The transformations have been confirmed in T_1 (seed generation) plants by PCR assays for the gene, leaf painting with Basta, and Southern blot analysis. T_2 generation plants are currently being bulked up for confirmation and virus testing.

This is the first report of the production of transgenic yellow lupins. It is significant in that there is interest in cultivation of *L. luteus* in Portugal, Spain and Eastern Europe (Poland, Russia), in addition to Australia. The species is capable of growing in acid soils not suited to the growth of *L. angustifolius*; it is also relatively tolerant to waterlogging (see Chapter 2) and the protein content of the harvested grain is 8–10% greater than that of *L. angustifolius*. There are 100,000 ha of soils suited to *L. luteus* in Western Australia alone, and with the possible generation of resistance to virus coupled with selections for resistance to fungal diseases (*Pleiochaeta* root rot and brown leaf spot, Erudu patch disease) and low alkaloid levels, it is likely that there will be a considerable expansion in growth of yellow lupin as a major grain legume.

Transformation to Confer Viral Resistance

Narrow-leafed lupins (*L. angustifolius*) are susceptible to two major virus diseases in Western Australia. These are cucumber mosaic cucumovirus (CMV) and bean yellow mosaic potyvirus (BYMV). CMV is a major threat to the lupin industry both in Western Australia and south-eastern Australia (Bowyer and Kiernan, 1981; Alberts *et al.*, 1985; Jones, 1987, 1988). Infection with CMV can cause individual crop failures and widespread losses, as well as disrupting breeding and trialling programmes and production of certified seed (Jones, 1988, 1991;

Jones and McLean, 1989). Although there is a limited degree of natural resistance to CMV in *L. angustifolius* populations, this is not the case for BYMV. For other lupin species (*L. luteus*, *L. albus*) useful natural resistance to CMV is available and BYMV is the major pathogen.

Cucumber mosaic virus

The major source of CMV infection in *L. angustifolius* crops is infected seed, where seed transmission rates of up to 34% have been recorded (Jones, 1994; Jones and Proudlove, 1991). The virus is spread non-persistently via aphids from seed-infected plants, to infect plants of the current season's crop. The symptoms of CMV infection include stunted plants, down-curling of leaflets and chlorosis. Although yield may be severely depressed, seed production can occur from infected plants, providing CMV-infected seed to continue the cycle. Three approaches have been used to combat CMV in *L. angustifolius*: development of a seed testing service to ensure that CMV-free seed is sown; management strategies to reduce aphid transmission; and the development of resistance to CMV (either natural or via molecular methods).

1. Seed testing service, CMV subgroup

Since 1988 a seed testing service has been provided by the Western Australian Department of Agriculture (now Agriculture WA), in which farmers' seed samples are analysed for presence of CMV. Initially this was carried out by enzyme-linked immunosorbent assay (ELISA), but a more sensitive reverse-transcriptase polymerase chain reaction (RT-PCR) method has been developed (Wylie *et al.*, 1993) which can routinely detect one CMV-infected seed in 1000 seed samples. This has been used as the primary screen for CMV in about 1500 seed samples per annum.

As part of the development of the RT-PCR assay, it was found that only subgroup II of CMV was present in lupins (from diverse geographical locations and different varieties), although CMV subgroup I is also widely found in Western Australia, in subterranean clover (*Trifolium subterraneum*) pastures in the south-west and in bananas in the Gasgoyne central region (Wylie *et al.*, 1993; Singh *et al.*, 1995).

2. CMV – molecular aspects

CMV is the type member of the cucumoviruses, and both subgroups have a positive sense tripartite genome (RNA1,2,3) with a subgenomic RNA4, satellites and defective sequences. There is about 95% sequence homology within CMV subgroups, but about 75% sequence homology between the subgroups (Singh *et al.*, 1995). Most published work has been done on subgroup I isolates of CMV, which tend to be more virulent than subgroup II isolates. Because of its diverse host range, several molecular approaches have been used to engineer resistance to CMV, mostly using subgroup I isolates (Singh *et al.*, 1998).

The approaches followed include expression of the coat protein gene of CMV-C (subgroup I) and a defective replicase from CMV-Fny (subgroup I) in transgenic tobacco plants. These plants have been challenged with CMV isolates of both subgroups from Western Australia (Singh *et al.*, 1998). It was found that although the coat protein-containing transgenic plants conferred resistance to subgroup I isolates from Western Australia, they did not confer resistance to the lupin isolates. Similarly, tobacco plants with the defective replicase gene were highly resistant to subterranean clover and banana isolates of CMV (subgroup I), but resistance to lupin isolates was limited (Singh *et al.*, 1998).

3. Constructs for engineered CMV resistance in lupins

Based on the information gained from published work, knowledge of the subgroup of CMV in lupins and tests on transgenic tobacco plants with CMV subgroup I-based constructs, a series of new constructs based on the most aggressive lupin isolate, LY (subgroup II), have been made. These include the coat protein gene (RNA3) and the movement protein gene (RNA3a) in sense, anti-sense, full length and shorter versions. In addition, full length and defective replicase (RNA2) constructs are being made. At present, each of these constructs is in a binary vector driven by the CaMV 35S promoter, with the latter driving the *bar* gene as a selectable marker. Transformation into *L. angustifolius* is in progress using the method outlined in the Genetic Engineering section above.

Bean yellow mosaic virus

BYMV causes a serious necrotic disease of *L. angustifolius*, both in the higher rainfall areas of Western Australia and in south-eastern Australia. It is not seed-borne in *L. angustifolius*, because infected plants die and do not produce seed. However, BYMV is seed-borne in other lupin species with up to 12% seed transmission in *L. albus* and 7% in *L. luteus* (Jones, 1994). BYMV is spread by migrant aphids, mainly from neighbouring subterranean clover pastures. When young plants are infected by aphids, a necrotic streaking occurs in younger shoots and the apical shoot usually bends over to give the characteristic 'shepherd's crook' symptom. The growing tip dies; the leaves become pale, wilt and abscise. The whole plant becomes blackened and dies (Jones, 1994) although infection of older plants allows some seed to be set. Infection is usually worst at crop margins and localized within the crop, but it can become generalized with resulting heavy losses in grain yield (Jones, 1994). No natural resistance to BYMV has been found in any of the major lupin species (*L. angustifolius, L. luteus, L. albus*). Although an agronomic package, based on methods to reduced aphid transmission to the crop, has been devised (Jones, 1994), the only effective approach to develop resistance appears to be using engineered viral genes.

BYMV – molecular aspects

BYMV is a typical potyvirus with an overall genome structure as shown in Fig. 3.1. It exists as flexuous, filamentous particles 750 nm long and 15 nm wide and has a single-stranded sense genomic RNA of about 10 kb with a VPg protein linked to the 5′ end and a poly A tail at the 3′ end. The viral genome has one open reading frame (ORF), which is translated as a polyprotein and subsequently cleaved into functional polypeptides by the sequential action of viral proteases. The coat protein gene is located at the 3′ end of the genome; 5′ to this is the NIb

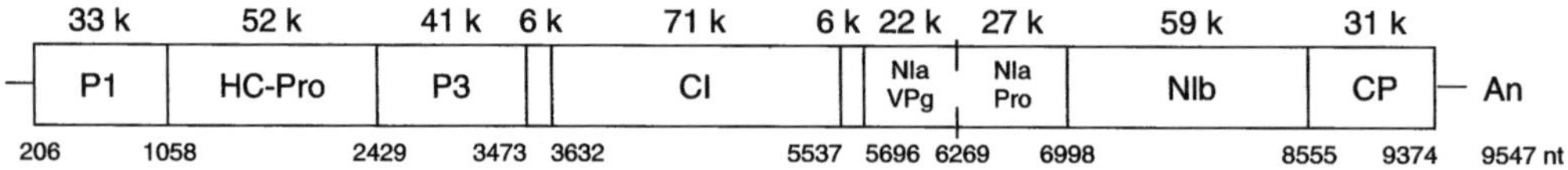

Fig. 3.1. Organization of the BYMV genome. The single lines represent 5′ and 3′ untranslated regions, whilst the box region indicates the viral polyprotein with cleavage sites indicated by vertical lines. The numbers below the polyprotein indicate the first nucleotide coding for each protein of BYMV-S. The protein products are P1, HC-Pro and NIa (viral proteases), P3 protein, CI (cytoplasmic inclusion), NIa-VPg and NIa-protease, NIa (nuclease inclusion protein a), NIb replicase (nuclease inclusion protein b), CP (coat protein) and 6 K proteins (from Guyatt *et al.*, 1996).

replicase and NIa (VPg/protease) gene. About half of the genome of the BYMV-MI isolate from Western Australia has been sequenced (Mathews *et al.*, 1995; EMBL accession number X81124 – coat protein gene, S. Wylie and G. Dwyer, Perth, 1996, unpublished). Recently a complete sequence of an Australian BYMV isolate (S) has also been reported (Guyatt *et al.*, 1996): this isolate appears to be somewhat different from lupin isolates of BYMV (G. Dwyer, Perth, 1996, unpublished).

Because BYMV is a major pathogen of all the cultivated lupin species, for which no natural resistance is available, and resistance to BYMV is the remaining major trait thought to be needed to enable widespread adoption of *L. luteus* in Western Australia, more effort has been made to generate and introduce resistance constructs against BYMV.

Constructs for engineered resistance to BYMV in lupins

A series of 34 potential resistance constructs, based on the coat protein, NIb and NIa (VPg/protease) genes of BYMV-MI, have been generated. These include full length or truncated sense, anti-sense, untranslatable and mutated genes, all driven at present by the CaMV 35S promoter, and linked to the CaMV 35S-*bar* selectable marker gene for Basta resistance. They have been introduced in *A. tumefaciens* strains AgL1, AgL0 and LBA4404, in the binary plasmids pTAB10 (L. Tabe, CSIRO Plant Industry, Canberra) or pPZB series (S. Wylie, Perth, 1996, unpublished).

Transformation

Transformation of *L. angustifolius* has been carried out essentially as described above, initially with *A. tumefaciens* strain AgL1 but subsequently with the more effective strain AgL0. In a series of experiments, transgenic plants (T_0, T_1, T_2) have been generated containing coat protein, NIb and NIa-based constructs, and the *bar* herbicide resistance gene. Transgenic lines of *L. angustifolius* generated include Merrit, Myallie, 83A:473, 84A-479 and 85L:460. These plants (T_1 generation) have been characterized by leaf painting tests with the herbicide Basta, by PAT assays, by PCR analysis for the gene and by Southern blotting. The plants are currently being bulked up for testing for possible resistance to BYMV.

Transgenic *L. luteus* plants of three cultivars have also been generated (outlined above) which contain either an NIa construct, an anti-sense NIa construct or the BYMV coat protein gene.

A current view of the mechanism of virus homologous gene-dependent resistance in transgenic plants is that it is probably based on an RNA-dependent degradation mechanism, which is initiated by overexpression of mRNA of the transgene, such that it and incoming viruses with the same sequence are degraded (Lindbo *et al.*, 1993; Mueller *et al.*, 1995). If this is the correct basis of resistance induced by expression of viral gene sequences in transgenic plants, it requires high levels of transgene mRNA transcription (but low levels of protein expression), and perhaps only one in ten or 20 transgenic events will provide a high level of virus resistance. Hence a series of transgenic events for each construct must be tested.

Perspectives for Yield Improvement of Lupins Through Genetic Engineering to Alter Hormone Balance

In addition to the first outcomes described above from the Lupin Genetic Engineering programme of the Centre for Legumes in Mediterranean Agriculture (CLIMA) in Western Australia, transformed lines are being generated in which changes to alter source/sink relations and harvest index, to enhance seed quality and increase tolerance to stress are sought.

A major limitation to realizing the yield potential of narrow-leafed lupin in the farmer's field is the loss of flowers and pods due to their excessive abortion and abscission. Under a wide range of environmental conditions as many as 94% of flowers and immature pods abscise (Farrington and Gladstones, 1974; Pigeaire *et al.*, 1992a,b) and harvest index is extremely variable (Delane *et al.*, 1988; Pigeaire *et al.*, 1990). Abiotic stresses such as drought, changes in temperature and nutrient deficiency reduce reproductive development and no doubt contribute to the variation in harvest index and grain yield (Chapter 8). While both nutritional and hormonal factors are likely to be involved there is evidence that specific changes in the balance of growth regulators in flowers or young pods have an impact on pod set and grain filling. Exogenous application of cytokinin to flowers at or close to anthesis prevents their abortion (Atkins and Pigeaire, 1993) and a similar effect can be achieved by application of an inhibitor of ethylene action (silver thiosulphate) to flowers at anthesis (Clements *et al.*, 1994). Field application of a range of growth-regulating chemicals (including cytokinin, gibberellins and auxin) to lupins prior to or during pod set had little effect on plant growth or grain yield in trials conducted over three seasons and at four sites (Seymour, 1996). Small yield increases were found with Cultar (paclobutrazol), Lontrel (cyclopyralid) or a mixture of Lexone and Brodal (metribuzin-diflufenican) and these were due to greater pod set. However, Seymour (1996) has indicated that cost and the need for precise timing of their application mitigate against their practical use.

Modifying the effects of growth regulators might be realized endogenously by inserting genes which alter their synthesis, transport and metabolism or which modify the plant's sensitivity to changes in their concentration. For example, stable enhanced synthesis of cytokinin has been achieved in a number of species by transferring the *ipt* (isopentenyl transferase) gene from *A. tumefaciens* (Medford *et al.*, 1989; Smigocki, 1991; Smigocki *et al.*, 1993). Enhanced cytokinin synthesis in these transgenics delayed senescence, altered source/sink relations and, in the case of *Nicotiana* transformed with *ipt* behind a proteinase inhibitor, also resulted in enhanced resistance to insect predation (Smigocki, 1991). Constructs of *ipt* coupled with a constitutive promoter (CaMV35S), a flower-specific promoter from *Arabidopsis thaliana* or a seed storage protein promoter from lupin have been used to transform *L. angustifolius* in an attempt to alter endogenous cytokinin synthesis (J. Barton, Perth, 1996, personal communication).

The rate of ethylene synthesis has been successfully manipulated in plants by expressing anti-sense constructs for ACC synthase or oxidase or by inserting a bacterial ACC deaminase gene (Kende, 1993). In tomato these changes delay fruit ripening and the onset of leaf senescence and could conceivably alter the stress response in a species like lupin. Although no clear relationship has been established between enhanced ethylene synthesis and the train of events which lead to the expression of components of the stress response of lupin (such as premature senescence, flower and pod abortion and reduced grain filling), preliminary studies of cultivar sensitivity to exogenous ethylene and the rate of endogenous synthesis following shock indicate a pivotal role for the gaseous regulator (J. Clements and C.A. Atkins, Perth, 1996, unpublished data).

It is unlikely that cytokinin and ethylene are the only regulators involved in the response of lupins to changes in environmental factors. Accumulation of abscisic acid (ABA) in different lupin species appears to be related to their origin in wet or dry regions (Chapter 8) and the response of stomatal conductance to ABA indicates a role in response to water deficit (Gallardo *et al.*, 1994). Similarly ABA, gibberellins and auxin together with cytokinin and ethylene have been implicated in the regulation of reproductive development. Strategies for directed modification of the balance among these regulators between organs or with time have not been developed. However, an increasing number of plant hormone mutants are being described (Reid, 1993), offering the potential for a more detailed molecular analysis of regulator transduction pathways and the possible cloning of genes and gene promoters which might alter their expression.

Detailed empirical models for the flow of assimilates, solutes of nitrogen and inorganic ions have been described for *L. albus* and, to some extent, for *L. angustifolius* (Chapter 7). These source/sink descriptions indicate sites likely to control solute flux (Layzell *et al.*, 1981) and highlight the possible role of 'transfer cells' in enhancing translocation and imposing specificity (Atkins, 1993). It is not yet possible to identify the transport proteins which operate at this level, let alone their encoding genes and the means of regulation. However the central role for asparagine in fashioning the composition of the transport channels, particularly in the specificity of transfer of N-solutes between xylem and phloem in leaf minor veins and nodes or between the filial and maternal tissues of the developing seed, predicts that over- or under-expression of amide/amino acid transporters in the transfer cell plasmamembrane offers a possible basis for genetic manipulation (Atkins, 1993).

Acknowledgements

We wish to acknowledge S. Carstairs, S. Gupta, Ms N. Fletcher, Dr J. Barton, Dr P.M.C. Smith, Dr M.G.K. Jones, S. Wylie, Ms L. Hua, V. Somsap and G. Dwyer for permission to cite unpublished data.

References

Alberts, E., Hanna, J. and Randles, J.W. (1985) An epidemic of cucumber mosaic virus in South Australian lupins. *Australian Journal of Agricultural Research* 36, 267–273.

An, G. (1986) Development of plant promoter expression vectors and their use for analysis of differential activity of nopaline synthase promoter in transformed tobacco cells. *Plant Physiology* 81, 86–91.

Atkins, C.A. (1993) Nitrogen assimilation and translocation in relation to plant growth. In: Buxton, D.R., Shibles, R., Forsberg, R.A., Blad, B.L., Asay, K.H., Paulsen, G.M. and Wilson, R.F. (eds) *International Crop Science. I.* Crop Science Society of America, Madison, WI, USA, pp. 807–811.

Atkins, C.A. and Pigeaire, A. (1993) Application of cytokinins to flowers to increase pod set in *Lupinus angustifolius* L. *Australian Journal of Agricultural Research* 44, 1799–1819.

Atkins, C.A., Smith, P.M.C. and Pigeaire, A. (1994) Direct genetic manipulation to alter processes which might enhance the utility and value of lupins. In: Dracup, M. and Palta, J. (eds) *Proceedings of the First Australian Lupin Technical Symposium.* Department of Agriculture, Perth, Western Australia, pp. 119–122.

Bowyer, J.W. and Kiernan, E. (1981) Cucumber mosaic virus in lupin. *Australasian Plant Pathology* 10, 27–29.

Buirchell, B.J. (1994) Domestication and potential of the rough-seeded lupins. In: Neves Martins, J.M. and Beirão da Costa, M.L. (eds) *Advances in Lupin Research, Proceedings of VIIth International Lupin Conference.* ISA Press, Évora, pp. 19–24.

Busmann-Loock, A., Dambroth, M. and Menge-Hartmann, U. (1992) Histological observations on interspecific crosses in the genus *Lupinus. Plant Breeding* 109, 82–85.

Carstairs, S.A., Buirchell, B.J. and Cowling, W.A. (1992) Chromosome number, size and interspecific crossing ability of three Old World lupins, *Lupinus princei* Harm., *L. atlanticus* Gladst. and *L. digitatus* Forsk., and implications for cyto-systematic relationships among the rough-seeded lupins. *Journal of the Royal Society of Western Australia* 75, 83–88.

Christou, P. (1994) The biotechnology of crop legumes. *Euphytica* 74, 165–185.

Clements, J.C., Atkins, C.A. and Fletcher, N. (1994) Ethylene and pod set in lupins. In: Dracup, M. and Palta, J. (eds) *Proceedings of the First Australian Lupin Technical Symposium.* Department of Agriculture, Perth, Western Australia, p. 261.

Darlington, C.D. and La Cour, L.F. (1969) *The Handling of Chromosomes*, 5th edn. George Allen & Unwin Ltd, London, 272 pp.

Daza, A. and Chamber, M.A. (1993) Plant regeneration from hypocotyl segments of *Lupinus luteus* L. cv. Aurea. *Plant Cell Tissue and Organ Culture* 34, 303–305.

Delane, R.J., Matthews, J. and Hamblin, J. (1988) Effect of genotype and environment on harvest index of *L. angustifolius.* In: *Proceedings of the Fifth Lupin Conference.* International Lupin Association, Poznań, Poland, p. 527.

Dunn, D.B. and Gillett, J.M. (1966) *The Lupines of Canada and Alaska.* Canadian Department of Agriculture Monograph 2, Ottawa.

Eichorn, A. (1949) À propos de la caryologie du *Lupinus tassilicus* Maire. *Revue de Cytologie et de Biologie Végétales* 11, 333–350.

Faluyi, M. (1980) Studies on the breeding system in lupin species. PhD thesis, University of Reading, UK.

Farrington, P. and Gladstones, J.S. (1974) Effect of genotype on yield of narrow-leafed lupin (*Lupinus angustifolius*) and sand plain lupin (*Lupinus cosentinii*). *Australian Journal of Experimental Agriculture and Animal Husbandry* 14, 742–748.

Fruwith, C. (1910) *Die Züchtung der landwirtschaftlichen Kulturpflanzen. Bd.* 3, 133.

Gallardo, M., Turner, N.C. and Ludwig, C. (1994) Water relations, gas exchange and abscisic acid content of *Lupinus cosentinii* leaves in response to drying different proportions of the root system. *Journal of Experimental Botany* 45, 909–918.

Gill, B.S., Friebe, B. and Endo, T.R. (1991) Standard karyotype and nomenclature system for description of chromosome bands and structural aberrations in wheat (*Triticum aestivum*). *Genome* 34, 830–839.

Gladstones, J.S. (1958) The naturalized and cultivated species of *Lupinus* (Leguminosae) recorded in Western Australia. *Journal of the Royal Society of Western Australia* 41, 29–33.

Gladstones, J.S. (1974) *Lupins of the Mediterranean Region and Africa.* Western Australian Department of Agriculture Technical Bulletin 26, Perth.

Gladstones, J.S. (1984) Present situation and potential of Mediterranean/African lupins for crop production. In: *Proceedings of the Third International Lupin Congress.* International Lupin Association, La Rochelle, France, pp. 18–37.

Gollmick, F. (1937) Über Artkreuzungen bei Lupinen. *Der Züchter* 9, 65–68.

Gorer, R. (1988) *The Development of Garden Flowers.* Eyre and Spottiswoode, London.

Gupta, S. and Buirchell, B.J. (1998) Genetic analysis of seed permeability, sweetness, non-shattering pods and early flowering in *Lupinus cosentinii* Guss, *L. atlanticus* Gladst. and their interspecific hybrids. In: Hill, G.D. (ed.) *Proceedings of 8th International Lupin Conference.* International Lupin Association, California, USA (in press).

Gupta, S., Buirchell, B.J. and Cowling, W.A. (1996) Interspecific reproductive barriers and genomic similarity among the rough-seeded *Lupinus* species. *Plant Breeding* 115, 123–127.

Guyatt, K.J., Proll, D.F., Menseen, A. and Davidson, A.D. (1996) The complete nucleotide sequence of Bean Yellow Mosaic Potyvirus RNA. *Archives of Virology* 141, 1231–1246.

Janssen, B.J. and Gardner, R.C. (1989) Localized transient expression of GUS in leaf discs following cocultivation with *Agrobacterium. Plant Molecular Biology* 14, 61–72.

Jaranowski, J. (1962) Fertilization and embryo development in the genus *Lupinus* Tourn. Part II. Fertilization and embryo development following reciprocal species hybridization. *Genetica Polonica* 3, 333–363.

Jones, R.A.C. (1987) Cucumber mosaic virus in lupins. *Western Australian Journal of Agriculture,* Fourth Series 28, 50–54.

Jones, R.A.C. (1988) Seed borne cucumber mosiac virus infection of narrow-leafed lupin (*Lupinus angustifolius*) in Western Australia. *Annals of Applied Biology* 113, 507–518.

Jones, R.A.C. (1991) Reflective mulch decreases the spread of two non-persistently aphid-transmitted viruses to narrow-leafed lupin (*Lupinus angustifolius*). *Annals of Applied Biology* 118, 79–85.

Jones, R.A.C. (1994) *Virus Diseases of Lupins. Department of Agriculture, Western Australia, Bulletin* 4294, 15pp.

Jones, R.A.C. and McLean, G.D. (1989) Virus diseases of lupins (a review). *Annals of Applied Biology* 114, 609–637.

Jones, R.A.C. and Proudlove, W. (1991) Further studies on cucumber mosaic virus infection of narrow-leafed lupin (*Lupinus angustifolius*): seed-borne infection, aphid transmission spread and effects on grain yield. *Annals of Applied Biology* 118, 319–329.

Kasten, W. and Kunert, R. (1991) A culture method for isolated immature embryos of different *Lupinus* species. *Biologishes Zentralblatt* 110, 290–300.

Kasten, W., Paradies, T., Kunert, R. and Straka, P. (1991) Progress in realization of interspecific hybrids in the genus *Lupinus* by means of embryo rescue technique. *Biologishes Zentralblatt* 110, 301–309.

Kazimierski, T. (1960) An interspecific hybrid in the genus Lupinus (*Lupinus albus* L. × *L. jugoslavicus* Kazim. & Nov.). *Genetica Polonica* 1, 3–60.

Kazimierski, T. (1961) Interspecific hybridization of *Lupinus*. *Genetica Polonica* 2, 97–102.

Kazimierski, T. (1964) Interspecific crossing in the Mediterranean and American lupin species. *Wiss Zeitschr Karl-Marx-University, Leipzig, Math-Naturw* 13, 697–698.

Kazimierski, T. (1980) Cytogenetics of species and hybrids in the Lutei section of the genus *Lupinus*. In: *Proceedings of the First International Lupin Conference*. International Lupin Association, Lima, Peru, pp. 51–68.

Kende, H. (1993) Ethylene biosynthesis. *Annual Review of Plant Physiology and Plant Molecular Biology* 44, 283–307.

Layzell, D.B., Pate, J.S., Atkins, C.A. and Canvin, D.T. (1981) Partitioning of carbon and nitrogen to root and shoot apex in a nodulated legume. *Plant Physiology* 67, 30–36.

Lazo, G.R., Stein, P.A. and Ludwig, R.A. (1991) A DNA transformation-competent *Arabidopsis* genomic library in *Agrobacterium*. *Plant Molecular Bio/Technology* 9, 963–967.

Lee, A.E. (1955a) Growth in culture of excised portions of lupine embryos. *Botanical Gazette* 116, 359–364.

Lee, A.E. (1955b) Potentially unlimited growth of lupine callus. *Botanical Gazette* 116, 364–368.

Lindbo, J.A., Silva-Rosales, L., Proebsting, W.M. and Dougherty, W.G. (1993) Induction of a highly specific antiviral state in transgenic plants: implications of regulation of gene expression and virus resistance. *Plant Cell* 5, 1749–1759.

Malheiros, N. (1942) Elementos para o estudo citologico do género *Lupinus*. *Agronomia Lusitana* 4, 231–236.

Mathews, A., Dwyer, G., Wylie, S. and Jones, M.G.K. (1995) Nucleotide and deduced amino acid sequence of the 3′ end of the BYMV-MI genome (Brief Report). *Archives of Virology* 140, 2269–2272.

Medford, J.I., Horgan, R., El-Sami, Z. and Klee, H.J. (1989) Alterations of endogenous cytokinins in transgenic plants using a chimeric isopentenyl transferase gene. *Plant Cell* 1, 403–413.

Molvig, L. (1992) Regeneration of lupin (*Lupinus angustifolius* L.) from tissue culture via organogenesis. In: *Proceedings of the Australian Society of Plant Physiologists 32nd Annual General Meeting, La Trobe University*, P3.

Molvig, L., Schroeder, H., Wardley-Richardson, T., Spencer, D. and Higgins, T.J. (1993) Histological comparison of regeneration and transformation in an amenable grain legume (*Pisum sativum* L.) and a recalcitrant species (*Lupinus angustifolius* L.). In: *Proceedings of the Australian Society of Plant Physiologists. 33rd Annual General Meeting*. University of Western Australia, Perth, P44.

Molvig, L., Schroeder, H.E., Tabe, L.M., Moore, A., Kortt, A.A., Spencer, D. and Higgins, T.J.V. (1994) Improvement of seed protein quality of lupin (*Lupinus augustifolius* cv. Ararran) by genetic engineering. In: Dracup, M. and Palta, J. (eds) *Proceedings of the First Australian Lupin Technical Symposium*. Department of Agriculture, Perth, Western Australia, p. 285.

Mueller, E., Gilbert, J., Davenport, G., Brigneti, G. and Baulcombe, D.C. (1995) Homology-dependent transgenic virus resistance in plants related to homology-dependent gene silencing. *Plant Journal* 7, 1001–1013.

Mugnier, J. (1988) Establishment of new axenic hairy root lines by inoculation with *Agrobacterium rhizogenes*. *Plant Cell Reports* 7, 9–12.

Murashige, T. and Skoog, F.J. (1962) A revised medium for rapid growth and bioassays with tobacco tissue cultures. *Physiologia Plantarum* 15, 473–497.

Nam, Y.I. and Zayakin, V.V. (1998) Approaches to the obtaining of interspecific lupin hybrids between species with different chromosome number. In: Hill, G.D. (ed.) *Proceedings of 8th International Lupin Conference*, International Lupin Association, California (in press).

Pazy, B., Heyn, C.C., Herrnstadt, I. and Plitmann, U. (1977) Studies in populations of the Old World *Lupinus* species. I. Chromosomes of the East Mediterranean Lupines. *Israel Journal of Botany* 26, 115–127.

Pazy, B., Plitmann, U. and Heyn, C.C. (1981) Genetic relationships between *Lupinus pilosus* and *L. palaestinus* (Fabaceae). *Plant Systematics and Evolution* 137, 39–44.

Pigeaire, A., Pate, J.S. and Atkins, C.A. (1990) Contrasting effects of rooting volume and aerial spacing on shoot morphology and reproductive performance of *L. angustifolius*. *Australian Journal of Agricultural Research* 41, 339–350.

Pigeaire, A., Delane, R.J., Seymour, M. and Atkins, C.A. (1992a) Predominance of flowers and newly-formed pods in reproductive abscission of *Lupinus angustifolius* (L.). *Australian Journal of Agricultural Research* 43, 1117–1129.

Pigeaire, A., Seymour, M., Delane, R.J. and Atkins, C.A. (1992b) Partitioning of dry matter into primary branches and pod initiation on the main inflorescence of *Lupinus angustifolius* (L.). *Australian Journal of Agricultural Research* 43, 685–696.

Pigeaire, A., Lu, C.Y. and Cornish, E. (1994) High frequency shoot regeneration from lupin seedling explants (*Lupinus angustifolius* L.) via organogenesis. In: *Proceedings of the VIII International Congress of Plant Tissue Culture. Florence, Italy*, p. 130.

Pigeaire, A., Abernethy, D., Smith, P.M.C., Simpson, K., Fletcher, N., Lu, C.-Y., Atkins, C.A. and Cornish, E. (1995) Routine transformation of *Lupinus angustifolius* L. via *Agrobacterium tumefaciens*-mediated gene transfer to shoot apices. In: *Proceedings of the Australian Society of Plant Physiologists. 35th Annual General Meeting. Sydney*, Abstr. 39.

Pigeaire, A., Abernethy, D., Smith, P.M.C., Simpson, K., Fletcher, N., Lu, C.-Y., Atkins, C.A. and Cornish, E. (1997) Routine transformation of a grain legume crop (*Lupinus angustifolius* L.) via *Agrobacterium tumefaciens*-mediated gene transfer to shoot apices. *Molecular Breeding* 3, 341–349.

Plitmann, U. and Heyn, C.C. (1984) Old World *Lupinus*: taxonomy, evolutionary relationships, and links with New World species. In: *Proceedings of the Third International Lupin Congress*. International Lupin Association, La Rochelle, France, pp. 56–66.

Plitmann, U. and Pazy, B. (1984) Cytogeographical distribution of the Old World *Lupinus*. *Webbia* 38, 531–539.

Plitmann, U., Heyn, C.C. and Pazy, B. (1980) Biological flora of Israel. 7. *Lupinus palaestinus* Boiss. and *L. pilosus* Murr. *Israel Journal of Botany* 28, 108–130.

Poirier, Y., Nawrath, C. and Somerville, C. (1995) Production of polyhydroxyalkanoates, a family of biodegradable plastics and elastomers, in bacteria and plants. *Bio/Technology* 13, 142–150.

Rahim M.A. and Caligari, P.D.S. (1998) Multiple shoot regeneration in *Lupinus mutabilis* Sweet. In: Hill, G.D. (ed.) *Proceedings of 8th International Lupin Conference*. International Lupin Association, California, USA (in press).

Reid, J.B. (1993) Plant hormone mutants. *Journal of Plant Growth Regulation* 12, 207–226.

Römer, P., Caligari, P.D.S., Huyghe, C., Neves Martins, J.M. and Sawicka-Sienkiewicz, E.J. (1998) Breeding perspectives of *Lupinus mutabilis* in middle Europe. In: Hill, G.D. (ed.) *Proceedings of 8th International Lupin Conference*. International Lupin Association, California, USA (in press).

Roy, N.N. and Gladstones, J.S. (1985) Prospects for interspecific hybridization of *Lupinus atlanticus* Gladst. with *L. cosentinii* Guss. *Theoretical and Applied Genetics* 71, 238–241.

Roy, N.N. and Gladstones, J.S. (1988) Further studies with interspecific hybridization among Mediterranean/African lupin species. *Theoretical and Applied Genetics* 75, 606–609.

Sator, C. (1985) Studies on shoot regeneration of lupins (*Lupinus* sp.). *Plant Cell Reports* 4, 126–128.

Sawicka-Sienkiewicz, E.J. (1998) Interspecific crossability of Andean lupin (*Lupinus mutabilis* Sweet). In: Hill, G.D. (ed.) *Proceedings of 8th International Lupin Conference*. International Lupin Association, California, USA (in press).

Schäfer-Menuhr, A. (1989) Isolation and culture of lupin protoplasts IV. Regeneration of plants from protoplasts. *Landbauforsschung Völkenrode* 39, 133–136.

Schäfer-Menuhr, A., Czerwinski, T. and Busmann, A. (1988) The use of embryo culture for the regeneration of interspecific hybrids from *Lupinus mutabilis* × *Lupinus hartwegii*. *Landbauforschung Völkenrode* 38, 173–177.

Schroeder, H.E., Schotz, A.H., Wardley-Richardson, T., Spencer, D. and Higgins, T.J.V. (1993) Transformation and regeneration of two cultivars of pea (*Pisum sativum* L.). *Plant Physiology* 101, 751–757.

Seymour, M. (1996) Response of narrow-leafed lupin (*Lupinus angustifolius* L.) to foliar application of growth regulators in Western Australia. *Australian Journal of Experimental Agriculture* 36, 473–478.

Siamasonta, M.B. (1996) Aspects of lupin breeding: interspecific hybridisation in lupins. PhD thesis, University of Reading, UK.

Siamasonta, M.B. and Caligari, P.D.S. (1998) Embryo growth rate *in vivo* in *L. albus* and *L. mutabilis* following self, intra- and inter-specific pollination. In: Hill, G.D. (ed.) *Proceedings of 8th International Lupin Conference*. International Lupin Association, California, USA (in press).

Singh, R.J. and Hymowitz, T. (1988) The genomic relationship between *Glycine max* (L.) Murr. and *G. soja* Sieb. and Zucc. as revealed by pachytene chromosome analysis. *Theoretical and Applied Genetics* 76, 705–711.

Singh, Z., Jones, R.A.C. and Jones, M.G.K. (1995) Identification of cucumber mosaic virus subgroup I isolates from banana plants affected by infectious chlorosis disease using RT-PCR. *Plant Disease* 79, 713–716.

Singh, Z., Jones, M.G.K. and Jones, R.A.C. (1998) Effectiveness of coat protein and defective replicase gene-mediated resistance against Australian isolates of cucumber mosaic cucumovirus. *Australian Journal of Experimental Agriculture* (in press).

Smigocki, A.C. (1991) Cytokinin content and tissue distribution in plants transformed by a reconstituted isopentenyl transferase gene. *Plant Molecular Biology* 16, 105–115.

Smigocki, A.C., Neal, J.W., McCanna, I. and Douglass, L. (1993) Cytokinin-mediated insect resistance in *Nicotiana* plants transformed with the *ipt* gene. *Plant Molecular Biology* 23, 325–335.

Snow, R. (1963) Alcoholic hydrochloric acid-carmine as a stain for chromosomes in squash preparations. *Stain Technology* 38, 9–13.

Somsap, V., Atkins, C. and Jones, M.G.K. (1993) Tissue culture for transformation of narrow- and broad-leafed lupins. *Proceedings of the Australian Society of Plant*

Physiologists. 33rd Annual General Meeting, University of Western Australia, Perth, p. 46.

Somsap, V., Cooper, J.I., Li, D. and Jones, M.G.K. (1994) Tissue culture and transformation of lupins. In: Dracup, M. and Palta, J. (eds) *Proceedings of the First Australian Lupin Technical Symposium*. Department of Agriculture, Perth, Western Australia, p. 312.

Sroga, G.E. (1987) Plant regeneration of two *Lupinus* spp. from callus cultures via organogenesis. *Plant Science* 51, 245–249.

Tuschnjakowa, M. (1935) Uber die Chromosomen einiger *Lupinus*-Arten. *Der Zuchter* 7, 169–174.

Van Der Leij, F.R. and Witholt, B. (1995) Strategies for the sustainable production of new biodegradable polyesters in plants: a review. *Canadian Journal of Microbiology* 41 (suppl.), 222–238.

Wetten, A., Bese, N., Masumbuko, L. and Caligari, P.D.S. (1994) Isolation of lupin protoplasts for the production of interspecific hybrids. In: Dracup, M. and Palta, J. (eds) *Proceedings of 1st Australian Lupin Technical Symposium*. Department of Agriculture, Perth, Western Australia, p. 321.

Wetten, A., Sinha, A. and Caligari, P.D.S. (1998) Electrofusion of lupin protoplasts for the production of interspecific hybrids. In: Hill, G.D. (ed.) *Proceedings of 8th International Lupin Conference*. International Lupin Association, California (in press).

Williams, W., Akhtar, M.A. and Faluyi, M. (1980) Cross compatibility between European and American *Lupinus* species. *Botanical Journal of Linnean Society* 81, 225–232.

Wolko, B. and Weeden, N.F. (1994) Linkage map of isozyme and RAPD markers for *Lupinus angustifolius* L. In: Neves Martins, J.M. and Beirão da Costa, M.L. (eds) *Advances in Lupin Research, Proceedings of the VIIth International Lupin Conference*. ISA Press, Evora, pp. 42–49.

Wylie, S., Wilson, C.R., Jones, R.A.C. and Jones, M.G.K. (1993) A polymerase chain reaction assay for cucumber mosaic virus in lupin seeds. *Australian Journal of Agricultural Research* 44, 41–51.

Young, S. (1994) A paintbox for plant genes. *New Scientist* 143, 28–31.

4 Lupin Breeding

W.A. Cowling[1], C. Huyghe[2] and W. Swiecicki[3]

[1]*Agriculture Western Australia, 3 Baron-Hay Court, South Perth, Western Australia 6151;* [2]*Station d'Amélioration des Plantes Fourragères, Institut National de la Recherche Agronomique (INRA), 86600 Lusignan, France;* [3]*Poznań Plant Breeders, 63-004 Tulce, Kasztanowa 5, Poland*

Domestication – First Steps for Modern Lupin Breeding (1928–1945)

Lupins did not fully join the ranks of domesticated grain legumes until the latter half of the 20th century – several thousand years after food legumes such as chickpea, field pea, lentil, faba bean or soybean. As shown in Chapter 1, *Lupinus albus* and *L. mutabilis* were partially domesticated around the first millennium BC in the Mediterranean region and South America, respectively. Lupins were well known for their beneficial properties to soil and several species appear to have been selected for human food with permeable seeds and sometimes non-shattering pods. However, their use was always hampered by the presence of bitter alkaloids. The first steps for modern lupin breeding were taken in 1928/29 by von Sengbusch in Germany, who selected the first low-alkaloid cultivars of *L. luteus* and *L. angustifolius*.

Von Sengbusch (1942), Hackbarth and Troll (1956), Gladstones (1970) and Hondelmann (1984) describe German breeding of these two species over the next few years, until the work was disrupted towards the end of World War II. Along with low alkaloid genes, von Sengbusch found a natural mutant of *L. luteus* with permeable seed coats in 1928 and a single non-shattering plant in a 5 ha field of yellow bitter lupins in 1935 (Hackbarth and Troll, 1956). These traits were found to be controlled by single genes (*w* for permeable seed coats and *invulnerabilis* for non-shattering), and the two genes were combined with white-seeded lines by Troll to develop the first fully domesticated *L. luteus* cv. Weiko II in 1943 (Hackbarth and Troll, 1956). Weiko is an abbreviation in German for *weisses korn* or white grain (Hanelt, 1960).

Von Sengbusch followed up his initial work on *L. luteus* and *L. angustifolius* by selecting low-alkaloid strains within *L. albus* during 1930 and 1931 (von Sengbusch, 1942). However, the first sweet selections of *L. albus* were too late maturing for northern Europe. Only after World War II did breeders develop cultivars with early enough maturity for spring sowing in northern Europe,

 Lupins as Crop Plants: Biology, Production and Utilization (eds J.S. Gladstones, C.A. Atkins and J. Hamblin)

notably Pflugs Gela and Pflugs Ultra in West Germany and Neutra in East Germany.

The first fully domesticated cultivar of *L. angustifolius* was not developed until the 1960s. Permeable seed coats became available in bitter *L. angustifolius* lines of uncertain origin in the 1930s. Troll crossed these with von Sengbusch's original hard-seeded sweet line to release Müncheberger Blaue Süsslupine II in 1944 (Hackbarth and Troll, 1956; Troll, 1964a). Despite earlier claims of improved non-shattering in *L. angustifolius*, this character proved to be unstable (Hondelmann, 1984). Little further progress was made until Gladstones in 1960 discovered two recessive mutants (*lentus* and *tardus*) with non-shattering pods in fields of bitter cultivated *L. angustifolius* in Western Australia (Gladstones, 1967). The two mutant genes were complementary in their action and, together, almost completely stopped shattering before harvest. They were crossed into white-flowered and white-seeded sweet types (the *leucospermus* gene). The first variety of sweet narrow-leafed lupins (cv. Uniwhite, with the *tardus* non-shattering gene) was released in 1967. Uniwhite was quickly followed by Uniharvest (with both *lentus* and *tardus* non-shattering genes) in 1970 (Gladstones, 1977). The gene *leucospermus* became a useful unlinked marker for sweetness (*iucundis*), allowing ready identification of blue-flowered and dark-seeded bitter contaminants or progeny of cross-pollination with bitter types in sweet crops of *L. angustifolius* (Gladstones, 1977).

Gladstones also proceeded to domesticate *L. cosentinii* with the aid of natural and induced mutants, and released cv. Erregulla (Anon., 1977) which had non-permeable seeds. Owing to agronomic problems, it did not become established commercially. A soft-seeded selection from Erregulla had deformed seeds and low seed yield, and was not released (J.S. Gladstones, Western Australia, 1997, personal communication).

Sweet mutants of *L. mutabilis* were also found by von Sengbusch in the 1930s (von Sengbusch, 1942; Hondelmann, 1984), but these were apparently lost. Recently a sweet cultivar of *L. mutabilis* (cv. Inti) was reported from a breeding programme in Chile (von Baer and von Baer, 1988), which marks the beginning of full domestication in this species. The names of domestication genes and early lupin varieties for several species are found in Gladstones (1970) and will not be expanded upon here.

Domestication in *L. angustifolius*, *L. albus*, *L. luteus*, *L. cosentinii* and *L. mutabilis* provides an excellent example of Vavilov's law of homologous series (Vavilov, 1951). Genes essential for domestication, such as non-shattering fruits and permeable seeds, are present in a wide range of crop species such as wheat, beans or maize. It was expected by the early German lupin breeders that such genes should also be available in lupins. Genes for low alkaloids, non-shattering pods and permeable seed coats have now been found in all of the above-mentioned lupin species, and are being assembled as well in *L. atlanticus* and *L. pilosus* (B.J. Buirchell, Western Australia, 1997, personal communication). This chapter focuses on the three major domesticated species, *L. albus*, *L. angustifolius* and *L. luteus*, with only minor reference to other species.

Defining the Domesticated Lupin

Although most lupin breeders have focused on improvements in sweet, fully domesticated lupins since the 1940s, strong disagreements continue among lupin breeders, researchers and promoters over the role of bitter types. The early German breeders realized that lupins would not reach their full potential as a crop plant until alkaloids were removed by breeding. The traditional role of bitter lupins as green manure crops, although valuable, would always be very limited. As sweet yellow lupin varieties quickly spread throughout Eastern Europe after World War II, Poland and other countries brought in regulations to prevent the growing of bitter yellow lupins which might cause contamination by admixture or cross-pollination. Such countries developed very high standards of sweet lupin breeding and production. Most lupin breeders have continued with the philosophy of eliminating bitter types from their breeding material (Gladstones, 1994).

In direct conflict with these breeding aims, a bitter lupin industry has been promoted recently in the European Community. A prototype refining factory was established where seed protein and fibre were separated and alkaloids purified as putative pharmaceutical or antimicrobial agents (W. Jaeggle, Mittex Anlagenbau GmbH, Ravesburg, Germany, unpublished abstract at the 7th International Lupin Conference, Portugal, 1993). However, no evidence exists that there is sufficient demand or value for the alkaloid products in the market-place. In Portugal, where it is very difficult to grow and maintain pure sweet yellow lupin crops, bitter yellow lupins are promoted for cropping and feeding to animals (Soares, 1994). However, modern animal feed industries will quickly turn away from lupins if one 'bad batch' causes feeding problems due to high alkaloids.

Clearly the two types of lupin industry are not compatible. The major barrier to the accreditation of lupins in human food is the presence of alkaloids (Jasper, 1994). Any presence of a bitter lupin industry will make sweet lupin seed more difficult to promote on international export markets as a valuable protein source for humans or animals. It appears that little progress has been made in resolving the alkaloid issue, at least on an international scale, since the first sweet mutants were discovered in Germany in 1928.

Difficulties remain for growing sweet lupins in some countries. In Spain, Portugal and Italy it is difficult to keep sweet cultivars of *L. albus*, *L. angustifolius* and *L. luteus* free of bitter contaminants. This is due to the common occurrence of landrace or wild bitter lupins of the same species inside or near crops of sweet lupin. The diversity of seed colours among these bitter types makes them especially hard to distinguish from sweet cultivars, no matter what colour markers the latter might have. Fencing is also essential to protect sweet lupin from herbivore pests, and this makes cropping sweet lupins more expensive (or impossible) in the extensive grazing systems of southern Europe.

There is no doubt that the relative success of the sweet lupin industries in countries such as Australia and Poland is due to their strict control over alkaloids during breeding and production. This is aided by a flower or seed colour marker for sweetness, and the relative absence of bitter weedy types. In Australia, alkaloids are controlled on an industry-wide basis all the way from the early stages of lupin breeding and release of new varieties, through to testing of farm-produced lupins at regional centres (Harris, 1994). Sweet lupins are receiving attention in France, the UK, the USA, Chile and other countries where wild weedy types of domesticated species are rare and the climate and soils are suitable for lupin production. The way forward for the lupin industry, in our opinion, is the further development of sweet lupins through breeding.

Impact of Domestication on Plant Performance

Many domesticated plant species such as maize are genetically isolated from their wild relatives after millennia of selection by humans. Others, such as garlic and ginger, are fully dependent on cultivation for their existence and their wild relatives are presumed to be extinct (Robinson, 1996). In lupins, domestication has occurred only recently, and domestication genes have only recently been introduced. The opportunity therefore exists in lupins to investigate the impact of genes for domestication on plant performance.

In *L. albus*, near-isogenic pairs of lines with either the dominant (bitter) or recessive (sweet) form of the gene *pauper* were subjected to a series of measurements during 1993 and 1994 in France (Table 4.1). Sweet lines were more frost-tolerant than bitter lines when tested in a growth chamber. In contrast, in 1994 (which was very cold and seedlings suffered frost damage and fungal diseases) field survival of bitter lines exceeded that of the sweet lines. On average, the yield of sweet lines was the same as bitter lines during the mild growing

Table 4.1. Performance of near-isogenic pairs of bitter and sweet *L. albus* in the field, including frost tolerance (as measured in growth chambers), 1994 winter survival (plant density in spring compared with the original sowing density of 22 plants m^{-2}), seed yield in winter 1993, seed yield in winter 1994, and seed weight (C. Huyghe, unpublished results).

Near-isogenic pair	Frost Tolerance (ratio[1])		1994 winter survival (spring plant density m^{-2})		1993 seed yield (t ha^{-1})		1994 seed yield (t ha^{-1})		Seed weight (mg)	
	Bitter	Sweet	Bitter	Sweet	Bitter	Sweet	Bitter	Sweet	Bitter	Sweet
1	1.31	1.51	19.10	7.20	4.38	4.36	4.12	2.57	284	278
2	1.38	1.80	18.50	9.50	4.81	3.71	4.81	3.23	286	298
3	1.63	1.97	21.00	10.10	4.33	4.21	4.59	3.24	315	322
4	1.17	1.66	18.80	9.80	4.08	4.36	3.92	3.17	365	285
5	1.06	1.42	17.90	13.10	4.72	4.30	4.73	4.29	339	401
6	1.09	1.41	19.80	16.80	4.43	5.20	4.77	5.14	363	376
7	1.57	2.13	20.50	13.20	5.17	5.06	4.77	4.06	316	291
8	1.54	1.75	16.70	11.00	3.94	4.33	4.42	3.34	330	323
9	1.57	2.79	18.90	11.30	5.54	4.59	4.54	3.04	331	321
10	—	—	19.40	11.30	4.36	4.17	4.31	3.20	340	336
11	1.47	1.69	19.20	11.80	4.34	3.90	4.03	3.33	336	317
12	1.49	1.85	15.90	11.60	4.79	4.21	3.72	3.66	335	333
13	—	—	17.80	12.30	4.28	4.49	3.77	3.08	312	316
Average	1.39	1.82	18.73	11.46	4.55	4.39	4.35	3.49	327	323
% (Sweet/Bitter)	131		61		96		80		99	

[1]Frost tolerance, measured in controlled conditions, is the ratio of mean weight of plants 15 days after cold stress/mean weight before the cold stress (average of three independent tests).

season of 1993, but in the more severe 1994 season they yielded less. The seed size of bitter and sweet lines was very similar. Overall, the bitter lines appeared to tolerate cold and disease stress better in the field than the sweet lines, mostly due to better winter survival.

Sweet *L. mutabilis* cv. Inti yielded less seed than comparable bitter lines (von Baer and von Baer, 1988). Near-isogenic lines of *L. angustifolius* with the dominant (bitter) form of the gene *iucundus* were on average 30% higher yielding than lines with the recessive (sweet) form of the gene in experiments of Oram (1983). Near-isogenic lines with either the dominant (blue flowers and grey testa) or recessive (white flowers and white testa) form of the gene *leucospermus* did not differ in yield. Sweetness in *L. angustifolius*, and to a lesser degree the gene *tardus* for non-shattering, also had negative pleiotropic effects that predisposed the plants to the split seed disorder caused by manganese deficiency (Walton and Francis, 1975; see Chapter 5). This was partly resolved in genotypes with the gene for early flowering, *Ku*, which allowed maturation under less stressful conditions.

Genes for non-shattering pods were also examined for their effects on dry matter accumulation, harvest index, seed yield, and pod and stem dry matter in *L. angustifolius* and *L. cosentinii* (Edwards, 1980). The recessive forms of the genes *lentus* in *L. angustifolius* and *macer* in *L. cosentinii* modified the endocarp of the pod wall and were associated with lower seed weights and lower pod, stem and total plant dry matter. In contrast, the recessive forms of the genes *tardus* in *L. angustifolius* and *coniunctus* in *L. cosentinii*, which caused fusion of the sclerenchyma strips in dorsal and ventral seams of pods, were associated with greater dry matter and seed yield, greater pod and stem dry matter and larger seeds. There was no disadvantage if both non-shattering genes were present; in fact domesticated *L. angustifolius* with both *le* and *ta* genes for non-shattering had greater dry matter and seed yield than their 'wild' counterparts.

There is also a negative impact on plant performance associated with a soft-seed mutation in *L. cosentinii* and other rough-seeded lupins that distorts the seed and reduces seed filling (B.J. Buirchell, Western Australia, 1997, personal communication).

The domestication genes that have consistently reduced yield are the recessive genes for low alkaloid content. Apart from their role in discouraging grazing by herbivores, high alkaloid genes appear to increase the tolerance of lupins to environmental stresses, and in some cases increase resistance to insects and diseases. However, this is by no means universal, as bitter lupins are susceptible to insects such as the lupin aphid *Macrosiphum albifrons* (von Eppler and Hinz, 1987) and diseases such as phomopsis stem blight (Cowling *et al.*, 1987), brown spot (Gladstones and Crosbie, 1978) and pleiochaeta root rot (Sweetingham *et al.*, 1994). Therefore, the physiological role of alkaloids that imparts improved seed yield (in the absence of herbivorous grazing) remains uncertain. Fortunately, there is strong evidence that breeding has succeeded in improving yield and disease resistance following the initial set-back caused by the low alkaloid genes in *L. angustifolius* (Oram, 1983; Gladstones, 1994; Cowling, 1996).

Genes that change the phenology of the plant to match the target environments are likely to improve plant performance. In *L. angustifolius*, the gene *Ku* brought forward flowering by 2 to 5 weeks and allowed narrow-leafed lupins to be grown as a successful commercial crop in the Mediterranean climate of Western Australia (Gladstones and Hill, 1969). Plant architecture in *L. albus* is the subject of a later section in this chapter.

Domesticated lupins still have a number of characters considered as 'wild'. Pod and seed walls are much thicker than in longer-established food legume species. For example, the pod wall is 34% of the mature fruit dry matter in *L. angustifolius*, 33% in *L. albus*, 46% in *L. luteus* and 46% in *L. mutabilis*, compared with 13% in field pea (M. Reader and M. Dracup, Western Australia, 1997, personal communication). This is an inefficient use of photosynthate that could be incorporated into the seeds. It is also a hindrance to rapid desiccation of pods and thus to grain harvest of *L. albus* in continental Europe (Hondelmann, 1984).

Similarly, seed coat in *L. albus* accounts for 18% of seed dry weight, 24% in *L. angustifolius*, 25% in *L. luteus* and 13% in *L. mutabilis* (M. Reader and M. Dracup, Western Australia, 1997, personal communication), compared with 9% in soybean (Hondelmann, 1984). There may be potential to reduce both the pod wall and seed coat thickness in the future breeding of lupins. Theoretically, this will allow improvements in overall seed yield and quality, with more protein and metabolizable energy and less coarse fibre.

The Era of Modern Lupin Breeding

The pioneers of modern lupin breeding (post 1945) include J. Hackbarth and H.-J. Troll in West and East Germany, respectively, M. Lenoble in France, W. Williams in England, S. Barbacki in Poland, E. von Baer in Chile, N.A. Maisurian and V.I. Golovchenko in the USSR, I. Forbes and H.D. Wells in the USA, and J.S. Gladstones in Australia. These breeders and others formed the new gene pool of sweet lupins upon which all future breeders will depend. As a result of their work, sweet lupins have become a significant crop in many countries. There are now more than 1 million ha of lupins grown annually in Australia (Cowling, 1996). Australian sweet lupin (*L. angustifolius*) is now recognized internationally as a valuable source of vegetable protein. The next few sections outline the bases for these breeding advances since 1945, and some possible future advances and directions for lupin breeding.

The importance of wild lupins in modern lupin breeding

Modern lupin breeding has benefited from the extensive collection of genetic material in the wild or in areas of traditional cultivation (in the case of landraces of *L. albus* and *L. luteus*) (see Chapter 2). As a result of close international collaboration among lupin breeders, these genetic resources have been duplicated and used in several breeding programmes throughout the world.

For instance, frost tolerance in *L. albus* was found in landrace types collected in northern Italy (J. Papineau and C. Huyghe, 1993, unpublished results). Sweetingham *et al.* (1994) found resistance to Pleiocheata root rot in *L. albus* collected in the Azores Islands. The urgency of such collections is highlighted by the situation on the Azores Islands, where traditional agricultural systems (including landraces of *L. albus*) are being replaced with agriculture that is more intensive. When J. Papineau collected in the Island of Faial in 1989, he found 20 populations of *L. albus* (Papineau and Huyghe, 1989), but in 1994 he found only one. Recently, resistance to anthracnose was noted in landraces of *L. albus* from Ethiopia (B.J. Buirchell, Western Australia, 1997, personal communication). Breeders are now using these landraces of *L. albus* in their breeding programmes to improve frost tolerance and disease resistance.

Resistance to phomopsis stem blight of *L. angustifolius* is present in accessions from throughout the natural habitat of this species in Morocco, Portugal, Spain and Italy (Cowling *et al.*, 1987). Primary crosses between cv. Illyarrie and wild types from Spain and Morocco resulted in the phomopsis-resistant cultivars Gungurru and Yorrel (Gladstones, 1989). The progeny of some crosses with wild *L. angustifolius* from Italy and Israel carried partial resistance to brown spot and root rot caused by *Pleiochaeta setosa* (Cowling *et al.*, 1997). Resistance to grey leaf spot and anthracnose and frost tolerance in *L. angustifolius* was found in progeny of crosses with a Portuguese wild type (Wells and Forbes, 1982). Field evaluation of wild populations of *L. angustifolius* from the Aegean region was useful to identify types with excellent pod set and large seeds for crossing into domesticated lupins (Clements and Cowling, 1994).

Wild lupins have also been important in breeding *L. luteus*. Resistance to fusarium wilt was found in Portuguese wild types (Lamberts, 1952) and was controlled by a dominant gene *Fus1* (Hackbarth and Troll, 1956). The first resistant domesticated cultivar was Refusa (bred in East Germany by Troll and Jagoda; released in 1965). The same resistance gene is present in more recent varieties such as Juno and Teo (from Poland), and Borsaja and Borselfa (from Germany), and remains very effective against fusarium wilt. Resistance to powdery mildew was also found in Portuguese wild types of *L. luteus* and the gene named *Er* (H. Lamberts, the Netherlands, 1975, personal communication to W. Swiecicki).

The importance of natural and induced mutants in modern lupin breeding

The recent history of lupin breeding is also marked by the use of chemical and irradiation mutagenesis (Micke and Swiecicki, 1988). With these techniques, mutants have rarely been directly released as new cultivars but some have been identified as potential parents.

In the Ukraine, various mutants were generated in an *L. albus* landrace from western Georgia in the late 1950s, some of which were included in V.I. Golovchenko's breeding programme and contributed to the release of cvv. Kiev Mutant and Horizont (Golovchenko, 1982). An early maturing mutant was generated in G.G. Gataulina's *L. albus* breeding programme in Moscow and gave rise directly to the low alkaloid cv. Start (Vavilov and Gataulina, 1984).

J.S. Gladstones and C.M. Francis in Western Australia artificially induced the genes for low alkaloid content (*sw*), early flowering and maturity (*xe*) and white flowers and seeds (*wfs*) in *L. cosentinii* cv. Erregulla, a mostly domesticated variety with non-permeable seeds (Anon., 1977). An induced mutant gene *efl* was used in breeding the moderately early flowering *L. angustifolius* cv. Chittick in Western Australia (Anon., 1982). The spontaneous dominant mutation *Ku* in

L. angustifolius, found in a crop of the late-flowering cv. Borre (Gladstones and Hill, 1969), completely removes the vernalization requirement for flowering. Crosses have shown *Ku* to be identical with the gene *Julius* used by Maissurian and others in Eastern Europe (J.S. Gladstones, Western Australia, 1997, personal communication). Tedin in Sweden selected early-flowering induced mutants in *L. luteus*, and these were later used in Poland to breed cvv. Aga and Reda (Micke and Swiecicki, 1988).

Spontaneous and induced mutants with a range of restricted branching patterns and dwarfism have appeared in all domesticated lupin species. These have especially been sought in continental Europe, where early growth cessation and rapid maturity are required for the short summer growing season. Polish breeders induced recessive 'epigonal' genes '*ep1*' in *L. albus* (Mikolajczyk *et al.*, 1984), and '*mut1*' in *L. angustifolius* (Bromberek *et al.*, 1984). The gene for self-completion (highly restricted branching) in *L. luteus* was found as a spontaneous mutant by Svab and was described by Troll (1967). W. Swiecicki in Poland developed *L. luteus* cv. Manru using this self-completing spontaneous mutant; Manru is semi-late flowering, but has restricted branching with the uppermost three to four leaf nodes replaced by flowers. In France, M. Lenoble and J. Papineau selected the *L. albus* dwarf mutant XA100 in a mutagenesis programme in 1981 and fixed it as a pure line in 1989. G.G. Gataulina's *L. albus* cv. Start in Russia has a mild restriction in the uppermost branches with fewer leaf nodes than normal.

Both dominant and recessive genes control restricted branching in *L. angustifolius*, as found in the Byelorussian lines Kupala (recessive) and Lanedesk (dominant) (Micke and Swiecicki, 1988). A restricted-branching type selected by I. Forbes in the USA in 1965, later used in the Australian *L. angustifolius* breeding programme, resulted from a natural mutation that appeared to be incompletely dominant (Gladstones, 1994). A large proportion of progeny from crosses between Forbes' restricted branching line and normal branching types were intermediate for restricted branching, but selection of selfed progeny resulted in some that appeared to be fixed for the intermediate character. The genetics of this intermediate character is being studied currently by M. Dracup, N. Galwey and W.A. Cowling in Western Australia. The genetic studies are complicated by the variable nature of the intermediate restricted branching phenotype, which is possibly due to an increase in the outcrossing rate in restricted branching types.

Recent varieties of restricted branching *L. angustifolius* include cv. Ladny, bred by V.I. Derbensky at NPO 'Podmoskovje' in Moscow and released in 1992. Ladny was derived from a cross of Lanedesk with the USA cv. Frost. It is early flowering and rapid maturing, which is an advantage in short continental summer growing seasons. The first Australian restricted branching variety of *L. angustifolius*, cv. Tallerack, was bred by W.A. Cowling at Agriculture Western Australia and is being released in 1997. This variety arose as a natural variant in a brown spot resistance breeding population, and has a very mild restriction with fewer leaf nodes in the uppermost branches than normal branching types. Early indications suggest that Tallerack has a relative advantage in high rainfall regions where pod set and not biomass limits seed yield.

Yield improvements in modern lupin breeding

The first step in a new lupin breeding programme is to identify the phenotype adapted to local climatic conditions and soil types (Wells, 1984). This usually provides large improvements in grain yield. Adaptation of *L. angustifolius* to the mediterranean climate of Western Australia was greatly improved by the *Ku* gene for early flowering in cv. Unicrop, as mentioned above. Similarly, thermoneutral types of *L. luteus* that mature rapidly have improved yields of this species in northern Europe. In France and the UK, large improvements in yield of *L. albus* are now being seen in dwarf, determinate and vernalizing types which may be sown early in the autumn and which resist freezing damage over winter (see Plant architecture changes section below). This is a completely different model of plant architecture from that of the early-maturing spring-sown cultivars developed in Eastern Europe over the past 40 years. In each case, induced and spontaneous mutants have been useful to overcome some major agronomic deficiencies of lupins.

The second step is to seek quantitative improvements in yield, quality and important agronomic characters, which are usually under polygenic control. Attempts should also be made to improve resistance to diseases of local importance. An example of parallel improvement in yield, lodging resistance, disease resistance and quality is provided by the narrow-leafed lupin breeding programme in Western Australia from 1973, when cv. Unicrop was released, to 1991, when the high yielding and disease resistant cv. Merrit was released.

A suite of eight cultivars, released from 1973 to 1991, were grown in historical variety trials at 11 locations in Western Australia in 1991 and 1992. The rate of yield improvement was 2.4% per year from cv. Unicrop (1973) to cv. Merrit (1991) at the high target plant density of 70 plants m^{-2} (Fig. 4.1) (Cowling and Speijers,

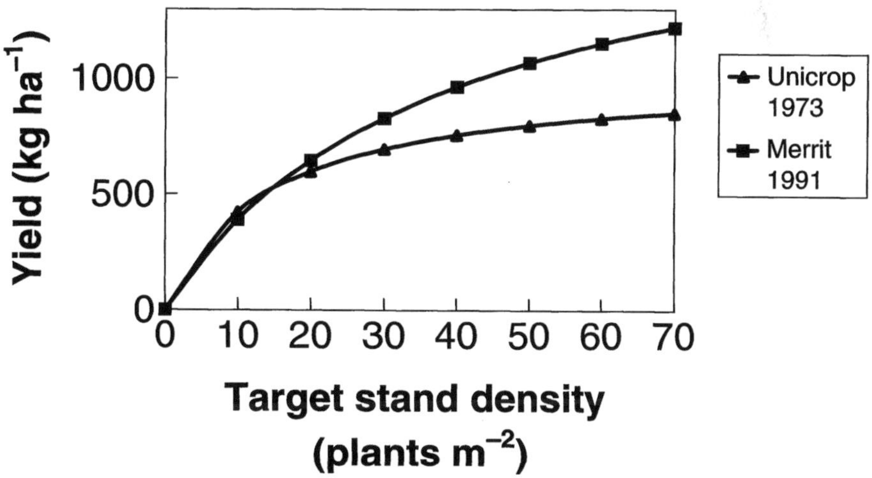

Fig. 4.1. Influence of target plant density (plants m^{-2}) on yield (kg ha^{-1}) of narrow-leafed lupin cultivars Merrit (released in 1991) and Unicrop (released in 1973) in historic variety trials in Western Australia (data from Cowling and Speijers, 1994).

1994). There was no difference in yield between Unicrop and Merrit at 10 plants m^{-2} (Fig. 4.1). The superior yield of *L. angustifolius* cv. Merrit is due to the improved ability of Merrit plants to set pods and seeds in dense populations. It follows that yield improvements in narrow-leafed lupins are most easily attained at high plant density, and that selection for yield should occur in the breeding programme at high plant density. This parallels experience with most other crops.

Yield improvements in *L. angustifolius* cultivars from 1973 to 1991 were associated with improvements in harvest index, but there was no increase in total biomass of cultivars (Tapscott *et al.*, 1994). Harvest index increased from 0.24 in cv. Unicrop to 0.29 in cv. Merrit, mostly due to more pods and seeds per plant. However, harvest index is still low compared with other grain legumes, and this gives grounds to hope for large future improvements in grain yield through breeding. This is verified by recent releases such as cv. Kalya, which is higher yielding on average than Merrit (Cowling 1997a).

Disease resistance improvements in modern lupin breeding

Resistance to fusarium wilt became essential for lupin cultivation in Europe from the 1960s and 1970s. The gene *Fus1* has remained effective against fusarium wilt in yellow lupin since its introduction from wild types in the 1950s. Methodology for breeding yellow lupin for resistance to fusarium was outlined by Troll (1964b) and Nowacki and Swiecicki (1973), and all cultivars of *L. luteus* in eastern Europe (such as Juno, Teo, Borsaja, Borselfa, Piast, Amulet) are resistant. Resistance to fusarium wilt has also been selected in *L. angustifolius* cultivars in Belarus (N.I. Kuptsov, Minsk, 1996, personal communication).

Production of narrow-leafed lupins increased in Australia in the late 1980s following the release of Phomopsis resistant cultivars (Gladstones, 1989). Resistance reduced the toxicity of lupin stubbles that caused lupinosis in sheep (Cowling *et al.*, 1988). This improved the profitability of including lupins in mixed grazing/cropping farming systems (Warren *et al.*, 1989).

Resistance to grey leaf spot (caused by *Stemphylium vesicarium*) and anthracnose (caused by *Glomerella cingulata*) was developed in *L. angustifolius* cv. Rancher in the southern USA in the 1960s (Wells and Forbes, 1982). Resistance to both diseases was transferred by recombination to the Australian cv. Marri (Gladstones, 1977) and later cv. Illyarrie (Gladstones, 1994).

Without doubt, the greatest challenge to lupin breeding in the 1990s is to improve resistance to anthracnose disease. A highly virulent strain known as VCG2 (see Chapter 9, this volume) has spread around the world in lupin seed since the late 1970s. Australian lupin breeders and pathologists are undertaking a large screening experiment in 1997 to select resistance in cultivated and wild forms of a range of lupin species. In addition, arrangements have been made for international testing of a Standard Set of 18 lines (including older cultivars which were originally resistant) to compare reactions across different countries.

The two main virus diseases of lupins are bean yellow mosaic virus (BYMV) and cucumber mosaic virus (CMV). *L. albus* is practically immune to CMV, and very high resistance to CMV was recently found in several *L. luteus* cultivars including cvv. Teo (Poland) and Motiv 369 (Belarus) (Jones and Latham, 1996).

Strong resistance to CMV in *L. angustifolius* is not known; however, cv. Danja and some advanced crossbred lines have very low seed transmission rates, most likely due to polygenic resistance (Jones and Cowling, 1995). *L. angustifolius* cv. Wandoo had to be withdrawn from release in Western Australia due to its high rates of CMV seed transmission (Gladstones, 1994). Resistance to BYMV in *L. luteus* is quantitative, and is expressed primarily as a reduction in seed transmission rates – from 15.8% in the most susceptible genotype 'TS' to 0.2% in strain 'V-1' (Schmidt, 1988).

Cowling *et al.* (1997) in Western Australia demonstrated quantitative improvements in resistance to brown spot and pleiochaeta root rot (caused by *Pleiochaeta setosa*) in *L. angustifolius*. Broad-sense heritability of resistance was very high (h^2 = 0.89–0.94) across environments for brown spot but much lower (h^2 = 0.00–0.53) for pleiochaeta root rot. In two environments there were genotypic correlations of 0.57 and 0.75 between resistance to brown spot and pleiochaeta root rot (Cowling *et al.*, 1997). Resistance to brown spot has increased in a recurrent selective breeding programme which aims also to improve yield and quality by transgressive segregation for these characters. Progeny are selected as F_2-derived lines and intercrossed every four years in closed populations. Three cycles of crossing and selection significantly improved resistance to brown spot and grain yield, while seed alkaloids decreased (Cowling, 1996). Myallie, released in 1995 from this recurrent selection programme, is the first *L. angustifolius* cultivar with moderate brown spot resistance (Cowling, 1997b).

Resistance to powdery mildew in *L. luteus* is controlled by the gene *Er* (H. Lamberts, the Netherlands, 1975, personal communication to W. Swiecicki). Serious losses due to powdery mildew occurred in the Western Cape Province of South Africa in the 1970s. Van Jaarsveld (1983) reported genetic resistance to powdery mildew in field and glasshouse experiments in *L. albus* cv. Ultra, *L. mutabilis*, some *L. luteus* lines, and several wild *L. angustifolius* accessions from Spain, Italy, Morocco and Israel. The latter were used in a breeding programme for resistance to powdery mildew in *L. angustifolius* (van Jaarsveld, 1983).

Plant architecture changes in modern lupin breeding

This section will focus on European autumn and spring-sown types of albus or white lupin (*L. albus*) and to a lesser extent on spring-sown yellow lupin (*L. luteus*). Plant development and structure, which controls dry matter accumulation and partitioning and ultimately seed yield, is similar across all lupin species cultivated for grain production. Recent developments in albus lupin in Europe demonstrate that plant architecture holds the key to new prospects for this species. Breeders have 're-designed' the development and structure of albus lupins to match better the environment of autumn sowing in Europe, and in doing so have greatly improved the prospects for albus lupins as a viable crop in this region.

Plant structure

Albus lupin, and most agricultural lupin species, classically shows an indeterminate growth habit with a modular structure, where each module on the main stem or branch is determinate since it carries a terminal inflorescence. Each module may support the elaboration of new modules, and if this occurs indefinitely the growth habit is said to be unrestricted or indeterminate. The modular growth habit of lupins, with its determinate branches, has caused some confusion in the literature when the word 'determinate' is also used to describe whole plant phenotype. In restricted branching phenotypes (also referred to as determinate phenotypes), modules cease to form at some point because floral initiation has occurred in all buds and there are no further vegetative buds produced. Highly restricted branching phenotypes produce few, if any, branch modules in the upper region of the main stem.

The main stem is the first module. The number of main stem leaf nodes varies with genotype, location and sowing date, as shown in Fig. 4.2 for two autumn-sown genotypes in different years, locations, and different sowing dates within each combination. More leaf nodes are produced on the main stem in autumn sowings than in spring sowings; more in some genotypes (Fig. 4.2a) than in others (Fig. 4.2b); and more at some locations and some years than at others.

The number of leaf nodes depends on the rate of leaf primordia production during the vegetative stage of the apex and on the date when it turns floral and stops producing leaf primordia (Huÿghe, 1991). The number of leaf primordia of a vegetative apex is a function of the thermal time from sowing and of the seed size (Huyghe, 1993), whether the origin of seed size variation is genotypic or environmental. A similar pattern of development of leaf primordia was observed in *L. mutabilis* (Hardy, 1997).

The apex becomes floral once the vernalization requirements of the genotype have been fulfilled. These requirements can be statistically quantified as the minimum amount of cold that an apex needs to receive before turning floral. This is calculated between 1 and 14°C according to Huyghe (1991) for autumn-sown genotypes and 1–17°C according to Duthion *et al.* (1994) for spring-sown genotypes of *L. albus*. These requirements are genotype dependent and are naturally greater for autumn-sown genotypes than for spring-sown genotypes. The extremely early flowering spring cv. Start, bred by G.G. Gataulina (Moscow, Russia), has been quantified as thermoneutral by Clapham and Willcott (1995).

The asymptotic curves in Fig. 4.2 suggest the existence, verified by experiment, of a minimum number of leaves for any genotype. This minimum value may be considered as the juvenility period, i.e. the period during which the apex is accumulating its cold units for vernalization, but is unable under any circumstances to become floral (Huyghe *et al.*, 1994c). The duration of the juvenility period and the thermal time of cold accumulation required for vernalization appear to be closely coupled, but results from Peel and Galwey (1998) suggest some variations around the relationship.

As a consequence of vernalization requirements, the number of main-stem leaves will increase for genotypes with greater vernalization requirements and when the weather is warm soon after sowing (for instance, early autumn sowings or late spring sowings). The number of main-stem leaves is therefore directly linked to the flowering date expressed in thermal time from sowing (Julier and Huyghe, 1993; Duthion *et al.*, 1994). The number of first-order lateral branches (the next group of modules) is also related to the number of main-stem leaves

(Huyghe *et al.*, 1994a). Because of the dry matter allocation pattern and within-plant competition for assimilates, the number of branches and thus the main-stem structure influences assimilate partitioning after flowering. It would be of major interest to stabilize the main-stem structure – either with genes for restricted branching or with modification of the relationship between vernalization requirements and juvenility period.

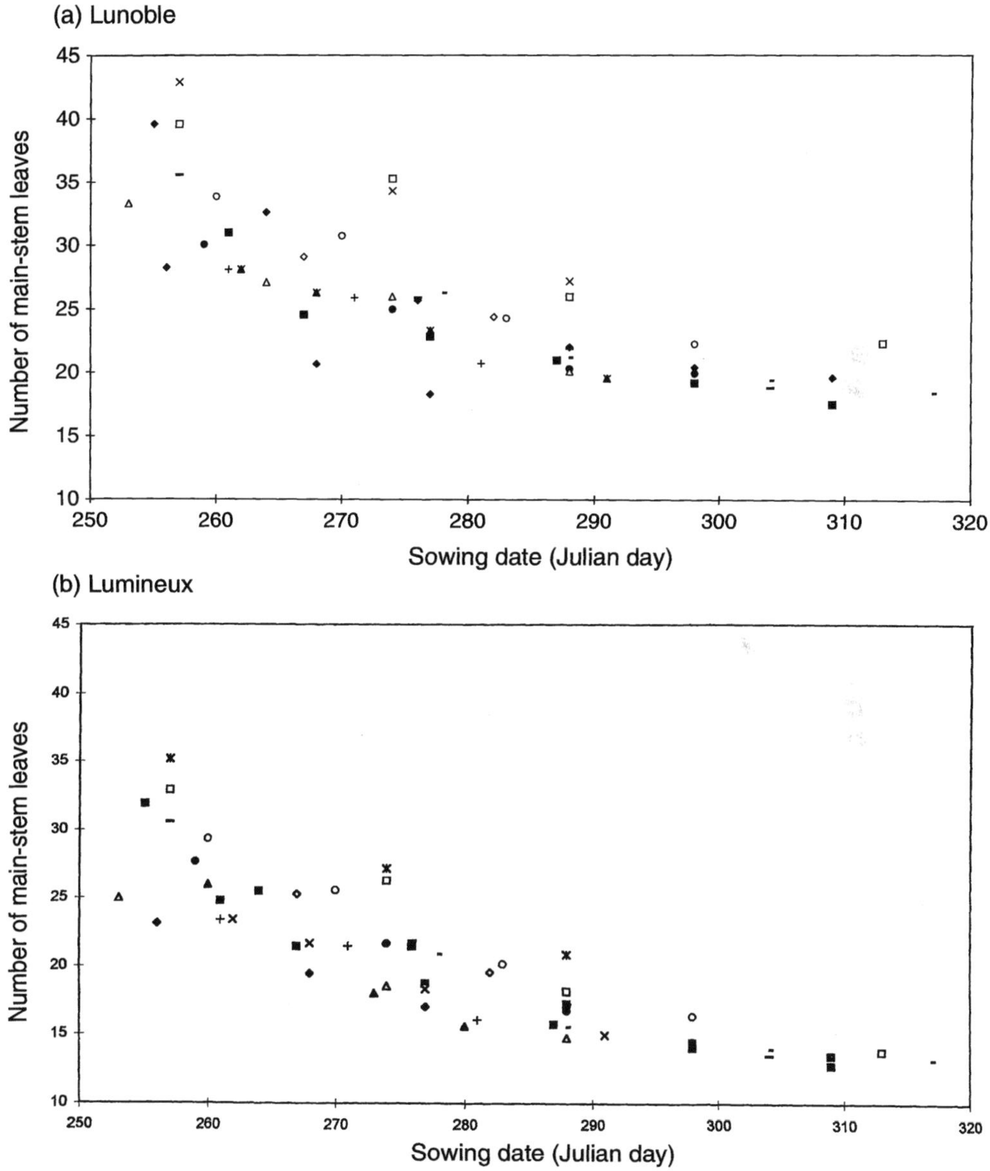

Fig. 4.2. Effect of the sowing date on the number of main-stem leaves for two autumn-sown varieties. Each set of symbols corresponds to one location × year combination. (a) Lunoble; (b) Lumineux.

Since the main-stem structure is temperature dependent, it is possible to predict stem structure for any location from the corresponding weather records. One can also use long-term weather databases to estimate the optimum sowing date to achieve a targeted number of main-stem leaves and to decide whether an area is suitable for lupin cultivation. Using such an approach, together with information on soil pH and soil drainage, Siddons *et al.* (1994) established a land suitability map for autumn-sown genotypes in Wales and England.

Primary or first-order branches grow in the axils of the upper leaf nodes of the main stem. These branches produce a terminal inflorescence under which, in turn, another order of branches develops. Pigeaire *et al.* (1985) found that the branch order n must reach a dry matter threshold to develop the order $n + 1$. Munier-Jolain *et al.* (1996) clearly showed that the number of branches of a given order was related to the growth rate of the plant at the beginning of elongation of the subtending order. Moreover, they showed that the number of axillary buds of an axis, which defines the potential number of branches of the next order, was related to the number of leaves on the axis.

Such a pattern of development results in severe competition between the new branches and the pods setting on the inflorescences of the previous orders. It can also lead to a long growing cycle, the duration of which will depend on the number of branch orders, and jeopardize plant maturation in cool and wet conditions. The mean duration of the cycle reached 1700°Cd (above 3°C) in spring-sown crops such as cv. Kali (Duthion *et al.*, 1987) and up to 2600°Cd for the autumn-sown indeterminate cv. Lunoble (Julier *et al.*, 1994).

Architecture of albus lupins is significantly modified by two major genetic systems: restriction of branching (determinacy) and dwarfism. As defined above, determinacy is the result of a change in growth pattern such that, at a given time in the growth cycle, all buds become floral, precluding the formation of new vegetative organs. The physiological mechanisms of this determinacy are unknown.

In *L. albus*, determinacy was first obtained after chemical mutagenesis in Poland by Mikolajczyk *et al.* (1984). The first-order branches of the original mutant were replaced with a single flower in leaf axils. This highly determinate character is monogenetically and recessively inherited (Julier and Huyghe, 1992). In France, this character was crossed into the autumn-sown genetic background. The structure of the branches of the determinates was very different from the indeterminate controls (Julier and Huyghe, 1993). Branches produced at the last node of the main stem were usually restricted to a small inflorescence. Progressing downwards along the main stem, the number of leaves per branch increased to a plateau. At the lowest branch position, there was a decrease in the frequency of branching. In contrast, the number of leaves per branch was constant in indeterminate genotypes at all branch positions.

For determinate genotypes, the number of primary branches is also closely related to the number of main-stem leaves (Julier and Huyghe, 1993). The leaf area index of determinate types is reduced because of a reduction of the total number of leaves on the primary branches and a reduction in the number of branch orders. Among the determinate genotypes, genetic variation was observed in the number of leaves on the longest branch, and to a lesser extent for the number of primary branches once corrected for the number of main-stem leaves. Determinacy reduced the number of branch orders down to one or two levels and induced earlier maturation, which was especially useful in northern and cool climates such as in England (Milford *et al.*, 1993).

Determinate mutants were also found in the other three lupin species cultivated for grain production. The mutants found in *L. luteus* and *L. mutabilis* (Römer, 1995) are under monogenic recessive control. In *L. angustifolius*, several determinate mutants have been reported, including an incompletely dominant natural mutant discovered by I. Forbes in the USA and several recessives that were either natural or induced by chemical mutagenesis (Gladstones, 1994). Bromberek *et al.* (1984) also reported a monogenic recessive gene. The determinate gene in *L. angustifolius* cv. Lanedesk is dominant (Micke and Swiecicki, 1988).

Dwarfism shortens the internodes. Several dwarf mutants have been described in *L. albus* (Muranyi and Sawicka, 1990; Harzic *et al.*, 1995). XA100 originated in an autumn-sown population after a chemical mutagenesis treatment. It shows a reduction of the internode length of 41% and 22%, respectively, on the main stem and the primary branches (Harzic and Huyghe, 1996). This source of dwarfism is controlled by two recessive genes (Harzic, 1996) and is responsive to the application of exogenous gibberellic acid. It modifies neither the number of leaves on the main stem and branches, nor the leaf size and leaf area (Harzic and Huyghe, 1996). It is a very interesting way to control the risk of lodging through reduction of plant height in autumn-sown lupins.

Other sources of dwarfism include a spontaneous Polish mutant, controlled by a single recessive gene, that results in a reduction in leaf size and petiole length. A similar mutant was identified in the same mutagenesis programme as XA100 and caused a reduction in the pod elongation rate.

Dry matter accumulation and partitioning – consequences for seed yield

The growth pattern of indeterminate types leads to a large leaf area index and high potential biomass production. More than 10 t dry matter ha^{-1} were reported for the spring-sown early flowering *L. albus* cv. Kali in France (Duthion *et al.*, 1987) while Julier *et al.* (1994) reported total biomass higher than 15 t ha^{-1} for the autumn-sown cv. Lunoble.

However, the partitioning of assimilates between growing branches and pods is very inconsistent and results in poor stability of seed yield. Consequently, strong correlations are often observed between seed yield and harvest index while correlations between seed yield and biomass are rarely significant. This is especially true for the very vigorous autumn-sown genotypes (Huyghe *et al.*, 1994b). The inconsistency of partitioning is a direct consequence of competition between the vegetative and reproductive compartments and of variation in the size of the vegetative compartment. To stabilize and possibly reduce intra-plant competition, it is necessary to modify the plant structure by the genetic systems previously described.

In the dwarf autumn-sown progenitor XA100, allocation of dry matter to the pods occurred at higher percentages than in non-dwarf types (Harzic *et al.*, 1995) but, because of numerous late pod and seed abortions, this preferential allocation did not always result in higher seed yield. When the number of pods set was reduced by pruning soon after flowering, the number of seeds per pod increased, as did the mean seed mass and also the seed yield per plant for intermediate intensities of pruning (Fig. 4.3). The numerous abortions observed on the control could be a direct consequence of competition for assimilates late in the growth cycle. In *L. angustifolius*, spraying cytokinins on flowers increased pod setting (Pigeaire and Atkins, 1992) but, because of heavy pod and seed abortions, the seed yields were reduced (Palta and Ludwig, 1998).

Determinacy modifies dry matter accumulation and partitioning more severely than dwarfism. After late autumn sowings, the leaf area index (LAI) may become too restricted and crops may not reach full light interception (Huyghe *et al.*, 1994c). It is not necessary to reach full light interception before flowering

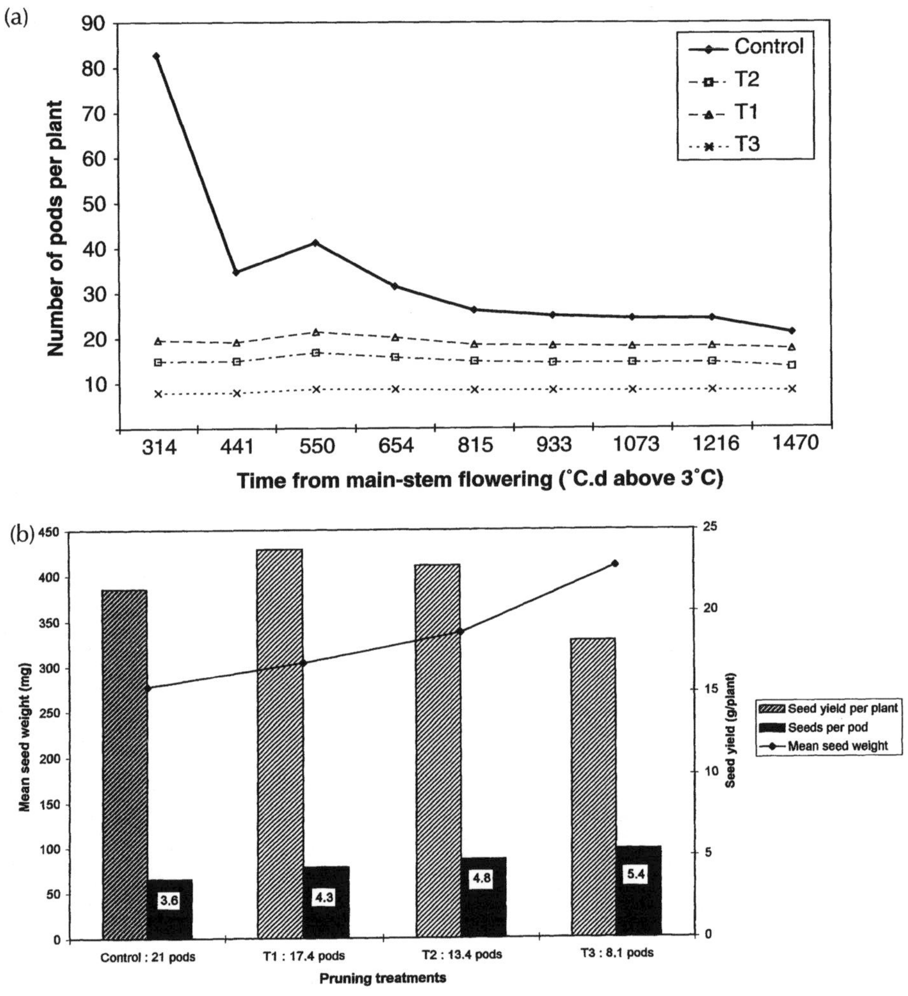

Fig. 4.3. Effect of three intensities of flower pruning on (a) the number of pods during the pod growth cycle and (b) the yield components at maturity.

because assimilates stored in the stems are remobilized at very low levels towards the growing fruits (Withers and Forde, 1979). In contrast, it is necessary to reach maximum light interception soon after flowering in order to increase biomass production, pod set and seed filling.

The choice of genotypes and of agronomic practices such as sowing date and density is important. The incidence of leaf diseases after flowering, such as rust, may be very deleterious to seed yield because of the reduction in the LAI. Julier *et al.* (1993a) showed on the first autumn-sown determinate genotypes that seed yield was highly correlated to the above-ground biomass at harvest ($r = 0.61$) but not to the harvest index. When considering the pattern of assimilate partitioning, it can be seen that the determinate genotypes allocate all the assimilates to the pods after 250°Cd from flowering, while on the indeterminates the pods and the stems are growing simultaneously during 1100°Cd from main-stem flowering. Thus, determinacy allows separation of vegetative growth and reproductive growth and results in a more stable yield (Julier *et al.*, 1993b).

The major limitation of the determinate lines is the risk of lodging. Indeed, in order to maximize dry matter accumulation and thus seed yield, early autumn sowings are recommended and this results in plants with numerous main-stem leaves and tall main stems. The use of dwarfism in combination with determinacy is possible. Fig. 4.4 shows the relationship between the number of the main-stem leaves and the main-stem height for two determinate genotypes and two dwarf determinates. The variation in the number of main-stem leaves was due to different years and different sowing dates within each year.

Dwarfism allows the creation of genotypes with numerous leaves and short main stems, which are resistant to lodging. Among the dwarf determinates, the

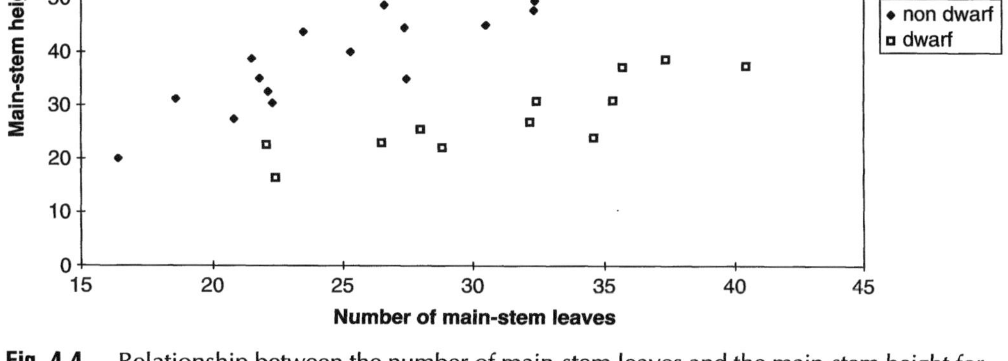

Fig. 4.4. Relationship between the number of main-stem leaves and the main-stem height for two autumn-sown dwarf (□) and two non-dwarf (♦) determinate genotypes. Variations in the main-stem leaf number are induced by variation in sowing date and year.

first studies over three years showed that the seed yield potential is high and seems to be correlated to harvest index and biomass. To reduce the impact of water deficits during the reproductive phase, breeding is now oriented towards the selection of early-flowering material which would produce a large primary branch compartment to ensure full light interception. While it not difficult to find early-flowering genotypes with branches carrying numerous leaves, it is more difficult but not impossible to have numerous primary branches. Such materials clearly result from a recombination between the sizes of the main stem and of the primary branch compartments.

The genetic parameters in different architectures

For breeding of different plant architectures, information is needed on their genetic parameters and heritabilities. Dwarfism and determinacy have been introduced, either alone or in combination, into diverse genetic backgrounds. The combination of both has so far been mainly in a late-flowering context. This is where genotypes with well-developed primary branches are most found, and where the selection for high yield seems most likely to be successful.

Although genetic backgrounds and testing environments differed, the genetic parameters and correlations are still interesting to compare. Here we compare results obtained in autumn-sown genotypes carrying either dwarfism (35 genotypes in five environments; Harzic *et al.*, 1996), determinacy (40 genotypes in five environments; Julier *et al.*, 1995) or determinacy with different intensities of dwarfism (105 genotypes in one environment; Huyghe *et al.*, 1998). In the first two populations, the genotype × environment interactions for yield were insignificant or small.

Table 4.2 shows for different characters the mean values, the genotypic variances and the broad-sense heritabilities in four different groups, including the values obtained by LeSech and Huyghe (1991) in a diallel mating design in spring-sown indeterminate white lupin. The single trial for evaluation of the dwarf determinate genotypes was planted early, which to some extent explains

Table 4.2. Mean values, genetic variance and broad sense heritabilities (h^2) of different morphological characters and yield components in three genetic backgrounds of autumn-sown albus lupin. The values obtained by LeSech and Huyghe (1991) in spring-sown albus lupin are given.

	Spring-sown		Dwarf		Determinate		Dwarf-determinate	
Characters	h^2	mean	h^2	mean	h^2	mean	h^2	mean
MS flowering date (°Cd > 3°C)	0.94	513	0.96	920	0.97	1030	0.91	1367
MS height (cm)	na	na	0.92	26	na	na	0.82	57
No. leaves on primary branch[1]	0.89	6.8	0.98	7.4	0.9	3.7	0.92	7.4
Yield (t ha^{-1})	0.45	5.22	0.68	3.77	0.78	3.22	0.62	4.22
Mean seed weight (mg)	0.76	433	0.92	372	0.94	246	0.72	322
No. seeds m^{-2}	na	na	0.82	1031	0.81	1321	0.62	1312

[1]Number of leaves on the second primary branch for the dwarfs and on the longest primary branch for the determinates and the dwarf determinates.
na = not available; MS = main stem.

the high temperature summation to flowering and the tall main stems. Notable in this table is, firstly, the mean seed weight, which was large in the dwarf population, small in the determinate, and intermediate in the dwarf determinate population that was created by crossing the two. This character is particularly important for the winter survival, as seedling growth is directly related to seed size and is a major component of winter survival (Huyghe, 1993). Secondly, there is an important shift in the branch structure from the normal determinate to the dwarf determinate, as a consequence of selection for greater branch leaf number in the latter.

Broad-sense heritabilities are high for all the morphological characters and mean seed mass and slightly lower for seed yield and the number of seeds m^{-2}. However, the heritabilities are high enough to expect rapid progress in breeding.

Table 4.3 shows the genetic correlations between the morphological characters, seed yield and yield components in the three populations of genotypes. The introduction of dwarfism in the determinate background (comparison of determinate and dwarf determinate) did not change the relationship between yield and its components. However, in the dwarf indeterminate background, the negative correlation between the mean seed weight and the number of seeds m^{-2} was much stronger. In the determinate background, leaf number per branch was highly correlated with yield and proved to be the major factor limiting yield. This was due to the average earliness of this population as shown by the positive genetic correlation between lateness of flowering and seed yield. The earliest genotypes were indeed epigonal. In the dwarf determinate population the correlation between branch leaf number and seed yield no longer existed, and

Table 4.3. Genetic correlations between morphological characters and yield in dwarf types (A), determinate types (B) and dwarf determinate types (C).

Characters	Plant type	MS height (cm)	No. leaves on primary branch[1]	Yield (t ha^{-1})	Mean seed weight (mg)	No. seeds m–2
MS flowering date	A	0.27	0.33	–0.07	–0.37	0.32
(°Cd > 3°C)	B	na	0.69	–0.61	–0.05	0.52
	C	0.13	–0.28	–0.46	–0.24	–0.25
MS height (cm)	A		0.75	0.57	0.42	–0.02
	B		na	na	na	na
	C		–0.29	–0.03	0.16	–0.08
No. leaves on primary	A			0.26	0.52	–0.28
branch[1]	B			0.73	–0.06	0.72
	C			0.12	0.05	0.06
Yield (t ha^{-1})	A				0.11	0.51
	B				0.22	0.76
	C				0.27	0.70
Mean seed weight (mg)	A					–0.80
	B					–0.45
	C					–0.47

[1]Number of leaves on the second primary branch for the dwarfs and on the longest primary branch for the determinates and the dwarf determinates.
na = not available; MS = main stem.

the genetic correlation between flowering date and seed yield had become negative. This shows that among this material some genotypes were too late flowering.

Three additional characters were scored on the dwarf determinate populations. Lodging was scored from 0 (no lodging) to 10 (100% of the plants lodged). It was highly heritable ($h^2 = 0.56$), and positively correlated ($r = 0.40$) to main-stem height. This suggests the need to reduce the main-stem height and to introduce dwarfism in the material under selection.

At the genotypic level, seed yield was positively correlated with both total biomass at harvest and harvest index (0.84 and 0.77, respectively). These characters were independent of each other.

These data show that the major problems of seed yield elaboration in white lupin under European conditions can be solved by modifying plant architecture. An optimum ideotype (dwarf determinate genotypes with early sowing) has been defined. Selection for yield will be efficient, especially in the dwarf determinate context where early maturity and lodging tolerance can be found. Selection must now be towards early-flowering genotypes with a large primary branch compartment.

Stem characters and growth rhythm in *L. luteus*

The main features of the stem of *L. luteus*, important for cultivar improvement, are thickness of the root neck and the type of branching. Stem architecture is controlled by plant density, development pattern and genetics. Wild yellow lupins are 'bushy' with a thick root neck and strong branching on lower stem leaf nodes. Most varieties have a rather slow growth rhythm, controlled by the gene *Cres* (*Crescens*). Plants are short, well branched and non-lodging.

The alleles $cres_{alt}$ (*crescens altus*) and $cres_{cel}$ (*crescens celer*) result in taller plants with rapid elongation, thin stems that tend to lodge, and light green leaves (Hackbarth and Troll, 1956). Their rosette phase of seedling growth is much reduced (2–3 weeks) compared with *Cres* genotypes (about 5 weeks). This allows the plants to compete better with weeds. A gene at a different locus *Rp* (*Rapidus*), epistatic to $cres_{cel}$ and $cres_{alt}$, results in moderately fast elongation, a thicker stem, and better resistance to lodging (Lamberts, 1953). However, lodging was experienced in Australia with *Rp* genotypes (J.S. Gladstones, Western Australia, 1997, personal communication). *Rp* does not cause a lighter leaf colour. In early Polish breeding material one more gene was described, the *pr* (*promptus*) gene, controlling fast growth without lighter colour of leaves (Mikolajczyk, 1962). In breeding *L. luteus*, the most popular genes are *Cres* and *Rp*. Other genes are available with clearly different plant architecture, such as the self-completing genotypes with only one, unbranched stem.

Seed quality improvements in modern lupin breeding

The major component of seed quality of concern to lupin breeders is alkaloid content. Lupin breeding in Australia adheres to the Australian food standard of 0.02% alkaloids as the upper limit for new varieties, and the *L. angustifolius* cv. Danja is used as the upper limit control in breeding and advanced variety testing.

Thus, Australian sweet lupin (*L. angustifolius*) cultivars may be used directly for human consumption.

The genetic control of alkaloid, protein and oil content of whole seeds and seed weight was evaluated across 482 *L. angustifolius* variety trials in Western Australia from 1984 to 1994 (L. Miguel, W.A. Cowling, A. Tarr and J. Hamblin, Western Australia, 1997, unpublished data). The variance component for genotypes was always larger than the variance components for genotype × site, genotype × year and genotype × site × year interactions for all seed quality traits. Therefore, these traits were under relatively strong genetic control. Danja had the highest average seed alkaloid content on an 'as received' basis (0.016%) and Yorrel the lowest (0.006%). Myallie had the highest average protein content on a dry basis (35.5%) and Danja the lowest (33.5%). As with many other crops, there was a negative association between oil and protein of seeds: Myallie had the lowest oil content on a dry basis (6.4%) and Danja had the highest (7.0%). Yorrel had the highest average seed weight (157 mg) and Merrit had the lowest (144 mg).

This information is very useful to breeders, as it indicates that selection for seed quality may be carried out in any 'normal' lupin experiment in Western Australia. In the case of alkaloids, breeders may confidently select early generation lines for alkaloid content in any environment where alkaloids are above detectable limits. However, data from several locations and years are required to demonstrate that advanced breeding lines have acceptable alkaloid content compared with the upper limit control variety Danja. The lupin breeding programme at Agriculture WA currently screens 3000 to 4000 breeding lines every year for alkaloid content. Measurements are carried out on ground seed samples by gas chromatography, which represents a cost of about 10% of the annual budget for lupin breeding. This investment has a very high return, through the early elimination of high alkaloid lines (above that in Danja) and therefore more efficient and faster selection for yield, disease resistance and agronomic traits required for improved varieties.

There are few reports of breeding for higher protein or oil within lupin species, though this is entirely possible based on results with soybean and other legumes. Williams (1979) believed that worthwhile advances in oil content were feasible in *L. albus* and *L. mutabilis*, although protein content showed little intraspecific variation. Apparently, von Sengbusch carried out selection for high oil in *L. albus* in the 1930s (Hackbarth and Troll, 1956), but this work did not lead to new varieties with higher oil. More important have been the broad quality traits of individual species – for example, the higher protein and Cys, Met and Lys contents of *L. luteus* seed compared with other lupin species (Petterson and Mackintosh, 1994). The high oil and protein content of *L. mutabilis* (Williams, 1979; Gladstones, 1980) makes it nearly equivalent to soybean for nutritive value.

If lupins develop a major role in human nutrition, breeders will be presented with a new suite of selection criteria for crop improvement. For example, most lupin species have high oligosaccharide content that may discourage widespread use in human food (Saini and Gladstones, 1986). In that study, enough variability was present in *L. angustifolius* to suggest the possibility of breeding for improved saccharide composition.

Lupin quality is severely scrutinized by modern intensive animal industries, especially the monogastric animal industries such as pigs and poultry, where nutrition experts study digestible energy and availability of protein and amino acids. There can be little doubt that scope exists for breeding to improve composition of lupins for target animal species. However, there are other, perhaps more important, criteria on which to base breeding programmes at the present developmental stage of lupins as crop plants.

Improvements in abiotic stress tolerance in modern lupin breeding

Breeders of *L. angustifolius* and *L. albus* in the USA in the 1950s and 1960s focused on developing frost tolerance (Wells and Forbes, 1982). This was necessary for the crop to withstand winter freezes in the southern USA, where lupins are sown in autumn. Despite the release of new varieties with improved frost tolerance, freezing damage and disease have continued to discourage growers from using lupins in this region. Further work in France is improving frost tolerance in *L. albus* (Papineau and Huyghe, 1992), which is vital for the success of autumn-sown dwarf determinate *L. albus* in that region.

The yellow lupin, *L. luteus*, is moderately resistant in the seedling stage to spring frosts (to −7°C), but sensitive to the same temperature in the autumn. Until recently, all European cultivars required vernalization during early growth (Laczynska-Hulewicz, 1954). This necessitated early spring sowing, which is not always feasible. Thermoneutrality is therefore very important in spring sowings of *L. luteus* in northern Europe, and this was found for the first time in 1977. The first thermoneutral cultivars were bred in Poland (Juno and Teo) and in Germany (Borsaja) (Swiecicki and Swiecicki, 1994). Thermoneutrality appeared to be a dominant character with an additional advantage of allowing plants to escape BYMV.

Yellow lupin is the most resistant of the species to drought under European conditions, and has the deepest root system. Ironically, it is the most sensitive to water deficit during flowering and pod setting. Variation for this character appears to be narrow. Yellow lupin is also strongly sensitive to free calcium in the soil. A pH above 6.5 leads to chlorosis of the young seedlings. Some authors have described a range of tolerance inside the species (Barbacki, 1952; Hackbarth and Troll, 1956).

Experience in Western Australia demonstrates that *L. luteus* is more tolerant of acid, aluminium toxic soils than *L. angustifolius* (Carr and Sweetingham, 1994). This provides another example of where breeding for abiotic stress tolerance has focused on differences among species rather than selecting within species.

Conclusions

Lupin breeders will continue to improve disease resistance, yield and quality in the domesticated lupin species. A yield improvement of 2.4% per year (1973–1991) in *L. angustifolius* exceeds that in most other crops around the world (Cowling, 1996). This has largely been possible because domestication has been so recent. Considerable advances are still being made through primary crosses with wild lupins – for example, in the development of *Phomopsis* resistance in *L. angustifolius* (Gladstones, 1989). Many 'primitive' traits remain in lupins, such as thick pod walls, which divert assimilate and probably reduce seed yields compared with other grain legumes that were domesticated in ancient times. Much opportunity therefore exists to manipulate the phenotype to achieve further improvements. Genetic engineering or other molecular technology may help (Chapter 3, this volume).

Disease resistance will remain important, and in many cases can be achieved in parallel with improvements in grain yield and quality by using recurrent selection (Cowling, 1996). Once an ideal plant architecture is achieved, breeders will need to create wider gene pools to allow improvements to continue. For example, an autumn-sown dwarf determinate *L. albus* breeding programme in Europe (Huyghe *et al.*, 1994c) will begin its breeding life with a very narrow gene pool that must be expanded by crossing into other phenotypes and re-extracting the desired phenotype. Similarly, the successful thermoneutral cultivars of *L. luteus* in Poland are narrowly based because they had to be developed in a background of strong resistance to *Fusarium* wilt (Swiecicki and Swiecicki, 1994). It is paramount that the correct phenotypes to suit particular environments are recognized as soon as possible, so that the necessary wider crossbreeding may begin. As with all crops, strong regionally based breeding programmes will be needed if the crop is to become widely adapted and grown commercially by farmers.

References

Anon. (1977) Registration of *Lupinus cosentinii* Guss. (sandplain lupin) cv. Erregulla. *Journal of the Australian Institute of Agricultural Science* 43, 87–89.

Anon. (1982) Registration of *Lupinus angustifolius* L. (narrow-leafed lupin) cv. Chittick. *Journal of the Australian Institute of Agricultural Science* 48, 122–124.

Barbacki, S. (1952) [*Lupin*] Panstwowe Wydawnietwo Rolniczne i Lesne, Warsaw, 172 pp. (in Polish).

Bromberek, S., Mikolajczyk, J. and Wiza, M. (1984) Self-completing mutation of blue lupin. In: *Proceedings of the Third International Lupin Conference.* International Lupin Association, La Rochelle, France, pp. 566–567 (in French).

Carr, S.J. and Sweetingham, M.W. (1994) Root elongation of *Lupinus angustifolius* and *L. luteus* differs in the presence of phytotoxic aluminium. In: Dracup, M. and Palta, J. (eds) *Proceedings of the First Australian Lupin Technical Symposium.* Department of Agriculture Western Australia, South Perth, p. 260.

Clapham, W.M. and Willcott, J.B. (1995) Thermosensitivity in spring white lupin. *Annals of Botany* 76, 349–357.

Clements, J.C. and Cowling, W.A. (1994) Patterns of morphological diversity and geographical origins of wild *Lupinus angustifolius* from the Aegean region. *Genetic Resources and Crop Evolution* 41, 109–122.

Cowling, W.A. (1996) Plant breeding for stable agriculture: Presidential Address 1994. *Journal of the Royal Society of Western Australia* 79, 183–194.

Cowling, W.A. (1997a) Register of Australian Grain Legume Cultivars. *Lupinus angustifolius* L. (narrow-leafed lupin) cv. Kalya. *Australian Journal of Experimental Agriculture* 37, 271–273.

Cowling, W.A. (1997b) Register of Australian Grain Legume Cultivars. *Lupinus angustifolius* L. (narrow-leafed lupin) cv. Myallie. *Australian Journal of Experimental Agriculture* 37, 274–275.

Cowling, W.A. and Speijers, E.J. (1994) Grain yield improvement in narrow-leafed lupin cultivars from 1973 to 1991 as predicted from plant density experiments. In: Dracup, M. and Palta, J. (eds) *Proceedings of the First Australian Lupin Technical Symposium*. Department of Agriculture Western Australia, South Perth, p. 262.

Cowling, W.A., Hamblin, J., Wood, P.McR. and Gladstones, J.S. (1987) Resistance to Phomopsis stem blight in *Lupinus angustifolius* L. *Crop Science* 27, 648–652.

Cowling, W.A., Allen, J.G. and Wood, P.McR. (1988) Resistance to Phomopsis stem blight reduces the lupinosis toxicity of narrow-leafed lupin stems. *Australian Journal of Experimental Agriculture* 28, 195–202.

Cowling, W.A., Sweetingham, M.W., Diepeveen, D. and Cullis, B.R. (1997) Heritability of resistance to brown spot and root rot of narrow-leafed lupins caused by *Pleiochaeta setosa* (Kirchn.) Hughes in field experiments. *Plant Breeding* 116, 341–345.

Duthion, C., Amarger, N. and Mariotti, A. (1987) Potential dry matter and nitrogen accumulation in spring white lupin (*Lupinus albus* L.). *Agronomie* 7, 585–593.

Duthion, C., Ney, B. and Munier-Jolain, N.M. (1994) Development and growth of white lupin: implications for crop management. *Agronomy Journal* 86, 1039–1045.

Edwards, C.S. (1980) Dry matter distribution and forage quality in selected lines of *Lupinus angustifolius* L. and *Lupinus cosentinii* Guss. PhD thesis, The University of Western Australia, Perth, Australia.

Gladstones, J.S. (1967) Selection for economic characters in *Lupinus angustifolius* and *L. digitatus*. 1. Non-shattering pods. *Australian Journal of Experimental Agriculture and Animal Husbandry* 7, 360–366.

Gladstones, J.S. (1970) Lupins as crop plants. *Field Crop Abstracts* 23, 123–148.

Gladstones, J.S. (1977) *The narrow-leafed lupin in Western Australia* (Lupinus angustifolius *L.*). Bulletin 3990, Western Australian Department of Agriculture, South Perth, 39 pp.

Gladstones, J.S. (1980) Recent developments in the understanding, improvement, and use of lupins. In: Summerfield, R.J. and Bunting, A.H. (eds) *Advances in Legume Science, Proceedings of the International Legume Conference, Kew 1978*. Royal Botanic Gardens, Kew, pp. 603–611.

Gladstones, J.S. (1989) Breeding Phomopsis-resistant lupins. *Journal of Agriculture Western Australia* 30, 3–7.

Gladstones, J.S. (1994) An historical view of lupins in Australia. In: Dracup, M. and Palta, J. (eds) *Proceedings of the First Australian Lupin Technical Symposium*. Department of Agriculture Western Australia, South Perth, pp. 1–38.

Gladstones, J.S. and Crosbie, G.B. (1978) *Lupin wild types introduced into Western Australia to 1973*. Technical Bulletin No. 43, Department of Agriculture, Western Australia, South Perth.

Gladstones, J.S. and Hill, G.D. (1969) Selection for economic characters in *Lupinus angustifolius* and *L. digitatus*. 2. Time of flowering. *Australian Journal of Experimental Agriculture and Animal Husbandry* 9, 213–220.

Golovchenko, V.I. (1982) Mutagenesis in selection of alkaloidless varieties of white and yellow lupine. In: *Proceedings of the Second International Lupin Conference*. Publicaciones Agrarias, Ministry of Agriculture, Madrid, Spain, pp. 43–45.

Hackbarth, J. and Troll, H.-J. (1956) [Lupins as grain legumes and fodder plants.] In: Kappert, H. and Rudorf, W. (eds) [*Handbook of Plant Breeding, Part IV*], 2nd edn. Verlag Paul Parey, Berlin and Hamburg, pp. 1–51 (in German).

Hanelt, P. (1960) [*The Lupins*]. A. Ziemsen, Wittenberg, 101 pp. (in German).

Hardy, A. (1997) Physiological and genetical study of architecture and growth in the Andean Lupin (*Lupinus mutabilis* Sweet). Consequences on the grain yield. PhD thesis, École Nationale Supérieure Agronomique de Rennes, 107 pp (in French).

Harris, D.J. (1994) Quality in production and breeding. In: Dracup, M. and Palta, J. (eds) *Proceedings of the First Australian Lupin Technical Symposium*. Department of Agriculture Western Australia, South Perth, pp. 98–105.

Harzic, N. (1996) Use of dwarfism for genetic improvement in winter white lupin. PhD thesis, INA Paris-Grignon, France, 90 pp (in French).

Harzic, N. and Huyghe, C. (1996) Dwarfism does not modify mean area per leaf and light interception in indeterminate autumn-sown white lupin. *Journal of Agricultural Science* 127, 337–345.

Harzic, N., Huyghe, C. and Papineau, J. (1995) Dry matter accumulation and seed yield of dwarf autumn-sown white lupin (*Lupinus albus* L.). *Canadian Journal of Plant Science* 75, 549–555.

Harzic, N., Huyghe, C., Papineau, J., Billot, C., Esnault, R. and Deroo, C. (1996) Genotypic variation for seed yield and architectural traits in dwarf autumn-sown white lupin. *Agronomie* 16, 309–319.

Hondelmann, W. (1984) The lupin – ancient and modern crop plant. *Theoretical and Applied Genetics* 68, 1–9.

Huyghe, C. (1991) Winter growth of autumn-sown white lupin (*Lupinus albus* L.): main apex growth model. *Annals of Botany* 67, 429–434.

Huyghe, C. (1993) Growth of white lupin seedlings during the rosette stage as affected by seed size. *Agronomie* 13, 145–153.

Huyghe, C., Harzic, N., Julier, B. and Papineau, J. (1994a) Comparison of determinate and indeterminate autumn-sown white lupins under the western European climate. In: Dracup, M. and Palta, J. (eds) *Proceedings of the First Australian Lupin Technical Symposium*. Department of Agriculture Western Australia, South Perth, pp. 123–128.

Huyghe, C., Julier, B., Harzic, N. and Papineau, J. (1994b) Yield and yield components of indeterminate autumn-sown white lupin (*Lupinus albus*) cv. Lunoble. *European Journal of Agriculture* 3, 145–152.

Huyghe, C., Julier, B., Harzic, N. and Papineau, J. (1994c) Breeding of *Lupinus albus*: new architectures for a further domestication. In: Neves Martins, J.M. and Beirão da Costa, L. (eds) *Advances in Lupin Research*. ISA Press, Lisbon, pp. 25–41.

Huyghe, C., Harzic, N., Papineau, J. and Pacault, D. (1998) Breeding lupin using physiological criteria. In: Hill, G. (ed.) *Proceedings of the Eighth International Lupin Conference, Asilomar, USA* (in press).

Jasper, R. (1994) The challenges confronting new lupin foods entering the food industry. In: Neves Martins, J.M. and Beirão da Costa, L. (eds) *Advances in Lupin Research*. ISA Press, Lisbon, pp. 492–499.

Jones, R.A.C. and Cowling, W.A. (1995) Resistance to seed transmission of cucumber mosaic virus in narrow-leafed lupins (*Lupinus angustifolius*). *Australian Journal of Agricultural Research* 46, 1339–1352.

Jones, R.A.C. and Latham, L.J. (1996) Natural resistance to cucumber mosaic virus in lupin species. *Annals of Applied Biology* 129, 523–542.

Julier, B. and Huyghe, C. (1992) Heredity of determinate growth in winter white lupin (*Lupinus albus* L.). Influence of the sowing time on architecture. In: Association Européenne des Protéagineux (ed.) *Proceedings of the First European Conference on Grain Legumes, Angers, France*, pp. 47–48.

Julier, B. and Huyghe, C. (1993) Description and model of the architecture of four genotypes of determinate autumn-sown white lupin (*Lupinus albus* L.) as influenced by location, sowing date and density. *Annals of Botany* 72, 493–501.

Julier, B., Huyghe, C. and Papineau, J. (1993a) Dry matter and nitrogen accumulation and seed yield in determinate autumn-sown white lupins (*Lupinus albus* L.). *Agronomie* 13, 877–888.

Julier, B., Huyghe, C., Papineau, J., Milford, G.F.J., Day, J.M., Billot, C. and Mangin, P. (1993b) Seed yield and yield stability of determinate and indeterminate autumn-sown white lupins (*Lupinus albus*) grown at different locations in France and the UK. *Journal of Agricultural Science* 121, 177–186.

Julier, B., Huyghe, C. and Papineau, J. (1994) Dry matter and nitrogen accumulation on autumn-sown white lupin (*Lupinus albus*) cv. Lunoble. *European Journal of Agronomy* 3, 153–160.

Julier, B., Huyghe, C., Papineau, J., Billot, C. and Deroo, C. (1995) Genetic and environmental variations for architecture and yield components in determinate autumn-sown white lupin (*Lupinus albus* L.). *Euphytica* 81, 171–179.

Laczynska-Hulewicz, T. (1954) Undersuchungen ueber ein Einfluss von Aussenafactoren auf die Entwicklung der Lupine. *Roczniki Nauk Rolniczych* 69, 189–242.

Lamberts, H. (1952) Resistance to *Fusarium oxysporum* in yellow lupins. *Euphytica* 1, 199–200.

Lamberts, H. (1953) A new type of rapid growth in yellow lupin (*Lupinus luteus*). *Euphytica* 2, 59–61.

LeSech, L. and Huyghe, C. (1991) Diallel analysis on white lupin: breeding consequences. *Agronomie* 11, 719–726.

Micke, A. and Swiecicki, W. (1988) Induced mutations in lupins. In: Twardowski, T. (ed.) *Proceedings of the Fifth International Lupin Conference*. Panstwowe Wydawnietwo Rolniczne i Lesne, Poznań, Poland, pp. 110–127.

Mikolajczyk, J. (1962) Genetical studies on yellow lupin (*Lupinus luteus* L.) Part 1. Spontaneous mutations in Polish yellow lupins. *Genetica Polonica* 3, 189–208.

Mikolajczyk, J., Stawinski, S. and Wiza, M. (1984) Present direction of improvement and present state of research on acclimation of white lupin in Poland. In: *Proceedings of the Third International Lupin Conference*. International Lupin Association, La Rochelle, France, pp. 570–571.

Milford, G.F.J., Day, J.M., Huyghe, C. and Julier, B. (1993) Floral determinacy in autumn-sown white lupins (*Lupinus albus*): the development of varieties for cooler European climates. *Aspects of Applied Biology* 34, 89–97.

Munier-Jolain, N.M., Ney, B. and Duthion, C. (1996) Analysis of branching in spring-sown white lupin (*Lupinus albus* L.). The significance of the number of axillary buds. *Annals of Botany* 77, 123–131.

Muranyi, R. and Sawicka, E. (1990) Mutants obtained in *Lupinus albus* after gamma irradiation. *Abstracts 6th International Lupin Conference, Pucon, Chile*, pp. 134.

Nowacki, E. and Swiecicki, W. (1973) Different biochemical reactions of lupine forms susceptible and resistant to *Fusarium. Genetica Polonica* 14, 375–382.

Oram, R.N. (1983) Selection for higher seed yield in the presence of the deleterious low alkaloid allele *iucundis* in *Lupinus angustifolius* L. *Field Crops Research* 7, 169–180.

Palta, J.A. and Ludwig, C. (1998) Improving seed yield in lupin: how many pods to be set? In: Hill, G. (ed.) *Proceedings of the Eighth International Lupin Conference, Asilomar, USA* (in press).

Papineau, J. and Huyghe, C. (1989) Collecting white lupin in the Azores. *FAO/IBPGR Plant Genetic Resources Newsletter*, 88/89, 77–78.

Papineau, J. and Huyghe, C. (1992) Measure of the frost tolerance of autumn-sown white lupin (*Lupinus albus* L.) under controlled conditions. In: Association Européenne des Protéagineux (eds) *Proceedings of the First European Conference on Grain Legumes, Angers, France*, pp. 33–34.

Peel, D.N.Y. and Galwey, N.W. (1998) Identification of a stable plant architecture: the key to development of *Lupinus albus* as a crop for Northern Europe. In: Hill, G. (ed.) *Proceedings of Eighth International Lupin Conference, Asilomar, USA* (in press).

Petterson, D.S. and Mackintosh, J.B. (1994) *The Chemical Composition and Nutritive Value of Australian Grain Legumes.* Grains Research and Development Corporation, Canberra, 68 pp.

Pigeaire, A. and Atkins, C.A. (1992) Effect of cytokinins applications to flowers on pod set in *Lupinus angustifolius* L. In: Association Européenne des Protéagineux (eds) *Proceedings of the First European Conference on Grain Legumes, Angers, France,* pp. 157–158.

Pigeaire, A., Reponty, H. and Duthion, C. (1985) White lupin plant building: relationship between level of dry matter accumulated and its partition. *Agronomie* 5, 936–947 (in French).

Robinson, R.A. (1996) *Return to Resistance: Breeding Crops to Reduce Pesticide Dependence.* agAccess, Davis, California, 480 pp.

Römer, P. (1995) New attempts to select early maturing *Lupinus mutabilis* for middle Europe. In: Association Européenne des Protéagineux (eds) *Proceedings of the Second European Conference on Grain Legumes, Copenhagen, Denmark,* p. 254.

Saini, H.S. and Gladstones, J.S. (1986) Variability in the total and component galactosyl sucrose oligosaccharides of *Lupinus* species. *Australian Journal of Agricultural Research* 37, 157–166.

Schmidt, H.E. (1988) Quantitative resistance of yellow lupin (*Lupinus luteus* L.) to bean yellow mosaic virus. In: Twardowski, T. (ed.) *Proceedings of the Fifth International Lupin Conference.* Panstowe Wydawnietwo Rolniczne i Lesne, Poznań, Poland, pp. 230-235.

Siddons, P.A., Jones, R.J.A., Hollis, J.M., Hallett, S.H., Huyghe, C., Day, J.M., Scott, T. and Milford, G.F.J. (1994) The use of a land suitability model to predict where autumn-sown, determinate genotypes of the white lupin (*Lupinus albus*) might be grown in England and Wales. *Journal of Agricultural Science* 123, 199–205.

Soares, M.C. (1994) Benefits of bitter yellow lupin (*Lupinus luteus* L.) crop promotion in Portugal. In: Neves Martins, J.M. and Beirão da Costa, L. (eds) *Advances in Lupin Research.* ISA Press, Lisbon, pp. 551–564.

Sweetingham, M.W., Henderson, J., Yang, H.A., Cowling, W.A. and Buirchell, B.J. (1994) Glasshouse screening of *Lupinus* species for resistance to brown spot and Pleiochaeta root rot. In: Dracup, M. and Palta, J. (eds) *Proceedings of the First Australian Lupin Technical Symposium.* Department of Agriculture Western Australia, South Perth, p. 313.

Swiecicki, W. and Swiecicki, W.K. (1994) Self completion and thermoneutrality for seed yield improvement. In: Dracup, M. and Palta, J. (eds) *Proceedings of the First Australian Lupin Technical Symposium.* Department of Agriculture Western Australia, South Perth, p. 314.

Tapscott, H.L., Cowling, W.A., Dracup, M. and Speijers, E.J. (1994) Effect of genotype, site and plant density on yield components of a historical set of narrow-leafed lupin cultivars. In: Dracup, M. and Palta, J. (eds) *Proceedings of the First Australian Lupin Technical Symposium.* Department of Agriculture Western Australia, South Perth, p. 317.

Troll, H.-J. (1964a) [Cultivation, breeding and seed problems in large seeded legumes as contribution to the remedy of protein fodder-deficiency in particular in reference to cultivation – competition between lupins and field peas]. *Wissenschaftliche Zeitschrift der Karl-Marx-Universität* 13, 651–663 (in German).

Troll, H.-J. (1964b) [Breeding for *Fusarium* resistance in *Lupinus luteus*]. *Wissenschaftliche Zeitschrift der Karl-Marx-Universität* 13, 727–732 (in German).

Troll, H.-J. (1967) [Phylogenetically interesting mutants of *Lupinus luteus*]. *Erwin Baer Gedaechtnissvorlesungen* 4, 131–138 (in German).

van Jaarsveld, A.B. (1983) Resistance of lupin accessions to powdery mildew (*Erisiphe poligoni*). *Phytophylactica* 15, 55.

Vavilov, N.I. (1951) The origin, variation, immunity and breeding of cultivated plants. Selected writings of N.I. Vavilov translated from the Russian by K Starr Chester. *Chronica Botanica* 13, 1–366.

Vavilov, P.P. and Gataulina, G.G. (1984) Growth and development pattern of different spring forms of *Lupinus albus*. In: *Proceedings of the Third International Lupin Conference.* International Lupin Association, La Rochelle, France, pp. 562–565.

von Baer, E. and von Baer, D. (1988) *Lupinus mutabilis*: cultivation and breeding. In: Twardowski, T. (ed.) *Proceedings of the Fifth International Lupin Conference.* Panstwowe Wydawnietwo Rolniczne i Lesne, Poznań, Poland, pp. 237–247.

von Eppler, A. and Hinz, U. (1987) [The lupin aphid, *Macrosiphum albifrons* Essig, a new insect pest and virus vector in Germany.] *Journal of Applied Entomology* 104, 510–518 (in German).

von Sengbusch, R. (1942) [Sweet lupins and oil lupins. The history of the origin of some new crop plants.] *Landwirtschaftliche Jahrbücher* 91, 719–880 (in German).

Walton, G.H. and Francis, C.M. (1975) Genetic influences on the split seed disorder in *Lupinus angustifolius* L. *Australian Journal of Agricultural Research* 26, 641–646.

Warren, J., Allen, J.G. and Cowling, W.A. (1989) Economic impact of growing Phomopsis-resistant lupins. *Journal of Agriculture – Western Australia* (4th Series) 30, 8–10.

Wells, H.D. (1984) Breeding aims and criteria for lupines. In: *Proceedings of the Third International Lupin Conference.* International Lupin Association, La Rochelle, France, pp. 166–178.

Wells, H.D., and Forbes, I. (1982) Genetics and breeding of *Lupinus* for winter production for the south of the United States of America. In: Gross, R. and Bunting, E.S. (eds) *Agricultural and Nutritional Aspects of Lupines. Proceedings of the First International Lupine Workshop.* GTZ GmbH, Eschborn, Germany, pp. 77–89.

Williams, W. (1979) Studies on the development of lupines for oil and protein. *Euphytica* 28, 481–488.

Withers, N.J. and Forde, B.J. (1979) Translocation of ^{14}C in *Lupinus albus*. *New Zealand Journal of Agricultural Research* 22, 261–269.

5 Lupin Nutrition

NANCY LONGNECKER[1], ROSS BRENNAN[2] AND ALAN ROBSON[1]

[1]Centre for Legumes in Mediterranean Agriculture (CLIMA) and Soil Science, Plant Nutrition and Faculty of Agriculture, University of Western Australia, Nedlands, Western Australia 6907; [2]Agriculture Western Australia, 444 Albany Highway, Albany, Western Australia 6330

Introduction

Lupins are known for their ability to thrive on soils of low fertility. This review examines characteristics of the genus *Lupinus* which facilitate this and also highlights differences in the nutrition of species which makes extrapolation pointless.

Most available information relates to west Australian conditions and to *L. angustifolius*. Western Australia is known for its nutritionally impoverished soils: for *L. angustifolius* to succeed as a crop, many nutritional challenges had to be overcome. This has required a sound understanding of plant nutrition and the transfer of this information into suitable fertilizer practices and agronomic management. There are still nutritional problems of efficient and sustainable lupin production to be solved.

While most studies involve *L. angustifolius* and/or *L. albus*, some discussion of other lupin species is included. First, an overview of nutrient uptake, transport and utilization is given, followed by the diagnosis and amelioration of deficiencies. Finally, manganese nutrition is examined as a case study which illustrates principles covered in the first two sections.

Nutrient Uptake and Utilization

An understanding of the factors causing nutritional problems improves our ability to diagnose and correct or predict and prevent nutrient deficiencies. This section provides that background information.

Root/soil interactions

Lupinus species characteristically grow on coarse-textured soils of acid to neutral pH. In their naturalized distribution, lupins grow on poor, sandy soils and rarely on calcareous or alkaline soils (Gladstones, 1970). The genus as a whole requires well drained soils which may be deep (450 cm) or shallow (< 20 cm) (Gladstones, 1970).

Generalizations can be made about the different soil types on which different *Lupinus* species thrive. *L. angustifolius* is cropped on infertile, deep sands in south-west and southern Australia. Baylis and Hamblin (1986) suggested that worldwide cultivation is restricted to sands or loams of pH 4 to 7.5 (measured in water at a ratio of 1 : 5). *L. albus* is adapted to moderately acid to neutral loamy sands of better nutritional status. In south-western Australia, *L. cosentinii* is cultivated or volunteers on coastal sandy soils overlying limestone. *L. luteus* is adapted to strongly to mildly acid soils of low fertility while *L. pilosus, L. atlanticus* and *L. cosentinii* generally grow on higher pH soils than does *L. angustifolius* (Clements and Cowling, 1990).

One mechanism that increases plant nutrient uptake from poor soils is infection of roots by mycorrhizal fungi. However, Robson (1986) pointed out that lupins differ from most agricultural plants in that they do not form mycorrhizal associations. In *L. angustifolius* and *L. cosentinii* collected from 20 locations, less than 10% of the root length was infected by vesicular-arbuscular mycorrhizal fungi (Trinick, 1977). Similarly, Gumley (1989) did not find mycorrhizal infection of *L. angustifolius, L. albus* or a perennial lupin species, *L. longifolius,* supplied with five rates of phosphorus. As discussed below, there are other characteristics of lupin roots which allow them to prosper in soils of poor nutrient status.

Depth of rooting

Lupins are deeper rooting than many other agricultural plants. In a deep loamy yellow earth, *L. angustifolius* and *L. cosentinii* reached a maximum depth of 2.2 m compared with 1.3 m for wheat and 0.5 m for peas (Hamblin and Hamblin, 1985). Also, the proportion of the total root length below the top 20 cm was greater (60%) for lupins than for the other species (< 30%).

Having a greater proportion of roots at depth enhances the ability of lupins to take up nutrients and water compared with wheat or other legumes. This is especially true in climates where crops suffer drought during grain filling and on soil types which have substantial nutrient reserves at depth. For example, potassium concentrations increase with depth in many of the duplex soils of Western Australia (Robson *et al.*, 1992). *L. angustifolius* is better able to extract potassium from these soils than cereals (Gladstones *et al.*, 1964) and can grow on soils where potassium deficiency limits the growth of *Trifolium subterraneum* (Toms and Fitzpatrick, 1961). The greater depth of rooting of lupins may also explain why deep banding superphosphate in Western Australia has been shown to increase fertilizer effectiveness more for *L. angustifolius* than for wheat (Jarvis and Bolland, 1990, 1991).

Lateral roots

Clements *et al.* (1993) found a wide range of rooting habit among 12 annual *Lupinus* species (Table 5.1). The root morphology of *L. angustifolius* (cv. Gungurru) was at one end of the spectrum of types for those species examined. Clements *et al.* (1993) speculated that the relatively thick roots and less extensive

Table 5.1. Root length and morphology, including occurrence of cluster or proteiod roots, secondary or tertiary lateral roots of twelve *Lupinus* species (taken from Clements *et al.*, 1993).

Species	Total root length (m per plant)	Cluster roots	Secondary	Tertiary
L. angustifolius	111	No	Yes	No
L. albus	357	Yes	Yes	No
L. luteus	137	No	Yes	No
L. cosentinii	177	Yes	Yes	No
L. pilosus	270	Yes	Yes	Yes
L. palaestinus	124	Yes	Yes	Yes
L. mutabilis	219	No	Yes	Yes
L. hispanicus	86	No	Yes	Yes
L. micranthus	145	Yes	Yes	Yes
L. digitatus	95	Yes	Yes	Yes
L. princei	151	Yes	Yes	No
L. atlanticus	220	Yes	Yes	Yes

lateral root system of *L. angustifolius* may partially explain its poor growth on fine-textured soils where fine lateral roots may be advantageous. The root system of *L. angustifolius* consists of a dominant taproot with many primary laterals but few secondary and tertiary laterals. The number of primary lateral roots decreased rapidly with depth for *L. angustifolius* and also for *L. mutabilis* but the distribution was more even for the other species. *L. angustifolius* and *L. luteus* had relatively thick roots compared with the other species.

Bishop *et al.* (1986) compared the rooting system of normal branching *L. angustifolius* (e.g. cv. Illyarrie) with reduced or restricted branching lines. Freely branching types had greater total root length density and greater length per primary lateral than the restricted branching types. This difference in rooting pattern could allow the freely branching types to take up soil water to support continued growth while the more determinate habit did not require such an extensive root system.

Total root length and surface area have a profound impact on nutrient uptake, particularly of nutrients such as phosphorus which are relatively immobile in soil. Thus, it is not surprising that root proliferation is affected by nutrient status. Lunt *et al.* (1994) showed that banded placement of superphosphate resulted in a greater number of lateral roots and greater total root length of *L. angustifolius* compared with top dressed or no added superphosphate. Roots of *L. angustifolius* (cv. Gungurru) supplied with either top dressed or banded superphosphate proliferated in the zone of enhanced phosphorus supply (Fig. 5.1).

Cluster roots

Cluster or proteoid roots (specialized lateral roots) in *L. albus* have attracted much attention by researchers (Gardner *et al.*, 1981, 1982a,b, 1983; Marschner *et al.*,

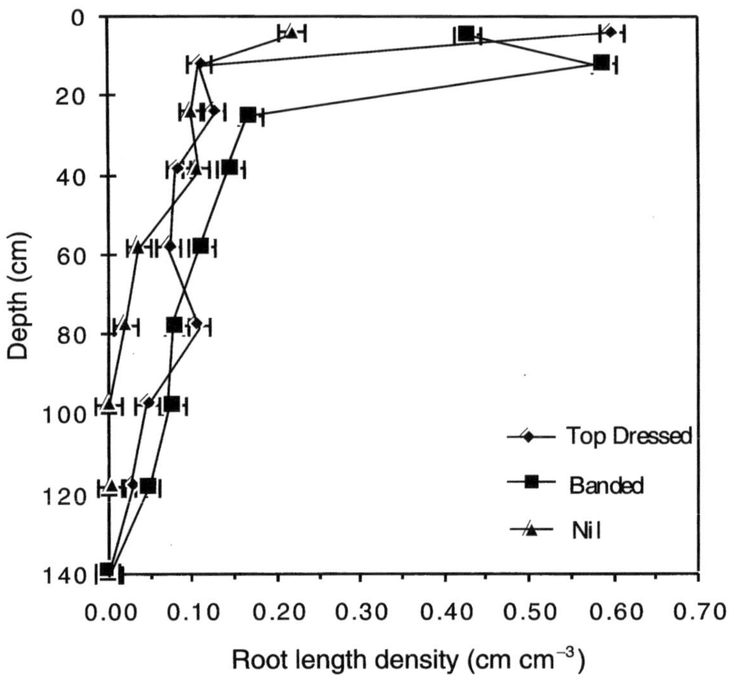

Fig. 5.1. Root length density (cm root per cm^3 soil) of *L. angustifolius* at depth (cm) from the soil surface. Lupins were grown in pots with three phosphorus fertilizer treatments: topdressed phosphorus, banded 7 cm below the seed and nil (no added phosphorus) in a highly phosphorus-deficient soil from South Carrabin, Western Australia (G. Knell, B. Bowden, A. Diggle and N. Longnecker, Perth, 1994, unpublished data).

1987; Dinkelaker *et al.*, 1989; Johnson *et al.*, 1994, 1996a,b). Other lupins which form cluster roots include *L. cosentinii, L. pilosus, L. palaestinus, L. micranthus, L. digitatus, L. princei* and *L. atlanticus* (Clements *et al.*, 1993). However they are not found in *L. angustifolius.* In the Clements *et al.* (1993) study, cluster roots were particularly numerous in *L. micranthus* and *L. albus.*

Development of cluster roots is enhanced under conditions of phosphorus deficiency and it has been postulated that they increase nutrient availability and uptake by a number of mechanisms. Dissolution of calcium, iron and aluminium phosphates and manganese dioxide was observed around cluster roots, with little dissolution elsewhere along the root except for areas of high density of lateral roots (Gardner *et al.*, 1982a). Exudation of protons, as well as reducing and chelating agents from cluster roots, could account for these increases in nutrient solubility (Gardner *et al.*, 1982b, 1983; Johnson *et al.*, 1996a,b). Phosphorus-deficient *L. albus* have been reported to exude up to 23% of their dry weight as citrate (Dinkelaker *et al.*, 1989) compared with only 0.3% exuded by phosphorus-deficient *Medicago sativa* (Lipton *et al.*, 1987). Non-photosynthetic CO_2 fixation apparently contributes carbon for enhanced citrate exudation (Johnson *et al.*, 1996a) with expression of phosphoenolpyruvate carboxylase being much higher in cluster roots of *L. albus* grown without phosphorus supplied (Johnson *et al.*,

1996b). Excretion of acid phosphatases from roots is also increased under phosphorus deficiency (Hirata *et al.*, 1982; Ozawa *et al.*, 1995). Acid phosphatase action can hydrolyse organic phosphorus in soils into inorganic orthophosphate (Tarafdar and Jungk, 1987) which is presumed to be the form absorbed by roots.

Adding phosphorus decreases growth of cluster roots (Marschner, 1986). Presumably there is a lower requirement for enhanced surface area for uptake when phosphorus supply is adequate, but the mechanisms that trigger or inhibit cluster root growth are not understood. Generally, those species that form cluster roots (Table 5.1) have greater nutrient uptake when grown in similar conditions than species that do not form cluster roots. For example, *L. cosentinii* appears to be more able to obtain phosphorus from soil than does *L. angustifolius* (Rahman and Gladstones, 1974; Pairunan, 1980). *L. cosentinii* and *L. albus* obtained more phosphorus for a given amount applied than *L. luteus* and *L. angustifolius* (Bolland, 1997). About twice as much of an insoluble rock phosphate was dissolved and taken up by *L. albus* as by *L. angustifolius* (Hinsinger and Gilkes, 1995). *L. albus* accumulates much greater concentrations of manganese than either *L. angustifolius* or *L. luteus* (Gladstones, 1962). Furthermore, *L. albus* and *L. cosentinii* are better able to obtain cobalt from soil than either *L. angustifolius* or *L. luteus* (Gladstones *et al.*, 1977). Finally, *L. cosentinii* had greater copper concentrations in its shoots than *L. angustifolius* (Gladstones, 1962). There are exceptions to the generalizations that species that form cluster roots have greater nutrient uptake than those that do not. In a study by Gladstones (1962), copper concentration in shoots of *L. luteus* was greater than that in *L. albus* and similar to that in *L. cosentinii*. Shenker *et al.* (1989) reported that *L. albus* and *L. palaestinus* were more susceptible to iron deficiency than *L. pilosus* even though all three species produce cluster roots. It seems probable that production of cluster roots is one of a number of mechanisms that increase nutrient uptake.

There are links between root growth and nutrient uptake under different conditions of nutrient availability, but the mechanisms triggering growth and/or uptake are not understood. For example, the signals that cause lateral roots of *L. angustifolius* to develop in zones of high phosphorus supply on the one hand and cause cluster roots of *L. albus* to develop in conditions of deficient phosphorus supply on the other hand are unknown.

Root hairs

For obvious reasons of sampling difficulty, there are few studies that examine root morphology and even fewer that include root hairs. Pairunan (1980), however, observed much greater total root surface area of *L. angustifolius* supplied with adequate phosphorus than with deficient phosphorus. There was little difference in total root surface area of the mycorrhizal species, *T. subterraneum*. The greater total root surface area was largely due to greater numbers and longer root hairs (Table 5.2). This contrasts with the phosphorus-induced root proliferation of *L. angustifolius* shown in Fig. 5.1, which was due to proliferation of lateral roots.

In *L. angustifolius* (which does not form cluster roots) Papavizas and Davey (1961) observed an extensive rhizosphere extending at least 18 mm from the root

Table 5.2. Root characteristics of *L. angustifolius* and subterranean clover (*Trifolium subterraneum*) supplied with no added phosphorus (–P) or an amount adequate for maximum shoot growth (+P) (taken from Pairunan, 1980).

	Lupinus angustifolius		*Trifolium subterrananeum*	
Characteristic	–P	+P	–P	+P
Root hair number (per mm root)	340	358	51	58
Root hair length (mm)	0.76	0.79	0.13	0.15
Root hair surface area (cm^2 per plant)	374	947	14	25
Total root surface area (cm^2 per plant)	533	1300	109	166

surface. Alleviating phosphorus deficiency of *L. angustifolius* resulted in half the amount of soil adhering to roots (Pairunan, 1980), supporting other observations that plant nutrient status affects root/soil interactions.

Nutrients and Nitrogen Fixation

Mineral nutrients play a key role in nitrogen fixation by legumes (see reviews by Robson, 1978, 1983; OHara *et al.*, 1988; Robson and Bottomley, 1991). There is little evidence that effects observed for other species do not also occur with lupins. Nodulation, nodule development and the rate of N_2 fixation can all be inhibited by high levels of available soil nitrogen (Atkins, 1984). Nodulation and nitrogen fixation appear to require more phosphorus than does host plant growth (Robson, 1986; Israel, 1987; Robson and Bottomley, 1991). Nitrogen fixation also requires greater levels of iron and cobalt than does plant growth (Dilworth *et al.*, 1979; Tang *et al.*, 1990b).

There has been relatively little study of the role of mineral nutrients in nitrogen fixation by lupins. A notable exception is the detailed and extensive studies of the role of cobalt. In fact, much of our current knowledge of the role of cobalt in nitrogen fixation is based on studies in Western Australia with narrow-leafed lupins.

Cobalt deficiency is likely to limit nitrogen fixation by effects on multiplication of both bradyrhizobial and nodule function. Narrow-leafed lupins are particularly sensitive to cobalt deficiency. Responses to cobalt were greater when seeds were not inoculated with root nodule bacteria and plants were reliant on low numbers of indigenous bradyrhizobia for nodulation (Robson *et al.*, 1979).

Cobalt deficiency decreased nitrogen fixation and initially decreased nodule number (Dilworth *et al.*, 1979). There were marked effects of cobalt deficiency in decreasing the number of bacteroids per gram nodule weight, and cobalamin and leghaemoglobin concentrations in nodules (Dilworth *et al.*, 1979). It is likely that proliferation of bradyrhizobia within the nodule is the most severely impaired process in cobalt-deficient plants, with the low leghaemoglobin contents a secondary effect (Riley and Dilworth, 1985a,b). Cobalt deficiency is also likely to decrease nitrogen fixation by decreasing the activity of enzymes known to be cobalamin dependent (Dilworth *et al.*, 1979; Dilworth and Bisseling, 1984).

Recently the role of copper in nitrogen fixation by *L. luteus* has been investigated (Seliga, 1993). Copper deficiency decreased nitrogen concentration in nodules but increased nitrogen concentrations in shoots and roots. Effects of copper deficiency on decreasing nodule weight was more marked than effects on shoot or root weight. It thus appears likely that copper deficiency limits nitrogen fixation in lupins in a similar way to effects in other legumes (Snowball *et al.*, 1980). The mechanisms are poorly understood but starch accumulation has been observed in the nodules of both lupins (Weryszko, 1992, cited by Seliga, 1993) and subterranean clover (Cartwright and Hallsworth, 1970) consistent with reduced rates of sugar utilization and respiration.

Iron deficiency decreases nodule number and specific nitrogenase activity (Tang, 1995). Tang *et al.* (1990b) used a split-root system to show that the decreased nodule number in *L. angustifolius* caused by iron deficiency was a direct effect; nodule numbers were decreased in half of the roots not supplied with iron but the half supplied with iron had nodules similar to control plants. The effects of iron deficiency on plants solely dependent on nitrogen fixation were greater than on plants supplied mineral nitrogen, and Tang *et al.* (1990a) concluded that the symbiosis has a greater iron requirement than host plant growth.

Utilization and Genotypic Differences

Because both phloem and xylem sap can be collected from *L. angustifolius* and *L. albus*, long-distance nutrient transport has been studied more extensively in *Lupinus* than in most genera. The *L. albus* system has been used to describe nutrient circulation in the plant, including transport from shoots to roots in phloem and re-entry into xylem (Jeschke *et al.*, 1985). Data illustrate the greater phloem mobility of K^+ and to a lesser extent Mg^{2+} and Na^+ compared with Ca^{2+}, for which the rate of transport to the roots via phloem was low (Jeschke *et al.*, 1985, 1987). For more detailed discussion, see Chapter 13, this volume.

The focus of this section is on nutrient utilization and genotypic differences between and within different lupin species. Phosphorus nutrition is examined in more detail in later sections on seed development and on diagnosis and amelioration of deficiencies.

Nitrogen

Management decisions based on a perceived yield penalty due to the cost of nitrogen fixation vary greatly in different legume crops around the world. Discussion of minimizing the cost of nitrogen fixation in terms of the carbon : nitrogen economy of the plant and maximizing the efficiency of nitrogen fixation (Atkins, 1984) is outside the scope of this chapter. Most lupin crops are used as a source of nitrogen in the farming system and nitrogen fertilizer application to a lupin crop is rare.

In addition to the input of nitrogen from fixation, some nitrogen is usually assimilated from the rooting medium. The proportion of fixed to assimilated nitrogen varies, depending on factors such as species, soil nitrogen levels and environmental conditions. Generally high levels of available nitrogen in soil will decrease nitrogen fixation, and lupins are no exception.

One intriguing puzzle in the nitrogen nutrition of symbiotic legumes involves signals that allow a plant to 'sense' its nutrient status. This is necessary in order for a symbiotic plant to regulate nitrogen fixation so that its nitrogen requirement is met efficiently. One school of thought is that ratios of particular amino acids in phloem may act as signals to the nodule, allowing it to sense the plant's nitrogen status and thus its requirement for the products of fixation (Parsons and Baker, 1996). Identification of the signals that trigger plant and nodule growth and activity in response to nutrient status will be a challenge in the coming decade.

Work on growth of lupin apices, which are fed primarily by phloem, has focused on a related hypothesis; namely, that changes in growth of the apex in response to nitrogen supply of the plant must be triggered by a signal other than nitrogen concentration of the apex. Interestingly, although asparagine is the predominant amino acid in both xylem and phloem of lupin (Pate *et al.*, 1979; Parsons and Baker, 1996), glutamine as a nitrogen source for cultured apices produced better growth than asparagine (J. Quealy, N. Longnecker and C. Atkins, 1994, Perth, unpublished data).

When nitrogen is absorbed as nitrate from the soil, reduction must occur prior to assimilation. Lupins differ from many legumes in that nitrate reduction takes place primarily in the roots (Atkins *et al.*, 1979). While shoot nitrate reductase activity was stimulated by increasing the nitrate supply to a very high level (30 mM), when *L. albus* was exposed to nitrate levels more commonly experienced the shoots contained only 25–40% of total plant nitrate reductase activity (Atkins *et al.*, 1979). Even in plants receiving 30 mM nitrate, NO_3^- accounted for less than 20% of xylem-borne nitrogen, with most transported as asparagine and less as glutamine (Atkins *et al.*, 1979).

Temporary nitrogen deficiency is common. It can occur, for example, when nodulation and/or nitrogen fixation are delayed and soil nitrogen levels are low or during periods of waterlogging in the growing season. There are effects of temporary nitrogen deficiency on lupin development which irreversibly affect yield potential and achievement of that potential. Development of *L. angustifolius* is sensitive to early, temporary periods of nitrogen deficiency (Ma *et al.*, 1997). When nitrogen fixation was temporarily inhibited by flushing the root system with oxygen for 30 minutes per day over a 10-day period, flower number, pod number and ultimately grain yield of *L. angustifolius* were lower than those of nitrogen-fixing plants that had no interruption of fixation, or plants continuously supplied with adequate levels of mineral nitrogen (Ma *et al.*, 1998). A two-week period of lowered nitrogen supply slowed the rate of leaf emergence and hence delayed flowering (Ma *et al.*, 1997).

Phosphorus

With a given rainfall, inadequate phosphorus nutrition is frequently the limiting factor for growth of lupins. In addition to limitations for vegetative growth and nodulation, phosphorus deficiency has been found to decrease the harvest index of *L. angustifolius* in both pot trials (M. Qifu and N. Longnecker, unpublished data, 1996) and field experiments (Jarvis and Bolland, 1991). The reason for a greater effect of phosphorus deficiency on reproductive than vegetative yield is unclear and warrants further investigation.

Iron

The poor growth of *L. angustifolius* in alkaline soils has been attributed to a number of factors (Jessop *et al.*, 1990; White, 1990; Atwell, 1991; see also Chapter 8, this volume), including susceptibility to iron deficiency (White, 1990; Tang *et al.*, 1995). In a comparison of the severity of iron chlorosis when grown on the same soils, *L. angustifolius, L. luteus* and *L. albus* were more affected than *L. pilosus, L. atlanticus* and *L. cosentinii* (White and Robson, 1989a; Tang *et al.*, 1995). There was a negative correlation ($P < 0.05$) between chlorosis rating five weeks after sowing and relative grain yield on both an alkaline clay ($r = -0.61$) and an alkaline sand ($r = -0.68$) (Tang *et al.*, 1995). The greater sensitivity of *L. angustifolius* to iron deficiency is not due to an inability to reduce Fe^{3+} (thereby increasing its availability and uptake) or to retranslocate iron (White and Robson, 1989b). In a comparison of lupin species for growth in nutrient solution with different levels of iron, *L. pilosus* was the most tolerant to low concentration, followed by *L. angustifolius* and *L. luteus* (Tang and Robson, 1993).

L. pilosus was more resistant to iron deficiency than were *L. palaestinus* and *L. albus* even though both *L. palestinus* and *L. albus* exhibited response mechanisms thought to be indicative of 'iron efficiency' (increased reducing activity and decreased pH along the roots) (Shenker *et al.*, 1989). *L. pilosus* showed the least chlorosis yet had no detectable reducing activity along the root, indicating that any definition of 'iron efficiency' needs to be based on plant growth response in the field rather than on laboratory measurements, unless their correlation is well documented (Longnecker, 1988).

Both *L. albus* and *L. angustifolius* exhibit genotypic variation in susceptiblity to iron deficiency (Tang *et al.*, 1995). However, Tang *et al.* (1996) determined that iron deficiency of *L. angustifolius* in Western Australia was generally a secondary effect, with decreased root growth being the main effect. Thus increased tolerance to alkaline soils would result from improved root growth in a substrate with high pH, and increased root growth in alkaline media is a better screening method than decreased chlorosis for alkaline soils of Western Australia.

Zinc

The wheat belt of Western Australia has one of the largest areas of zinc-deficient soils in the world (Gilkes, 1994). It has been assumed that if zinc is adequate for cereal production, it will also be adequate for *L. angustifolius* within the rotation and that zinc deficiency is not a constraint to lupin production. Yet Riley *et al.* (1992) observed an increase in shoot growth, grain yield and grain zinc concentration of *L. angustifolius* in response to zinc fertilizer on the Eradu sand plain in Western Australia. *L. albus* is more sensitive to zinc deficiency than *L. angustifolius* and exhibited severe deficiency symptoms and a 40% increase in grain yield when zinc was applied, while *L. angustifolius* and *Vicia faba* did not respond to zinc fertilization (R. Brennan, N. Longnecker and S. Loss, 1995, Perth, unpublished data). In fact *L. albus* has exhibited severe zinc deficiency at that particular site for many years, although no symptoms have been seen in *L. angustifolius* (B. Buirchell and G. White, Perth, personal communication, 1995). The difference does not appear to be related to increased uptake of zinc, since Pakendorf *et al.* (1970) observed lower concentrations of zinc in *L. albus* than in *L. angustifolius* grown at one site.

L. albus exhibits genotypic differences in susceptibility to zinc deficiency (B. Buirchell, personal communication, 1996). This could be an important trait in breeding programmes if *L. albus* is being recommended for alkaline soils where zinc availability could be limiting. Zinc deficiency has not been observed in *L. atlanticus* and *L. pilosus* (B. Buirchell, personal communication, 1996).

Copper

Gladstones and Drover (1962) speculated that the good growth of *L. cosentinii* (blue lupin) on coastal sandy soils of Western Australia was due to a low requirement for copper. Copper concentration in whole plant tops was generally *L. cosentinii* > *L. luteus* > *L. angustifolius* > *L. albus* (Gladstones, 1962; Gladstones and Drover, 1962). This might indicate that the success of *L. consentinii* is due to a better ability to take up copper rather than a lower requirement. Within *L. luteus*, the bitter form contained more copper than the sweet variety (Weiko III). Copper deficiency of *L. angustifolius* is rarely observed in field-grown crops, even when other crops show marked response to copper fertilizer (Table 5.3).

Cobalt

As mentioned previously, *L. angustifolius* may be particularly sensitive to cobalt deficiency. Gladstones *et al.* (1977) demonstrated a 50% response of vegetative yield in *L. angustifolius* to cobalt where *L. cosentinii, L. luteus* and *T. subterraneum* showed no response. Grain yield of *L. angustifolius* on a deep yellow sand in Western Australia without cobalt application was about 40% of that on cobalt-fertilized plots (J. Gartrell, 1978, Perth, unpublished data). Cobalt deficiency results in decreased nitrogen concentrations in shoots.

Table 5.3. Response to copper and zinc fertilizer by *L. angustifolius* and wheat (*Triticum aestivum*) grown on copper and zinc deficient soil in Western Australia (R. Brennan, Perth, 1990, unpublished data).

	Grain yield (kg ha^{-1})	
Treatment	*L. angustifolius*	Wheat
–Cu	1900	2100
+Cu	1950	2800
–Zn	1870	2190
+Zn	1850	2690

Impact of Nutrition on Seed Quality

Nutrient accumulation in the seed has important implications for feed and food quality, for removal of nutrients from the system at harvest and for growth of the next generation. Lupin species differ markedly in the nutrients accumulated in their seeds (Table 5.4). Copper and especially manganese concentrations are higher in *L. albus* than in *L. angustifolius* (Gladstones, 1962; Pakendorf *et al.*, 1970; White *et al.*, 1981). Copper and molybdenum concentrations are higher in *L. cosentinii* and *L. luteus* than in either *L. albus* or *L. angustifolius* (Gladstones and Drover, 1962). Zinc concentration tends to be higher in *L. angustifolius* than in *L. albus* while sulphur concentrations are similar in seeds of the two species (White *et al.*, 1981).

Seeds of all legumes are low in the sulphur amino acids required by animals for a balanced diet. Thus there have been ongoing efforts to increase the sulphur content of legume seeds for many years. Insertion of a sunflower albumin seed gene into *L. angustifolius* has resulted in a 94% increase in methionine content and a 12% decrease in cysteine content of the transgenic (T_4 generation) seeds compared with their non-transgenic parents (Molvig *et al.*, 1996). This approach should dramatically improve the feed and food quality of lupin.

For micronutrients, which are by definition needed in small quantities, seed nutrient content can have a significant impact on crop growth. In the case of cobalt, greater content in the seed can provide enough cobalt to eliminate fertilizer response in the growing crop. There is a strong inverse relationship between cobalt concentration in seeds and response to applied cobalt (Robson and Mead, 1980; Robson and Snowball, 1987). Grain yield of *L. angustifolius* can be affected by seed manganese concentration (Fig. 5.2). As shown in the next section, seed macronutrient content can also affect crop growth and yield.

There is some evidence that nutrient status of the plant affects the content of anti-nutritionals in seeds. Lupins and other legumes in the *Papilionoideae* subfamily appear to take particular advantage of their nitrogen fixing symbiosis by producing nitrogen-rich secondary metabolites (Waterman, 1994). The alkaloids and non-protein amino acids in lupins are effective allelochemicals against pathogens and pests, including vertebrate herbivores. For example, alkaloid-rich 'bitter' lupins are less susceptible to herbivores than 'sweet' lupins, which are low in alkaloids (Wink, 1987). In addition to their role as allelochemicals, the levels of accumulation of secondary metabolites in seeds (as much as 0.043% in *L. angustifolius*; Petterson and Mackintosh, 1994) leads to speculation that they also serve as storage pools for the next generation (Johns, 1994).

Since most alkaloids are synthesized in young, actively growing tissues, it is not surprising that nutrient deficiencies which restrict growth also affect alkaloid accumulation (Waller and Nowacki, 1978). High levels of nitrogen fertilizer resulted in increased concentration of alkaloids from 0.9 to 1.3% in *L. angustifolius* plants before flowering (Nowacki *et al.*, 1976). Lupins receiving low potassium supply had higher alkaloid levels (Andreeva, 1952; Mironenko, 1965, 1975; all reported in Waller and Nowacki, 1978), possibly because of greater levels of catabolic enzymes (Waller and Nowacki, 1978). The effect of phosphorus fertilizer is less marked and the opposite to that of potassium: increased phosphorus fertilizer resulted in increased levels of alkaloids (Mironenko, 1965, 1975, reported in Waller and Nowacki, 1978). Mironenko (1965, reported in Waller and Nowacki, 1978) observed that addition of boron, molybdenum, manganese and copper decreased alkaloid levels in seed of three lupin species. Clearly, more data from experiments in which detailed seed analyses are

Table 5.4. Critical nutrient concentrations in whole shoots and nutrient concentrations in seeds of *L. angustifolius* and *L. albus* (little information is available for other species).

Species	Nutrient	Critical concentration in whole shoot	Ref.	Seed concentration (kg^{-1})	Ref.
L. angustifolius	Nitrogen	—			
	Potassium	1.5%*	e	11.5 g	a
	Phosphorus	0.23%	d	3.1 g	a
	Magnesium	—		3.1 g	a
	Sulphur	—		2.5 g	c
	Calcium	—		2.1 g	a
	Iron	—		61 mg	a
	Manganese	40–55 $\mu g\ g^{-1}$		32 mg	c
	Zinc	—		30–34 mg	a,c
	Boron	—		16.2 mg	a
	Copper	< 4 $\mu g\ g^{-1}$		3.7 mg	c
	Selenium	—		0.9 mg	c
	Cobalt	—		0.7 mg	c
L. albus	Nitrogen	—			
	Potassium	—		9.8 g	b
	Phosphorus	0.22%	f	3.6 g	b
	Magnesium	—		1.4 g	b
	Sulphur	—		2.4 g	b
	Calcium	—		2.0 g	b
	Iron	—		26.2 mg	b
	Manganese	—		835 mg	b
	Zinc	—		29.7 mg	b
	Boron	—		—	
	Copper	—		5.0 mg	b
	Selenium	—		—	
	Cobalt	—		0.3 mg	b

*At flowering

References:

a Pakendorf *et al.*, 1970

b Petterson and Mackintosh, 1994

c White *et al.*, 1981

d Rahman and Gladstones, 1974

e Cox, 1979

f K. Snowball and A.D. Robson Perth, 1985, unpublished data

available are needed before we can understand the effects of nutrient status on alkaloid metabolism.

Phosphorus

Seedlings of *L. angustifolius* grown from seed with low phosphorus concentration (< 2.1 g P kg^{-1}, Thomson *et al.*, 1992) had decreased early growth whether or not there was an adequate supply of phosphate fertilizer (Thomson *et al.*, 1992). Bolland *et al.* (1989) also showed that grain yield of *L. angustifolius* grown from seed with low phosphorus concentration (2.0 g P kg^{-1}) was depressed compared with plots sown with 2.8 g P kg^{-1} seed when phosphorus in the soil was limiting (37 mg bicarbonate-soluble P g^{-1}).

Increasing availability of seed phosphorus reserves can increase root growth of the seedling (Thomson *et al.*, 1992; G. Knell, N. Longnecker, B. Bowden, R. Lunt and A. Diggle, 1996, Perth, unpublished data). This may account for the increase in shoot growth and ultimate grain yield since greater root growth may allow greater volume of soil to be explored and hence increase subsequent phosphorus uptake.

Copper

Seed of *L. albus* generally has a higher copper concentration than seed of *L. angustifolius* when grown on the same soil (Gladstones, 1962; Pakendorf *et al.*, 1970; White *et al.*, 1981). In glasshouse studies using defined nutrient solution, more copper was translocated to the plant top of *L. angustifolius* than in *L. albus*,

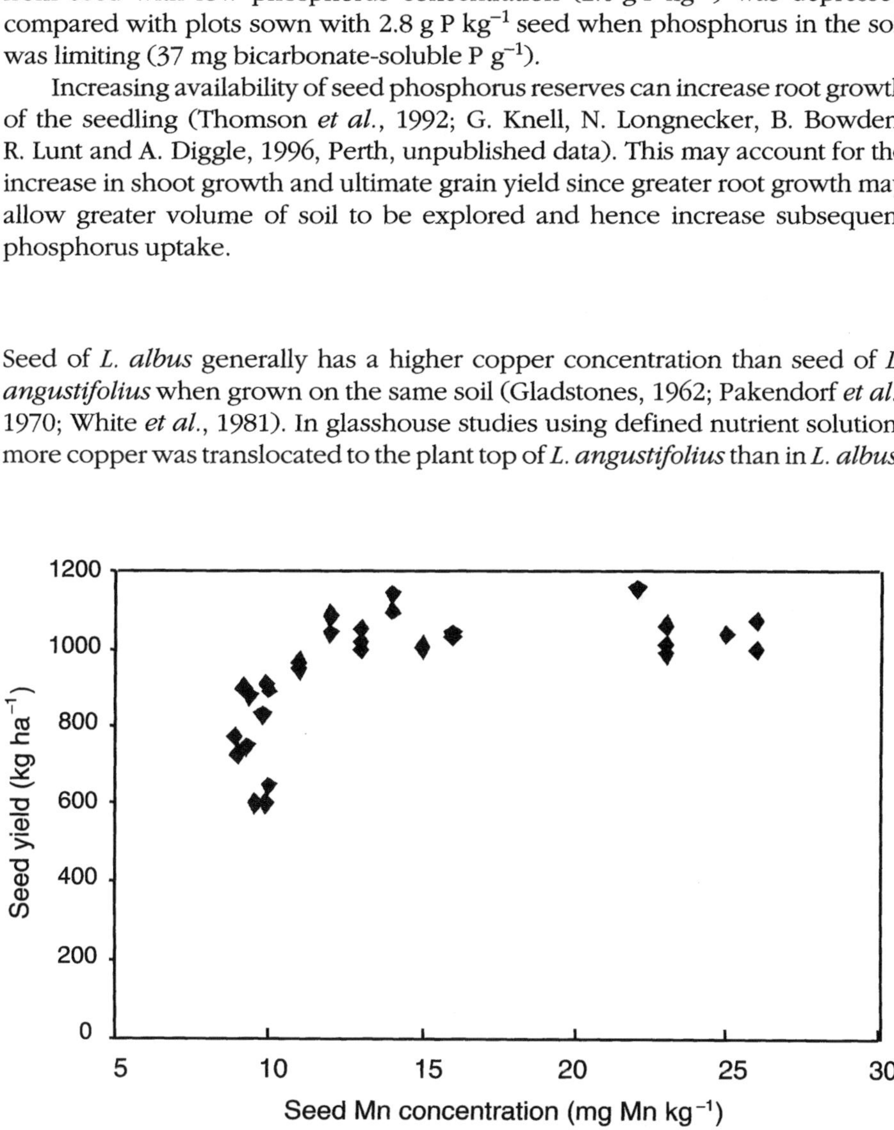

Fig. 5.2. Seed yield (kg ha^{-1}) of *L. angustifolius* grown from seed with different manganese concentrations (mg Mn kg^{-1} seed) (R. Brennan, Perth, 1995, unpublished data).

but *L. albus* still accumulated a greater copper concentration in its seed (Hocking and Pate, 1978). This implies that there is greater loading of copper into phloem of *L. albus* than *L. angustifolius*.

Diagnosis and Amelioration

One of the biggest priorities for correcting nutrient deficiencies of lupins has been phosphorus and much of the discussion in this section reflects its importance. Manganese, for which prognostic tests and reliable recommendations have also been developed, is discussed in the following section.

Diagnosis

Nutrient deficiencies can be diagnosed or predicted using soil tests and/or plant analysis. Predicting or diagnosing nutritional disorders in lupins by either means requires a diagnostic framework that integrates local knowledge and takes into account factors known to promote nutrient deficiencies.

Soil testing

Numerous studies have established growth and grain yield responses of *L. angustifolius* to phosphorus applications (Jarvis and Bolland, 1990, 1991; Bolland and Barrow, 1991; Bolland, 1992a,b,c). At suboptimal levels of application, *L. angustifolius* required two to three times more phosphorus than did wheat or barley to produce the same yield, because of a sigmoidal yield response in *L. angustifolius* compared with an exponential response in the cereals (Bolland and Barrow, 1991). However, all three species required similar amounts of added phosphorus to achieve maximum yield.

Bolland (1992a,b,c) has established a relationship between bicarbonate-extractable phosphorus (the soil test for phosphorus commonly used in southern Australia) and grain yield of *L. angustifolius*. The relationship between yield and soil test values had similar form, but varied between years, making a general recommendation difficult.

Soil testing for potassium has been calibrated in Western Australia and generally predicts a potential deficiency before the lupin crop is sown. However, often only the top 10 cm is sampled for chemical analysis, so the test values may be misleading for deep-rooted lupin species if the subsoil contains markedly different levels of potassium (Robson *et al.*, 1992). For example, the deep sands of the Esperance district of Western Australia often have low levels of extractable potassium (25–30 mg K kg^{-1}) in the surface and a response to potassium fertilizer might be expected. These soils often have similar values to depths of 30 cm and hence response to potassium fertilizer is not always observed (N. Edwards, Perth, 1994, personal communication). If the 0 to 10 cm soil sample from deep grey sands of Badgingarra–Eneabba region of Western Australia had less than 30 mg K g^{-1} of soil, a response to potassium occurred (N. Edwards, Perth, 1994, unpublished data). If the surface soil test value was between 30 and 40 mg K g^{-1}, the response of lupin to potassium fertilizer was influenced by the available potassium in the subsoil. If the surface soil extractable potassium was greater than 40 mg K g^{-1} for both the deep grey sands of the Badgingarra or

Esperance districts, it was unlikely that lupins would respond to potassium fertilizer.

Plant analysis

A large amount of information is necessary to develop fertilizer recommendations based on soil test results and consequently plant analysis is frequently used to diagnose nutritional status and determine adequacy of fertilizers for new crops or for crops with a small sown area. However, critical plant levels cannot be extrapolated from one species to another and they are not constant throughout the life cycle of a plant.

In order to develop procedures for diagnosing nutrient deficiencies by plant analysis, understanding factors affecting distribution of the nutrient in plants is essential. Differences in concentration of a nutrient between plant parts result from absorption, transport in xylem and retranslocation via phloem. The relative activities of these processes can differ between species. Such information determines which plant parts should be sampled for diagnosis of nutrient deficiency and assists in interpretation of analytical results.

In copper-deficient *L. angustifolius* there is relatively little difference between the copper concentration of leaves of different ages (D. Vetterlein and A.D. Robson, 1985, unpublished data). This contrasts with copper in young and old leaves of wheat (Loneragan *et al.*, 1980) and subterranean clover (Reuter *et al.*, 1981) where the concentration is lower in young leaves of deficient plants. Zinc concentration in leaves of different ages depends on zinc supply, both for subclover and *L. angustifolius*. In zinc-deficient plants, young leaves tend to have higher zinc concentrations than older leaves (Reuter *et al.*, 1982; Longnecker and Robson, 1993). With low to adequate zinc supply, the concentration in reproductive plant parts tends to be higher than that of vegetative parts (Hocking and Pate, 1978).

L. angustifolius has the ability to concentrate molybdenum in the nodules of both molybdenum-deficient and molybdenum-adequate plants. However, under adequate supply, molybdenum accumulated in the stems of lupins rather than in leaves (M. Riley, 1986, unpublished data).

Data are available for critical levels of potassium (Cox, 1979) and phosphorus (Rahman and Gladstones, 1974) and for concentrations of other nutrients in *L. angustifolius* and *L. albus* with and without nutrient deficiency symptoms (Table 5.4). Potassium deficiency of lupins is often observed on deep sand with low organic matter and clay content. N. Edwards (1994, Perth, unpublished data) examined the possibility of using the potassium concentration in the youngest shoot tips of *L. angustifolius* instead of whole tops to diagnose deficiency. The concentration of potassium in lupin tips was higher than in whole tops, reflecting its mobility in the plant (Fig. 5.3a). The tips had a narrower range of potassium concentrations than whole tops for the same level of potassium fertilizer applications. There was a linear relationship between the potassium concentration in tips and the concentration in whole tops of *L. angustifolius*. However, the relationship between potassium concentration and grain yield was steeper and less sensitive to added potassium than were the concentrations in whole tops (Fig. 5.3b).

Sulphur concentrations in young leaves, stem and whole shoot and N : S ratios in young leaves are all useful indicators of sulphur deficiency in *L. angustifolius* (Robson *et al.*, 1995). Concentrations in the youngest open leaves were higher than in older leaves at all rates of sulphur support. However, the critical sulphur values were unaffected by plant maturity (Robson *et al.*, 1995), differing in this respect from other legumes such as subterranean clover (Spencer *et al.*, 1977). These results verify the usual practice of analysing whole shoots, which many farmers prefer since these shoots can be meaningfully analysed for a broader spectrum of nutrients.

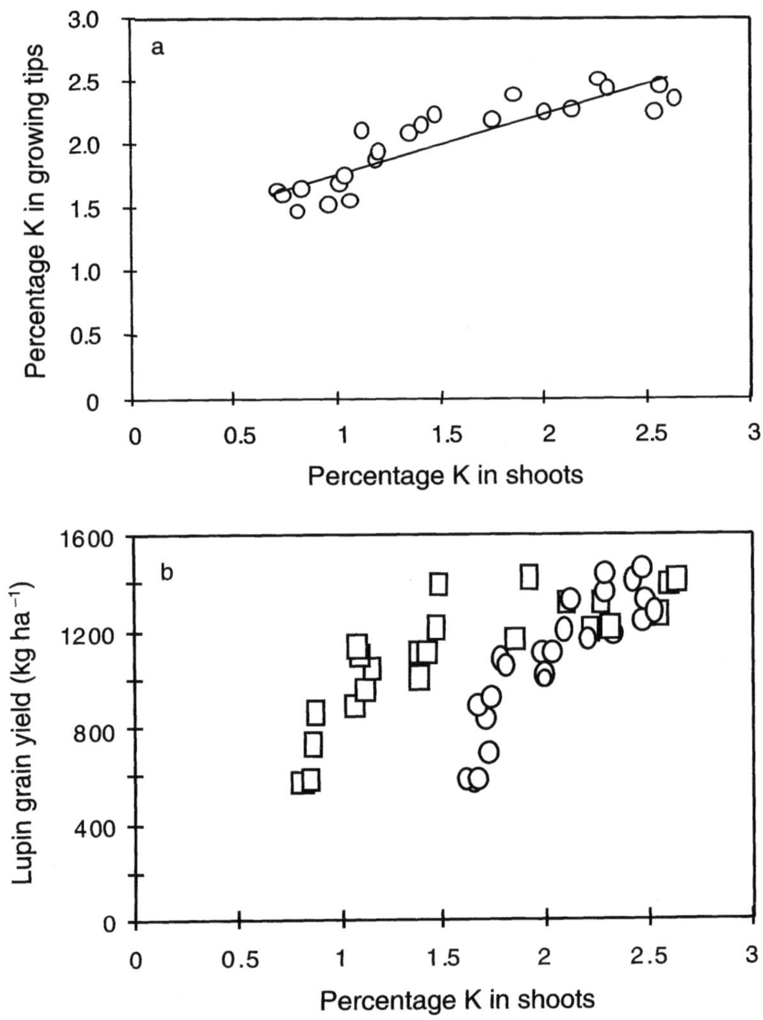

Fig. 5.3. Relationships for *L. angustifolius* between (a) potassium concentration (% dry weight) in the whole shoots and growing tips and (b) the potassium concentration (% dry weight) in the whole shoots (□) or growing tips (○) and grain yield (kg ha^{-1}) (N. Edwards, Perth, 1994, unpublished data).

Amelioration

Soil application

Comparing the amount of phosphorus fertilizer required for maximum growth, *L. angustifolius* required more than wheat (*Triticum aestivum*) while *L. cosentinii* required less (Bolland, 1992c). This is consistent with results observed by Ozanne *et al.* (1976), but differs from that of Bolland and Barrow (1991).

Jarvis and Bolland (1990) found that superphosphate (9.1% P) placed 5 to 13 cm deep during a wheat cropping year was 30 to 60% more effective for the following year's *L. angustifolius* crop than superphosphate freshly drilled with the lupin seed. Jarvis and Bolland (1991) then examined the effectiveness for lupin production when superphosphate was applied in the lupin year of the rotation: either drilled with the seed, topdressed to the soil surface or banded below the seed. This work found that regardless of *L. angustifolius* cultivar or phosphorus status of the soil, superphosphate effectiveness was increased anywhere from 10 to 90% by banding below the seed, compared with the drilled treatment. Topdressing was 30 to 60% less effective than drilling with the seed. Residual value of superphosphate fertilizer is predictably low, with effectiveness decreasing by 70–80% after one year on a yellow sandplain soil in Western Australia (Bolland, 1992b). The recommendation from this work was that phosphorus fertilizer be banded below the seed on the sandy-textured soils in south-west Australia.

Similarly, deep-placed (12–15 cm below the soil surface) manganese fertilizer significantly increased seed yield and decreased the numbers of split seed produced by *L. angustifolius* compared with topdressed manganese fertilizer (Brennan, 1996). This occurred in all seasons and across all experimental sites in Western Australia that were responsive to manganese. Deep-placed manganese fertilizer was also significantly better than manganese fertilizer placed with the seed in seasons with a dry spring when the top 10 cm of soil dried.

Response to micronutrient fertilizers can depend on the nutrient concentration in the seed, as noted in the example of seed cobalt which was discussed earlier. In that work, Robson and Snowball (1987) showed that when *L. angustifolius* seed concentrations of cobalt exceeded 128 mg kg^{-1}, there was little or no response to added cobalt fertilizer.

Foliar application

Seed filling of lupins occurs in late spring in south-western Australia. At this time, water is becoming limiting and air temperature is rising. In dry soil, nutrients become unavailable to the lupin plant and this may lead to deficiencies affecting both plant growth and seed development. Brennan and Seymour (1995) applied nitrogen, phosphorus, potassium, sulphur, manganese, magnesium and boron as foliar sprays to *L. angustifolius* grown in the Esperance district of Western Australia. Applied individually, no single nutrient increased the seed yield of lupins. Mixtures of nitrogen, phosphorus, potassium, sulphur and boron were applied at three growth stages (at flowering and two or six weeks after flowering) but no combination applied at any of the three growth stages increased the seed yield. A single application of 15 kg nitrogen ha^{-1} at flowering or two weeks after flowering and 4 kg sulphur ha^{-1} two weeks after flowering reduced the seed yield by 10% (300 kg ha^{-1}).

The lack of positive response to the above treatments is probably more related to the stage of development of the crop at the time of application rather than the mode of application; the data are consistent with findings of Q. Ma, N. Longnecker and C. Atkins (1995, Perth, unpublished data) in which nitrogen applications to soil at or after flowering had no effect on yield of *L. angustifolius*, while low amounts of nitrogen applied at sowing increased grain yield. In contrast, foliar applications of manganese fertilizer to *L. angustifolius* crops at early pod fill is a successful means of preventing or minimizing split seed (Brennan, 1995; see section on Manganese, below).

Seed coating

Seed coating is a method prone to complications, but has been used successfully to supply nutrients and increase yields in some circumstances. Little work has been done with lupins.

Shoot dry matter has been shown to increase in response to coating seed of *L. angustifolius* with molybdenum in soils with pH < 4.4, but no grain yield responses have been observed (M. Riley, 1986, Perth, unpublished data). Coating *L. angustifolius* seed with a solution of sodium molybdate has resulted in nodulation problems, especially in situations where the crop is sown for the first time, possibly due to the molybdenum salt killing the rhizobium inoculum. Also, the sodium molybdate solution (5 to 6% molybdenum) required for seed dressing is alkaline (pH 9 to 10), causing breakdown of the Rovaral™ seed dressing used to control *Pleiochaeta*. As discussed in the next section, preliminary work on manganese seed coating of *L. angustifolius* resulted in delayed emergence (Longnecker *et al.*, 1996).

Manganese – a Case Study

We have selected manganese as a case study because it presents one of the most complete stories of lupin nutrition. As will become apparent, there are still some mysteries to be solved.

Manganese deficiency in lupin leads to the split seed disorder (Walton and Francis, 1975; Perry and Gartrell, 1977; Brennan, 1993). *L. angustifolius* is more susceptible to split seed than *L. albus, L. luteus* and *L. cosentinii*. Low alkaloid cultivars of *L. angustifolius* are particularly susceptible (Walton and Francis, 1975; Walton, 1978), possibly because of their greater functional requirement for manganese than high alkaloid cultivars (Radjagukguk, 1980).

L. albus can accumulate excess levels of manganese in its seed, up to 100 times that of *L. angustifolius* even when grown on the same soil (Table 5.4). Seeds of low alkaloid lines of *L. albus* had greater manganese concentration than high alkaloid lines (Oram *et al.*, 1979). High levels of manganese in *L. albus* can result in loss of appetite, mild toxicity and reduced growth rate in pigs and sheep (Batterham, 1979; Grace, 1973).

Uptake

Low manganese availability is a common problem in soils used for production of *L. angustifolius* in Western Australia. Manganese deficiency is seldom observed in the vegetative stage of development of *L. angustifolius*, but affects seed filling (causing the split seed disorder). This contrasts with *L. luteus* (Radjagukguk, 1980) and *L. cosentinii* (Walton, 1978), which exhibit manganese deficiency symptoms in leaves but are less susceptible to split seed.

Drought stress during seed filling, low manganese uptake at that time and poor remobilization from leaves to seeds of *L. angustifolius* contribute to low manganese concentration in much of the lupin seed produced in Western Australia (White *et al.*, 1981; Crosbie *et al.*, 1994). A genotype of *L. angustifolius* which has high alkaloid content had consistently greater (up to 50% greater) manganese in its shoots compared with a closely related low alkaloid gentoype when grown in soil (Radjagukguk, 1980). However, there were no differences in their rates of uptake from a nutrient solution containing labelled manganese. In the same experiment, *L. luteus* had greater rates of manganese uptake.

As mentioned previously, *L. albus*, which produces cluster roots, tends to have higher concentrations of manganese than *L. angustifolius*. In one study with soil-grown plants, *L. albus* had twice the manganese concentration in its shoot compared with *L. angustifolius*, but eight times the manganese concentration in its seeds (Gladstones, 1962). In contrast, the two species grown in nutrient solution with the same level of added manganese showed a smaller difference between the manganese concentration in their seed (e.g. 54 mg Mn kg^{-1} for *L. albus* compared with 30 mg Mn kg^{-1} for *L. angustifolius*; Hocking *et al.*, 1977). Thus, a large part of the difference in manganese accumulation in seeds must be due to differences in manganese uptake from soil as well as differences in transport and remobilization of manganese to the seed in phloem.

Transport and utilization

Distribution patterns of manganese in mature plants indicate that manganese is transported freely by xylem (Hocking *et al.*, 1977). There is genotypic variability in manganese mobility in the phloem (Hocking *et al.*, 1977; Radjagukguk, 1980), being lower in those genotypes susceptible to split seed (e.g. *L. angustifolius).* There may also be variability in the amount of manganese retained in the stem (Radjagukguk, 1980). This may be an important factor in determining lower susceptibility to split seed.

Hannam *et al.* (1985a) showed that manganese stored in the stem is available to filling seeds while manganese in leaves is not retranslocated. If there is not an adequate store of manganese in the stems or adequate manganese uptake during grain filling, *L. angustifolius* is susceptible to split seed (Hannam *et al.*, 1985b).

The lack of manganese retranslocation from leaves of *L. angustifolius* is difficult to explain since manganese in leaves can be leached out with water (Hannam *et al.*, 1985a). The problem appears to be poor phloem loading of manganese in leaves rather than manganese being trapped in an insoluble form. This invites further investigation.

In a comparison of sweet and bitter *L. angustifolius* and *L. luteus,* Radjagukguk (1980) showed that those genotypes susceptible to split seed had lower proportions of the manganese in the plant transported to the shoot tip (which would be mainly fed by the phloem). Hocking *et al.* (1977) observed lower manganese concentrations in phloem sap of *L. angustifolius* than *L. albus* when external supply was low to adequate. At high levels, the differences were smaller. There is a close relationship between the concentration of manganese in phloem sap collected from the pod and incidence of split seed (Hocking *et al.*, 1977; Radjagukguk, 1980; Hannam *et al.*, 1985a).

Seed nutrition

Little is known about the physiological cause of abnormal seed development in manganese-deficient *L. angustifolius.* The seed coats of immature seeds inside the pods split along the edge furthest from their attachment to the pod and the cotyledons continue to grow and protrude from the seed coat (Brennan, 1993). Many seeds which do not split cease developing, dry off and shrivel if immature. However, other seeds develop normally. Normal seed has greater manganese concentration than split or discoloured seed (Hocking *et al.*, 1977; Radjagukguk, 1980). If normal seeds are sieved into different sizes, there is no difference in their manganese concentrations (J. Crosbie, N. Longnecker and A.D. Robson, 1992, unpublished data).

Emergence can be impaired from seed of *L. angustifolius* which appears normal but has low manganese concentrations (Longnecker *et al.*, 1996). Whether or not manganese fertilizer was supplied, Longnecker *et al.* (1996) observed emergence of 35 to 52% in pots of sand sown with seed together with 7 mg Mn kg^{-1} seed compared with 88 to 100% emergence in pots sown with seed containing 11 mg Mn kg^{-1} or higher. When adequate manganese was supplied, seedling growth was not affected by the manganese concentration of the seed (Crosbie *et al.*, 1994). However, when supply was deficient, seedling growth was positively correlated with the seed manganese concentration (Crosbie *et al.*, 1994).

Diagnosis and amelioration

When manganese concentration in leaflets of the youngest fully emerged leaves was less than 30 mg g^{-1}, vegetative growth of *L. angustifolius* was inhibited (Hannam *et al.*, 1985a). Manganese concentration in the stems at anthesis was closely correlated with the incidence of split seed in *L. angustifolius* (Hannam *et al.*, 1985b). If the manganese concentration in the main stem of *L. angustifolius* at anthesis is greater than 20 mg kg^{-1}, there is usually no problem with split seed (Brennan, 1993). However if manganese concentration is less than 20 mg kg^{-1}, a foliar spray of $MnSO_4$ is recommended, to be applied when pods on the main stem are about 2–3 cm long (Brennan, 1995).

Seed manganese coatings increased shoot manganese concentrations in *L. angustifolius* from 42 mg kg^{-1} in the nil manganese treatment to 104 ($MnSO_4$) and 225 (MnO_2), and increased shoot and root growth in the MnO_2 treatment (Longnecker *et al.*, 1996). Coating with $MnSO_4$ resulted in delayed emergence, perhaps because of high osmotic potential, but did not decrease the final percentage emergence. Because of its lower solubility, MnO_2 may have better utility as a seed coating with lower toxicity to the germinating seed.

Concluding Remarks

A sound understanding of factors affecting nutrient availability and use has been critical for the development of the lupin industry. Research methodologies which are now available offer possibilities for solving intractable problems. For example, transformations of *L. angustifolius* with sunflower genes to increase seed sulphur amino acids (Molvig *et al.*, 1996) may solve a long-standing problem of imbalanced legume seed protein for animal nutrition.

Lupin genotypes with unusual and varied nutritional physiology have been identified which provide useful systems for studying fundamental physiology. This will allow identification of mechanisms controlling plant growth, crop yield and seed quality. For example, the difference between *L. albus* and *L. angustifolius* in seed accumulation of manganese provides a useful system for the study of seed development and nutrient transport into seeds.

Rapid changes in plant growth in response to nutritional status allow studies of the signals that control plant growth regulation. For example, a useful system may arise from the observations that lateral roots of *L. angustifolius* proliferate in zones of high phosphorus while cluster roots of *L. albus* proliferate in deficient phosphorus conditions. Studies using systems such as these will improve our understanding of how plants partition their resources and may give clues about how to increase lupin yield potential and reliability of yield.

Fundamental knowledge of how plants work will provide the basis for future improvements in efficient and sustainable lupin production.

Acknowledgements

The authors thank Dr Ma Qifu and Mr Nick Edwards for help with figures and Professor Craig Atkins for constructive review of the manuscript.

References

Atkins, C.A. (1984) Efficiencies and inefficiencies in the legume/Rhizobium symbiosis – a review. *Plant and Soil* 82, 273–284.

Atkins, C.A., Pate, J.S. and Layzell, D.B. (1979) Assimilation and transport of nitrogen in non-nodulated (NO_3^--grown) *Lupinus albus* L. *Plant Physiology* 64, 1078–1082.

Atwell, B.J. (1991) Factors which affect the growth of grain legumes on a solonized brown soil. II. Genotypic responses to soil chemical factors. *Australian Journal of Agricultural Research* 42, 107–119.

Batterham, E.S. (1979) *L. albus* cv Ultra and *L. angustifolius* cv Unicrop as protein concentrates for growing pigs. *Australian Journal of Agricultural Research* 30, 369–375.

Baylis, J.M. and Hamblin, J. (1986) Lupins in the farming system. A survey of production. In: *Proceedings of the Fourth International Lupin Conference, Geraldton.* Agriculture Western Australia, Perth, p. 161.

Bishop, A., Delane, R. and Hamblin, J. (1986) Root characteristics in branching and reduced branching lupins. In: *Proceedings of the Fourth International Lupin Conference, Geraldton*. Agriculture Western Australia, Perth, p. 313.

Bolland, M.D.A. (1992a) The current and residual value of superphosphate for lupins grown in rotation with oats and wheat on a deep sandy soil. *Fertilizer Research* 31, 319–329.

Bolland, M.D.A. (1992b) Residual value of superphosphate for wheat and lupin grain production on a uniform yellow sandplain soil. *Fertilizer Research* 31, 331–340.

Bolland, M.D.A. (1992c) The phosphorus requirement of different crop species compared with wheat on lateritic soils. *Fertilizer Research* 32, 27–36.

Bolland, M.D.A. (1997) Comparative phosphorus requirement of *Lupinus luteus, L. cosentinii, L. angustifolius* and *L. albus*. *Journal of Plant Nutrition* 20, 1239–1253.

Bolland, M.D.A. and Barrow, N.J. (1991) The effect of level of application on the residual value of superphosphate on a sandy soil in south-western Australia. *Fertilizer Research* 29, 163–172.

Bolland, M.D.A. and Jarvis, R.J. (1996) Effectiveness of different methods of applying superphosphate for lupins grown on sandplain soils. *Australian Journal of Experimental Agriculture* 36, 707–715.

Bolland, M.D.A., Paynter, B.H. and Baker, M.J. (1989) Increasing phosphorus concentration in lupin seed increased grain yields on phosphorus deficient soil. *Australian Journal of Experimental Agriculture* 29, 797–801.

Brennan, R. (1993) *Symptoms of Mn Deficiency (Split Seed) in Narrow Leafed Lupins*. West Australian Department of Agriculture Farmnote 70/93.

Brennan, R. (1995) *Treatment of Mn Deficiency (Split Seed) in Narrow Leafed Lupins*. West Australian Department of Agriculture Farmnote 71/93.

Brennan, R. (1996) Effectiveness of different sources of manganese foliar sprays in alleviating manganese deficiency of *Lupinus angustifolius* L. grown on manganese deficient soils in Western Australia. *Journal of Plant Nutrition* 19, 293–304.

Brennan, R. and Seymour, M. (1995) Nutrient sprays applied to the foliage of narrow leafed lupins (*Lupinus angustifolius*) during flowering and podding do not increase seed yields. *Australian Journal of Experimental Agriculture* 35, 381–385.

Cartwright, B. and Hallsworth, E.G. (1970) Effects of copper deficiency on root nodules of subterranean clover. *Plant and Soil* 33, 685–698.

Clements, J.C. and Cowling, W. (1990) Summary of wild and semi-domesticated lupin accessions introduced to Australia from 1974 to 1989. *Australian Plant Introduction Review* 21, 1–14.

Clements, J.C., White, P.F. and Buirchell, B. (1993) The root morphology of *Lupinus angustifolius* in relation to other *Lupinus* species. *Australian Journal of Agricultural Research* 44, 1367–1375.

Cox, W.J. (1979) The relationship between grain yield and plant potassium concentration in *Lupinus angustifolius* cv. Unicrop. *Communications in Soil Science and Plant Analysis* 10, 747–761.

Crosbie, J., Longnecker, N.E. and Robson, A.D. (1994) Seed manganese affects the early growth of lupins in manganese-deficient conditions. *Australian Journal of Agricultural Research* 45, 1469–1482.

Dilworth, M.J. and Bisseling, T. (1984) Cobalt and nitrogen fixation in *L. angustifolius* L. III. DNA and methionine in bacteroids. *New Phytologist* 98, 311–316.

Dilworth, M.J., Robson, A.D. and Chatel, D.L. (1979) Cobalt and nitrogen fixation in *L. angustifolius* L. II. Nodule formation and functions. *New Phytologist* 83, 63–79.

Dinkelaker, B., Römheld, V. and Marschner, H. (1989) Citric acid excretion and precipitation of calcium citrate in the rhizosphere of white lupin (*Lupinus albus* L.). *Plant Cell and Environment* 12, 285–292.

Gardner, W.K., Parbery, D.G. and Barber, D.A. (1981) Proteoid root morphology and function in *Lupinus albus*. *Plant and Soil* 60, 143–147.

Gardner, W.K., Parbery, D.G. and Barber, D.A. (1982a) The acquisition of phosphorus by *L. albus* L. I. Some characteristics of the soil/root interface. *Plant and Soil* 68, 19–32.

Gardner, W.K., Parbery, D.G. and Barber, D.A. (1982b) The acquisition of phosphorus by *L. albus* L. II. The effect of varying phosphorus supply and soil type on some characteristics of the soil/root interface. *Plant and Soil* 68, 33–41.

Gardner, W.K., Barber, D.A. and Parbery, D.G. (1983) The acquisition of phosphorus by *L. albus* L. III. The probable mechanism by which phosphorus movement in the soil/root interface is enhanced. *Plant and Soil* 70, 107–124.

Gilkes, R.J. (1994) The availability of zinc in fertilisers for Western Australian soils. *CSBP Productivity Focus* 11, 1–3.

Gladstones, J.S. (1962) The mineral composition of lupins. 2. A comparison of the copper, manganese, molybdenum and cobalt contents of lupins and other species at one site. *Australian Journal of Experimental Agriculture and Animal Husbandry* 2, 213–220.

Gladstones, J.S. (1970) Lupins as crop plants. *Field Crop Abstracts* 23, 123–148.

Gladstones, J.S. and Drover, D.P. (1962) The mineral composition of lupins. 1. A survey of the copper, molybdenum, and manganese contents of lupins in the south-west of Western Australia. *Australian Journal of Experimental Agriculture and Animal Husbandry* 2, 46–53.

Gladstones, J.S., Leach, B.J., Drover, D.P. and Asher, C.J. (1964) Field studies on the potassium and phosphorus nutrition of lupins. *Australian Journal of Experimental Agriculture and Animal Husbandry* 4, 158–164.

Gladstones, J.S., Loneragan, J.F. and Goodchild, N.A. (1977) Field responses to cobalt and molybdenum by different legume species, with inferences on the role of cobalt in legume growth. *Australian Journal of Agricultural Research* 28, 619–628.

Grace, N.D. (1973) Effect of high dietary Mn levels on the growth rate and the level of mineral elements in the plasma and soft tissues of sheep. *New Zealand Journal of Agricultural Research* 16, 177–180.

Gumley, J. (1989) Can a comparison of root characteristics between perennial and annual lupins be related to their relative responses to applied soil phosphorus? Honours thesis. The University of Western Australia, Nedlands, Western Australia.

Hamblin, A.P. and Hamblin, J. (1985) Root characteristics of some temperate legume species and varieties on deep, free draining entisols. *Australian Journal of Agricultural Research* 36, 63–72.

Hannam, R.J., Graham, R.D. and Riggs, J.L. (1985a) Redistribution of manganese in maturing *Lupinus angustifolius* cv Illyarrie in relation to levels of previous accumulation. *Annals of Botany* 56, 821–834.

Hannam, R.J., Graham, R.D. and Riggs, J.L. (1985b) Diagnosis and prognosis of manganese deficiency in *Lupinus angustifolius* L. *Australian Journal of Agricultural Research* 36, 765–777.

Hinsinger, P. and Gilkes, R.J. (1995) Root-induced dissolution of phosphate rock in the rhizosphere of lupins grown in alkaline soil. *Australian Journal of Soil Research* 33, 477–489.

Hirata, H., Hisaka, H. and Hirata, A. (1982) Effects of phosphorus and potassium deficiency treatment on root secretions of wheat and rice seedlings. *Soil Science and Plant Nutrition* 28, 543–552.

Hocking, P.J. and Pate, J.S. (1978) Accumulation and distribution of mineral elements in annual lupins *Lupinus albus* and *Lupinus angustifolius* L. *Australian Journal of Agricultural Research* 29, 267–280.

Hocking, P.J., Pate, J.S., Wee, S.C. and McComb, A.J. (1977) Manganese nutrition of *Lupinus* spp. especially in relation to developing seeds. *Annals of Botany* 41, 677–688.

Hung, T.V., Handson, P.D., Amenta, V.C., Kyle, W.S.A. and Yu, R. S.-T. (1987) Content and distribution of manganese in lupin seed grown in Victoria and in lupin flour, spray-dried powder and protein isolate prepared from the seeds. *Journal of the Science of Food and Agriculture* 41, 131–139.

Israel, D.W. (1987) Investigation of the role of phosphorus in symbiotic dinitrogen fixation. *Plant Physiology* 84, 835–840.

Jarvis, R. and Bolland, M.D.A. (1990) Placing superphosphate at different depths in the soil changes its effectiveness for wheat and lupin production. *Fertiliser Research* 22, 97–107.

Jarvis, R. and Bolland, M.D.A. (1991) Lupin grain yields and fertilizer effectiveness are increased by banding superphosphate below the seed. *Australian Journal of Experimental Agriculture* 31, 357–366.

Jeschke, W.D., Atkins, C.A. and Pate, J.S. (1985) Ion circulation via phloem and xylem between root and shoot of nodulated white lupin. *Journal of Plant Physiology* 117, 319–330.

Jeschke, W.D., Pate, J.S. and Atkins, C.A. (1987) Partitioning of K^+, Na^+, Mg^{++} and Ca^{++} through xylem and phloem to component organs of nodulated white lupin under mild salinity. *Journal of Plant Physiology* 128, 77–93.

Jessop, R.S., Roth, G. and Sale, P. (1990) Effects of increased levels of soil $CaCO_3$ on lupin (*Lupinus angustifolius*) growth and nutrition. *Australian Journal of Soil Research* 28, 955–962.

Johns, T. (1994) Defense of nitrogen-rich seeds constrains selection for reduced toxicity during the domestication of the grain legumes In: Sprent, J.I. and McKey, D. (eds) *Advances in Legume Systematics 5: The Nitrogen Factor*. Royal Botanic Gardens, Kew, pp. 151–167.

Johnson, J.F., Allan, D.L. and Vance, C.P. (1994) Phosphorus stress-induced proteoid roots show altered metabolism in *Lupinus albus*. *Plant Physiology* 104, 657–665.

Johnson, J.F., Allan, D.L., Vance, C.P. and Weiblen, G. (1996a) Root carbon dioxide fixation by phosphorus-deficienct *Lupinus albus*. Contribution to organic acid exudation by proteoid roots. *Plant Physiology* 112, 19–30.

Johnson, J.F., Vance, C.P. and Allan, D.L. (1996b) Phosphorus deficiency in *Lupinus albus*. Altered lateral root development and enhanced expression of phosphoeneolpyruvate carboxylase. *Plant Physiology* 112, 31–41.

Lipton, D.S., Blanchar, R.W. and Blevins, D.G. (1987) Citrate, malate and succinate concentration in exudates from P-sufficient and P-stressed *Medicago sativa* L. seedlings. *Plant Physiology* 85, 315–317.

Loneragan, J.F., Snowball, K. and Robson, A.D. (1980) Copper supply in relation to content and redistribution of copper among organs of the wheat plant. *Annals of Botany* 45, 621–632.

Longnecker, N.E. (1988) Iron nutrition of plants. In: *ISI Atlas of Science: Animal and Plant Sciences*, Vol. 1. ISI Press, Philadelphia, PA, pp. 143–150.

Longnecker, N.E. and Robson, A.D. (1993) Distribution and transport of zinc in plants. In: Robson, A.D. (ed.) *Zinc in Soils and Plants*. Kluwer Academic Publishers, Dordrecht, pp. 79–91.

Longnecker, N.E., Crosbie, J., Davies, F. and Robson, A. (1996) Low seed manganese concentration and decreased emergence of *Lupinus angustifolius*. *Crop Science* 36, 355–361.

Lunt, R.J., Diggle, A.J. and Bowden, J.W. (1994) Response by lupin roots to placed phosphorus. *Proceedings of the First Australian Lupin Technical Symposium*. Agriculture Western Australia, Perth, p. 283.

Ma, Q., Longnecker, N. and Dracup, M. (1997) Nitrogen deficiency slows leaf development and delays flowering in narrow-leafed lupin. *Annals of Botany* 79, 403–409.

Ma, Q., Longnecker, N., Emery, N. and Atkins, C. (1998) Growth and yield in *Lupinus angustifolius* are depressed by early transient nitrogen deficiency. *Australian Journal of Agricultural Research* (in press).

Marschner, H. (1986) *The Mineral Nutrition of Higher Plants*. Academic Press, London, p. 461.

Marschner, H., Römheld, V. and Cakmak, I. (1987) Root-induced changes of nutrient availability in the rhizosphere. *Journal of Plant Nutrition* 10, 1175–1184.

Molvig, L., Tabe, L.M., Eggum, B.O., Spenser, D. and Higgins, T.J.V. (1996) Increased nutritional quality of seeds of transgenic lupins (*L. angustifolius*) expressing a sunflower seed albumin seed gene. In: *Abstracts of the 1996 Meeting of the Australian Society for Biochemistry and Molecular Biology Incorporated and Australian Society of Plant Physiologists Incorporated*, Canberra, Australia.

Nowacki, E., Jurzysta, M., Gorski, P., Nowacka, D. and Waller, G.R. (1976) Effect of nitrogen nutrition on alkaloid metabolism in plants. *Biochemie und Physiologie von Pflanzen* 169, 231–240.

O'Hara, G.W., Dilworth, M.J., Boonkerd, N. and Parkpian, P. (1988) Iron-deficiency specifically limits nodule development in peanut inoculated with *Bradyrhizobium* sp. *New Phytologist* 108, 51–57.

Oram, R.N., David, D.J., Green, A.G. and Read, B.J. (1979) Selection in *Lupinus albus* L. for lower seed manganese concentration. *Australian Journal of Agricultural Research* 30, 467–476.

Ozanne, P.G., Howes, K.M.W. and Petch, A. (1976) The comparative phosphate requirements of four annual pastures and two crops. *Australian Journal of Agricultural Research* 27, 479–488.

Ozawa, K., Osaki, M., Matsui, H., Honma, M. and Tadano, T. (1995) Purification and properties of acid phosphatase secreted from lupin roots under phosphorus-deficiency conditions. *Soil Science and Plant Nutrition* 41, 461–469.

Pairunan, A.K. (1980) Factors affecting the availability of poorly soluble phosphorus sources to plants. PhD thesis, The University of Western Australia. Nedlands, Western Australia.

Pakendorf, K.W., Coetzer, F.J. and van Schalkwyk, D.J. (1970) Genotypic and species variation in mineral element concentration in the genus *Lupinus*. I. Seeds and plant data. *Zeitshrift für Pflanzenzuechtung* 63, 125–136.

Papavizas, G.C. and Davey, C.B. (1961) Extent and nature of the rhizosphere of *Lupinus*. *Plant and Soil* 14, 215–236.

Parsons, R. and Baker, A. (1996) Cycling of amino compounds in symbiotic lupin. *Journal of Experimental Botany* 47, 421–429.

Pate, J.S., Atkins, C.A., Hamel, K., McNeil, D.L. and Layzell, D.B. (1979) Transport of organic solutes in phloem and xylem of a nodulated legume. *Plant Physiology* 63, 1082–1088.

Perry, M.W. and Gartrell, J.W. (1977) Lupin 'split seed': a disorder of seed production in sweet, narrow-leafed lupins. *Journal of Agriculture, Western Australia* 17, 20–25.

Petterson, D.S. and Mackintosh, J.B. (1994) *The Chemical Composition and Nutritive Value of Australian Grain Legumes.* Grains Research and Development Corporation, Canberra, Australia.

Radjagukguk, B. (1980) Manganese nutrition in lupins: plant response and the relationship of manganese supply to its distribution. PhD thesis, The University of Western Australia, Nedlands, Western Australia.

Rahman, M.S. and Gladstones, J.S. (1974) Differences among *Lupinus* species in field response to superphosphate. *Australian Journal of Experimental Agriculture and Animal Husbandry* 14, 214–223.

Reuter, D.J., Robson, A.D., Loneragan, J.F. and Tranthim-Fryer, D.J. (1981) Copper nutrition of subterranean clover (*Trifolium subterraneum* L. cv. Seaton Park). II. Effects of copper supply on distribution of copper and the diagnosis of copper deficiency by plant analysis. *Australian Journal of Agricultural Research* 32, 267–282.

Reuter, D.J., Loneragan, J.F., Robson, A.D. and Plaskett, D. (1982) Behavior of zinc in subterranean clover (*Trifolium subterraneum* L. cv. Seaton Park). I. Effects of zinc supply on distribution of zinc and dry weight among plant parts. *Australian Journal of Agricultural Research* 33, 989–999.

Riley, I.T. and Dilworth, M.J. (1985a) Cobalt status and its effects on soil populations of *Rhizobium lupini*, rhizosphere colonization and nodule initiation. *Soil Biology and Biochemistry* 17, 81–85.

Riley, I.T. and Dilworth, M.J. (1985b) Cobalt requirement for nodule development and function in *Lupinus angustifolius* L. *New Phytologist* 100, 361–365.

Riley, M.M., Gartrell, J.W., Brennan, R.F., Hamblin, J. and Coates, P. (1992) Zinc deficiency in wheat and lupins in Western Australia is affected by the source of phosphate fertilizers. *Australian Journal of Experimental Agriculture* 32, 455–463.

Robson, A.D. (1978) Mineral nutrition limiting nitrogen fixation in legumes. In: Andrew, C.S. and Kamprath, E.J. (eds) *The Mineral Nutrition of Legumes on Tropical and Subtropical Soils.* CSIRO, Melbourne, pp. 277–293.

Robson, A.D. (1983) Mineral nutrition. In: Broughton, W.J. (ed.) *Nitrogen Fixation*, Vol. 3. Clarendon Press, Oxford, pp. 36–55.

Robson, A.D. (1986) The mineral nutrition of lupins. In: *Proceedings of the First Australian Lupin Technical Symposium.* Agriculture, Western Australia, Perth, Western Australia, pp. 84–95.

Robson, A.D. (1988) Nutrient requirement of pulses. In: Summerfield, R.J. (ed.) *World Crops: Cool Season Legumes.* Kluwer Academic Publishers, Dordrecht, pp. 869–881.

Robson, A.D. and Bottomley, P.J. (1991) Limitations in the use of legumes in agriculture and forestry. In: Dilworth, M.J. and Glenn, A.R. (eds) *The Biology and Biochemistry of Nitrogen Fixation.* Elsevier, Amsterdam, pp. 320–349.

Robson, A.D. and Mead, G.R. (1980) Seed cobalt in *Lupinus angustifolius Australian Journal of Agricultural Research* 31, 109–116.

Robson, A.D. and Snowball, K. (1987) Response of narrow-leafed lupins to cobalt application in relation to cobalt concentration in seed. *Australian Journal of Experimental Agriculture* 27, 657–660.

Robson, A.D., Dilworth, M.J. and Chatel, D.L. (1979) Cobalt and nitrogen fixation in *Lupinus angustifolius* L. I. Growth, nitrogen concentrations and cobalt distribution. *New Phytologist* 83, 53–62.

Robson, A.D., Longnecker, N.E. and Osborne, L.D. (1992) Effects of heterogeneous nutrient supply on root growth and nutrient uptake in relation to nutrient supply on duplex soils. *Australian Journal of Experimental Agriculture* 32, 879–886.

Robson, A.D., Osborne, L.D., Snowball, K. and Simmons, W.J. (1995) Assessing sulfur status in lupins and wheat. *Australian Journal of Experimental Agriculture* 35, 79–86.

Seliga, H. (1993) The role of copper in nitrogen fixation in *Lupinus luteus* L. In: Barrow N.J. (ed.) *Plant Nutrition – from Genetic Engineering to Field Practice* Kluwer Academic Publishers, Dordrecht, pp. 459–462.

Shenker, M., Chen, Y. and Dovrat, A. (1989) Uptake mechanisms of iron: the lupin case. In: *Lupin Production and Bio-processing for Feed, Food and Other By-products. Proceedings of the Joint CEC-NCRD Workshop.*

Snowball, K. and Robson, A.D. (1986) *Symptoms of Nutrient Deficiencies: Lupins.* School of Agriculture, University of Western Australia, Perth, Western Australia.

Snowball, K., Robson, A.D. and Loneragan, J.F. (1980) The effect of copper on nitrogen fixation in subterranean clover (*Trifolium subterraneum*). *New Phytologist* 85, 63–72.

Spencer, K., Jones, M.B. and Freney, J.R. (1977) Diagnostic indices for *S* status of subterranean clover. *Australian Journal of Agricultural Research* 28, 401–412.

Tang, C. (1995) Iron in symbiotic nitrogen fixation in legumes. In: Hemantaranjan, A. (ed.) *Advancements in Iron Nutrition Research.* Scientific Publishers, Jodhpur, India.

Tang, C. and Robson, A.D. (1993) *Lupinus* species differ in their requirements for iron. *Plant and Soil* 157, 11–18.

Tang, C., Robson, A.D. and Dilworth, M.J. (1990a) A split-root experiment shows that iron is required for nodule initiation in *Lupinus angustifolius* L. *New Phytologist* 115, 61–67.

Tang, C., Robson, A.D. and Dilworth, M.J. (1990b) The role of iron in nodulation and nitrogen fixation in *Lupinus angustifolius* L. *New Phytologist* 114, 173–182.

Tang, C., Robson, A.D., Longnecker, N.E. and Buirchell, B.J. (1995) The growth of *Lupinus* species on alkaline soils. *Australian Journal of Agricultural Research* 46, 255–268.

Tang, C., Adams, H., Longnecker, N.E. and Robson, A.D. (1996) A method to identify lupin species tolerant of alkaline soils. *Australian Journal of Experimental Agriculture* 36, 595–601.

Tarafdar, J.C. and Jungk, A. (1987) Phosphatase activity in the rhizosphere and its relation to the depletion of soil organic phosphorus. *Biology and Fertility of Soils* 3, 199–204.

Thomson, B.D., Bell, R.W. and Bolland, M.D.A. (1992) Low seed phosphorus concentration depresses early growth and nodulation of narrow-leafed lupin (*Lupinus angustifolius* cv. Gungurru). *Journal of Plant Nutrition* 15(8), 1193–1214.

Toms, W.J. and Fitzpatrick, E.N. (1961) Potassium deficiency in medium rainfall areas. *Journal of the Department of Agriculture for Western Australia* 2, 763–773.

Trinick, M.J. (1977) Vesicular-arbuscular infection and soil phosphorus utilization in *Lupinus* spp. *New Phytologist* 78, 297–304.

Waller, G.R. and Nowacki, E.K. (1978) Environmental influences on alkaloid production In: *Alkaloid Biology and Metabolism in Plants.* Plenum Press, New York, pp. 85–119.

Walton, G.H. (1978) The effect of manganese on seed yield and the split seed disorder of sweet and bitter phenotypes of *L. angustifolius* and *L. cosentinii*. *Australian Journal of Agricultural Research* 29, 1177–1189.

Walton, G.H. and Francis, C.M. (1975) Genetic influences on the split seed disorder in *Lupinus angustifolius* L. *Australian Journal of Agricultural Research* 26, 641–646.

Waterman, P.G. (1994) Costs and benefits of secondary metabolites to the Leguminosae In: Sprent, J.I. and McKey, D. (eds.) *Advances in Legume Systematics 5: The Nitrogen Factor.* Royal Botanic Gardens, Kew, pp. 129–149.

White, C.L., Robson, A.D. and Fisher, H.M. (1981) Variation in nitrogen, sulfur, selenium, cobalt, manganese, copper and zinc contents of grain from wheat and two lupin species grown in a range of Mediterranean environments. *Australian Journal of Agricultural Research* 32, 47–49.

White, P.F. (1990) Soil and plant factors relating to the poor growth of *Lupinus* species on fine-textured alkaline soils – a review. *Australian Journal of Agricultural Research* 41, 871–890.

White, P.F. and Robson, A.D. (1989a) Lupin species and peas vary widely in their sensitivity to Fe deficiency. *Australian Journal of Agricultural Research* 40, 539–547.

White, P.F. and Robson, A.D. (1989b) Rhizosphere acidification and Fe^{3+} reduction in lupins and peas: iron deficiency in lupins is not due to a poor ability to reduce Fe^{3+}. *Plant and Soil* 119, 163–175.

Wink, M. (1987) Chemical ecology of quinolizidine alkaloids. In: Waller, G. (ed.) *Allelochemicals.* American Chemical Society, Washington, DC, pp. 524–533.

6 Nodulation, Nitrogen Fixation and Nitrogen Balance

John G. Howieson[1,6*], Ian R.P. Fillery[1,3], Andrzej B. Legocki[4], Michal M. Sikorski[4], Tomasz Stepkowski[4], Frank R. Minchin[5] and Michael J. Dilworth[1,2]

*[1]Centre for Legumes in Mediterranean Agriculture, University of Western Australia, Nedlands, Western Australia 6907; [2]Murdoch University, Murdoch, Western Australia 6150; [3]CSIRO Private Bag, Wembley, Western Australia 6019; [4]Institute of Bioorganic Chemistry, Polish Academy of Sciences, Noskowskiego 12/14, 61-704 Poznań, Poland; [5]Institute of Grassland and Environmental Research, Plas Gogerddan, Aberystwyth, Dyfed SY23 3EB, UK; [6]Agriculture Western Australia, 3 Baron-Hay Court, South Perth, Western Australia 6151; *present address: Centre for Rhizobium Studies, Murdoch University, South Street, Murdoch, Western Australia 6150, Australia.*

Introduction

The *Lupinus* genus is nodulated by the soil microorganism *Bradyrhizobium* sp. (*Lupinus*) (Jordan, 1982). Although we have learnt an enormous amount about root-nodule bacteria in the last two decades, this particular microorganism has not been as extensively studied as its faster growing root-nodulating relatives, either in the laboratory or in the field. The reason for this, perhaps, lies in its slow growth rate (7–10 days to form 1 mm colonies on yeast-mannitol agar), which is a deterrent to rapid experimentation. Also, its symbiosis with the lupin is exceedingly robust under normal conditions for lupin cultivation and, as such, has not warranted intervention. Importantly though, world production of lupins has been from only three major species – *L. angustifolius, L. luteus* and *L. albus* – whose microsymbiotic requirements have been adequately addressed. There is a current effort to seek adaptation to specific niches from minor species such as *L. pilosis, L. atlanticus* and *L. princei*. Of these, *L. princei* has been shown to have rather specific requirements for its microsymbiont (Howieson *et al*., 1994), which has implications for future inoculant selection and commercialization. To obviate the potential for future suboptimal nodulation in major species, an inoculant strain selected for *L. princei* would ideally be capable of effective nodulation (i.e. high nitrogen fixation) with important commercial lupin species and cultivars likely to be grown on the same land or nearby. The selection of commercial inoculant

 Lupins as Crop Plants: Biology, Production and Utilization (eds J.S. Gladstones, C.A. Atkins and J. Hamblin)

strains that combine a broad host range with adaptation to specific edaphic conditions is not a trivial undertaking (Howieson, 1995).

Entry of the lupin microsymbiont into the plant is unusual. Cracks or openings in the root epidermis allow the bacteria to move through intercellular spaces in the root. This is followed by penetration of cells in the root cortex, possibly facilitated by degradation of the host cell wall (Tang *et al.*, 1993a). This mode of entry, then, differs substantially from the entry of root-nodule bacteria at the junction of lateral and tap roots in *Arachis hypogaea*. There is also no evidence of a substantial infection thread carrying the lupin microsymbiont to the cortical cells, as is found with many other legumes (Sprent, 1989). Lupin nodules are indeterminate and possess active meristems, but their morphology is unlike that of indeterminate nodules in other legumes, being described by Corby (1981) as 'lupinoid'. As this chapter unfolds it will become clear that, from the enhanced survival of the lupin microsymbiont in acidic soils through the nodulation process to high nitrogen fixation by the lupin, there are many facets of the symbiosis which are peculiar to lupins and *Bradyrhizobium* sp. (*Lupinus)*. This chapter covers ecological, physiological and molecular aspects of the symbiosis, commencing with the ecology of the lupin microsymbiont through to the fixation of nitrogen and its fate in the environment.

Ecology of *Bradyrhizobium* sp. (*Lupinus*)

Bradyrhizobia are rarely encountered as microsymbionts of legumes of Mediterranean origin. They are limited to nodule occupation in only a few herbaceous genera such as *Lotus, Ornithopus* and *Argyrolobium* as well as a few shrub genera such as *Chaemycytisus, Lupinus* and *Acacia* (Allen and Allen, 1981). Parenthetically, this generalization is reversed when considering tropical agriculture (Norris, 1965). Amongst the *Bradyrhizobium* genus, the lupin-nodulating form appears to be a taxonomic outlier on the basis of fatty acid composition (Graham *et al.*, 1995).

The microsymbiont of the *Lupinus* genus is thus a unique organism when compared with the root-nodule bacteria associated with most other legumes in temperate or mediterranean agriculture. This distinctiveness arises initially because it is one of a few representatives of the *Bradyrhizobium* genus with an agricultural host in mediterranean agriculture. Secondly, many lupin-nodulating *Bradyrhizobium* strains also have a dual host in agriculture: they may effectively nodulate the herbaceous genus *Ornithopus*. In fact, the lupin inoculant strain in Australia, WU425, was recommended as the commercial inoculant for *Ornithopus* for nearly 30 years, until 1996 when strain WSM471 became available.

Associated with this distinctiveness, *Bradyhizobium* sp. (*Lupinus*) can be considered highly acid- and aluminium-tolerant and perhaps even an acidophile. This assertion arises from studies conducted *in vitro* and *in vivo*, as well as from anecdotal evidence dating from the ancient use of lupins, where practitioners were warned of poor performance of the lupin on soils containing limestone (Gladstones, 1970). The preference of the lupin microorganism for low pH conditions has been frequently recorded in studies of its growth at low pH (Graham and Parker, 1964; Norris, 1965; Rerkasem, 1977; Khan *et al.*, 1984), in its sensitivity to factors associated with high pH such as high Ca concentration (Norris, 1956; Parker and Oakley, 1965; Rerkasem, 1977), as well as in modern comparative nodulation studies (e.g. Jessop and Mahoney, 1982; Subramaniam *et al.*, 1982; Tang and Robson, 1995). Further, the ability of the organism to persist in high numbers in acidic field soils in Australia has been well documented (Chatel and Parker, 1973; Slattery and Coventry, 1989; Ballard, 1996). In fact survival of the organism as a saprophyte in acid soils is so effective that 'second crop' inoculation is not recommended. On the highly acidic and aluminium-saturated soils of the low rainfall regions of Western Australia and New South Wales, where pH values between 3.5 and 4.0 are common, the symbiosis between *Bradyrhizobium* sp. (*Lupinus*) and either lupins or serradella is never visibly challenged by acidity-related nutritional stress. This experience is repeated in Oregon, where representatives of *Bradyrhizobium* sp. (*Lupinus*) were also found to persist well in soils despite few agricultural legume hosts being commonly grown in that region (Bottomley *et al.*, 1994).

Jordan (1982) separated fast and slow growing rhizobia into the genera *Rhizobium* and *Bradyrhizobium*, respectively, and it is contended that the ecologies of the two are substantially different (Bushby, 1982; Bottomley, 1992). Apart from differential tolerance to acidity, different growth rates and C-utilization pattern *in vitro*, the contrast between the lupin bradyrhizobia and the genus *Rhizobium* is sustained when examination is made of the general effect of calcium on the dynamics of the attachment of root-nodule bacteria to roots and root hairs prior to nodule initiation. Whereas the proportion of rhizobial cells that attached to roots generally increased with increasing calcium concentration (Caetono Anolles *et al.*, 1989; Howieson *et al.*, 1993), the proportion of bradyrhizobial cells that attached to lupin roots was decreased by increasing calcium concentration (Wisniewski and Delmotte, 1996). The reason for this may be found in the markedly different surface electrophoretic properties of rhizobia in comparison with some bradyrhizobia (Marshall, 1967; Bushby, 1990), or in the sensitivity of the lupin microsymbiont to calcium in aqueous solution *per se* (Rerkasem, 1977).

Opinion is divided on the question of the propensity of bradyrhizobial inoculant to survive on seed. It has been described as a poor survivor when applied to lupins (Lange and Parker, 1960; Roughley *et al.*, 1993), but survived very well on naked serradella seed (Bowman *et al.*, 1995). This discrepancy may give reason to search for compounds inhibitory to bradyrhizobia in the lupin seed coat (e.g. Thompson, 1960), or stimulatory ones emanating from serradella (e.g. Hartwig *et al.*, 1991).

The *Lupinus* genus is exotic to Australia and this has implications for its nodulation in that country, where it has emerged as an important agricultural legume (Chapter 1, this volume). Some of the earliest ecological studies of lupin rhizobia, mainly from Western Australia, noted the ease with which *L. cosentinii* (then *L. digitatus*) was nodulated by soil-borne bacteria (Adams and Riches, 1930). Some 30 years later, investigations by Lange (1961a) in the same region confirmed the effective nodulation of this species by several bradyrhizobial isolates from 39 endemic species of legume. In contrast, both *L. luteus* and *L. angustifolius* were not nodulated by any of 85 strains of bacteria cultured from the root-nodules of native legumes (Lange, 1961b). From these observations, it follows that there are substantial host/strain interactions in specificity for nodulation in the symbiosis between lupins and their root-nodule bacteria. It was somewhat strange that, in their investigations, Lange and Parker (1960) could find no evidence of specificity within a small group of root-nodule bacteria isolated from the three lupin species they examined, or later with serradella species (Parker and Oakley, 1963). However, subsequent investigations have reported rather substantial variation in host/strain compatibility for nitrogen fixation when a broader germplasm base has been evaluated in Australia (Gault *et al.*, 1986; Howieson *et al.*, 1994) and in North America (Bottomley *et al.*, 1994). When combined with the knowledge that only some lupin bradyrhizobia nodulate *Ornithopus* spp. (Lange and Parker, 1960; Graham and Parker, 1964; Bottomley *et al.*, 1994), this specificity for both nodulation and nitrogen fixation has implications when the molecular characterization of the genes responsible for these functions is attempted (see below).

Early Australian authors (e.g. Adams and Riches, 1930) and some more recently (Corbin, 1978) emphasized the low level of nodulation in uninoculated controls of both *L. angustifolius* and *L. luteus* in field experiments. Some 30 years after the comprehensive studies of Lange and Parker (1960), the situation has changed entirely, at least in Western Australia. In contemporary field experimentation on land that has never been sown to lupins or serradella it is impossible to keep uninoculated treatments of *L. angustifolius* nodule free for more than six weeks (J. Howieson, Perth, 1997, unpublished data), except on alkaline fine-textured earths unsuitable for lupin production (Tang and Robson, 1995).

The reasons for this gradual evolution in reaction of lupins to the endemic bradyrhizobial flora in south-west Australia are probably twofold. Firstly, inoculated *L. angustifolius* and *O. compressus* have been sown over approximately 4 million hectares of agricultural land in the years since 1970. Secondly, the inoculant strain for almost this entire area was WU425, selected on the basis of its remarkable ability to persist in, and colonize, acid sandy soils (Chatel and Greenwood, 1973; Chatel and Parker, 1973). This same strain has since been shown to be highly persistent in the acid soils of North America (Ayisi *et al.*, 1992) and it is possible that through its widespread use in south-west Australia it has now become established as a common component of the soil microflora.

To determine whether WU425 has, in fact, broadly colonized soils of Western Australia, there has been some recent assessment of bradyrhizobial biodiversity in the region using a method of identification based on PCR (polymerase chain reaction) (Harrison *et al.*, 1992; Richardson *et al.*, 1995). WU425 was a major occupant (37%) of serradella nodules three years after inoculation at Carrabin, in the eastern wheatbelt of Western Australia, but had disappeared five years after inoculation from a site at Albany, on the south coast of WA (McInnes *et al.*, 1996). These results suggest an intriguing and rapid increase in biodiversity of *Bradyrhizobium* sp. *(Lupinus)* capable of nodulating lupins and serradella in the region since 1930. Effective strains from native legumes other than those sampled by Lange (1961b) may have increased in the presence of the exotic hosts *Ornithopus* and *Lupinus* or, alternatively, bradyrhizobia from indigenous legumes may have acquired genes from WU425 that enable them to nodulate the exotic hosts. An analogous situation has been reported in New Zealand (Sullivan *et al.*, 1995) where the genetic diversity of rhizobia nodulating *Lotus* has increased over time. A possible source of bradyrhizobia closely related to the lupin bacteria, and therefore susceptible to *in situ* gene transfer, may be native legumes, especially *Acacia* spp. In a recent investigation of bradyrhizobial isolates from *Acacia saligna* growing in a range of soils from the south-west of Western Australia, half showed their highest 16*S* rRNA homology to *Bradyrhizobium* sp. (*Lupinus*) (N. Marsudi, Perth, 1997, personal communication).

Nodulation on alkaline, fine-textured soils

Commercial cultivars of *L. angustifolius* do not grow well on alkaline, fine-textured soils (Gladstones, 1970; White, 1990) and this can be related to poor nodulation under conditions present on these soil types. Evidence, available for some time, indicates detrimental effects of calcium carbonate application on lupin nodulation and growth, variously associated with the process of lime pelleting (Norris, 1956; Parker and Oakley, 1965), of applying lime to soil to overcome soil acidity (Jessop and Mahoney, 1982) and direct effects of calcium on the survival of lupin bradyrhizobia (Rerkasem, 1977). Lime pelleting reduced the nodulation of both lupins and serradella (Parker and Oakley, 1965), two legumes nodulated by the same microorganism but with substantially different modes of infection (crack entry vs. root-hair curling) and nodule physiology. This suggests a direct effect of lime, or high pH, on the behaviour of the microsymbiont itself. Cultured *in vitro* there is evidence of a preference for low pH growth conditions by the organism (Graham and Parker, 1964). More recently, Cobley (1993) demonstrated that WU425 grew and survived much better in buffered solutions at pH 6 than at pH 7 or 8. Although surveys have alluded to the poor survival of the lupin microsymbiont under alkaline conditions (e.g Amarger *et al.*, 1984), there is also evidence of very good survival of WU425 over many years under alkaline and sandy conditions (Mock and Gibson, 1994). These contrasting data indicate that the survival and physiological response of bradyrhizobia to high pH, as a factor limiting nodulation of lupins under alkaline conditions, requires additional research.

The reduced growth of lupins on alkaline soils is, in part, due to direct effects of the root chemical environment on the nodulation process, as elucidated in soil experiments (Subramaniam *et al.*, 1982; Tang *et al.*, 1995) as well as in solution culture (Tang and Robson, 1993). Poor nodulation at high pH occurs in commercial cultivars of *L. angustifolius* despite the species having been recorded as growing on infertile soils of pH 9 in its native habitats (Cowling and Clements, 1993). However, germplasm collection reports rarely provide accompanying data to indicate the level of nodulation in such extreme environments. Poor nodulation under alkaline conditions has also been recorded for *L. luteus, L. albus* and *L. pilosus* (Tang and Robson, 1995) although both *L. pilosus* and *L. albus* are considered to be better adapted than *L. angustifolius* to soil profiles that contain free lime (Pacault *et al.*, 1994).

The underlying mechanisms leading to the decrease in nodulation of lupins at high pH are complex and difficult to quantify because of a number of interactive effects. Thus, pH can affect root growth (Tang *et al.*, 1992a) and root surface integrity (Tang *et al.*, 1993a), bicarbonate concentration can affect nodulation (Blight, 1989), iron availability can influence shoot growth (White and Robson, 1989) and nodule initiation (Tang *et al.*, 1991, 1992b) and high calcium content can control plant growth and nodulation (Jessop *et al.*, 1990), root extension (Tang *et al.*, 1993b) and survival of bradyrhizobia (Rerkasem, 1977). The effect of soil texture almost certainly provides a second-order interaction with many of these factors because of the associated effect of soil buffering capacity. Lupins release organic acids from their roots (Chapter 5), which acidify the rhizosphere and may serve to overcome transient nutrient deficiencies or toxicities associated with alkalinity in that zone. However, the net effect of this acidification is reduced as the buffering capacity of the soil is increased (B. Cobley, Perth, 1997, unpublished data). In this context, it should be remembered that some species of *Lupinus*, e.g. *L. cosentinii*, make excellent growth on quite alkaline, coarse-textured soils and their nodulation appears to be entirely satisfactory.

Molecular Characterization of the Lupin – *Bradyrhizobium* sp. (*Lupinus*) symbiosis

In the nodulation process, communication is initiated between symbiotic partners when they exchange diffusible chemical signals in the rhizosphere. In the legume–rhizobial symbiosis, the first signals derived from the host plant that initiate symbiotic events are plant flavonoids, released initially from the seed coat (Phillips *et al.*, 1994, 1995) and subsequently as root exudates (Long, 1989). Root-nodule bacteria in the rhizosphere recognize specific flavonoids and respond to them with the production of '*Nod* factors'. The reception of this rhizobial-mediated *Nod* factor at the root surface stimulates the plant responses characteristic for early stages of the symbiosis, including root cortical cell divisions, root hair deformation and changes in the transcription level of a number of plant genes. There have been substantial reviews of this process for legumes in general and the reader is referred to Pueppke (1996) and Long (1989).

The flavonoid compounds present in the genus *Lupinus*, mainly isoflavonoids, have been identified as derivatives of genistein (4′,5,7-trihydroxyisoflavone), 2′-hydrogenistein and genistin (7-*O*-β-D-glucosyl genistein; Tahara *et al.*, 1989; Stobiecki and Wojtaszek, 1990; Ingham *et al.*, 1993; Gagnon *et al.*, 1995a). The generic *Nod* factors, whose transcription is potentiated by these flavonoids, are a family of lipo-chitinoligosaccharide molecules formed by the sequential activities of *NodC*, *NodB* and *NodA* (Spaink *et al.*, 1991). These genes are highly conserved in rhizobia and bradyrhizobia. Structurally, the generic *Nod* factor consists of an *N*-acylated chitin backbone. Specificity for host nodulation is conferred by addition of fatty acyl group decorations to the chitin backbone mediated by the actions of the products of additional genes such as *NodE, NodF* in rhizobia (Spaink *et al.*, 1991), and *NodS, Nod Z* in bradyrhizobia (Downie and Young, 1996). The latter authors have noted a strong correlation between the type of nodule morphogenesis and the presence of polyunsaturated acyl groups on the cognate *Nod* factors. For example, bacteria which nodulate legumes that form indeterminate nodules (e.g. *Rhizobium leguminosarum, Sinorhizobium meliloti*) make *Nod* factors with polyunsaturated fatty acyl groups, whereas those which nodulate legumes that form determinate nodules (e.g. *B. japonicum, B. elkanii, Rhizobium tropici, Sinorhizobium fredii*) produce predominantly mono-unsaturated fatty acyl group decorations. On this basis, one would expect the *Nod* factor from *Bradyrhizobium* sp. (*Lupinus*), which forms indeterminate nodules on both lupins and serradella, to contain polyunsaturated decorations. This would set it apart somewhat from bradyrhizobia studied to date (e.g Carlson *et al.*, 1994).

The majority of *Rhizobium* spp. are distinguished by a narrow host range, whilst the genus *Bradyrhizobium* generally nodulates a much broader spectrum of legume species. Some hypotheses explaining this generic division in host promiscuity on an evolutionary basis have been presented (Parker, 1957; Norris, 1965). Previous studies have noted the substantial host/strain specificity for nodulation and nitrogen fixation that exists between strains of *Bradyrhizobium* sp. (*Lupinus*) and genotypes of both *Lupinus* spp. and *Ornithopus* spp. (Gault *et al.*, 1986; Bottomley *et al.*, 1994; Howieson *et al.*, 1994). Thus, the *NodS* or *NodZ* gene products of infecting *Bradyrhizobium* could bear some strain specificity related to their cognate hosts and, as a consequence, the lipo-chitinoligosaccharide molecules active with different hosts may not be identical.

The ability of lupin *Bradyrhizobium* to nodulate *Lotus corniculatus* implies that in the infection process the *Nod* factors with structures similar to those reported for *Rhizobium loti* could be involved (Lopez-Lara *et al.*, 1995). Lupins were additionally infected by *R. loti, R. etli, R. tropici* and *Rhizobium* strain GRH2, but not by *R. fredii* or *Sinorhizobium* strain NGR234 (A.B. Legocki *et al.*, Poznan, 1997, unpublished). The main difference between the *Nod* factors of nodulating and non-nodulating bacteria of lupin is the presence of an *N*-methyl group on the non-reducing end, as well as modification of the reducing *N*-acetylglucosamine residue (Poupot *et al.*, 1993). Genetic analysis indicates that *Nod* factors of lupin-nodulating bacteria are *N*-methylated and carbamoylated and that they carry 4-*O*-acetyl-fucose at C-6 of *N*-acetylglucosamine at the reducing end. However, the precise identification of lupin *Nod* factors remains to be determined. Some known rhizobial and bradyrhizobial *Nod* factors are illustrated in Fig. 6.1, together with a predicted *Nod* factor from lupin-nodulating bradyrhizobia. We would expect the 'R' moiety to be a polyunsaturated fatty acyl group that may vary in structure depending on the particular lupin host it effectively nodulates.

Lupin Genes Symbiotically Regulated During Root Nodule Development

Reception of the (bradyrhizobial) *Nod* factor at the legume root surface stimulates transcription of plant genes which encode a group of proteins called 'nodulins', the functions of many of which remain obscure. A contemporary list of genes identified in nodule development of *Lupinus* is given in Table 6.1. The most abundant nodulins are leghaemoglobins (Wittenberg and Wittenberg, 1990). In *L. luteus* two major leghaemoglobins, LbI and LbII, have been identified and their primary structures determined (Egorov *et al.*, 1976, 1978). They are composed of 154 amino acid residues (including methionine) and their full-length coding cDNA clones have been identified (Strózycki and Legocki, 1995). The predicted amino acid sequences of lupin leghaemoglobins reveal 88% identity with each other and 55–67% identity with other analysed plant leghaemoglobin sequences. Northern hybridization analysis indicated that the transcription of *Lb* genes in lupin starts 12–13 days post infection, with the strongest signals observed 24 days post infection. This is consistent with the time-course for leghaemoglobin appearance reported following microscopic examination of nodules of *L. luteus* by Dilworth (1969). Both lupin leghaemoglobins are highly expressed in root nodules and their mRNAs are not detected by Northern hybridization in other lupin tissues. Southern analysis of lupin genomic DNA revealed that each

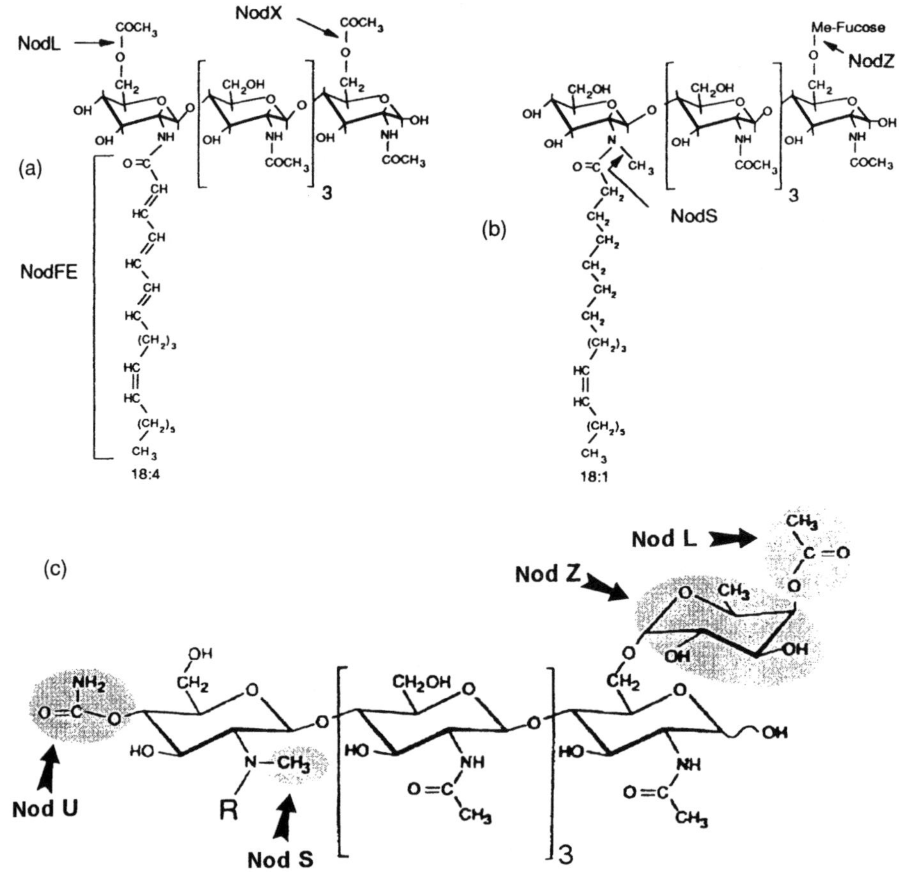

Fig. 6.1. (a) *R. leguminosarum* bv. *viciae Nod* factor showing the sites of action of the acetyl-transferases *Nod*L and *Nod*X, and also the polyunsaturated fatty acid made by *Nod*F and *Nod*E. (b) *Bradyrhizobium Nod* factor illustrating the sites of action of *Nod*S and *Nod*Z. Note the mono-unsaturated nature of the fatty acyl moiety (Downie and Young, 1996). (c) The predicted generic structure of the *Nod* factors of lupin nodulating bradyrhizobia (Legocki *et al.*, 1997).

Table 6.1. Nodule-specific genes and symbiotically regulated PR10 protein genes of *Lupinus* species.

Nodulin	Source	Name of the clone	References
Leghaemoglobin I	*L. luteus*	cDNA – *pSP25*	Strózycki and Legocki (1995)
Leghaemoglobin I	*L. luteus*	genomic – *LlIbI*	Strózycki *et al.* (1994)
Leghaemoglobin II	*L. luteus*	cDNA – *pSP22*	Strózycki and Legocki (1995)
Nodulin-45	*L. luteus*	cDNA – *Nod45*	Szczyglowski et al. (1989)
	L. angustifolius	cDNA – *Nod45*	Rice *et al.* (1993)
	L. angustifolius	genomic – *Nod45*	Macknight *et al.* (1995)
Glutamine synthetase	*L. luteus*	cDNA – *pcLINGS1* (partial)	Boroñ *et al.* (1989)
	L. luteus	genomic – *gcLINGS1*	Boroñ and Legocki (1993)
	L. angustifolius	cDNA – *pGS5*	Grant *et al.* (1989)
Early nodulin LIENOD40A	*L. luteus*	cDNA – *Llenod40A* (partial)	Podkowiñski *et al.* (1996)
Early nodulin LIENOD40B	*L. luteus*	genomic – *Llenod40B*	Podkowiñski *et al.* (1996)
Proline-rich protein LENOD2	*L. luteus*	cDNA – *LENOD2*	Szczyglowski and Legocki (1990)
Proline-rich protein PRP2	*L. luteus*	genomic – *Llprp2*	Karlowski *et al.* (in preparation)
Late nodulins of unknown function	*L. luteus*	cDNA – *pLN21-61* (partial)	Konieczny *et al.* (1988)
	L. luteus	– *pLN281* (partial)	Konieczny *et al.* (1988)
	L. luteus	– *pLN50* (partial)	Konieczny *et al.* (1988)
Late nodulins of unknown function*	*L. luteus*	cDNA – *pDD2/15T*	Swiderski and Legocki (1996)
	L. luteus	– *pDD2/18A* (partial)	Swiderski and Legocki (1996)
	L. luteus	– *pDD4/9A*	Swiderski and Legocki (1996)
	L. luteus	– *pDD3/20A*	Swiderski and Legocki (1996)
Proteins of PR10 class: LIPR10.1A	*L. luteus*	cDNA – *pLlpr10.1a*	Sikorski *et al.* (1996)
	L. luteus	genomic – *gLlpr10.1a*	Sikorski *et al.* (unpublished)
LIPR10.1B	*L. luteus*	cDNA – *pLlpr10.1b*	
	L. luteus	genomic – *gLlpr10.1b*	

*Found by differential display method (DDRT-PCR).

leghaemoglobin gene is represented by a single copy in the lupin genome. The organization of the nodule-specific region in lupin leghaemoglobin promoters seems to be significantly different from that reported for other legumes (Strózycki *et al.*, 1994).

The nodule-specific cDNA clone encoding glutamine synthetase (GS, EC 6.3.1.2) has also been characterized from yellow lupin (Boroñ and Legocki, 1993). In higher plants, GS together with glutamate synthase (GOGAT) is involved in the main pathway of ammonia assimilation (Lea *et al.*, 1990). In legumes GS is coded by a small family of differentially regulated nuclear genes which are expressed in roots at a very low level but are strongly induced in root nodules close to the onset of nitrogen fixation (Forde and Cullimore, 1989). The genomic clone harbouring the full copy of this gene (GS) has been selected from a yellow lupin genomic library (Boroñ and Legocki, 1993). It consists of 12 exons and 11 introns within the region of 3.5 kbp. A comparison of predicted GS polypeptides as well as 5′-untranslated regions of *L. luteus* and *L. angustifolius* (Grant *et al.*, 1989) revealed high sequence identity.

Another DNA sequence expressed in lupin nodules during symbiosis is a gene encoding nodulin-45. A full-length copy of the nodule-specific cDNA clone was selected from a yellow lupin genomic library (Szczyglowski *et al.*, 1989). Its mRNA appears prior to the onset of nitrogenase activity (11 days after infection). Two putative asparagine-linked glycosylation sites and an N-terminal signal peptide suggests that nodulin-45 is synthesized on membrane-bound polysomes. The *L. angustifolius* nodulin-45 shows high amino acid sequence similarity to the *L. luteus* nodulin-45 (Rice *et al.*, 1993; Macknight *et al.*, 1995).

An interesting group of nodule-specific sequences are early nodulin genes coding for ENOD40, known from soybean (Kouchi and Hata 1993; Yang *et al.*, 1993), pea (Matvienko *et al.*, 1994), alfalfa (Asad *et al.*, 1994; Crespi *et al.*, 1994) and *Vicia sativa* (Vijn *et al.*, 1995). Two different sequences homologous to the *Medicago truncatula* ENOD40 have been identified in yellow lupin (Podkowiñski *et al.*, 1996).

A group of genes expressed in early stages of root nodule development, and highly conserved among legumes, codes for (hydroxy) proline-rich proteins (ENOD2). The cDNA and genomic sequences for ENOD2 were selected from *L. luteus* libraries (Szczygowski and Legocki, 1990; W. Karlowski *et al.*, 1997, Poznań, personal communication). They code for protein 2½ times longer than ENOD2 sequences from soybean and alfalfa. The genomic sequence shows no introns in the coding region and a signal peptide closely similar to other ENOD2 sequences. The promoter region shows one element characteristic of all ENOD2 promoter genes. The presence of an OSE (organ specific element) homologous sequence makes the ENOD2 gene of yellow lupin different from others and, in general, from early nodulins. The gene coding for lupin leghaemoglobin I lacks such an element whereas the OSE in the GS gene is in a different position from its location in other ENOD2 genes.

Further nodule-specific sequences of unknown function have been identified from a yellow lupin library (Konieczny *et al.*, 1988), including the clones pLN21-67, pLN281 and pLN50, all of which are expressed as late nodulin genes (11–18 days after inoculation) suggesting roles in nitrogen fixation or assimilation rather than in nodule morphogenesis. Similarly, when four nodule-specific probes, amplified by DDRT-PCR (differential display reverse-transcribed technique), were used to screen a lupin cDNA library, the four clones identified and sequenced were all expressed exclusively in nodules 13 days after inoculation.

During effective nodule formation in *L. luteus,* the synthesis of two abundant, low molecular weight (M_r 18,000) root proteins is down-regulated (Sikorski *et al.*, 1989). When full-length cDNA copies (Sikorski *et al.*, 1996) and genomic clones (M. Sikorski, Poznań, 1997, personal communication) for each were identified and sequenced, the deduced amino acid sequences were of two 156 amino acid polypeptides with 76% sequence identity (90% similarity). They showed 40–80% similarity to pathogenesis-related (PR) and stress-induced proteins of the PR10 class (van Loon *et al.*, 1994), 40% similarity to major pollen allergens and 86% similarity to ABA-responsive proteins in soybean. The accumulation of mRNA transcripts of both genes continued at high levels in the roots of uninfected plants and in the hypocotyls, cotyledons and leaves of mature nodulated plants. Both identified lupin proteins represent a class of defence-related proteins which may be involved in the process of effective symbiosis development.

Analysis of the molecular response of lupins to nodulation should help us understand the complex signalling between host and microsymbiont that precedes nodule function and the ultimate export of fixed nitrogen. Whilst understanding of legume nodule physiology is more mature, specific investigation of the lupin nodule is revealing some variations on the accepted norm. Of particular interest is the diffusion of oxygen in the 'lupinoid' nodule.

Oxygen Diffusion Regulation in Lupin Nodules

Legume nodules have a large demand for O_2 for high rates of oxidative phosphorylation to fuel the energy-demanding N_2 fixation process. However, an excess of free O_2 within the infected cells will lead to loss of nitrogenase activity (Witty *et al.*, 1986), whilst an inadequate O_2 supply will initially curtail N_2 fixation and can cause eventual oxidant-induced nodule senescence (Escuredo *et al.*, 1996). Thus, a careful control of O_2 supply is essential for the efficient functioning of legume nodules. This is achieved through a variable O_2 diffusion barrier which balances O_2 ingress against consumption by metabolic activity so that levels of free O_2 within infected cells remain very low (Hunt and Layzell, 1993; Minchin, 1997). Efficient distribution of this O_2 within the infected cells is achieved through the O_2 carrier leghaemoglobin (Bergersen, 1994). Evidence for the existence of a variable O_2 diffusion barrier comes from a combination of physiological measurements and mathematical analyses (e.g. Witty *et al.*, 1986; Hunt and Layzell, 1993). Much of this data concerns soybean, but there is also a substantive body of work on white lupin. In both species the O_2 diffusion barrier responds to reduced root temperatures, detopping and prolonged darkness (Minchin *et al.*, 1986; Layzell *et al.*, 1990; Diaz del Castillo *et al.*, 1992; Minchin *et al.*, 1992). However, recent studies on white lupin have suggested several differences from soybean in the operation of the diffusion barrier. A limitation of these studies is that they largely focus on one symbiosis, between *Lupinus albus* cv. Multolupa and *Bradyrhizobium* sp. (*Lupinus*) strain ISLU 16, and the results may not be typical for lupins. Nevertheless, they provide an interesting contrast with soybean in terms of structures and mechanisms of operation of O_2 diffusion control and perhaps provide some insight into the unique morphology of 'lupinoid' nodules (Corby, 1981).

Diffusion control in the mid-cortex

Although the existence of a cortical diffusion barrier is still a matter of some controversy (Van Cauwenberghe *et al.*, 1994), structural studies with white lupin have provided strong supporting evidence. This is based on the production of intercellular occlusions of the spaces between the large, thick-walled cells of the mid-cortical region, accompanied by cell expansion and consequent shrinkage of the intercellular spaces (De Lorenzo *et al.*, 1993; Iannetta *et al.*, 1993a). Since gaseous diffusion is much faster (*c.* 10,000 times) through air than through water, it is reasonable to assume that the network of intercellular spaces will be the main route for gas diffusion through the nodule cortex. Therefore, the operation of a variable cortical barrier is most likely to involve alterations to these spaces (Witty *et al.*, 1986).

The intercellular occlusions can be immuno-labelled using monoclonal antibodies raised against the matrix glycoprotein of infection threads from pea and common bean nodules (MAC 236 and MAC 265; Rae *et al.*, 1991; Vanden-Bosch *et al.*, 1994). Such labelling of occlusions has also been reported for soybean nodules, but the published literature is limited to long-term treatments (James *et al.*, 1991; Iannetta *et al.*, 1993b) suggesting that the occlusions could be morphological adaptations. Within lupin nodules the labelled occlusions can be induced within a few hours and levels of glycoprotein in nodules (as determined by ELISA) correlate with measured changes in O_2 diffusion resistance in response to physiological stress treatments (De Lorenzo *et al.*, 1993; Iannetta *et al.*, 1993a). Indeed, the occlusions can be induced within 15 minutes in response to exposure of nodulated roots to 50% O_2 (Iannetta *et al.*, 1995).

A fascinating addition to the occlusion story has been provided by reports that diprenylated isoflavones also occur in the cortical spaces of lupin nodules (Gagnon *et al.*, 1995b; Grandmaison and Ibrahim, 1995). Furthermore, these share the same cortical origins and extrusion pathways (via intracellular vesicles) as the glycoproteins (Iannetta *et al.*, 1995). Grandmaison and Ibrahim (1995) suggested that the two compounds interacted within the intercellular spaces to produce a rigid 'gel' in a manner similar to that of a cell wall pathogen defence response. This concept of the occlusions being modified defence responses is strongly supported by the occurrence of the defence-related polygalacturonase inhibitor protein (PGIP) within the cortical intercellular spaces of lupin nodules (P.P.M. Iannetta and E.K. James, Dundee, 1996, personal communication).

Diffusion control in the inner cortex

The arrangement of the small inner cortical cells of lupin nodules is different from soybeans. In the latter they can be divided into a distribution zone, with large intercellular spaces (Witty *et al.*, 1987), and a boundary layer of closely packed cells with no apparent traversing intercellular spaces towards the outside

(Parsons and Day, 1990). The boundary layer is also present in lupin nodules but there is no obvious distribution zone (Iannetta *et al.*, 1993b). This difference can be explained by the recent demonstration of a continuous network of open pores within soybean, but not lupin, nodules (Witty and Minchin, 1994). Such an arrangement requires some open pores to cross the boundary layer of soybean nodules, with an effective distribution zone to avoid locally high concentrations of O_2 at the edge of the infected zone. The absence of such pores in lupin nodules precludes the need for a distribution zone but raises the possibility of negative pressure in the central tissue zone (Witty and Minchin, 1994). Negative pressure raises the possibility of gaseous movement facilitated by convection as well as diffusion.

The presence of open pores in soybean nodules also requires a rapid mechanism for increasing diffusion resistance within the inner cortex (Witty and Minchin, 1994). Recent anatomical evidence suggests that the inner cortical cells of soybean nodules are osmocontractile and may collapse to reduce the size of intercellular spaces and/or expel water into them (Serraj *et al.*, 1995). This idea is supported by microelectrode evidence for rapid O_2-induced membrane depolarizations (Denison and Kinraide, 1995) and X-ray microanalysis evidence for disturbance-induced shifts of Ca^{2+} and K^+ ions, occurring within 10 to 15 min (Minchin *et al.*, 1995). However, Ca^{2+}/K^+ shifts have not been observed in the inner cortex of lupin nodules (F.R. Minchin and B.J. Thomas, Aberystwyth, 1997, unpublished data), suggesting that these cells may not be osmocontractile. A possible explanation for this (at least for the Multolupa/ISLU 16 symbiosis) is that the inner cortical cells are involved in the exclusion of Na^+ and Cl^- in this salt-tolerant symbiosis (Fernandez-Pascual *et al.*, 1996). If these cells also operated as an osmocontractile diffusion barrier they could be 'triggered' by salt (Serraj *et al.*, 1995), causing an excessive reduction in O_2 diffusion. This problem could be avoided by eliminating osmocontractile diffusion regulation and replacing it with the rapid production of glycoprotein occlusions.

Early expression of intercellular glycoproteins

It is now well established that there is a need for O_2 regulation to create a microaerobic environment during the early stages of nodule production to allow expression of *nif* and *fix* genes (Sheehy and Thornley, 1988; Soupene *et al.*, 1995). Thus, if intercellular occlusions are an essential feature of O_2 diffusion control in lupin nodules, the matrix glycoproteins (MAC 236/MAC 265 antigens) should be expressed in developing nodules prior to the onset of nitrogenase activity. This has now been demonstrated in white lupin, where the glycoprotein was immunolocalized in the intercellular spaces of nodules eight days after initial infection whilst nitrogenase enzyme expression and nitrogenase activity were not detected until 17 days after infection (James *et al.*, 1997).

An infected region diffusion barrier?

This body of data provides strong evidence for a cortical O_2 diffusion barrier within white lupin nodules which operates by the production of intercellular occlusions. However, it is difficult to envisage how such a mechanism could, by itself, achieve sufficient fine control of O_2 supply to the infected region. Recent mathematical models for determinate nodules (soybean, cowpea) suggest that O_2 regulation could also occur within the infected zone (Bergersen, 1994; Thumfort *et al.*, 1994). Although such models do not yet exist for lupin nodules, where the infection zone has an almost complete absence of non-infected interstitial cells (Atkins *et al.*, 1991) it seems reasonable to conclude that a level of O_2 diffusion control does occur within this zone.

While complete knowledge of the mechanism of control of O_2 diffusion within the lupin nodule is yet to be revealed, nodule function remains robust and productive under a broad range of edaphic conditions. As a consequence, nitrogen fixation from lupins in mainstream agriculture is a major factor potentiating global food production on soils otherwise depauperate in nitrogen. The nitrogen from this symbiosis is harvested directly as high protein pulse grains, in addition to that portion which remains in the soil, in various plant components, to boost production from subsequent non-leguminous crops.

Nitrogen in Plant Components

Accumulation of N in lupin is typically positively correlated with biomass production (Evans *et al.*, 1989). Shoot mass yields of *L. angustifolius* grown in broad-acre agriculture can vary from as little as 2 t ha^{-1} (Smith *et al.*, 1987: Evans *et al.*, 1989) to in excess of 14 t ha^{-1} (Evans *et al.*, 1989; Unkovich *et al.*, 1994). Consequently, reported peak N in shoot biomass of *L. angustifolius* ranges from a low of 48 kg N ha^{-1} (Smith *et al.*, 1987) to as much as 372 kg N ha^{-1} (Unkovich *et al.*, 1994). The average shoot N content for the 48 measurements on *L. angustifolius*, summarized in Table 6.2, is 201 kg N ha^{-1}, whereas *L. albus*, on average, accumulated 333 kg N ha^{-1}, based on 21 studies. Better soil fertility and higher rainfall probably contributed to the higher average yield of N for *L. albus*, compared with *L. angustifolius*. Nevertheless, Armstrong *et al.* (1997) found *L. albus* to accumulate more N (419 kg N ha^{-1}) than *L. angustifolius* (303 kg N ha^{-1}) when these species were compared in the same environment.

The distribution of shoot N between stem, pod walls and seed has important implications for N harvest index, and the value of legume residues as sources of N for subsequent crops. A typical pattern of accumulation of N in components of *L. angustifolius* is shown in Fig. 6.2. N concentration in vegetative organs declines progressively throughout the growth of the plant, with a rapid decline evident after the onset of flowering when N is transported to reproductive organs. At maturity, lupin stem material can contain 0.7 to 1.3% N, leaf material *c.* 2% N, pod valves 0.4 to 0.6% N, and seed 4.5 to 5.8% N (Russell, 1996). The sharp contrast in N concentration between shoot components and seed at maturity is evidence of the efficiency of remobilization of N from vegetative tissue to seed in lupin. However, the combination of an indeterminate growth habit in domesticated lupin, and variable growing season rainfall, particularly during grain filling, has resulted in a wide range of values for N harvest index (11 to 80%).

Few studies have been made of the biomass of root material under lupin crops, and the content of N in this biomass. Unkovich *et al.* (1994) recovered 2.4 t DM ha^{-1} of root at peak biomass, which was equivalent to 24% of the total

Table 6.2. Quantities of nitrogen accummulated and rates of nitrogen fixation in cultivated lupin.

Lupin species	Region/rainfall	Experimental detail	Total N (kg N ha^{-1})	Measurement of N fixation	Nitrogen fixed (kg N ha^{-1})	Proportion N fixed (%)	Reference
L. angustifolius	SE Australia, 799 mm	1 site; 3 sowing dates	123–168	NA	247–253	77–85	Armstrong *et al.* (1997)
	SE Australia, 307–518 mm	2 sites over 2 years; 2 sowing dates	159–306	ID, NA	79–254	49–85	Evans *et al.* (1987a)
	SE Australia, 307–518 mm	2 sites over 2 years; 3 rates fert. N; straw retention	184–306	ID, NA	78–255	41–85	Evans *et al.* (1987b)
	SE Australia, 267–646 mm	14 sites over 3 years; 2.4–3.4 kg N ha^{-1} fert. N	57–297	ID	26–288	20–97	Evans *et al.* (1989)
	Southern Australia, 350 mm	2 sites over 2 years	225–226	$15N_2$ assay	136–155	61–64	Gibson and Alston (1987)
	SE Australia, 305–527 mm	4 sites over 2 years	121–231	ND, ARA, XS	72–223	59–97	Herridge and Doyle (1988)
	Canterbury, New Zealand, 680 mm	1 site in 2 seasons; innoc; 2.4 kg N ha^{-1} fert. N	130–347	ID	22–222	9–36	Haynes *et al.* (1993)
	Iceland, 469 mm	1 site; ± innoc., 30 kg N ha^{-1} fert. N	73–228	ID, ND	70–213	91–97	Palmason *et al.* (1992)
	Washington State, USA, ≈ 500 mm	2 sites over 2 years; disease ↓ yield at 1 site	48–121	ID	19–60	40–50	Smith *et al.* (1987)
	SE Australia, 350 mm†	1 site; 25 kg N ha^{-1} fert. N	306	ID	242	79	Smith *et al.* (1992)
	Western Australia, ≈ 400 mm	7 sites over 3 seasons	199–372	NA	181–327	78–91	Unkovich *et al.* (1994)
L. albus	SE Australia, 799 mm	1 site, 3 sowing dates	370–472	NA	268–400	68–85	Armstrong *et al.* (1997)
	Germany, 550–1100 mm	10 sites over 2 seasons	50–400	ND	190–360	44–92	Hartmann and Aldag (1989)
	California, USA, 256*–677 mm	1 site, 2 seasons; ± innoc; 4 cultivars	239–345	ND	147–247		Larson *et al.* (1989)
	France	1 site; 5 kg N ha^{-1}	293	ID	234	80	Duthion *et al.* (1987)

NA, natural abundance; ID, isotopic dilution; ARA, acetylene reduction assay; XS, xylem sap.
*Unspecified amount of irrigation applied.
†Including 150 mm of irrigation.

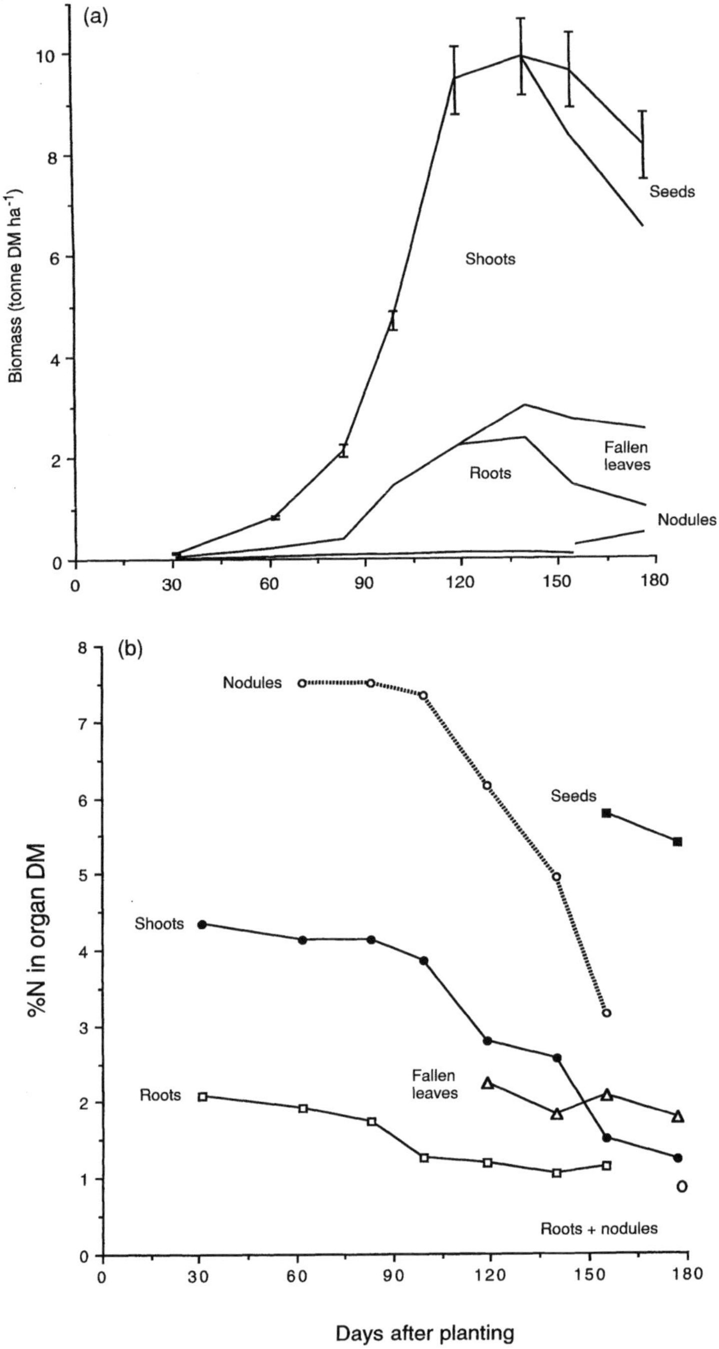

Fig. 6.2. Changes with age in (a) above- and below-ground harvestable dry matter and (b) concentrations of nitrogen in a crop of narrow-leafed lupin (*Lupinus angustifolius* L. cv. Illyarrie) at East Chapman Research Station, near Geraldton, in 1986. (Taken from Unkovich *et al.*, 1994).

lupin biomass. This biomass contained 31 kg N ha^{-1} (13% of total crop N). Farrington *et al.* (1977) found root N to be 30 to 40% of total lupin N from two to 14 weeks after seeding. By peak biomass, 18 weeks after seeding, root N comprised only 20% of total N. The efficiency of recovery of root material following soil coring and wet sieving procedures has not been established for lupin. These techniques probably do not recover all root material and they cannot recover N in decayed root or rhizodeposition material. Russell and Fillery (1996a,b) used ^{15}N dilution techniques to estimate the quantities of lupin below-ground biomass N after using a wick procedure to feed a known quantity of ^{15}N into the stems of growing lupin plants. The ^{15}N enrichment of clean root material, other root material and bulk soil was measured and the content of root-derived N in each fraction calculated by ^{15}N dilution procedures. About 28% of the total biomass, or 3 t ha^{-1}, and 28% of the total N content, or 91 kg N ha^{-1}, was estimated to be in the below-ground lupin biomass at maturity. This estimate of N in the below-ground biomass is appreciably higher than reported previously – indirect evidence that standard root recovery techniques could substantially underestimate the quantity of this below-ground biomass N under lupin crops.

Measurement of nitrogen fixation

In a review of alternative sources of N for crops, Nutman (1976) concluded that lupin fixed between 145 and 208 kg N ha^{-1}. No information was given on the species studied, the specific method used to assess N_2 fixation, and whether the lupin was grown for green manure or for grain production. Considerable progress has been made with techniques to measure N_2 fixation in legumes since Nutman's review. Ledgard and Peoples (1988) and Peoples *et al.* (1989) discussed the strengths and weaknesses of the different methods used. In brief, the ^{15}N isotope dilution technique (Chalk, 1985) and the ^{15}N natural abundance (Bergersen *et al.*, 1985; Pate *et al.*, 1994) are the preferred methods for estimating N_2 fixation in lupins, which are amide-producing plants (Chapter 12). Uncertainty over the calibration of the acetylene reduction assay (ARA), and problems associated with extrapolation of short-term assay procedures to estimate N inputs over a full crop-growth cycle, largely preclude the use of information from ARA to estimate N_2 fixation. Nevertheless, ARA measurements continue to be used to assess N_2 fixation in dispersed lupins in natural ecosystems. Several studies, summarized in Table 6.2, used total N uptake by a non-fixing reference plant to estimate the quantity of soil-derived N in lupin (Rennie, 1979). The accuracy of this simple procedure is very dependent on the closeness of the match in N uptake between the legume and the non-N_2 fixing plant.

The ^{15}N isotope dilution technique and the ^{15}N natural abundance both rely on the enrichment of ^{15}N in soil, compared with air, to proportion N in legume biomass between that derived from soil and that derived from atmospheric N_2. Use is made of the natural ^{15}N enrichment of soil N to discriminate the soil-derived N from atmospheric N in the natural abundance technique. In contrast, the isotope dilution technique depends on the enrichment of soil ^{15}N using exogenous N sources. A non-fixing reference plant is used in both cases to measure the $^{15}N/^{14}N$ ratio in soil. Ideally, the reference plant should have a comparable phenology to the legume under study, possess a similar root architecture and take inorganic N from the same soil layers during growth. In soybean, non-nodulating isolines (rji), which are comparable to normal fixing lines (Delves *et al.*, 1986), have proved very useful as samplers of soil N. However, these kinds of mutants are not available for lupin. The more uniform the distribution of enriched ^{15}N in soil, both spatially and temporally, the less critical are the requirements for synchronizing the pattern of uptake of N between the legume and reference plant. Stratification of ^{15}N in soil by depth, and rapid changes in the $^{15}N/^{14}N$ ratio in inorganic N after the application of ^{15}N-enriched N sources, complicate the choice of reference plant in the case of isotopic dilution techniques. An alternative isotopic dilution procedure uses the ^{15}N enrichment in the inorganic pool, total N and ^{15}N accumulation in lupin biomass to estimate the proportion of soil-derived N (Hamilton *et al.*, 1992). This technique has advantages over the traditional isotopic dilution methods that involve reference plants (Smith *et al.*, 1992), but the method has not been extensively evaluated.

The quantities of N fixed by lupin species in different regions, and ecosystems, are summarized in Table 6.2. Most measurements have been made in southern Australia on the domesticated, narrow-leafed, sweet lupin, *L. angustifolius*. This species can fix between 180 and 320 kg N ha^{-1} when grown on the deep sandplain soils of Western Australia (Unkovich *et al.*, 1994). A wider range of values for N_2 fixation (26–288 kg N ha^{-1}) has been noted in south-east Australian and New Zealand studies where *L. angustifolius* has been grown in a wider range of soil and climatic conditions. Rates of N_2 fixation in *L. angustifolius* in regions outside of Australasia are also variable, with 19 to 60 kg N ha^{-1} reported for crops in Washington State (Smith *et al.*, 1987), and 70 to 213 kg N ha^{-1} in Iceland (Palmason *et al.*, 1992).

Fewer measurements of N_2 fixation are available for *L. albus*. Under European conditions, this lupin has been found to fix up to 360 kg N ha^{-1} but, as is the case for *L. angustifolius*, rates of N_2 fixation as low as 20 kg N ha^{-1} have been reported (Hartmann and Aldag, 1989). In south-east Australia, *L. albus* has been found to fix between 268 and 400 kg N ha^{-1} (Armstrong *et al.*, 1997). No measurements of N_2 fixation have been undertaken on *L. albus* grown on sandy soils of Western Australia. Hamblin *et al.* (1993) report that both biomass and harvested grain in *L. albus* were less than for narrow-leafed lupin, implying that N content and N_2 fixation rates are likely to be lower than noted previously for *L. angustifolius* for this region.

No measurements of N_2 fixation have been made on *L. cosentinii* (sandplain blue lupin), which is widespread as a self-regenerating volunteer 'pasture' species on deep infertile sands in Western Australia (Unkovich *et al.*, 1997). The ability of *L. cosentinii* to colonize sands with very low N status and to produce large biomass is indirect evidence of its capacity to fix N. Another lupin species, *L. arboreus*, has been used during the establishment of *Pinus radiata* forest plantations on coastal sands in New Zealand. Rates of N_2 fixation for the species grown with establishing pine, averaged over a period of five years, were about 80 kg N ha^{-1} (Baker *et al.*, 1986). Double these rates of N_2 fixation were found by Gadgil (1971) for open stands of *L. arboreus*. Shading, fruit set and water stress all reduce N_2 fixation in *L. arboreus* (Sprent and Silvester, 1973).

Several attempts have been made to quantify the N_2 fixation in indigenous lupins in the understorey of lodgepole pine in North America. Compared with the introduced *L. arboreus* in New Zealand forest plantations, lupins in lodgepole pine forests fix < 2 kg N ha^{-1} per annum (Fahey *et al.*, 1985; Hendrickson and Burgess, 1989). Although the N_2 fixing capacity of individual wild lupin plants can be substantial when these first colonize new land, their low density results in a negligible contribution to N inputs when expressed on a per hectare basis (Halvorson *et al.*, 1992).

Factors affecting N_2 fixation in the field

The wide range in nitrogen fixation values reported in Table 6.2 demonstrate that soil and crop factors have a significant effect on the N_2 fixation process. Nevertheless, few studies have systematically evaluated the effect of soil properties on the rate of N_2 fixation in lupin. In a survey of N_2 fixation in *L. angustifolius* crops grown on a range of soils with different rainfall, Evans *et al.* (1989) found N_2 fixation (N_{fix}) to be explained by the relationship $N_{fix} = 9.2 + 19.79 \times DM - 0.507 \times N_{cer}$, ($r^2 = 0.78$); where DM was the lupin above-ground biomass production and N_{cer} is the mineral N extracted by an adjacent cereal crop. The dependence of N_{fix} on biomass could not be fully explained by the greater depletion of soil-derived mineral N by a larger biomass, leading Evans *et al.* (1989) to suggest that lupin, in some way, avoided use of inorganic N in soil. Addition of fertilizer N can reduce N_2 fixation, with inorganic N substituting for fixed N_2, since the total N in lupin is largely unaffected by N fertilizer addition (Evans *et al.*, 1987a). Increased sensitivity of the N_2 fixation process to inorganic N has been reported for late sown compared with early sown lupin crops (Evans *et al.*, 1989).

Soil constraints, including nutrition and alkaline pH, low rainfall, late sowing time and disease, reduce the potential for yield and appear to contribute to the low nitrogen fixation values reported in Table 6.2. Waterlogging and water stress at flowering have been reported to reduce ARA (Farrington *et al.*, 1977). N accumulation appears to be more affected by low root-zone temperature when lupins are dependent on N_2 fixation compared with lupin fed inorganic N (Legros and Smith, 1994).

Net N inputs

Except for the studies by Gibson and Alston (1987), Doyle *et al.* (1988) and Unkovich *et al.* (1994), estimates of the net input of N into soil have been derived by deducting N harvested in grain from estimates of fixed N based on above-ground biomass N only, and therefore exclude the contribution of the below-ground biomass N. As will be shown below, the focus on above-ground biomass N may have resulted in substantial underestimates of the contribution grain lupin makes to the N budgets in rotations. In south-east Australia (Evans *et al.*, 1987a,b, 1989; Doyle *et al.*, 1988; Armstrong *et al.*, 1997) the average net input from *L. angustifolius* is 67 kg N ha^{-1} with a range from −41 kg N ha^{-1} to +141 kg N ha^{-1}. The range in net input of N from *L. angustifolius* in Western Australia (Unkovich *et al.*, 1994) is much narrower (+32 to +96 kg N ha^{-1}) but the average net input (+65 kg N ha^{-1}) is similar to that observed in crops grown in south-east Australia.

A negative net N input (–18 kg N ha^{-1}) and a small positive net N input (+14 kg N ha^{-1}) were estimated by Smith *et al.* (1987) for *L. angustifolius* grown in two seasons in California. *L. angustifolius* either fixed N equivalent to that removed in grain or exhibited a small negative N input (–16 kg N ha^{-1}) in a New Zealand study (Haynes *et al.*, 1993).

There are sharp differences in net N inputs from *L. albus* between studies conducted in Australia (+53 kg N ha^{-1}) and Europe (+40 kg N ha^{-1}) and the single North American study (–41 kg N ha^{-1}). Exclusion of fallen leaf material in the estimate of total lupin N in the North American study (Larson *et al.*, 1989) contributed to the large negative estimate of net N input. Based on the ratio of leaf-N to stem- and pod-N in lupin biomass (Russell and Fillery, 1996), and the N content in fallen lupin leaf (Unkovich *et al.*, 1994), it seems reasonable to conclude that N in fallen leaf may have been equivalent to N in standing biomass, excluding grain N. Inclusion of N in fallen leaf in the estimate of N_2 fixation resulted in an average net N input that was closer to –17 kg N ha^{-1}, with a range between –61 and +24 kg N ha^{-1}. In the only common study of N inputs between lupin species, *L. albus* contributed less to soil N (+50 kg N ha^{-1}) than did *L. angustifolius* (+95 kg N ha^{-1}), indicating that while total N_2 fixation may be greater in *L. albus*, the higher harvest N index in this species reduced the net N input compared with *L. angustifolius* (Armstrong *et al.*, 1997). Recently isolated mutants of *L. angustifolius* cv. Danja which show normal patterns of leaf senescence but completely lack abscission (Atkins and Pigeare, 1993) provide an opportunity to recover full above-ground biomass at final harvest in the field.

Inclusion of lupin below-ground N in estimates of N turnover is imperative. Herridge and Doyle (1988) included root and nodule material in their estimate of N_2 fixation but did not state the ratio of below-ground N to above-ground N. Unkovich *et al.* (1994) used a ratio of 6.4 to 1 shoot N to root N to calculate root N. Farrington *et al.* (1977) observed root N to range between 0.3 and 0.4 of whole-plant N for 14 weeks after sowing. Thereafter this ratio declined steadily to be 0.2 at leaf drop and 0.1 at grain harvest. These estimates are much lower than the grain harvest ratio of below-ground N to total plant N (0.28) obtained by ^{15}N dilution techniques after *in situ* labelling plants with ^{15}N using stem feeding (Russell and Fillery, 1996). The use of ^{15}N dilution techniques to proportion N in soil between lupin and soil-derived N enabled inorganic N, fine root, decayed root/nodule material and root rhizodeposits, which are typically discarded with soil during root washing, to be included in the estimate of below-ground N. Only a third of the N estimated to be in the below-ground biomass was recovered in root separated from soil (Russell and Fillery, 1996).

Published values for fixed N (Table 6.2) have been adjusted for below-ground N by assuming this N was equivalent to 0.4 of above-ground biomass N or 0.28 of total N. With this adjustment, average net N inputs for *L. angustifolius* are +136 kg N ha^{-1} for south-east Australian studies (Evans *et al.*, 1987b, 1989; Doyle *et al.*, 1988; Armstrong *et al.*, 1997), +175 kg N ha^{-1} for Western Australia (Unkovich *et al.*, 1994), +70 kg N ha^{-1} for New Zealand (Haynes *et al.*, 1993) and +14 kg N ha^{-1} for the USA (Smith *et al.*, 1987). In the case of *L. albus*, the revised net N inputs are +145 kg N ha^{-1} for Europe (Duthion *et al.*, 1987; Hartmann and Aldag, 1989), +181 kg N ha^{-1} for south-east Australia (Armstrong *et al.*, 1997) and

+73 kg N ha^{-1} for Californian conditions (Larson *et al.*, 1989). These revised estimates of net N inputs for *L. angustifolius* and *L. albus* are two to three times higher than those determined from above-ground biomass N only. Given the importance of below-ground legume N in the transfer of N to subsequent crops (Russell and Fillery, 1996), further work is needed to establish the range in values of total lupin below-ground N for different lupin species grown worldwide, to improve the confidence in estimates of net N inputs as undertaken here.

Larger postharvest quantities of mineral N in soil have been found under lupin, compared with non-N-fixing reference plants, in a number of studies (Evans *et al.* 1987b, 1989; Chalk *et al.* 1993). The origin of this N is not clear. Lupin may potentially add to the soil inorganic N pool because of N_2 fixation, and it may also remove less inorganic N from soil because part of its N requirement is met by N_2 fixation. The latter phenomenon has been described as N 'sparing', implying that lupin does not utilize soil-derived mineral N in the same quantities used by non-N_2-fixing reference plants. Unkovich *et al.* (1997) advance a number of arguments against the thesis that lupin, and other legumes, use soil-derived mineral N sparingly, or that legume roots stimulate mineralization of soil organic matter. They note that in the study of Chalk *et al.* (1993) the build-up in mineral N in soil, compared with wheat, was only 14 kg N ha^{-1} whereas the apparent sparing of soil and fertilizer N was estimated, using ^{15}N dilution techniques, to be 136 kg N ha^{-1}. However, Chalk *et al.* (1993) report mineral N in the top 15 cm of soil only, whereas the cereal crop used by them to determine the quantity of spared N carried forward likely recovered inorganic N from deeper layers in the soil. Leaching of nitrate below 15 cm is not uncommon in the sandy soils widely used to grow lupin, highlighting the need for full soil profile assessments of inorganic N when the N budgets for lupin-based rotations are being determined (G.C. Anderson, I.R.P. Fillery, P.J. Dolling and F.X. Dunin, Perth, 1997, unpublished data).

Another plausible explanation for 'spared' N is that legume root and nodule decomposition and rhizodeposition provide readily mineralizable substrate which contributes to the within-season build-up of soil mineral N under lupin (Russell and Fillery, 1996; G.C. Anderson, I.R.P. Fillery and P.J. Dolling, unpublished data). If this is the case, the existence of the so-called spared N pool is further evidence of the importance of the below-ground N in the N balance under lupin.

Summary

Several aspects of the *Lupinus–Bradyrhizobium* sp. (*Lupinus*) symbiosis make it unusual, perhaps even unique, in comparison with many other well-researched legume symbioses. Its exceedingly robust nature under conditions of considerable edaphic stress, such as extreme soil acidity, immediately sets it apart from most other symbioses exploited in mediterranean-type agriculture. This symbiosis is initiated by the production of bacterial *Nod* factors speculated to possess monounsaturated fatty acyl group decorations. These are unlike those polyunsaturated side chains produced by known strains of bradyrhizobia which form determinate nodules on legumes other than the lupin. Perhaps in association with this, the mode of entry of the bacterium into lupins is uniquely peculiar – through epidermal cracks or wounds in the root. The nodule then formed has been described as 'lupinoid' in deference to its special character. Although the physiology of this nodule type requires further examination, research to date is strongly suggestive of an alternative means of O_2 diffusion control to other well-researched nodule types. Finally, the lupin microsymbiont, as exemplified by the Australian commercial inoculant strain WU425, is a taxonomic outlier in relation to other bradyrhizobia as revealed by analysis of its fatty acid content.

Despite these substantial unusual characteristics, too few strains of *Bradyrhizobium* sp. (*Lupinus*) have been examined for us to be dogmatic in our appraisal of the symbiosis as unique. It remains to be seen whether ongoing research reinforces this viewpoint.

References

Adams, A.B. and Riches, J.H. (1930) Root nodules on lupins. *Journal of the Department of Agriculture, Western Australia* 2(7), 556.

Allen, O.N. and Allen, E.K. (1981) *The Leguminosae: A Source Book of Characteristics, Uses and Nodulation*. Macmillan, London.

Amarger, N., Sommer, G., Giraud J.J. and Lagacherie B. (1984) *Rhizobium lupini*: presence et survie dans les sols, selection des souches et inoculation des semences des lupin. In *Proceedings of the 3rd International Lupin Conference, La Rochelle, France*, pp. 599–600.

Armstrong, E.L., Heenan, D.P., Pate, J.S. and Unkovich, M.J. (1997) Potential and realized nitrogen benefits of lupins, field pea and chickpea to wheat production in south-eastern Australia. *Australian Journal of Agricultural Research* 48, 39–47.

Asad, S., Fang, Y., Wycoff, K.L. and Hirsh, A.M. (1994) Isolation and characterization of cDNA genomic clones of MsENOD40; transcripts are detected in meristematic cells of alfalfa. *Protoplasma* 183, 10–23.

Atkins, C.A. and Pigeare, A. (1993) Applications of cytokinins to flowers to improve pod set in *Lupinus angustifolius*. *Australian Journal of Agricultural Research* 44, 1799–1819.

Atkins, C.A., Storer, P.J. and Young, E.B. (1991) Translocation of nitrogen and nodule-specific expression of uricase (nodulin-35) in *Robinia pseudoacacia*. *Physiologia Plantarum* 83, 483–491.

Ayisi, K.K., Putnam, D.H., Vance, C.P. and Graham, P.H. (1992) Dinitrogen fixation, nitrogen and dry matter accumulation and nodulation in white lupine. *Crop Science*, 32, 1197–1202.

Baker, T.G., Oliver, G.R. and Hodgkiss, P.D. (1986) Distribution and cycling of nutrients in *Pinus radiata* as affected by past lupin growth and fertiliser. *Forest Ecology and Management* 17, 169–187.

Ballard, R.A. (1996) Assessment of strains of Bradyrhizobium sp. (Lupinus) for serradellas (*Ornithopus* spp.). *Australian Journal of Experimental Agriculture* 36, 63–70.

Bergersen, F.J. (1994) Distribution of O_2 within infected cells of soybean root nodules: a new simulation. *Protoplasma* 183, 49–61.

Bergersen, F.J., Turner, G.L. Gault, R.R., Chase, D.L. and Brockwell, J. (1985) The natural abundance of ^{15}N in an irrigated soybean crop and its use for the calculation of nitrogen fixation. *Australian Journal of Agricultural Research* 36, 411–423.

Blight, A. (1989) Sensitivity of iron uptake, plant growth and nodulation of narrow leafed lupin genotypes to bicarbonate. BSc (Agric.) thesis, The University of Western Australia, Perth, Australia.

Boroñ, L. and Legocki, A.B. (1993) Cloning and characterization of a nodule-enhanced glutamine synthetase-encoding gene from *Lupinus luteus*. *Gene* 136, 95–102.

Boroñ, L., Szczyglowski, K., Konieczny, A. and Legocki, A.B. (1989) Glutamine synthetase in *Lupinus luteus*. Identification and preliminary characterization of nodule-specific cDNA clone. *Acta Biochimica Polonica* 36, 296–301.

Bottomley, P.J. (1992) Ecology of *Bradyrhizobium* and *Rhizobium*. In: Stacey, G., Burris, R.H. and Evans, H.J. (eds) *Biological Nitrogen Fixation*. Chapman & Hall, New York, pp. 293–348.

Bottomley, P.J., Cheng, H.H. and Strain, S.R. (1994) Genetic structure and symbiotic characteristics of a *Bradyrhizobium* population recovered from a pasture soil. *Applied and Environmental Microbiology* 60, 1754–1761.

Bowman, A.M., Hebb, D.M., Munnich, D., Rummery, G.L. and Brockwell, J. (1995) Field persistence of *Bradyrhizobium* sp. (*Lupinus*) inoculant for serradella (*Ornithopus compressus* L.). *Australian Journal of Experimental Agriculture* 35, 357–365.

Bushby, H.V.A. (1982) Ecology. In: Broughton, W.J. (ed.) *Nitrogen Fixation. Volume 2: Rhizobium*. Clarendon Press, Oxford, pp. 35–75.

Bushby, H.V.A. (1990) The role of bacterial surface charge in the ecology of root-nodule bacteria: an hypothesis. *Soil Biology and Biochemistry* 22, 1–9.

Caetano-Anolles, G., Lagares, A. and Favelukes, G. (1989) Adsorption of *Rhizobium meliloti* to alfalfa roots: dependence on divalent cations and pH. *Plant and Soil* 117, 67–74.

Carlson, R.W., Price, N.P.J. and Stacey, G. (1994) The biosynthesis of rhizobial lipo-oligosaccharide nodulation signal molecules. *Molecular Plant–Microbe Interactions* 7, 684–695.

Chalk, P.M. (1985) Estimation of N_2 fixation by isotope dilution: an appraisal of techniques involving ^{15}N enrichment and their application. *Soil Biology and Biochemistry* 17, 389–410.

Chalk, P.M., Smith, C.J., Hamilton, S.D. and Hopmans, P. (1993) Characterization of the N benefit of a grain legume (*Lupinus angustifolius* L.) to a cereal (*Hordeum vulgare* L.) by an *in situ* ^{15}N isotope dilution technique. *Biology and Fertility of Soils* 15, 39–44.

Chatel, D.L. and Greenwood, R.M. (1973) The location and distribution in soil of rhizobia under senesced annual legume pasture. *Soil Biology and Biochemistry* 5, 799–808.

Chatel, D.L. and Parker, C.A. (1973) Survival of field-grown rhizobia over the dry summer period in Western Australia. *Soil Biology and Biochemistry* 5, 415–423.

Cobley, B.T. (1993) Mechanisms of the impairment of lupin nodulation by high pH. BSc (Agric.) thesis, The University of Western Australia, Perth, Australia.

Corby, H.D.L. (1981) The systematic value of leguminous root nodules. In: Polhill, R.M. and Raven, P.H. (eds) *Advances in Legume Systematics, Part 1 and 2, Vol. 2 of the Proceedings of the International Legume Conference*. Royal Botanic Gardens, Kew, pp. 657–670.

Corbin, E.J. (1978) Lupin growing increasing in southern N.S.W. *Agriculture Gazette N.S.W.* 88, 9–11.

Cowling, W.A. and Clements, J.C. (1993) Association between collection site soil pH and chlorosis in *Lupinus angustifolius* induced by a fine textured alkaline soil. *Australian Journal of Agricultural Research* 44, 1821–1836.

Crespi, M.D., Jurkevitch, E., Poiret, M., d'Auberton-Carafa, Y., Petrovics, G., Kondorosi, E. and Kondorosi, A. (1994) Enod40, a gene expressed during nodule organogenesis, codes for a non-translatable RNA involved in plant growth. *European Molecular Biology Organisation Journal*, 21, 5099–5112.

De Lorenzo, C., Iannetta, P.P.M., Fernandez-Pascual, M., James, E.K., Lucas, M.M., Sprent, J.I., Witty, J.F., Minchin, F.R. and De Felipe, M.R. (1993) Oxygen diffusion in lupin nodules II. Mechanisms of diffusion barrier operation. *Journal of Experimental Botany* 44, 1469–1474.

Delves, A.C., Matthews, A., Day, D.A., Carter, A.S., Carroll, B.J. and Gresshoff, P.M. (1986) Regulation of the soybean–*Rhizobium* nodule symbiosis by shoot and root factors. *Plant Physiology* 82, 588–590.

Denison, R.F. and Kinraide, T.B. (1995) Oxygen-induced membrane depolarizations in legume root nodules: possible evidence for an osmoelectrical mechanism controlling gas permeability. *Plant Physiology* 108, 235–240.

Diaz del Castillo, L., Hunt, S. and Layzell, D.B. (1992) O_2 regulation and O_2-limitation of nitrogenase activity in root nodules of pea and lupin. *Physiologia Plantarum* 86, 269–278.

Dilworth, M.J. (1969) The plant as the genetic determinant of leghaemoglobin production in the legume root nodule. *Biochimica Biophysica Acta* 184, 432–441.

Downie, J.A. and Young, J.P.W. (1996) Evolution of specificity in legume nodulating rhizobia. In: Pate, J.S. (ed.) *Recent Progress in Research on Symbiotic Nitrogen Fixation*. CLIMA Occasional Publication 14, CLIMA, Nedlands, WA, 15 pp.

Doyle, A.D., Moore, K.J. and Herridge, D.F. (1988) The narrow-leafed lupin (*Lupinus angustifolius* L.) as a nitrogen-fixing rotation crop for cereal production. III. Residual effects of lupins on subsequent cereal yields. *Australian Journal of Agricultural Research* 39, 1029–1037.

Duthion, C., Amarger, N. and Marriotti, A. (1987) Accumulation potentielle de matière seche et d'azote chez le lupin blanc de printemps (*Lupinus albus* L.). *Agronomie* 7, 585–593.

Egorov, Ts.A., Feigina, M.Yu., Kazakov, V.K., Shakhparonov, M.I., Mitaleva, S.I. and Ovchinnikov, Yu.A. (1976) The complete amino acid sequence of leghemoglobin I from yellow lupin. *Bioorganicheskaya Khimiya* 2, 125–128.

Egorov, Ts.A., Kazakov, V.K., Shakhparonov, M.I., Ovchinnikov, Yu.A. and Feigina, M.Yu. (1978) The complete amino acid sequence of leghemoglobin II from yellow lupin (*Lupinus luteus*). *Bioorganicheskaya Khimiya* 4, 476–480.

Escuredo, P.R., Minchin, F.R., Gogorcena, Y., Iturbe-Ormaetxe, I., Klucas, R.V. and Becana, M. (1996) Involvement of activated oxygen in nitrate-induced senescence of pea root nodules. *Plant Physiology* 110, 1187–1195.

Evans, J., O'Connor, G.E., Turner, G.L. and Bergersen, F.J. (1987a) Influence of mineral nitrogen on nitrogen fixation by lupin (*Lupinus angustifolius*) as assessed by ^{15}N isotope dilution methods. *Field Crops Research* 17, 109–120.

Evans, J., Turner, G.L., O'Connnor, G.E. and Bergersen, F.J. (1987b) Nitrogen fixation and accretion of soil nitrogen by field-grown lupins (*Lupinus angustifolius*). *Field Crops Research* 16, 309–322.

Evans, J., O'Connor, G.E., Turner, G.L., Coventry, D.R., Fettell, N., Mahoney, J., Armstrong, E.L. and Walsgott, D.N. (1989) N_2 fixation and its value to soil N increase in lupin, field pea and other legumes in south-eastern Australia. *Australian Journal of Agricultural Research* 14, 791–805.

Fahey, T.J., Yavitt, J.B., Pearson, J.A. and Knight, D.H. (1985) The nitrogen cycle in lodgepole pine (*Pinus contorta* spp. *latifolia*) ecosystems, southeastern Wyoming. *Ecological Monographs* 53, 51–72.

Farrington, P., Greenwood, E.A.N., Titmanis, Z.V., Trinick, M.F. and Smith, D.W. (1977) Fixation, accumulation and distribution of nitrogen in a crop of *Lupinus angustifolius* cv. Unicrop. *Australian Journal of Agricultural Research* 28, 237–248.

Fernandez-Pascual, M., de Lorenzo, C., de Felipe, M.R., Rajalakshmi, S., Gordon, A.J., Thomas, B.J. and Minchin, F.R. (1996) Possible reasons for relative salt stress tolerance in nodules of white lupin. *Journal of Experimental Botany* 47, 1709–1716.

Forde, B.G. and Cullimore, J.V. (1989) The molecular biology of glutamine synthetase in higher plants. *Oxford Surveys of Plant Molecular and Cell Biology* 6, 247–296.

Gadgil, R. (1971) The nutritional role of *Lupinus arboreus* in coastal sand dune forestry. III. Nitrogen distribution in the ecosytem before tree planting. *Plant and Soil* 35, 113.

Gagnon, H., Tahara, S. and Ibrahim, R. (1995a) Biosynthesis, accumulation and secretion of isoflavonoids during germination and development of white lupin (*Lupinus albus* L.). *Journal of Experimental Botany* 197, 635–645.

Gagnon, H., Grandmaison, J. and Ibrahim, R.K. (1995b) Phytochemical and immuno-cytochemical evidence for the accumulation of 2′-hydroxylupalbigenin in lupin nodules and bacteroids. *Molecular Plant–Microbe Interactions* 8, 131–137.

Gault, R.R., Corbin, E.J., Boundy, K.A. and Brockwell, J. (1986) Nodulation studies on legumes exotic to Australia: *Lupinus* and *Ornithopus*. *Australian Journal of Experimental Agriculture* 26, 37–48.

Gibson, P.R. and Alston, A.M. (1987) Nitrogen balances in field-grown *Lupinus angustifolius* L. In: Bacon, P.E., Evans, J., Storrier, R.R. and Taylor, A.C. (eds) *Nitrogen Cycling in Temperate Agricultural Systems*, volume 1, Australian Society of Soil Science, Riverina Branch, Wagga Wagga, NSW, Australia, pp. 44–51.

Gladstones, J.S. (1970) Lupins as crop plants. *Field Crop Abstracts* 23, 123–148.

Graham, P.H. and Parker, C.A. (1964) Diagnostic features in the characterisation of root-nodule bacteria of legumes. *Plant and Soil* 20, 383–396.

Graham, P.H., Sadowski, M.J., Tighe, S.W., Thompson, J.A., Date, R.A., Howieson, J.G. and Thomas, R. (1995) Differences amongst strains of *Bradyrhizobium* in fatty acid–methyl ester (FAME) analysis. *Canadian Journal of Microbiology* 41, 1038–1042.

Grandmaison, J. and Ibrahim, R.K. (1995) Ultrastructural localization of a diprenylated isoflavone in *Rhizobium lupini–Lupinus albus* symbiotic association. *Journal of Experimental Botany* 46, 231–237.

Grant, M.R., Carne, A., Hill, D.F. and Farnden, K.J.F. (1989) The isolation and characterization of cDNA clone encoding *Lupinus angustifolius* root nodule glutamine synthetase. *Plant Molecular Biology* 13, 481–490.

Halvorson, J.J., Franz, E.H., Smith, J.L. and Black, R.A. (1992) Nitrogenase activity, nitrogen fixation, and nitrogen inputs by lupines at Mount St Helens. *Ecology* 73, 87–98.

Hamblin, J., Delane, R., Bishop, A. and Adam, G. (1993) The yield of wheat following lupins: effects of different lupin genotypes and management. *Australian Journal of Agricultural Research* 44, 645–659.

Hamilton, S.D., Smith, C.J. Chalk, P.M. and Hopmans, P. (1992) A model based on measurement of soil and plant ^{15}N enrichment to estimate N_2 fixation by soybean (*Glycine max* L. Merrill) grown in pots. *Soil Biology and Biochemistry* 24, 71–78.

Harrison, S.P., Mytton, L.R., Skot, L., Dye, M. and Cresswell, A. (1992) Characterisation of *Rhizobium* isolates by amplification of DNA polymorphisms using random primers. *Canadian Journal of Microbiology* 38, 1009–1015.

Hartmann, C. and Aldag, R. (1989) N_2 fixation and yield structure of white lupin (*Lupinus albus* L.) in comparison to *Vicia faba* L. and *Glycine max* (L.) Merr. on different sites. *Journal of Agronomy and Crop Science* 1633, 210–211.

Hartwig, U.A., Joseph, C.M. and Phillips, D.A. (1991) Flavonoids released naturally from alfalfa seeds enhance growth rate of *Rhizobium meliloti*. *Plant Physiology* 95, 797–803.

Haynes, R.J., Martin, R.J. and Goh, K.M. (1993) Nitrogen fixation, accumulation of soil nitrogen and nitrogen balance for some field-grown legume crops. *Field Crops Research* 35, 85–92.

Hendrickson, O.Q. and Burgess, D. (1989) Nitrogen-fixing plants in a cut-over lodgepole pine stand of southern British Columbia. *Canadian Journal of Forestry Research* 19, 936–939.

Herridge, D.F. and Doyle, A.D. (1988) The narrow-leafed lupin (*Lupinus angustifolius* L.) as a nitrogen-fixing rotation crop for cereal production. II Estimates of fixation by field-grown crops. *Australian Journal of Agricultural Research* 39, 1017–1028.

Howieson, J.G. (1995) Rhizobial persistence and its role in the development of sustainable agricultural systems in Mediterranean agriculture. *Soil Biology and Biochemistry* (Special Volume) 27, 603–610.

Howieson, J.G., Ewing, M.A., Robson, A.D. and Abbott, L.K. (1993) External phosphate and calcium concentrations, but not the products of rhizobial nodulation genes, affect the attachment of *Rhizobium meliloti* to roots of annual medics. *Soil Biology and Biochemistry* 25, 567–573.

Howieson, J.G., Reeve, N. and Yates, R. (1994) The selection of effective inoculant *Bradyrhizobium* sp. (*Lupinus*) for new lupin and serradella species. In: Dracup, M. and Palta, J. (eds) *Proceedings 1st Australian Lupin Technical Symposium*, Western Australian Department of Agriculture, Perth, pp. 270–273.

Hunt, S. and Layzell, D.B. (1993) Gas exchange of legume nodules and the regulation of nitrogenase activity. *Annual Review of Plant Physiology and Plant Molecular Biology* 44, 483–511.

Iannetta, P.P.M., De Lorenzo, C., James, E.K., Fernandez-Pascual, M., Sprent, J.I., Lucas, M.M., Witty, J.F., De Felipe, M.R. and Minchin, F.R. (1993a) Oxygen diffusion in lupin nodules. I. Visualisation of diffusion barrier operation. *Journal of Experimental Botany* 44, 1461–1467.

Iannetta, P.P.M., James, E.K., McHardy, P.D., Sprent, J.I. and Minchin, F.R. (1993b) An ELISA procedure for quantification of relative amounts of intercellular glycoprotein in legume nodules. *Annals of Botany* 71, 85–90.

Iannetta, P.P.M., James, E.K., Sprent, J.I. and Minchin, F.R. (1995) Time-course of changes involved in the operation of the oxygen diffusion barrier in white lupin nodules. *Journal of Experimental Botany* 46, 565–575.

Ingham, J.L., Tahara, S. and Harborne, J.B. (1993) Fungitoxic isoflavones from *Lupinus albus* and other *Lupinus* species. *Zeitschrift für Naturforschung* 38c, 194–200.

James, E.K., Sprent, J.I., Minchin, F.R. and Brewin, N.J. (1991) Intercellular localisation of glycoprotein in soybean nodules: effects of rhizosphere oxygen concentration. *Plant, Cell and Environment* 14, 467–476.

James, E.K., Minchin, F.R., Iannetta, P.P.M. and Sprent, J.I. (1997) Temporal relationships between nitrogenase and intercellular glycoprotein in developing white lupin nodules. *Annals of Botany* 79, 493–503.

Jessop, R.S and Mahoney, J. (1982) Effects of lime on the growth of four grain legumes. *Australian Journal of Soil Research* 20, 265–268.

Jessop, R.S., Roth, G. and Sale, P. (1990) Effects of increased levels of soil $CaCO_3$ on lupin (*Lupinus angustifolius*) growth and nodulation. *Australian Journal of Soil Research* 28, 955–962.

Jordan, D.C. (1982) Transfer of *Rhizobium japonicum* Buchanan 1980 to *Bradyrhizobium* gen. nov., a genus of slow-growing, root nodule bacteria from leguminous plants. *International Journal of Systematic Bacteriology* 32, 136–139.

Khan, S.I., Robson, A.D. and Parker, C.A. (1984) Comparison of the effects of acid soil factors on the growth of fast and slow growing species of *Rhizobium*. In: Kennedy, I.R. and Copeland, L. (eds) *Proceedings of 7th Australian Legume Nodulation Conference, Sydney*, pp. 151–152.

Konieczny, A., Szczyglowski, K., Boroñ, L., Przybylska, M. and Legocki, A.B. (1988) Expression of lupin nodulin genes during root nodule development. *Plant Science* 55, 145–149.

Kouchi, H. and Hata, S. (1993) Isolation and characterization of novel nodulin cDNAs representing genes expressed at early stages of soybean nodule development. *Molecular and General Genetics* 238, 106–119.

Lange, R.T. (1961a) Effective nodulation of *Lupinus digitatus* by native rhizobia in south-western Australia. *Plant and Soil* 15, 193–198.

Lange, R.T. (1961b) Nodule bacteria associated with the indigenous Leguminosae of south-western Australia. *Journal of General Microbiology* 61, 851–859.

Lange, R.T. and Parker, C.A. (1960) The symbiotic performance of lupin bacteria under glasshouse and field conditions. *Plant and Soil* 13, 137–146.

Larson, K.J., Cassman, K.G. and Phillips, D.A. (1989) Yield, dinitrogen fixation and aboveground nitrogen balance of irrigated white lupin in a mediterranean climate. *Agronomy Journal* 81, 538–543.

Layzell, D.B., Hunt, S. and Palmer, G.R. (1990) Mechanisms of nitrogenase inhibition of soybean nodules. Pulse-modulated spectroscopy indicates that nitrogenase activity is limited by O_2. *Plant Physiology* 92, 1101–1107.

Lea, P.J., Robinson, S.A. and Stewart, G.R. (1990) The enzymology and metabolism of glutamine, glutamate and asparagine. In: Miflin, B.J. (ed.) *The Biochemistry of Plants*, Volume 16. Academic Press, New York, pp. 121–159.

Ledgard, S.F. and Peoples, M.B. (1988) Measurement of nitrogen fixation in the field. In: Wilson, J.R. (ed.) *Advances in Nitrogen Cycling in Agricultural Systems*. CAB International, Wallingford, UK, pp. 351–367.

Legocki, A.B., Karlowski, W., Podkowiñski, J., Sikorski, M. and Stêpkowski, T. (1997) Advances in molecular characterization of the yellow lupin – *Bradyrhizobium* sp. (*Lupinus*) symobiotic model. In: Legocki, A.B., Bothe, H. and Pühler, A. (eds) *Biological Fixation of Nitrogen for Ecology and Sustainable Agriculture*. Springer Verlag, Heidelberg, NATO ASI Series, pp. 263–266.

Legros, T. and Smith, D.L. (1994) Root zone temperature sensitivity of nitrogen fixing and nitrate-supplied soybean [*Glycine max* (L.) Merr. cv Maple Arrow] and lupin (*Lupinus albus* L. cv. Ultra) plants. *Environmental and Experimental Botany* 34, 117–127.

Long, S.R. (1989) *Rhizobium* – legume nodulation: life together in the underground. *Cell* 56, 203–214.

Lopez-Lara, I.M., van den Berg, J.D.J., Thomas-Oates, J.E., Glushka, J.E., Lugtenberg, B.J.J. and Spaink, H.P. (1995) Structural identification of the lipo-chitin oligosaccharide signals from *Rhizobium loti*. *Molecular Microbiology* 14, 567–578.

Macknight, R.C., Reynolds, P.H.S. and Farnden, K.J.F. (1995) Analysis of the lupin Nodulin-45 promoter: conserved regulatory sequences are important for promoter activity. *Plant Molecular Biology* 27, 457–466.

Marshall, K.C. (1967) Electrophoretic properties of fast- and slow-growing species of *Rhizobium*. *Australian Journal of Biological Sciences* 20, 429–438.

Matvienko, M., van de Sande, K., Yang, W.-C., van Kammen, A., Bisseling, T. and Fransen, H. (1994) Comparison of soybean and pea ENOD40 cDNA clones representing genes expressed during both early and late stages of nodule development. *Plant Molecular Biology* 26, 487–493.

McInnes, A., Abbott, L.K., Howieson J.G. and Thies, J.E. (1996) Field populations of bradyrhizobia associated with serradella are diverse and effective. In: *Proceedings of the 11th Australian Nitrogen Fixation Conference, Perth*, pp. 11–12.

Minchin, F.R. (1997) Regulation of oxygen diffusion in legume nodules. *Soil Biology and Biochemistry* 29, 881–888.

Minchin, F.R., Sheehy, J.E. and Witty, J.F. (1986) Further errors in the acetylene reduction assay: effects of plant disturbance. *Journal of Experimental Botany* 37, 1581–1591.

Minchin, F.R., Iannetta, P.P.M., Fernandez-Pascual, M., de Lorenzo, C., Witty, J.F. and Sprent, J.I. (1992) A new procedure for the calculation of oxygen diffusion resistance in legume nodules from flow-through gas analysis data. *Annals of Botany* 70, 283–289.

Minchin, F.R., Iannetta, P.P.M., James, E.K., Sprent, J.I., Thomas, B.J. and Witty, J.F. (1995) Mechanisms for short-term (15 min.) changes in the oxygen diffusion barrier of legume nodules. In: Tikhonovich, I.A., Provorov, N.A., Romanov, V.I. and Newton, W.E. (eds) *Nitrogen Fixation: Fundamentals and Applications*. Kluwer Academic, Dordrecht, p. 600.

Mock, I.T. and Gibson, A.H. (1994) Evaluation of *Lupinus* species and strains of *Bradyrhizobium lupini* for improved production potential on calcareous soils. In: Dracup, M. and Palta, J. (eds) *Proceedings 1st Australian Lupin Technical Symposium*, Western Australian Department of Agriculture, Perth, pp. 182–186.

Norris, D.O. (1956) Legumes and the *Rhizobium* symbiosis. *Empire Journal of Experimental Agriculture* 24, 247–270.

Norris, D.O. (1965) Acid production by *Rhizobium* – a unifying concept. *Plant and Soil* 22, 143–166.

Nutman, P.S. (1976) Alternative sources of nitrogen for crops. *Journal of the Royal Agriculture Society of England* 137, 86–94.

Pacault, D., Roi, S., Huyghe, C. and Aksland, G. (1994) Lime tolerance in white lupin: definition of a screeening test under controlled conditions. In: Dracup, M. and Palta, J. (eds) *Proceedings 1st Australian Lupin Technical Symposium*, Western Australian Department of Agriculture, Perth, p. 287.

Palmason, F., Danso, S.K.A. and Hardarson, G. (1992) Nitrogen accumulation in sole and mixed stands of sweet-blue lupin (*Lupinus angustifolius* L.), ryegrass and oats. *Plant and Soil* 142, 135–142.

Parker, C.A. (1957) Evolution of nitrogen fixing symbiosis in higher plants. *Nature* 179, 593–594.

Parker, C.A. and Oakley, A.E. (1963) Nodule bacteria for two species of serradella – *Ornithopus sativus* and *Ornithopus compressus*. *Australian Journal of Experimental Agriculture and Animal Husbandry* 3, 9–10.

Parker, C.A. and Oakley, A.E. (1965) Reduced nodulation of lupins and serradella due to lime pelleting. *Australian Journal of Experimental Agriculture and Animal Husbandry* 5, 144–146.

Parsons, R. and Day, D.A. (1990) Mechanism of soybean nodule adaptation to different oxygen pressures. *Plant, Cell and Environment* 13, 501–512.

Pate, J.S., Unkovich, M.J., Armstrong, E.L. and Sanford, P. (1994) Selection of reference plants for ^{15}N natural abundance assessment of N_2 fixation by crop and pasture legumes in southwest Australia. *Australian Journal of Agricultural Research* 45, 133–147.

Peoples, M.B., Faizah, A.W., Rerkasem, B. and Herridge, D.F. (1989) *Methods for Evaluating Biological Nitrogen Fixation by Nodulated Legumes in the Field.* Monograph 11, Australian Centre for International Agricultural Research, Canberra.

Phillips, D.A., Dakora, F.D., Sande, E., Joseph, C.M. and Zon, J. (1994) Synthesis, release and transmission of alfalfa signals to rhizobial symbionts. *Plant and Soil* 161, 69–80.

Phillips, D.A., Wery, J., Joseph, C.M., Jones, A.D. and Teuber, L.R. (1995) Release of flavonoids and betaines from seeds of seven *Medicago* species. *Crop Science* 35, 805–808.

Podkowiñski, J., Karlowski, W.M., Kondorosi, A., Legocki, A.B. and Crespi, M. (1996) Two ENOD40 sequences from *Lupinus luteus.* In: *Abstracts of the 2nd European Nitrogen Fixation Conference.* Scientific Publishers OWN, Poznań, Poland, p. 94.

Poupot, R., Martinez-Romero, E. and Prome, J.-C. (1993) Nodulation factors from *Rhizobium tropici* are sulfated or nonsulfated chitopentasaccharides containing an *N*-methyl-*N*-acylglucosaminyl terminus. *Biochemistry* 32, 10430–10435.

Pueppke, S.G. (1996) The genetic and biochemical basis for nodulation of legumes by rhizobia. *Critical Reviews in Biotechnology* 16, 1–51.

Rae, A.L., Perotto, S., Knox, J.P., Kannenberg, E.L. and Brewin, N.J. (1991) Expression of extracellular glycoproteins in the uninfected cells of developing pea nodule tissue. *Molecular Plant–Microbe Interactions* 4, 563–570.

Rennie, R.J. (1979) Comparisons of ^{15}N aided methods for determining symbiotic dinitrogen fixation. *Revue d'Ecologie et de Biologie du Sol* 16, 453–463.

Rerkasem, B. (1977) Differential sensitivity to soil acidity of legume–*Rhizobium* symbioses. PhD thesis, The University of Western Australia, Perth, Australia.

Rice, S.J., Grant, M.R., Reynolds, P.H.S. and Farnden, K.J.F. (1993) DNA sequence of Nodulin-45 from *Lupinus angustifolius. Plant Science* 90, 155–166.

Richardson, A.E., Viccars, L.A., Watson, J.M. and Gibson, A.H. (1995) Differentiation of *Rhizobium* strains using the polymerase chain reaction with random and directed primers. *Soil Biology and Biochemistry* 27, 515–524.

Roughley, R.J., Gemmell, L.G., Thompson, J.A. and Brockwell, J. (1993) The number of *Bradyrhizobium leguminosarum* sp. (*Lupinus*) applied to seed and its effect on rhizosphere colonisation, nodulation and yield of lupin. *Soil Biology and Biochemistry* 25, 1453–1458.

Russell, C.A. (1996) Mineralisation of nitrogen from crop-lupin residue components. PhD thesis, University of Western Australia, Nedlands, Perth, Western Australia.

Russell, C.A. and Fillery, I.R.P. (1996) Estimates of lupin below-ground biomass nitrogen dry matter and nitrogen turnover to wheat. *Australian Journal of Agricultural Research* 47, 1047–1059.

Serraj, R., Fleurat-Lessard, P., Jaillard, B. and Drevon, J.J. (1995) Structural changes in the inner-cortex cells of soybean root nodules are induced by short-term exposure to high salt or oxygen concentrations. *Plant, Cell and Environment* 18, 455–462.

Sheehy, J.E. and Thornley, J.H.M. (1988) Oxygen, the *NifA* gene, nodule structure and the initiation of nitrogen fixation. *Annals of Botany* 61, 605–609.

Sikorski, M.M., Szybiak-Strŏzycka, U., Strŏzycki, P., Goliñska, B., Mldrzak, C.J., Kamp R.M., Wittmann-Liebold, B. and Legocki, A.B. (1989) Coordinated synthesis of leghemoglobin and root protein R18 in yellow lupin. *Acta Biochimica Polonica*, 36, 63–72.

Sikorski, M.M., Szlagowska, A.E. and Legocki, A.B. (1996) cDNA sequences encoding for two homologs of *Lupinus luteus* (L.) proteins (Accession nos. X79974 and X79975 for LlR18A and LlR18B mRNAs, respectively). *Plant Physiology* 110, 335.

Slattery, J.F. and Coventry, D.R. (1989) Populations of *Rhizobium lupini* in soils used for cereal–lupin rotations in north-eastern Victoria. *Soil Biology and Biochemistry* 21, 1009–1010.

Smith, C.J., Chalk, P.M., Hamilton, S.D. and Hopmans, P. (1992) Estimating N_2 fixation by field-grown lupins (*Lupinus angustifolius* L.) using soil and plant enrichment. *Biology and Fertility of Soils* 13, 235–241.

Smith, S.C., Bezdicek, D.F., Turco, R.F. and Cheng, H.H. (1987) Seasonal N_2 fixation by cool-season pulses based on several ^{15}N methods. *Plant and Soil* 97, 3–13.

Soupene, E., Foussard, M., Boistard, P., Truchet, G. and Batut, J. (1995) Oxygen as a key development regulator of *Rhizobium meliloti* N_2-fixation gene expression within the alfalfa root nodule. *Proceedings of the National Academy of Sciences (USA)* 92, 3759–3763.

Spaink, H.P., Sheeley, D.M., Van Brussel, A.A.N., Glushka, J.N., York, W.S., Tak, T., Geiger, O., Kennedy, E.P., Reinhold, V.N. and Lugtenberg, B.J.J (1991) A novel, highly unsaturated, fatty acid moiety of lipo-oligosaccharide signals determines host specificity of *Rhizobium leguminosarum*. *Nature* 354, 125–130.

Sprent, J. (1989) Which steps are essential for the formation of functional legume nodules? Tansley Review 15. *New Phytologist* 111, 129–153.

Sprent, J.I. and Silvester, W.B. (1973) Nitrogen fixation by *Lupinus arboreus* grown in the open and under different aged stands of *Pinus radiata*. *New Phytologist* 72, 991–1003.

Stobiecki, M. and Wojtaszek, P. (1990) Application of gas chromatography–mass spectrometry to the indenfication of isoflanoids in lupin root extracts. *Journal of Chromatography* 508, 391–398.

Strózycki, P.M. and Legocki, A.B. (1995) Leghemoglobins from an evolutionarily old legume, *Lupinus luteus*. *Plant Science* 110, 83–93.

Strózycki, P.M., Karlowski, W.M. and Legocki, A.B. (1994) *Lupinus luteus* leghemoglobin genes. Structure and regulation of expression. In: Kiss, G.B. and Endre, G. (eds) *Proceedings of the 1st European Nitrogen Fixation Conference*. Officina Press, Szeged, Hungary, p. 342.

Subramaniam, P., Paradinas, R.M. and Rodriguez-Barrueco, C. (1982) Nodulation and growth of *Lupinus angustifolius* L. as influenced by various pH levels of the rooting medium. In: *Proceedings of the 2nd International Lupin Conference, Torremolinos, Spain*, pp. 134–138.

Sullivan, J.T., Patrick, H.N., Lowther, W.L., Scott, D.B. and Ronson, C.W. (1995) Nodulating strains of *Rhizobium loti* arise through chromosomal gene transfer in the environment. *Proceedings of the National Academy of Sciences (USA)* 92, 8985–8989.

Swiderski, M. (1997) Biological fixation of nitrogen for ecology and sustainable agriculture. In: Legocki, A.B., Bothe, H. and Puhler, A. (eds) *NATO ASI Series*, Springer-Verlag, Berlin.

Swiderski, M. and Legocki, A.B. (1996) New cDNA clones coding nodulins from yellow lupin. In: *Abstracts of the 2nd European Nitrogen Fixation Conference*. Scientific Publishers OWN, Poznań, Poland, p. 101.

Szczyglowski, K. and Legocki, A.B. (1990) Isolation and nucleotide sequence of cDNA clone encoding nodule-specific (hydroxy)proline-rich protein LENOD2 from yellow lupin. *Plant Molecular Biology* 15, 361–363.

Szczyglowski, K., Boroñ, L., Szybiak-Stròzycka, U. and Legocki, A.B. (1989) Characterization of cDNA clone coding for Nodulin-45 from yellow lupin (*Lupinus luteus*). *Plant Science* 6, 87–95.

Tahara, S., Orihara, S., Ingham, J.L. and Mizutani, J. (1989) Seventeen isoflavonoids from *Lupinus albus* roots. *Biochemistry* 28, 901–911.

Tang, C. and Robson, A.D. (1993) pH above 6.0 reduces nodulation in *Lupinus* species. *Plant and Soil* 152, 269–276.

Tang, C. and Robson, A.D. (1995) Nodulation failure is important in the poor growth of two lupin species on an alkaline soil. *Australian Journal of Experimental Agriculture* 35, 87–91.

Tang, C., Robson, A.D. and Dilworth, M.J. (1991) A split root experiment shows that iron is required for nodule initiation in *Lupinus angustifolius* L. *New Phytologist* 115, 61–67.

Tang, C., Longnecker, N.E., Thomson, C.J., Greenway, H. and Robson, A.D. (1992a) Lupin (*Lupinus angustifolius* L.) and pea (*Pisum sativum*) roots differ in their sensitivity to pH above 6.0. *Journal of Plant Physiology* 140, 715–719.

Tang, C., Robson, A.D. and Dilworth, M.J. (1992b) The role of iron in the *(Brady)rhizobium*–legume symbiosis. *Journal of Plant Nutrition* 15, 2235–2252.

Tang, C., Kuo, J., Longnecker, N.E., Thomson, J.C. and Robson, A.D. (1993a) High pH causes disintegration of the root surface in *Lupinus angustifolius* L. *Annals of Botany* 71, 201–207.

Tang, C., Robson, A.D., Longnecker, N.E. and Greenway, H. (1993b) Physiological responses of lupin roots to high pH. *Plant and Soil* 155/6, 509–512.

Tang, C., Robson, A.D., Longnecker, N.E. and Buirchell, B.J. (1995) The growth of *Lupinus* species on alkaline soils. *Australian Journal of Agriculture* 46, 225–268.

Thompson, J.A. (1960) Inhibition of nodule bacteria by an antibiotic from legume seed coats. *Nature* 187, 619.

Thumfort, P.P., Atkins, C.A. and Layzell, D.B. (1994) A re-evaluation of the role of the infected cell in the control of O_2 diffusion in legume nodules. *Plant Physiology* 105, 1321–1333.

Unkovich, M.J., Pate, J.S. and Hamblin, J. (1994) The nitrogen economy of broadacre lupin in southwest Australia. *Australian Journal of Agricultural Research* 45, 149–164.

Unkovich, M.J., Pate, J.S. and Sanford, P. (1997) Nitrogen fixation by annual legumes in Australian mediterranean agriculture. *Australian Journal of Agricultural Research* 48, 267–293.

Van Cauwenberghe, O.R., Hunt, S., Newcomb, W., Canny, M.J. and Layzell, D.B. (1994) Evidence that short-term regulation of nodule permeability does not occur in the inner cortex. *Physiologia Plantarum* 91, 477–487.

Van Loon, L.C., Pierpoint, W.S., Boller, T.H. and Canejero W. (1994) Recommendations for naming plant pathogenesis-related proteins. *Plant Molecular Biology Reporter* 12, 245–264.

VandenBosch, K.A., Bradley, D.J., Knox, J.-P., Perotto, S., Butcher, G.W. and Brewin, N.J. (1989) Common components of the infection thread matrix and the intercellular space identified by the immunocytochemical analysis of pea nodules and uninfected roots. *European Molecular Biology Organization Journal* 8, 335–342.

Vijn, I., Yang, W.-C., Pallisgard, N., Jensen, E.Ø., van Kammen, A. and Bisseling, T. (1995) *VsENOD5*, *VsENOD12* and *VsENOD40* expression during *Rhizobium*-induced nodule formation on *Vicia sativa* roots. *Plant Molecular Biology* 28, 1111–1119.

White, P.F. (1990) Soil and plant factors relating to the poor growth of *Lupinus* species on fine-textured, alkaline soils – a review. *Australian Journal of Agricultural Research* 41, 871–890.

White, P.F. and Robson, A.D. (1989) Lupin species and peas differ widely in their sensitivity to iron deficiency. *Australian Journal of Agricultural Research* 40, 539–547.

Wisniewski, J.P. and Delmotte, F.M. (1966) Modulation of carbohydrate binding capacities and attachment ability of *Bradyrhizobium* sp. (*Lupinus*) to white lupin roots. *Canadian Journal of Microbiology* 42, 234–242.

Wittenberg, B.J. and Wittenberg, B.A. (1990) Mechanism of cytoplasmic leghaemoglobin and myoglobin function. *Annual Review of Biophysics and Biophysical Chemistry* 19, 217–241.

Witty, J.F. and Minchin, F.R. (1994) A new method to detect the presence of continuous gas-filled pathways for oxygen diffusion in legume nodules. *Journal of Experimental Botany* 45, 967–978.

Witty, J.F., Minchin, F.R., Skot, L. and Sheehy, J.E. (1986) Nitrogen fixation and oxygen in legume root nodules. *Oxford Surveys of Plant Molecular and Cell Biology* 3, 275–314.

Witty, J.F., Skot, L. and Revsbeck, N.P. (1987) Direct evidence for changes in the resistance of legume root nodules to O_2 diffusion. *Journal of Experimental Botany* 38, 1129–1140.

Yang, W.-C., Katinakis, P., Hendriks, P., Smolders, A., de Vries, F., Spee, J., van Kammen, A., Bisseling, T. and Fransen, H. (1993) Characterization of GmENOD40, a gene showing novel pattern of cell-specific expression during soybean nodule development. *Plant Journal* 3, 573–585.

7 Transport Physiology and Partitioning

J.S. PATE, R.J.N. EMERY AND C.A. ATKINS

Centre for Legumes in Mediterranean Agriculture (CLIMA), University of Western Australia, Nedlands, Western Australia 6907

Introduction

Transport processes, and the related partitioning of water and essential solutes, provide the mechanistic base and resources through which the processes of uptake, assimilation, growth, storage and reproduction are consummated and continuously integrated during plant development. Within this background, routes for transport are extensive at both macro and micro scaling; absolute and relative amounts of solutes transported in different channels change dramatically through plant growth and, with each time-dependent change in supply and demand of constituent parts, partitioning processes alter qualitatively and quantitatively according to highly precise programmes. The ultimate aim of the student of plant transport should be to determine how these processes and events, and the structural, physiological and biochemical features that underpin their operation, are organized and regulated at the molecular level. As will be seen from the ensuing account, we remain far from achieving these somewhat grandiose objectives as far as lupins are concerned.

The major thrust of this chapter is to assemble information on specific transport phenomena which occur in lupins. Firstly the complexities of structure relating to vascular transport will be considered in terms of bulk long-distance transport of water and solutes in xylem and phloem and the equally important short-distance transport phenomena which load, unload or exchange solutes at various locations along the two conducting systems. There then follows a detailed series of case studies of transport of two vital macronutrients, carbon (C) and nitrogen (N), to illustrate the highly complex manner in which their distribution in phloem and xylem impacts upon the growth and vital functioning of key source or sink organs of the plant. The case studies examined comprise the photosynthesizing leaf, growing fruit and N_2-fixing nodule.

Gleaning information from the above approach, the chapter then examines some of the principal mechanisms underlying the partitioning of C and N at whole plant level, using an empirical modelling approach in which transport exchanges between plant parts during specific intervals of growth are related quantitatively to the assimilatory inputs and demands relating to this pair of elements. The same modelling procedure is then extended to piece together how certain mineral nutrients are distributed and utilized, selecting as an example the ability displayed by lupin for cation discrimination at sites of loading and unloading of phloem and xylem. Such selectivity engenders highly characteristic differences in partitioning of Na^+ versus K^+, and between this pair of cations and Mg^{2+} and Ca^{2+}. Particular emphasis is placed in this section on the pivotal role of stems in differential distribution of element resources within the plant body.

In a further part of the chapter we return to C and N partitioning to examine how lupin plants adapt to artificially induced pertubations in their assimilatory inputs of one element relative to the other. Responses in variously treated and control plants are followed in terms of short-term changes in composition of phloem and xylem streams at different locations or as longer-term repercussions in organ composition and growth and eventual C : N balance of seeds. Appended to this section are results of a study in which $^{14}CO_2$ labelling of field-grown plants at different stages in the life cycle is used to monitor the phasing of transfer of C and N to fruits and specific reserves of seeds.

The chapter then switches to considerations of the transport and partitioning of growth substances within the plant. Basic difficulties involved in such studies are highlighted, especially in relation to prescribing definitively where such components are synthesized and transported and defining possible roles played by cytokinins, gibberellins and abscisic acid in the coordination of growth, activity and functional longevity of certain vegetative organs, reproductive parts and the rest of the plant body.

A final section of the chapter highlights what was said earlier regarding the inadequacy of knowledge regarding the programming and regulation of transport processes within target tissues of source and sink organs of the system, especially in relation to how differential loading and unloading of major organic solutes are continuously modulated during plant development. Radical new approaches and techniques applied at the molecular level will be needed for such study and future prospects for these will be discussed against the background of what is already known at a macro level concerning the transport arrangements of lupins.

Somewhat regrettably, transport processes of *Lupinus* have been studied exhaustively in only two species, *Lupinus albus* (white lupin) and *Lupinus angustifolius* (narrow-leafed lupin), and research on these has been mostly channelled towards partitioning of C and N. The relevant work has been conducted mostly in Perth and so the chapter contains an embarrassingly large proportion of parochially oriented references and data which may be viewed as idiosyncratic of Western Australian conditions. The programme of work presented derives from the fortuitous discovery some 25 years ago that certain annual species of lupin, including the principal grain crop species *Lupinus albus* and *L. angustifolius*, bleed prolifically from cut phloem of their leaf stalks, stems and fruits. As far as we are aware, *Lupinus* spp. are unique in this respect amongst herbaceous grain legumes. They thus provide unrivalled opportunities for assessing coordinated transport processes in phloem and xylem and the relationship of these to organ and whole plant growth. Whilst the data bank for lupins remains mostly descriptive and lacking in content at cell level, it has already provided physiologists with new insight as to how the processes of photosynthesis, N_2 fixation and nutrient uptake interact collectively during the growth cycle of a herbaceous crop legume.

Vascular Architecture and Long-distance Pathways of Solute Transport

Knowledge of the vascular anatomy of a plant, particularly of the vascular connections between leaves, roots, stems and reproductive parts, has obvious relevance to a general appreciation of how assimilates of various kinds might be distributed. Arrangement of leaves on shoots of *L. albus* and *L. angustifolius* is spiral with a phyllotaxis of 2/5, i.e. with an angle of approximately 72° between points of insertion of successive pairs of leaves on the stem (O'Neill, 1961). As might be expected of any randomly inherited attribute, close to half of the plants in a population have their leaves borne spirally in an anticlockwise manner, the other half in a clockwise arrangement. Patterns of insertion of flowers and fruits on each terminal inflorescence subscribe to the same spiral arrangement exhibited by leaves on the stem immediately below.

Xylem and phloem accompany one another throughout virtually all parts of the vascular network of the plant, but sectioning of vascular tissue serving different organs reveals variation with location in the proportion of conducting sieve elements of phloem relative to vessels and tracheids of xylem. For example, in stems and roots of *L. albus* the transectional area of conducting xylem greatly exceeds that of phloem, as might be expected of organs providing all of the xylem-based water utilized by foliar surfaces of the shoot. By contrast, fruits of the same species transpire weakly and depend mostly on phloem intake for supplying the bulk of their assimilates; and, again as might be predicted, they have vascular tissue in the fruit stalk which contains a 2–4 times greater transectional area of phloem sieve tubes than xylem-conducting elements (Pate *et al.*, 1978). Nevertheless the total xylem region of the fruit stalk is large, relative to phloem, because of the massive development of non-conducting xylem fibres to support the weight of the subtended fruit. As a final example the peripheral vascular strands surrounding the N_2-fixing bacterial tissue of nodules possess almost equal cross-sectional areas of conducting xylem and phloem (Pate *et al.*, 1969), a feature consistent with heavy traffic of phloem sugar and other assimilates into the nodule and a matching return flow in xylem of fixed N in the form of amino compounds.

Although interconnecting vascular networks between different ages and classes of organ may be seen to provide collectively a potential for transport from top to very bottom of the plant, appropriately chosen labelling studies are obviously needed to test which pathways are currently in use and which are of major importance in transporting specific classes of solutes between specific source and sink regions. $^{14}CO_2$ feeding of single leaves, coupled with study of distribution of exported ^{14}C assimilates, comprises the most widely used technique for providing such information. Studies of this kind in lupins (Pate *et al.*, 1980a; Pate and Layzell, 1981) have shown that upper leaves generally feed young leaves, vegetative apices and reproductive organs at the top of the branch on which they are located, whereas lower leaves feed lower stem, root and nodules, and mid-level leaves donate assimilates to organs both above and below their points of insertion. Effects due to phyllotaxis come into play at finer scale, as shown by the highly defined patterns of photosynthate flow which occur in *L. angustifolius* from upper main-stem leaves to individual flowers of the adjacent terminal inflorescence (Pate and Farrington, 1981). Assimilate flow is then shown to be confined largely to flowers inserted vertically above – that is, of the same orthostichy as the fed leaf (Fig. 7.1). However, once secondary thickening of the upper stem and inflorescence takes place during fruiting, affinities based on phyllotaxis become blurred since the capacity then exists for lateral movement of assimilates through the ring of secondary thickening in the fruit stalk. A fruit which has lost its original nurse leaf through disease, grazing or other debilitating agency may thus be 'rescued' nutritionally by cross-transfer of assimilates from leaves not normally supplying the fruit in question (Pate and Farrington, 1981).

Short-distance Transport and its Relationships with Partitioning through Xylem and Phloem

Before any long-distance channel can operate to full capacity, solutes must be loaded rapidly at one end and removed equally effectively at destinations along

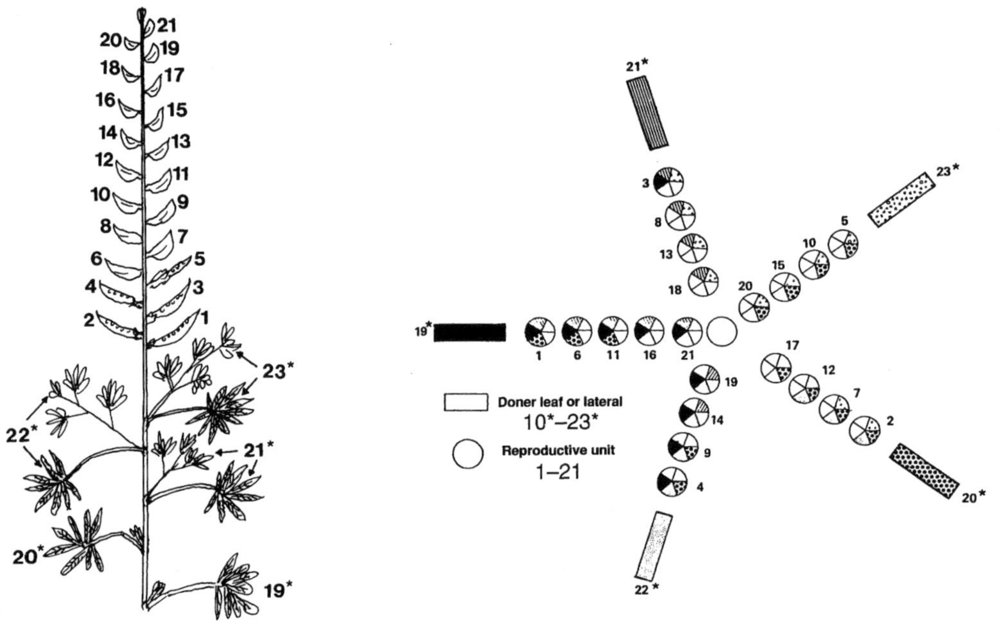

Fig. 7.1. Pattern of flow of ^{14}C assimilates to flowers and young fruits on the main-stem inflorescence of *Lupinus angustifolius* after feeding the topmost five leaves of the same stem with $^{14}CO_2$. Donor leaves (rectangles numbered upwards from 19 to 23) and flowers and fruits (circles numbered from 1 up to 21) are arranged according to the 2/5 phyllotaxis shown by the species. The extents to which reproductive units received photosynthate from different leaves are indicated using the coding system indicated. (From Pate, 1984b.)

or at the ends of the route in question. Added to this, phloem sieve tubes and xylem vessels represent long-distance transport conduits operating within two physically separate compartments of the plant – phloem transport occurring through the membrane-bound cytoplasmic or 'symplast' compartment, and xylem transport involving bulk flow of water and dilute solutes through the surrounding extracellular or 'apoplast' compartment. Hence, for effective mixing to occur between solute streams generated by the root in the xylem and by the mesophyll of photosynthesizing leaves, appropriately programmed mechanisms must exist for short-distance exchanges between apoplast and symplast across a whole series of traffic stations within the plant system (Pate and Gunning, 1972; Pate and Jeschke, 1995).

As is also the case for many other herbaceous plants, exchanges of the aforementioned nature are accomplished in lupin mostly through the agency of specialized parenchyma called transfer cells (Gunning and Pate, 1969; Pate and Gunning, 1972). Each such cell carries unique finger-like ingrowths of cellulose on strategically located parts of its wall, thus increasing greatly the surface area of plasma membrane exposed at the apoplast : symplast junction concerned. Because of this amplified membrane and associated ultrastructural features which promote highly effective solute uptake and transfer within its cytoplasm, the transfer cell is viewed as providing especially high efficiency in the short-distance transport of a diverse range of solutes at an equally diverse number of locations within or at the surface boundaries of the plant. Activity of the cell type in this respect has been shown to be highly selective, regardless of whether the fluxes involved comprise secretion from or absorption into the symplast of the compartment of the plant in which the transfer cells are located (Pate and Gunning, 1972; Gunning and Pate, 1974).

Transfer cells occur in legumes such as lupin at a range of potential 'bottlenecks' in the transport system. For example, they promote exchange between xylem and phloem in minor veins of leaves; they load xylem with fixation products in nodules; and they engage in transfers of materials between different nuclear generations of mother sporophyte tissues, nurse tissues of endosperm and the growing embryo of the developing seed (Pate, 1979). Figure 7.2 indicates principal locations for transfer cells in the vegetative body of a herbaceous legume such as lupin and depicts how respective sites and presumed functions relate to the apoplast and symplast compartments of the plant, its xylem and phloem channels and its primary access zones for water, minerals and fixation of CO_2 and N_2.

Transport of Nitrogen in Xylem and Phloem

Lupins may be regarded as classic amide plants, synthesizing large quantities of asparagine, and to a lesser extent glutamine, when N_2 is fixed in nodules or when nitrate or ammonium is assimilated in roots (Pate, 1980, 1984a; Pate and Atkins, 1983a). These same two compounds predominate in transport of N through xylem and phloem and in virtually all situations together carry 70% or more of the N moving in a specific transport pathway or exchanged between one channel and another (Pate *et al.*, 1979a,b; Atkins *et al.*, 1980b; Pate, 1989c). Even when nitrate is available at high levels in the rooting substrate it is reduced principally in the root and lower stem, so that reduced N as amide rather than free nitrate constitutes the major form of N ascending into the shoot via xylem (Pate *et al.*, 1975; Sharkey and Pate, 1975, 1976; Atkins *et al.*, 1979). As is the case in other plants, nitrate is absent or present at only trace levels in phloem. This applies even to sand-cultured plants whose roots are being fed concentrations of nitrate ten or more times those ever likely to be experienced by field-grown plants (Atkins *et al.*, 1979).

A number of other amino compounds, particularly aspartate, glutamate, threonine, serine, alanine, glycine and valine, accompany amides in both xylem and phloem. As a general rule the ratio of amide-N to nitrogen in the form of

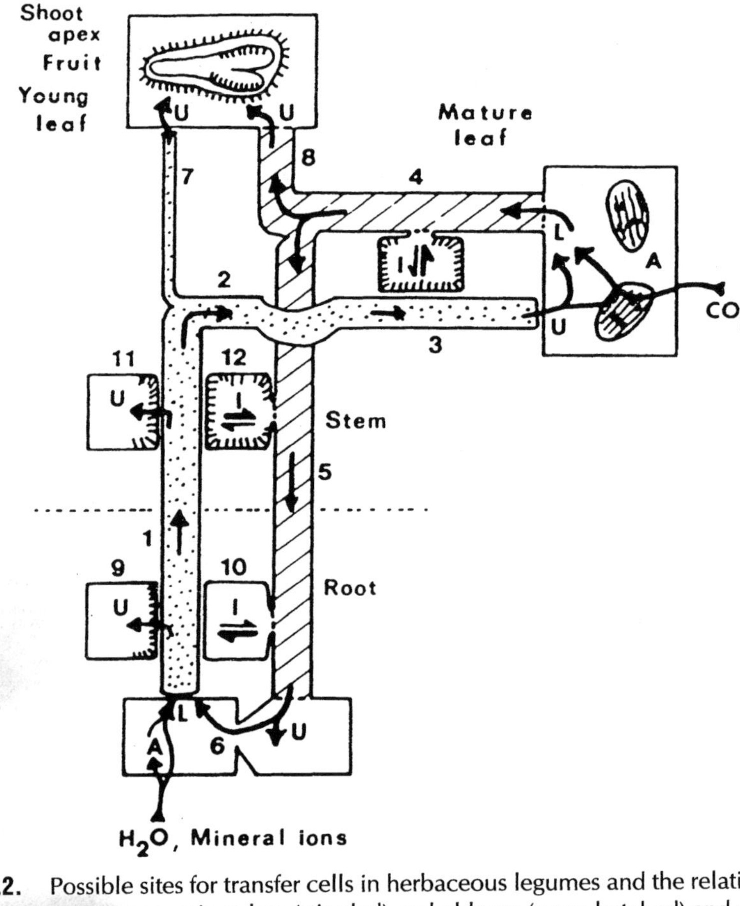

Fig. 7.2. Possible sites for transfer cells in herbaceous legumes and the relationships of these sites to transport in xylem (stippled) and phloem (cross-hatched) and organs providing, assimilating or accumulating carbon, nitrogen, minerals and water. U, unloading and L, loading of vascular channels; A, sites of assimilation; 1–6 with arrows denote the circulatory system of the plant; 7, 8, unidirectional flow from mature organs to sinks; 9, loading of fixation products into xylem in nodule; 10–12, short-distance lateral exchanges of metabolites. (With modifications from Pate, 1972.)

these other amino compounds is higher in xylem than in phloem. As shown using studies following the fate of ^{14}C- or ^{15}N-labelled amino compounds fed singly to cut transpiring shoots in the xylem (Sharkey and Pate, 1975; McNeil *et al.*, 1979; Atkins *et al.*, 1980b), asparagine is broken down in substantial amounts on arrival in the leaf and the N of its amide group can be traced to other amino acids, particularly those formed during photosynthesis – as identified by $^{14}CO_2$ feeding studies. A comprehensive picture has thus been obtained (Pate, 1979) of the relative extents to which the incoming xylem stream as opposed to the mesophyll of the leaf contribute amino compounds to the phloem translocate exiting from the leaf. The scheme proposed (Fig. 7.3) depicts the relative extent to which amino compounds involved in transfer reactions in the leaf vein donate carbon to specific amino compounds of phloem translocate. Direct xylem-to-phloem transfer, presumably within the minor veins of the leaf, emerges as the dominant activity in phloem loading of asparagine, glutamine, threonine and valine, whereas photosynthesis associated with metabolism of amides provides exportable amino acid-related N attached mostly to serine, alanine, aspartic acid, glutamic acid and glycine.

A complication not considered so far is that not all of the N moving into the shoot through xylem ever reaches the leaf and cannot therefore participate in the reactions portrayed in Fig. 7.3. The relevant N is abstracted mostly by lower stem tissue, particularly at the stem nodes, as has been demonstrated vividly by autoradiography of shoots of *L. albus* following feeding of different ^{14}C-labelled amino acids (McNeil *et al.*, 1979). Such studies show that the basic amino acid arginine is adsorbed most strongly by stems and very little reaches the leaves. This may account for the much lower proportional amount of arginine in phloem than in ascending xylem sap (Pate *et al.*, 1979a,b). Significant amounts of asparagine, glutamine, valine and serine are similarly absorbed laterally from the xylem by stem tissue, but heavy deposition in leaves is consistent with the exchange reactions already indicated in Fig. 7.3. In marked contrast to the situation described above for arginine, autoradiography of shoots fed with the acidic amino acids ^{14}C-glutamate or ^{14}C-aspartate exhibit very weak retention in vascular tissue of stem or leaf veins, indicating that the bulk of the label runs the gauntlet of the vascular system to accumulate in unusually great amount in interveinal (mesophyll) regions of the leaf. A special function of these two dicarboxylic amino acids in nourishing leaf tissues has been suggested (Pate *et al.*, 1981). Xylem sap becomes unusually high in aspartate relative to other amino compounds when plants experience N deficiency, indicating involvement of the compound in maintaining N balance of the leaf under such conditions.

A second feature relating to N transfer reactions of stem tissues concerns the xylem-to-phloem transfer processes which take place in uppermost regions of the shoot. Evidence for the efficacy of such transfers comes from the finding that phloem sap collected from tops of vegetative shoots or inflorescence stalks of flowering or fruiting shoots is very much richer in amino compounds relative to sugars than is the phloem sap obtained from petioles of the leaves from which such phloem streams originated (Pate *et al.*, 1979a; Pate, 1986a). Comparisons of the amino acid spectra of the related phloem streams of parent leaf and upper stem of plants of *L. albus* or *L. angustifolius* have consistently indicated that xylem-to-phloem flux in upper stems throughout the day mainly involves direct enrichment of phloem with the amides asparagine and glutamine, whereas absolute and relative levels of sucrose and non-amide amino compounds remain

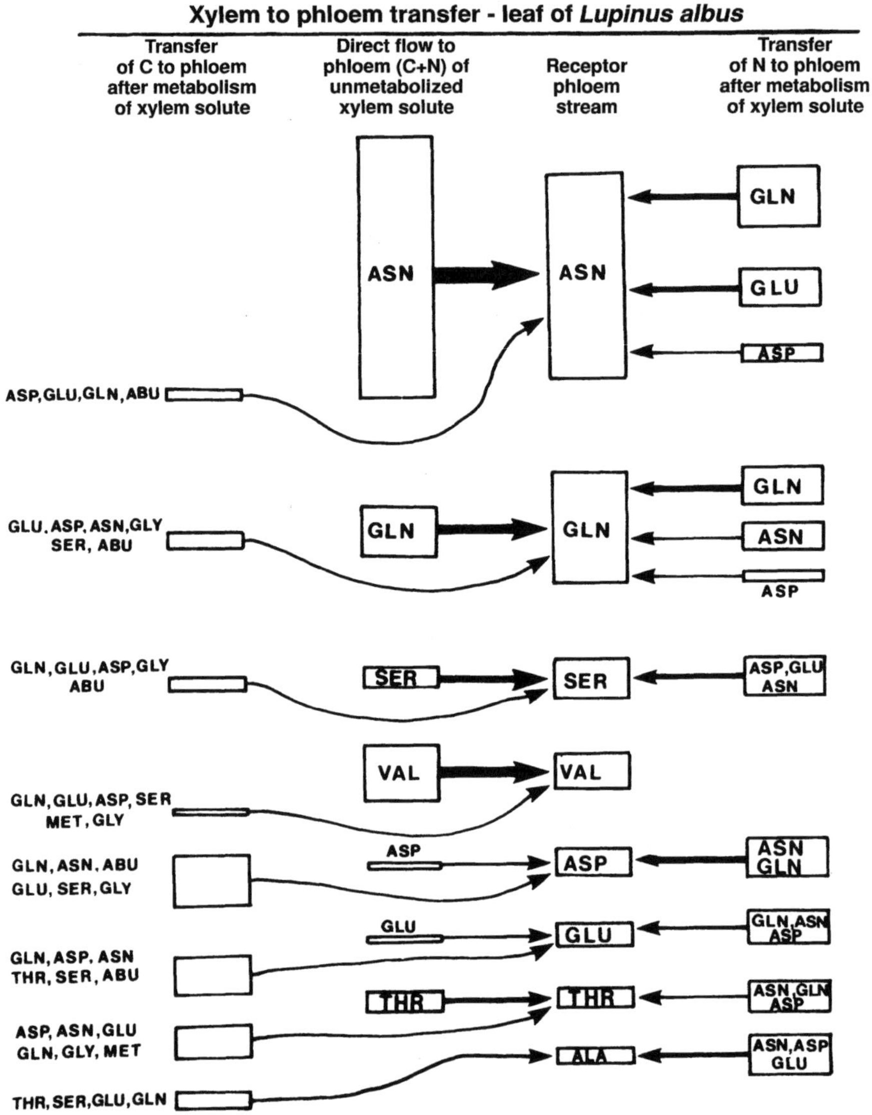

Fig. 7.3. Scheme depicting the relative extents to which each amino acid of the phloem stream translocated from a leaf acquires C from amino acids arriving from the root in the xylem or from photosynthetic fixation of C in the leaf. The extent of traffic via each amino acid is indicated by width of the rectangles representing the various amino acids in the upper and lower horizontal rows of the figure. The central block of the figure shows amino acids drawn with rectangles of size in proportion to abundance of each compound in the phloem on a carbon basis. (With modification from Pate, 1972.)

closely similar during transit to the apex (Fig. 7.4). As a result of such intake of amides, the C : N ratio of the upper stem phloem stream is lowered to a value of as little as 20 : 1 from a starting value of 60 : 1 or higher in the sugar-rich parent stream exported from the upper nurse leaves (Pate *et al.*, 1979a; Pate and Layzell, 1981; Pate and Jeschke, 1995). The significance of this to C and N partitioning in the whole plant will be discussed later.

Transport of Carbon in Phloem and Xylem

Wherever phloem sap is collected in lupins, sucrose comprises 98% or more of the translocated sugar component (Pate *et al.*, 1974, 1975, 1979a; Sharkey and Pate, 1976). C is also carried in phloem as amino acids and organic acids and a host of other minor organic solutes, but these rarely add up to more than 10% of the C in a sample of translocate. Thus sucrose transport mechanisms concerned with loading and unloading of phloem, working in concert with those relating

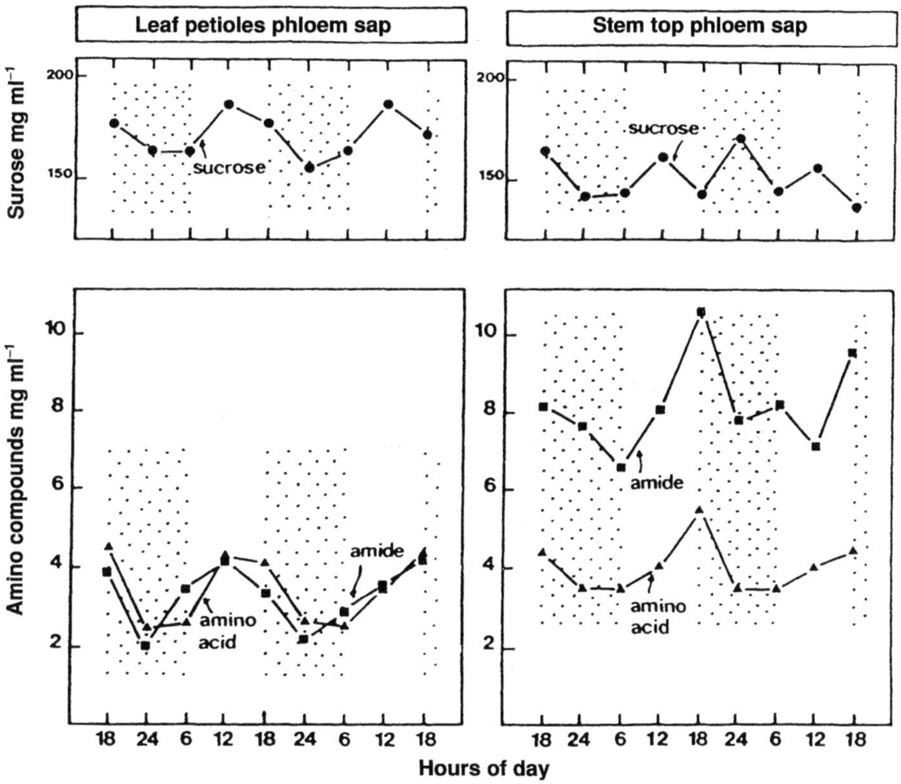

Fig. 7.4. Diurnal profiles of changes in sucrose, amide and amino acid concentrations in phloem sap collected from petioles of upper stem feeding the shoot apex of *L. albus* and in the same phloem stream intercepted by collecting sap from the subterminal parts of the stem. Note enrichment of phloem with amide but not appreciably with amino acids as it passes to the shoot apex. (From Pate 1986b, using data of Pate *et al.*, 1979b.)

to asparagine, comprise pivotal systems for determining phloem sap composition at different locations within the plant. The balance in activity between these two loading processes varies across very broad limits, judging from the wide range of ratios of sugar to amino compounds that can be encountered in phloem streams sampled at various locations at different times in the life of a plant (Pate *et al.*, 1979a). The full implications of such variations are explored below where the nutritional relationships of single organs or the whole plant are discussed.

Case Studies of Phloem and Xylem Transport in Organ Development and Function

C and N balance of a single leaf

Knowledge of how a leaf interacts with its neighbours, and the manner in which its demands for and subsequent export of C and N relate to the nutrition of the whole plant, is obviously paramount to understanding how transport exchanges continuously shape organ and whole plant growth and development. A study of this kind (Pate and Atkins, 1983b) conducted throughout the life of the uppermost main-stem leaf of white lupin has followed net day and night exchanges of C as CO_2 between leaf and atmosphere, net changes in C and N of leaf dry matter and changes in C : N weight ratios of the solute streams of xylem and phloem passing between leaf and parent plant. The resulting analysis is summarized diagrammatically in Fig. 7.5 for each of three cardinal periods in the life of the leaf, namely (A) the Importing phase, (B) the Principal Growth phase and (C) the Mature phase.

During the Importing phase (0–11 days, Fig. 7.5A), ending with the leaf having reached approximately one-third of its final maxima in dry weight C and N contents and area, the C and N balances of the leaf depict continued import of both elements through xylem and phloem. The bulk of the C incorporated into dry matter is imported at this time since photosynthetic gains of C only slightly exceed respiratory losses of C at night. Phloem imports much more C than xylem, whereas the reverse applies to N import. More than 90% of the asparagine entering the leaf is metabolized, with the amide group being released through the activity of asparaginase (Atkins *et al.*, 1983).

The Principal Growth phase (Fig. 7.5B) (11–20 days) commences with the leaf graduating to become a net exporter of C and N through phloem. It grows very rapidly throughout this period, reaching clear-cut maxima in area, C and N content and photosynthetic rate at or near the end of the phase. Import of N through xylem is especially rapid and most is immediately incorporated into chloroplasts and hence towards augmenting the photosynthetic potential of the leaf. With asparagine the principal imported N solute, key enzymes (asparaginase and to a lesser extent asparagine aminotransferase) involved in its utilization are particularly active in making both amino and amide groups available for amino acid and protein synthesis (Atkins *et al.*, 1983). Phloem export peaks at a C : N ratio of over 100 : 1 during the phase, indicative of escalating sugar production in photosynthesis but limited cross-transfer of amino N from xylem to phloem. Mean net daily intake of C from the atmosphere occurs at 20 times the rates of C loss in night respiration, and this is associated with highest rates of photosynthetic activity being recorded during this stage. The bulk of the net C gained photosynthetically during the phase is exported, rather than incorporated into leaf dry matter.

The Mature phase of leaf functioning (Fig. 7.5C) (20–66 days) is marked by progressively declining rates of photosynthesis but denotes a time when virtually all of the net photosynthate of the leaf is continuously exported. Due to loss of N from leaf dry matter, slightly more N cycles back out of the leaf in the phloem than enters through xylem. C : N ratios in phloem sap decline throughout the phase as photosynthesis wanes to a relatively faster extent than does xylem import of N. Losses of N from the fabric of the leaf continue to provide additional

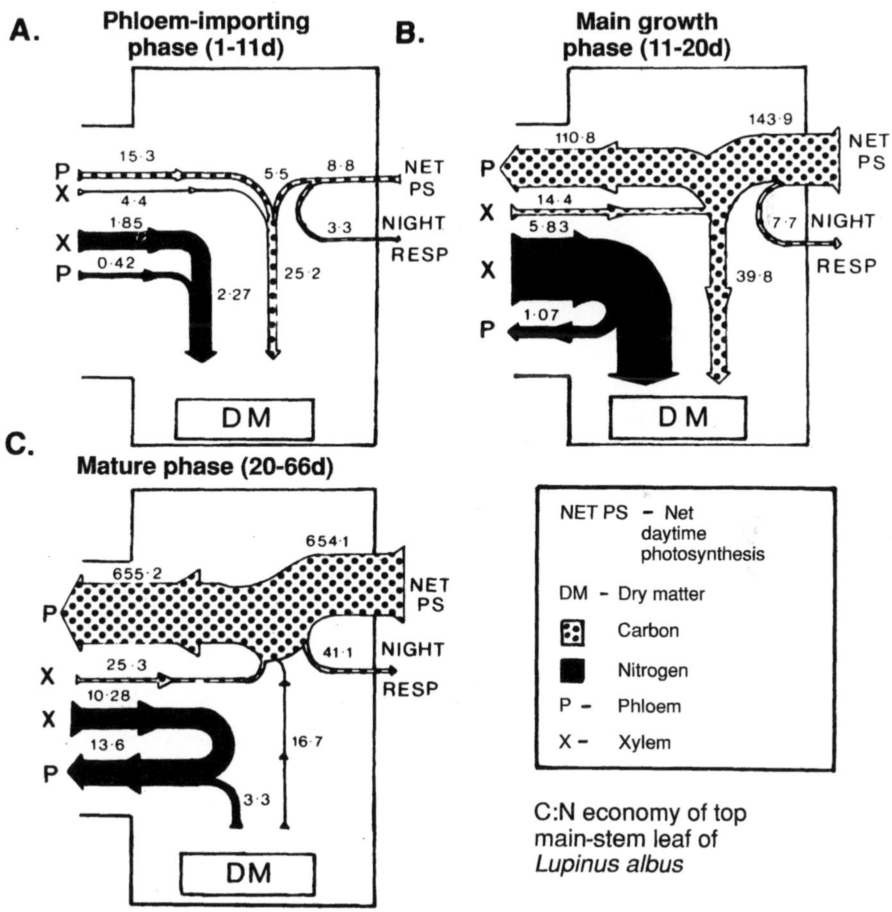

Fig. 7.5. Economies of C and N in the uppermost main-stem leaf of nodulated N_2-dependent plant of *L. albus*. Values for gaseous exchanges of C, increments or losses in C and N in leaf dry matter and exchanges of C and N in phloem and xylem between leaf and parent plant are recorded as mg C or mg N per leaf for each prescribed sequential period of leaf development (A–C). (From Pate, 1989b.)

supplies of N for phloem loading. During this time the activity of asparaginase declines to almost zero (Atkins *et al.*, 1983). However, the leaf retains the potential for asparagine metabolism through maintenance of high levels of an alternative enzymic system comprising asparagine aminotransferase and amino acid oxidase.

By combining the data from all three of the above three phases of leaf growth and functioning, one arrives at a summary picture of functioning of the whole life of the leaf. This indicates that, of the 7 mg N which the leaf initially invests into its structure and predominantly into its photosynthetic apparatus, 94% is imported through the xylem and only 6% through phloem. During its life the leaf imports a total of 18.4 mg N, and of this 11.4 mg cycles immediately back out of the leaf by xylem-to-phloem exchange, a further 3.3 mg N is mobilized prior to senescence and the remaining 3.7 mg remains as non-recoverable N when the leaf dies or abscises. The corresponding C budget for the leaf shows export of a massive 807 mg C as net photosynthate compared with a loss of only 52 mg C in night respiration. A final residue of 65 mg C remains as skeletal C upon death of the leaf. The data as a whole highlight the efficiency with which leaf surfaces of lupin operate in both fixing C and processing and redistributing the fixed N which they receive from the nodulated root system. The collective functioning of the differently aged strata of leaves in these respects is considered further when C and N partitioning within the whole plant is discussed below.

C, N and water relations of developing fruits

Fruits of plants possess generally low ratios of surface area to volume and are capable of extremely high rates of deposition of dry matter into the reserve materials of their seeds. Where fruits are chlorophyllous, as in the case of lupin, they show much lower densities of stomates and commensurately lower rates of transpiration and CO_2 fixation than do comparable foliar surfaces of the same plant (Pate and Kuo, 1981; Pate *et al.*, 1985; Atkins *et al.*, 1986; Pate, 1988, 1989a). An assessment of the combined economy of C, N and H_2O in fruits of white lupin (Pate *et al.*, 1977) has demonstrated how the above features collectively impact on the transport physiology of the fruit. Using information obtained during fruit development on the C and N increments in dry matter of seed and pod, the CO_2 exchanges which occur between fruit and surrounding atmosphere and the transpiration losses and gains or losses of water effected by seed and pod, a picture can first be obtained of the net weekly influxes of C, N and H_2O into pod and seed from the parent plant. Then, using parallel information on the C : N ratios of the xylem and phloem streams serving the fruit and absolute concentrations of C and N in these same streams, one arrives at the model depicted in Fig. 7.6. This essentially indicates what proportions of the C, N and H_2O acquired by the fruit enter by xylem and phloem. The assessments are based on the mixture of phloem and xylem streams which meet precisely the measured increments of C and N and corresponding CO_2 exchanges, transpiration losses and tissue water budgets of the fruit.

Figure 7.6 shows that, for every 100 units of C imported by the parent fruit, 98% enters by phloem and only 2% by xylem. For corresponding N import, 89%

arrives by phloem and 11% by xylem, while for water 40% enters as mass flow of phloem translocate and the major fraction (60%) through xylem. As is to be expected, intake of water by xylem matches closely the transpiration loss of the fruit.

The fruits of lupin, like those of most legumes, contain chlorophyll and the question arises as to whether they engage in sufficient photosynthesis to

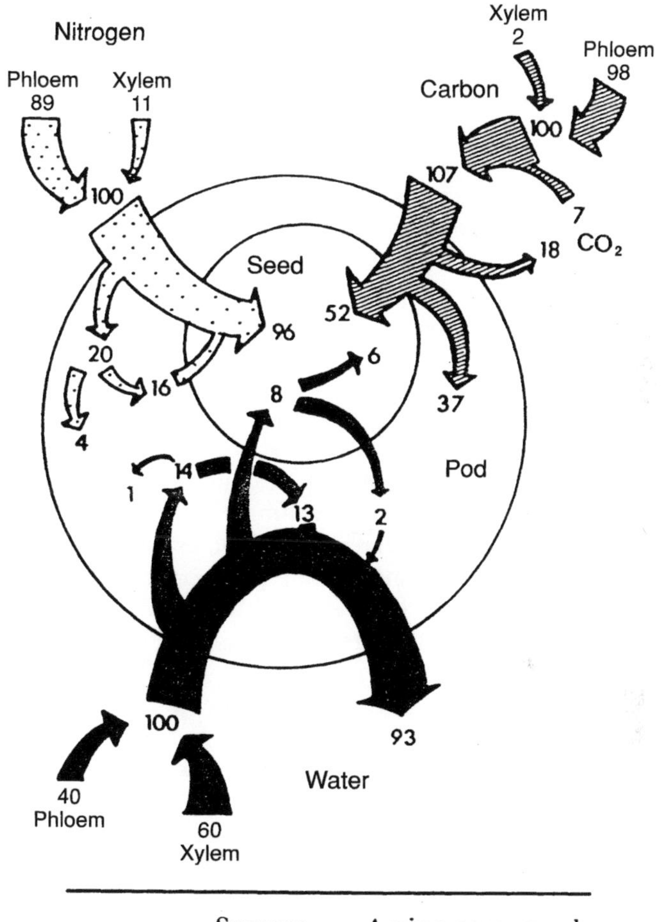

	Sucrose	Amino compounds
Xylem	0	1
Phloem	42	9

Fig. 7.6. Economy of C, N and H_2O for the whole 12-week growth cycle of a fruit of white lupin. Proportional intake by weight of C : N : H_2O by the whole fruit was 12 : 1 : 600 and the percentage intake of C, N and H_2O by fruit and import ratios of the seed by xylem and phloem are indicated relative to amounts of these commodities by weight incorporated into the seed or transpired or lost as respired CO_2 by the fruit. (From Pate, 1983, using data of Pate *et al.*, 1977.)

contribute significantly to seed yield. Measurements of gas exchange of *L. albus* fruits exposed to full sunlight in the field indicate net daytime gains in CO_2 from one week to ten weeks of their development. This accounts for about 7% of the C gain over the fruit's life (Fig. 7.6). However, darkening the fruit shows that photosynthesis serves to refix some of the CO_2 which would otherwise be lost from the fruit – sufficient, in fact, to reduce potential respiratory losses in the photoperiod by about 50%. Taken together, the conservation of translocated C plus the small net gain due to photosynthesis accounts for about one seed per pod if it is well illuminated (Pate *et al.*, 1977). Similar studies in peas highlighted the inner pod wall epidermis as the principal site of refixation, involving both RuBISCO and phosphoenolpyruvate (PEP) carboxylase (Atkins *et al.*, 1977; Flinn *et al.*, 1977). In lupin the developing cotyledons contain high levels of chlorophyll as well as the two carboxylases and it seems likely that the seeds themselves also conserve respired CO_2 through photosynthetic fixation (Atkins and Flinn, 1978).

Within the above budgets, phloem sucrose provides the bulk source of C for synthesis of reserves of carbohydrate, oil and fibrous components of seed and pod, while asparagine provides 50–70% of the N for synthesis of seed protein. Just as in the leaf, asparaginase is the key enzyme providing N for amino acid synthesis and reserve protein deposition in cotyledons (Atkins *et al.*, 1975). The overall ratio of C : N : H_2O imports for the 12-week period of fruit growth of *L. albus* turns out to be 12 : 1 : 600 (Pate *et al.*, 1977) and this translates to an efficiency of water use equivalent to a transpiration loss of 22.5 ml g^{-1} DM gain of the fruit (Pate, 1988, 1989a). Exceptionally high water use efficiency of this order of course reflects bulk intake of sucrose and other organic solutes, the transpirational penalty for manufacture of which has been incurred elsewhere in photosynthesizing parts of the parent plant. Expressed in another way, the overall effectiveness of the fruit in converting imported solutes into seed components may be determined once the weight of the solutes imported and the corresponding amounts of dry matter, protein, oil and soluble carbohydrate materials laid down in the seed are known (Pate, 1984a,b, 1985b). Table 7.1 presents such information for white lupin and two other legumes: pea and cowpea. Note the lower effectiveness of lupin in terms of producing seed dry matter per unit weight of all imported solutes but highest effectiveness of the three species when expressed in terms of protein production from imported solutes.

As a negative side to the excellent water economy of the fruit, low net intake of xylem fluid may result in the fruit running critically short of mineral elements mobile in xylem but absent or only sparingly mobile in phloem (Hocking and Pate, 1977, 1978; Pate and Hocking, 1978; Hocking *et al.*, 1978). Ca and Mn are classic examples of this and a specific Mn deficiency disorder known as 'split seed' of lupins has been documented for sweet (low alkaloid) cultivars of *L. angustifolius* (Hocking *et al.*, 1977). The details of Mn nutrition and the split seed disorder are discussed fully in Chapter 5.

The C and N relationships of N_2-fixing nodules

Nodules of lupin comprise at most 10–15% of the total fresh weight of the plant, yet in fully symbiotic plants they are responsible for fixing and exporting virtually

Table 7.1. Indices for conversion of translocated organic solutes into seed products by fruit of three species of grain legumes (from Pate, 1984b).

Item	Species		
	Garden pea (*Pisum sativum* cv. Greenfeast)	Cowpea(*Vigna unguiculata* cv. Vita 3)	White lupin (*Lupinus albus* cv. Ultra)
1. Organic solutes entering fruits in xylem and phloem (mg per fruit)			
Sucrose, organic acids	2404	3437	1756
Amides, amino acids	363	660	384
Ureides	30	72	—
Total	2797	4169	2140
2. Reserve materials laid down in seeds (mg per fruit)			
Total dry matter	1930	3220	1070
Protein	387	719	412
Oil	19	30	132
Perchloric acid-soluble carbohydrate	857	1803	110
3. Conversion efficiency by weight of imported solutes into seed components (%)			
Into total dry matter	69	77	50
Into total protein	14	17	19
Into total protein, oil and $HClO_4^-$ soluble carbohydrate	45	61	31

the entire N requirement of the rest of the plant. During their functioning, actively fixing nodules will attract through phloem the equivalent of up to 20% of the photosynthate currently produced by the shoot (Pate, 1996). To function to this effectiveness in terms of utilizing C and generating reduced N, they must possess transport systems engaging very intensively in exchange of organic solutes with the host plant. Furthermore, these exchanges are accomplished through vascular tissues which comprise only a few per cent of the total volume of the nodule (Pate *et al.*, 1969; Gunning *et al.*, 1974) and so efficiencies of operation of transport per unit volume or surface area of nodule must be inordinately high. Moreover, nodules are poor absorbers of water from the surrounding soil and cannot rely to any great extent, if at all, on surface uptake of water to flush nitrogenous solutes into the plant transpiration stream. Indeed, it has been suggested that the bulk of the water required in xylem export of N is actually supplied by the phloem stream to the nodule. As testimony to their extreme frugality in water use in xylem, sap bleeding from the cut xylem elements of freshly detached nodules such as lupin, pea and broad bean is found to contain asparagine and other nitrogenous compounds at concentrations close to their limits of solubility (Pate *et al.*, 1969, 1980b; Gunning *et al.*, 1974; Pate and Atkins, 1983a).

Turning more generally to the C and N transport arrangements of the nodule in relation to nodule growth and functioning, a modelling procedure based on information on N_2 fixation rates, respiratory outputs of CO_2 and gains of C and N in dry matter of growing nodules of white lupin has been used to compare the economies of functioning in nodules of cowpea and lupin. The resulting scheme is illustrated in Fig. 7.7, with values for lupin marked with asterisks below those for cowpea.

Note that for every 100 units of C imported through phloem by nodules of lupin, over half (55%) is lost in nodule respiration, 19% incorporated into new nodule tissue and the remaining 26% exported in the form of amino acids formed during assimilation of the ammonia generated by N_2-fixing bacteroids. The weight ratio of C imported to N fixed is 100 : 15 (i.e. an import of 6.6 mg C, or the rough equivalent of 17 mg carbohydrate mg^{-1} N fixed) and the model attempts to relate the observed respiration rate of the nodule to a range of functions which would be expected to utilize varying amounts of ATP and reductant. Corresponding data for cowpea nodules in Fig. 7.7 show greater economy in terms of C utilized per unit of N fixed, principally because of the use

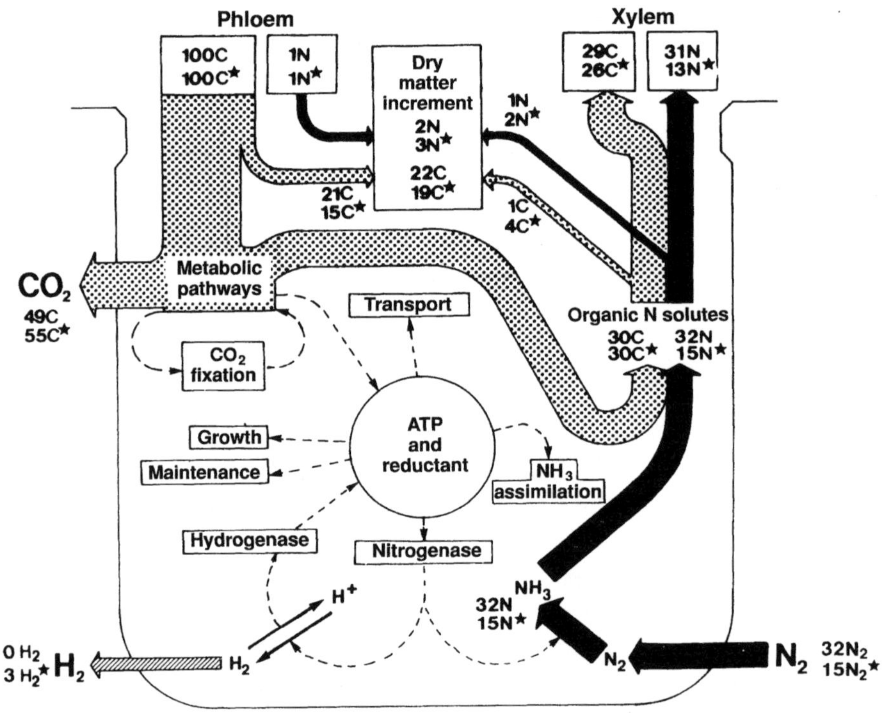

Fig. 7.7. Relationships between phloem intake of carbon and N_2 fixation of nodules of two grain legumes: cowpea and white lupin. Items of budgets for C and N for each species are expressed on a weight basis relative to an intake by the nodule of 100 units by weight of carbon as translocated carbohydrate. Values for cowpea are given above those for lupin (marked by stars). Items of the budget of the nodule which would be expected to use or generate ATP and reductant are indicated in relation to the observed respiratory loss of C by the nodule. Note that the nodules of the lupin association illustrated evolved H_2, while those for cowpea did not do so. (Redrawn from Layzell *et al.*, 1979.)

of ureides (C : N ratio 1 : 1) for export of N as opposed to asparagine (C : N ratio 2 : 1) in lupin.

Budgets of the above kind have proved important to understanding how efficiently nodules of different legume species operate in respect of C utilization. One can, for example, determine whether differences exist between nodules formed on the one legume species by different strains of nodule bacteria. Similarly, at a more detailed level, the modelling procedure allows one to assess to what extent differences in the nitrogenase, hydrogenase and H_2 evolving activities of nodules might impact upon their C economy (Layzell *et al.*, 1979; Pate *et al.*, 1981; Rainbird *et al.*, 1983a,b). With this background in mind it becomes possible to appreciate how costings relating to nodules of symbiotically reliant plants differ from those of non-nodulated counterparts subsisting solely on nitrate. Studies of this nature have been made in our laboratory for lupin (Atkins *et al.*, 1979; Pate *et al.*, 1979b), certain other herbaceous grain legumes (Atkins *et al.*, 1980a; Pate, 1986b, 1996; Pate and Layzell, 1990) and other species (Pate, 1986a). For *L. albus* there is a measurably increased C cost for N derived from fixation compared with N derived from NO_3 reduced primarily in the roots; 58–59% of the plant's total net photosynthate compared with 50% (Pate *et al.*, 1979b). This was reflected in a greater loss of respired CO_2 and was due to an increased cost for N reduction, for the growth and maintenance of nodules and for a greater rate of respiration of the supporting roots of nodulated plants compared with those supplied NO_3.

Partitioning Within the Whole Plant

Exchanges of C, N and inorganic cations between shoot and root

The simplest way of portraying partitioning within the whole plant is to determine over a specific interval of growth how much of a specific nutrient element passes between shoot and root in phloem and xylem, and to express how these flow attributes relate to the total amounts of that element absorbed from the external environment and incorporated into dry matter of root and shoot (Pate *et al.*, 1979c; Jeschke *et al.*, 1985, 1987; Pate and Jeschke, 1995). Figure 7.8 provides such information for symbiotically fed plants of white lupin for C, an element fixed by the shoot, N, an element provided exclusively by N_2 fixing nodules, and four nutrient cations (K^+, Na^+, Mg^{2+}, Ca^{2+}), all absorbed by the root but partitioned between shoot and root very differently from one another. A number of features emerge. The model for C shows (Fig. 7.8) the large sink associated with respiration of roots and nodules but highlights the appreciable fraction of this downward-directed C which returns back to the shoot via the xylem. As we showed earlier, this cycling component is attached mostly to amino compounds synthesized in and subsequently exported from the nodule. The related model for N partitioning (Fig. 7.8) depicts the shoot as a major sink for N and indicates that growing roots of symbiotically dependent plants gain their N indirectly in the form of phloem translocated from the shoot, rather than having N transferred directly from adjacent nodules (Oghoghorie and Pate, 1972).

The four flow profiles for cations (Fig. 7.8) rank these in descending order of phloem mobility as $K^+ > Na^+$, $Mg^{2+} > Ca^{2+}$ and also differentiate between the cations in terms of marked ability to cycle between shoot and root by phloem-to-xylem transfer in the root. The latter is clearly evident for K^+ and Mg^{2+} but not to any appreciable extent for Ca^{2+} and Na^+.

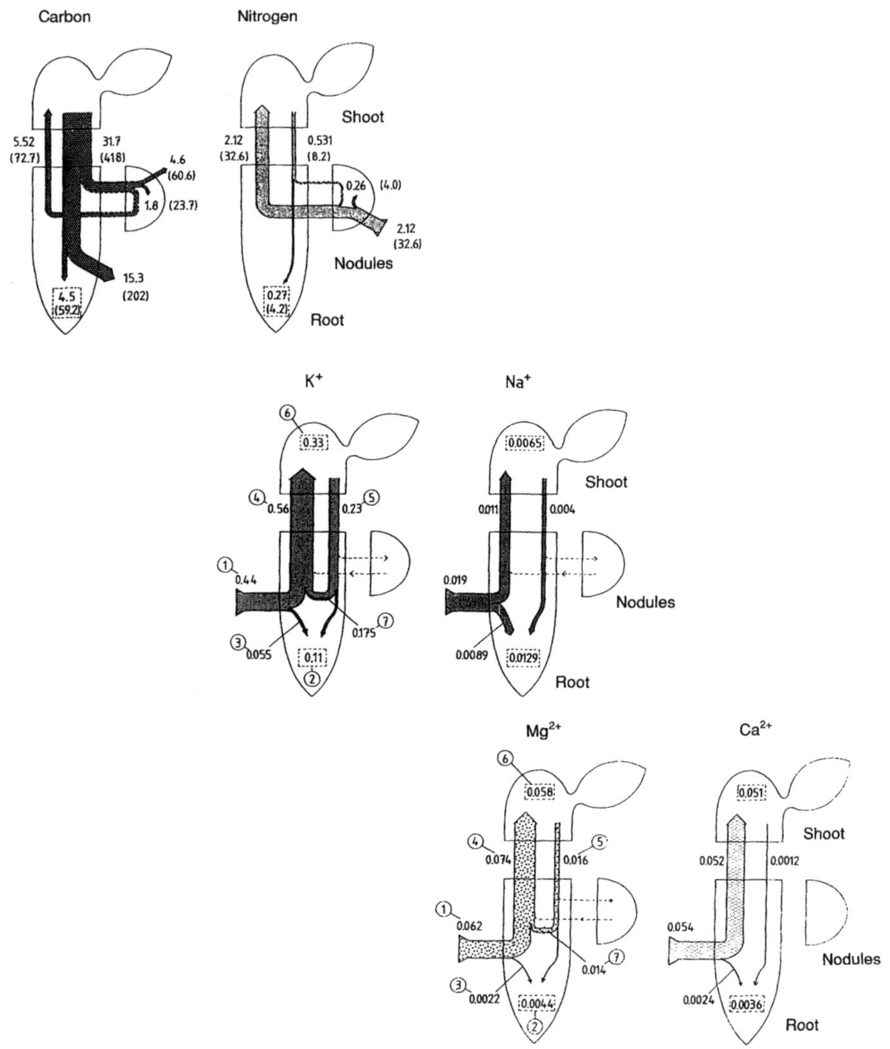

Fig. 7.8. Root : shoot exchange profiles for C, N and four nutrient cations for 39–49-day-old symbiotically dependent white lupin. Amounts of each quantity passing between different parts in xylem and phloem are indicated as μ mol g fresh weight h^{-1}, and also in the case of C and N (figures in brackets) as mg C or N over the 10-day period. Note exchanges with nodules are depicted only for C and N. Thicknesses of arrowed pathways are drawn proportional to exchange rates in each system. The seven numbered components in the four schemes for mineral cations are: 1, net uptake from medium; 2, accumulation by root; 3, direct uptake by root from medium; 4, transport in xylem; 5, transport in phloem; 6, uptake by shoot; 7, phloem-to-xylem transfer in root. (From Jeschke *et al.*, 1985.)

Based upon the modelling procedures outlined above for root : shoot exchanges, a series of more sophisticated profiles of partitioning of macronutrient elements has been developed to depict multiple exchanges between an age series of leaf and stem segments, shoot apex, roots and nodules in white lupin. Models relating to a specific interval of growth are then based on measurements of increments or losses of C and N shown by the unit parts, matching data on CO_2 exchanges of the same parts, and measurements of the C : N ratios of xylem and phloem sap collected at a series of strategic interchange points within the system (Pate *et al.*, 1980c; Layzell *et al.*, 1981; Pate and Layzell, 1981; Pate and Jeschke, 1995).

Flow profiles for C and N for a ten-day period following anthesis in N_2-dependent white lupin are shown in Figs 7.9 and 7.10. As might be expected, the model for partitioning of 1000 units of C (Fig. 7.9) of net photosynthate is dominated by phloem transfer of C from source leaves to root and growing organs

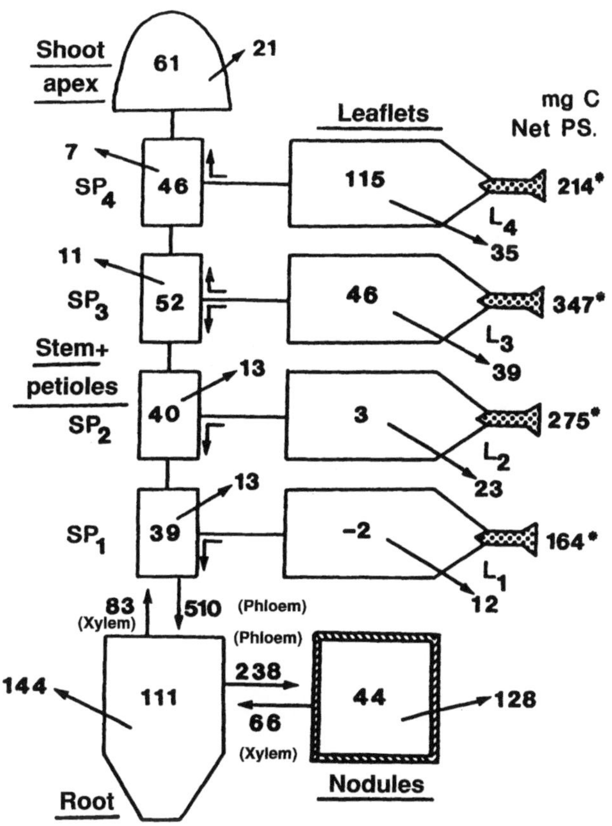

Fig. 7.9. Detailed partitioning profile for C of net photosynthate in symbiotically dependent white lupin over 51–58 days after sowing. Amounts of C lost in respiration, moved in xylem and phloem and incorporated into or lost from dry matter of plant parts are shown in relation to a net intake by the plant of 1000 units by weight of fixed C. (Pate, 1996, redrawn from Layzell *et al.*, 1981.)

at the top of the shoot. As mentioned above, xylem flow of C relates mostly to export of organic solutes of N synthesized by nodules. Approximately 8% of the total C donated from leaves returns to the shoot in this manner. The model also separates quantitatively the somewhat larger carbon budget of the root from that of the nodules. It also shows that over half of the 24% of the C of net photosynthate acquired by nodules is respired, the rest being incorporated into new nodule dry matter or cycled back to the shoot with fixed N.

Turning to the complementary model for partitioning of 1000 units of fixed N (Fig. 7.10), dominating features are bulk export of fixed N from nodules in

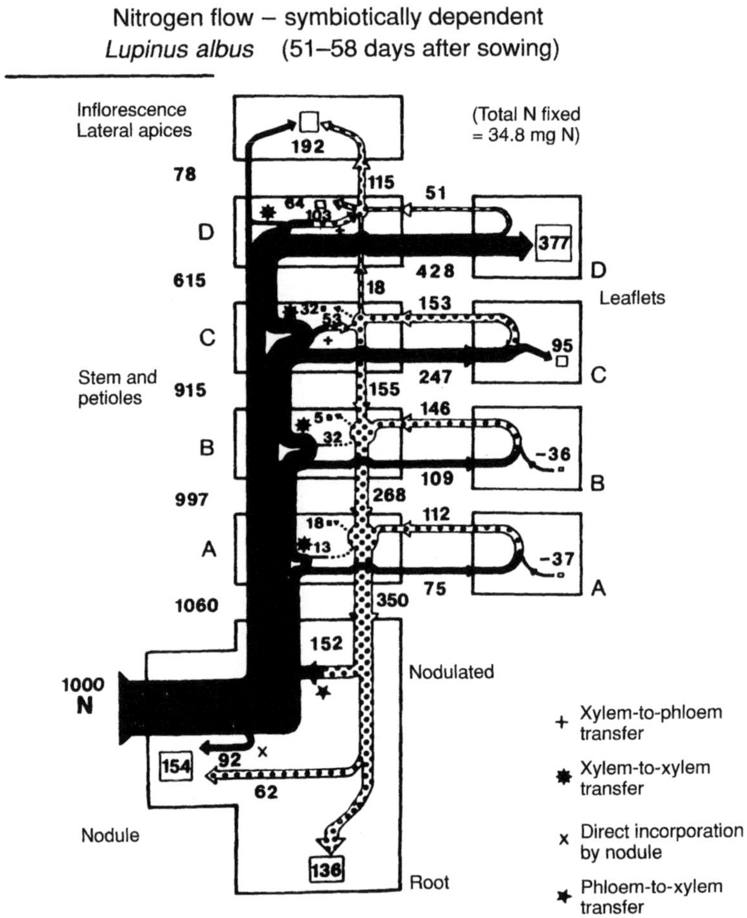

Fig. 7.10. Uptake by N_2 fixation, transport and utilization of N by symbiotically dependent white lupin in the 10-day period following anthesis. Exchanges and incorporations or losses of N are given in amounts relative to an uptake of 1000 units of fixed N by nodules. Xylem transport shown in black, phloem translocation shown with dotted pathways and short-distance exchanges between transport pathways are designated as essential components of the partitioning process. A–D refer to four successive strata of leaves and related stem regions on the shoot. (From Pate, 1986b.)

xylem and the role of transpiring leaves as principal destinations and sites of processing of N derived from the nodulated root. Massive transfer of N from xylem to phloem occurs within leaf veins, as already indicated earlier for the N economy of a single leaf in Fig. 7.8. Following such transfer the exiting N is distributed very much as shown for C; that is, with N cycled through lower leaves becoming mostly available for root and nodule growth, and that from upper leaves nourishing young apical regions of the shoot.

The model for N portrays lower nodes of the stem as participating in a series of xylem-to-xylem transfers that withdraw N from xylem traces supplying the lower leaves and pass this N back into xylem traces moving further up the stem. These transfers take place through the agency of xylem- and phloem-located transfer cells, concentrated especially within departing leaf traces in the stem (Gunning *et al.*, 1970; Pate *et al.*, 1970; Kuo *et al.*, 1980). This 'shuttle' system effectively enriches the xylem stream progressively as it ascends the stem, and the extent to which this happens can be assessed experimentally by simply measuring increases in N concentrations in xylem fluid extracted by mild vacuum from successive regions up a shoot (Layzell *et al.*, 1981). There are two important and related outcomes from this stem xylem-enrichment process. Firstly, lower leaves are substantially short-changed for N, resulting in less N becoming available for export back to the root than would be expected from an equal sharing of N between leaves on the basis of their respective transpiration losses. Secondly, due to the above effect, upper leaves receive proportionately more N than might be expected from their transpirational attraction of xylem water. They are thus able to build up large amounts of N particularly rapidly during their early growth (i.e. during the Principal Growth phase, Fig. 7.5B) and accordingly then commence upward export of N earlier and to greater relative extent than would be possible in the absence of the stem system for enriching xylem N. According to the results of modelling exercises on other herbaceous plants (castor bean, barley) and certain woody plants (*Banksia*: Pate and Jeschke, 1995; Jeschke and Pate, 1995; and *Eucalyptus*: J.S. Pate, Perth, 1997, unpublished data), shuttle systems of a similar nature also operate in non-leguminous species and may well prove a universal feature of partitioning processes favouring enrichment of translocation streams in N towards newly formed but weakly transpiring organs.

Visible also from the N model of Fig. 7.10 is the xylem-to-phloem transfer process in the upper stem which enriches with N the translocate stream moving from upper leaves to apex. Young leaves, shoot meristems, inflorescences and fruits are all weakly active in transpiration and this xylem-to-phloem transfer process, together with the shuttle system for the lower stem mentioned above, comprises elements absolutely vital to the differential partitioning of N towards shoots rather than roots (Layzell *et al.*, 1981; Pate, 1986b; Pate and Jeschke, 1995).

Once base models for partitioning of C and N have been constructed for a specific interval in the life of the plant, profiles for uptake, transport and accumulation of a range of other major or minor constituents can be constructed. In all of these cases one again needs primary information on net synthesis, loss or accumulation of the commodity in question in each component part of the plant and the models are then generated simply by using ratios of C or N to the specified component in xylem and phloem. One can thereby prescribe the mixtures of phloem and xylem inputs or outputs between each part of the plant which meet precisely the observed losses or gains recorded for that constituent in the plant part under study. Mention will be made later concerning synthesis and distribution of growth substances. At this point consideration is confined to a greater sophistication of the already mentioned differences in root : shoot partitioning of Na^+ and K^+ in *L. albus*.

The comparisons drawn up for this pair of cations in Fig. 7.11 are based on the same plants and time intervals over which the C and N budgets shown in Figs 7.9 and 7.10 were developed. The experimental situation involved a threefold higher concentration of Na^+ than K^+ in the external medium, and moderate ability to discriminate in favour of K^+ is evident from the net uptake of about as much K^+ (1.08 mmol per plant) as Na^+ (1.25 mmol per plant) during the study interval. The model for absorbed K^+ depicts the top regions of the shoot as major sinks for xylem-derived K^+, and xylem-to-phloem cycling through leaves as responsible for supplying major amounts of K^+ to other regions of the plant. Significant amounts of K^+ translocated downwards to the root via phloem are

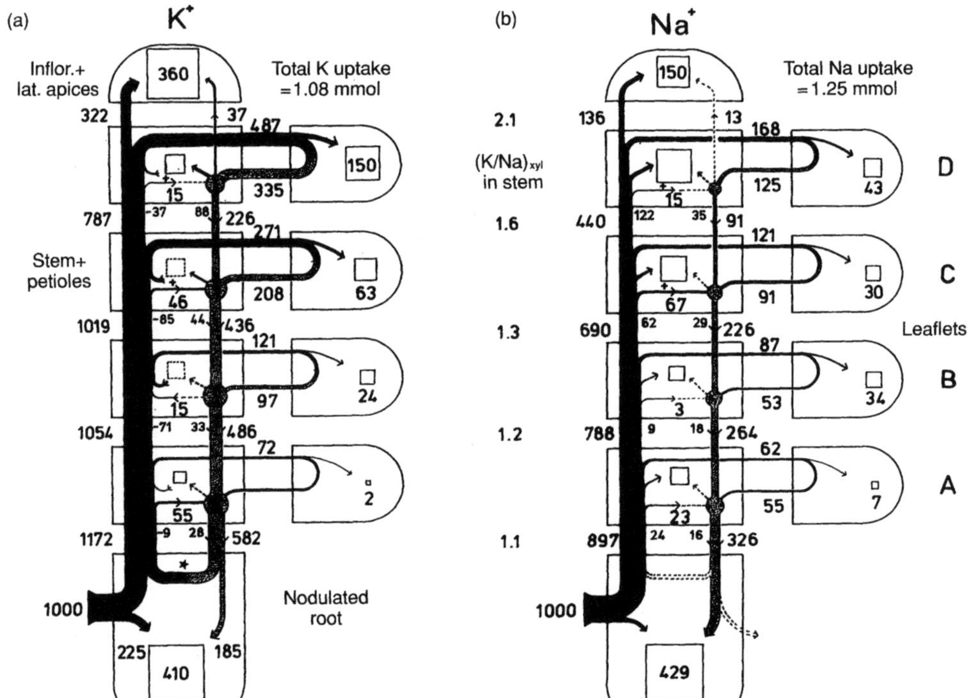

Fig. 7.11. Contrasting partitioning profiles for partitioning of K^+ and Na^+ in white lupin. Organs are represented and pathways drawn as shown in Fig. 7.10 for N, but with no information relating to nodules. Note that the progressively increasing values for concentration ratios of K : Na, depicted as the xylem stream ascends the stem, bear evidence of discrimination between the two ions and a resulting flow of less Na^+ into salt-sensitive upper parts of the shoot. (From Pate and Jeschke, 1995.)

seen to pass back to the shoot after phloem-to-xylem transfer in the root, thus completing the circulatory system for this element within the plant. Judging from the high levels of K^+ in both xylem and phloem sap, a substantial part of the current K^+ resource of the plant is committed to transport phenomena rather than merely residing in the tissue pools of organs.

Comparisons between the above partitioning profile for K^+ (Fig. 7.11a) and that for Na^+ (Fig. 7.11b) indicate relatively greater return flow of Na^+ from shoot to root and this is viewed as a mechanism for preventing Na^+ from reaching young, salt-sensitive regions of the shoot. Discrimination clearly occurs between Na^+ and K^+ at a range of interchange points in the system, especially in the stem. This is shown in Fig. 7.11 as substantial and progressive increases in K : Na ratio in xylem fluid from base to top of shoot. The model for Na^+ also contrasts with that for K^+ in showing little or no return back to the shoot of Na^+ which has reached the root through phloem. Excretion of this Na^+ back to the root medium is suggested as a means of ridding the plant of previously absorbed salt, and this might well comprise a regular outcome in saline soils after rain has leached salt out of upper horizons of the root profile.

Experimental perturbation of C and N partitioning through alterations of plant inputs of fixed CO_2 and N_2

So far we have restricted ourselves to considering how differential partitioning of fixed C and N is accomplished within symbiotically dependent white lupin where the plant is literally able to self-regulate the balance between the autotrophic processes of leaf and nodule. Understanding of the broadness or otherwise of the limits within which such coordinative processes adjust to redress imbalances in the systems have been obtained experimentally by elevating or decreasing inputs of N relative to C, and monitoring the ensuing processes through which the plant subsequently responds to perturbation of its C and N metabolism.

Our programme of research in this connection (Atkins *et al.*, 1984; Pate *et al.*, 1984a; Pate, 1986b) has chronicled both the short-term repercussions resulting from altered inputs of C and N in mid vegetative growth and much longer-term effects in which the C : N balance in seed is monitored after protracted exposure of reproductive parent plants to grossly reduced assimilation of C or N. Techniques include decreasing C inputs through shading or exposure to low-CO_2 atmospheres, greatly reducing N_2 fixation by enclosing roots of water-cultured plants in a mixture of Ar : O_2 (80 : 20 v/v) or exposing the nodulated roots of sand-cultured plants to 20 minutes daily flushings with 100% O_2, or by 'washing out' the nutrient solution surrounding roots of non-nodulated plants and replacing it with solution free of NO_3. Ar : O_2 treatment proves to be reversible since it simply denies nitrogenase of its normal substrate N_2 and allows N_2 fixation to return to its initial rate once the foreign gas mixture is replaced by air. During treatment with Ar, nitrogenase continues to function at a high rate, producing H_2 gas and maintaining a near normal sink for photosynthate (Pate *et al.*, 1984a); while there are rapid changes to the nature of N metabolic pathway enzyme levels in nodules, N export as asparagine is promptly resumed on transfer back to air

(Atkins *et al.*, 1984). By contrast, O_2 treatment irreversibly inactivates the nitrogenase system of the bacteroids, although daily application is required in this case to prevent N_2 fixation building up again as new crops of bacterial tissue grow in the same or newly formed nodules.

Experiments studying the effects of short-term exposures of Ar : O_2 on N transport in xylem and phloem have demonstrated declines within one day of commencement of treatment to xylem sap levels of less than 10% of those of control plants in air. Comparing the N composition of root-bleeding sap collected above and below the nodule zone of these Ar : O_2-treated and control plants, amide export from nodules is found to be specifically inhibited by the treatment, whereas cycling of N through the lower non-nodulated part of the root continues unaltered over a four-day period of study. Changes in phloem sap composition over the first day of treatment show sharp increases in C : N ratios of petioles, stem top and stem base of treated plants, as fully expected from the above-mentioned decline in N levels in xylem serving as N source for xylem-to-phloem transfer. During the next three days, stem phloem sap N levels remain low, reflecting continued unavailability of xylem N for xylem-to-phloem transfer in this part of the shoot. Concomitantly, however, lower stem phloem sap and that of petioles of lower leaves supplying roots progressively decrease in C : N ratio of transported solutes, consistent with N deficiency-induced onset of senescence of lower leaves and 'attendant' mobilization therefrom of appreciable amounts of N.

A parallel study of the effect of daily flushing of roots of nodulated 40-day-old white lupin with 100% O_2 has shown the progression of changes in dry matter and N contents of the upper eight of the 17 leaves on the main shoot, over an eight-day period after commencement of O_2 treatment (Pate, 1986b). The data indicate relatively small differences in dry matter content of the eight uppermost leaves of O_2-treated and control plants, but, as is to be expected, virtually no increase in N content of these leaves, and net decreases in N content of the lowermost three of these eight leaves under O_2 treatment. These changes occur despite a threefold collective increase of N in the corresponding set of leaves of control plants in which N_2 fixation continued unabated. Somewhat surprisingly, the data also showed N gains in nodulated roots of O_2-treated plants which almost equalled those of controls, implying how effectively N had become mobilized from shoot to root under N deficiency. The findings again agree fully with what would be predicted from the observed changes in C : N ratios of xylem and phloem of the Ar : O_2-treated plants.

Similar findings resulted when the nutrients surrounding the roots of non-nodulated *L. albus* grown with 10 mM NO_3 were washed out and replaced with N-free nutrients (Pate *et al.*, 1979a,b). In this case upper stem phloem exudate changed from a C : N of around 20 : 1 to 50 : 1, while that of the phloem serving the root system was unchanged. The model of N flow in the whole plant indicated that low xylem N precluded normal xylem : phloem exchange in the upper stem and that, as a consequence, growth of the shoot apex but not the root system was arrested by transient N deficiency.

Turning to the long-term effects of altered inputs of C and N on seed size and N content (J.S. Pate, C.A. Atkins, M. Peoples and P. Sanford, Perth, 1997, unpublished data), deprivation of C by shading or low CO_2 during flowering and fruiting reduced seed size by almost 50%, but increased significantly the seed's concentration of N from a normal value up from 5.5% of dry matter to one of 7.1%. This treatment thus elevated the ratio of protein to carbohydrate-based reserves of the seeds. Conversely, O_2 treatment of roots applied frequently enough to virtually eliminate N_2 fixation after anthesis had relatively small impact (10% reduction) on seed size, but produced seeds with N levels at less than 4%. These perturbations exceeded by a large margin the range of variation in seed size and composition that one encounters within *L. albus* under normal field conditions. The nature of partitioning of C and N to different seed reserve materials has not been determined, but the wide range of C : N obtained highlights the possible impact of environmental factors on seed quality (Chapters 12 and 13).

Carbon partitioning to fruits and seed – the significance of assimilation before and after anthesis

A major thrust of research on the nutrition of both cereals and grain legumes has been to identify which major sources of C are responsible for the filling of fruits and seeds, and the overall importance and respective time courses of contribution of each of these of sources to grain filling processes (Pate, 1976, 1989a,b; Pate *et al.*, 1983, 1988). Most information of this nature relates to cereals, where it has been generally shown that C laid down in vegetative parts of the plant before anthesis can be of substantial importance in implementing grain yield. This is particularly the case under dry-land agricultural conditions, where lack of soil water during late season may cause premature leaf senescence and accordingly detract from the performance of flag leaves and reproductive parts in providing net photosynthate during post-anthesis stages of crop development.

Regardless of the species of crop plant being considered, a basic difference invariably exists in respect of how much less effectively pre-anthesis reserves of C are mobilized to seeds in comparison with those of N (Pate and Minchin, 1980). This distinction becomes critically important when considering grain filling of legumes of high seed protein content. For example, in a study of C and N relationships of reproducing pea plants (Pate and Flinn, 1973) $^{14}CO_2$ labelling prior to anthesis led to only a few per cent of the assimilated labelled carbon being eventually donated to the fruits, whereas N given as pulses of $^{15}NO_3$ at a range of times during the growth cycle of the crop was in every instance mobilized to fruits with upwards of 70% efficiency. This marked difference, apparent also in *L. angustifolius* (Unkovich *et al.*, 1994), reflects the dynamic nature of the N pools of vegetative parts of the plant, and the correspondingly uniform effectiveness with which they can be channelled into reproductive parts, as contrasted with the general inability of the plant to degrade the C reserves of its tissues, particularly those bound into cellulose and lignin of cell walls. The same distinction applies of course to the C and N reserves of pod walls and seed coats of fruits and the respective effectivenesses with which they provide C and N to growing embryos (Pate *et al.*, 1977; Peoples *et al.*, 1983, 1985).

An example illustrating the extremely limited transfer of C from early formed photosynthate to developing fruits comes from a ^{14}C feeding study of narrow-leafed lupin (*L. angustifolius* cv. Unicrop) in the field (Pate *et al.*, 1980a). $^{14}CO_2$ was supplied to different sets of selected plants at a specific week in the 12-week period after anthesis and transfer of label, assayed periodically and non-destructively thereafter by sampling fruit phloem sap until reproductive maturity. All fed plants were then harvested to assay distribution of remaining ^{14}C between vegetative parts, fruits and seeds. Analyses were also conducted specifically on the distribution of seed ^{14}C between protein, oil, water-soluble and non-soluble, carbohydrate materials of the embryos of seeds formed following each occasion of $^{14}CO_2$ feeding.

A principal finding of the study (Pate *et al.*, 1980a) was that photosynthate fixed by the plant just before and up to 3 weeks after flowering provided a mere 5% of the final C accumulated by the fruits. A further 18% of the C came from photosynthesis 3–6 weeks after anthesis, followed by the remaining C from net photosynthesis during the critically important final 6 weeks of fruit filling. The inordinately low contribution of C fixed before flowering was not surprising, in view of the above-mentioned inefficiency of retrieval of C committed to skeletal materials, and the fact that anthesis in the study cultivar of *L. angustifolius* occurred at a time when the plant contained less than one-sixth of its maximum content of C and was therefore far from being fully effective in photosynthesis.

A further series of outcomes derived from comparisons between the mean specific radioactivities of phloem sap samples obtained from fruits with eventual specific ^{14}C activities of the total C of fruits of the same plants at final harvest. As might have been anticipated, specific radioactivities of phloem and final fruit C were closely similar when $^{14}CO_2$ was fed early, but with later times of feeding, fruits were less intensely labelled than expected from the ^{14}C activity of the C they received in phloem. The logical explanation was that late-arriving ^{14}C had become diluted with unlabelled C already accumulated by the fruit.

Finally, further information was forthcoming from examinations of the distribution of ^{14}C between different organic constituents of phloem sap of plants fed early or late with $^{14}CO_2$ (Pate *et al.*, 1980a). The phloem translocate supplied to fruits of early ^{14}C-fed plants showed an average 60% of its label in amino acids versus only 40% in non-amino solutes and this was regarded as consistent with the ^{14}C in question having been mobilized mostly from the protein of senescent leaves. Conversely, the phloem sap of plants fed 9 weeks after anthesis showed only 22% of its ^{14}C in amino compounds and the remaining 78% in non-amino compounds, principally sucrose. Equating these differences with the distribution of ^{14}C amongst seed embryo reserves for each feeding time of the study, the marked time-related drift in labelling patterns illustrated in Fig. 7.12 was obtained. This showed that almost 80% of the ^{14}C from photosynthate formed at or shortly after anthesis had ended up in protein, declining progressively to less than 20% in protein for ^{14}C-photosynthate synthesized during the penultimate stages of fruiting.

In general terms, then, the filling of a fruit and seed of a particular genotype of lupin or other grain legume with C comprises a series of highly complex step-wise processes in which each age group of leaves comes on stream sequentially to supply characteristic amounts and proportions of different assimilates. Modulations of this basic pattern would then be expected in response to cropping

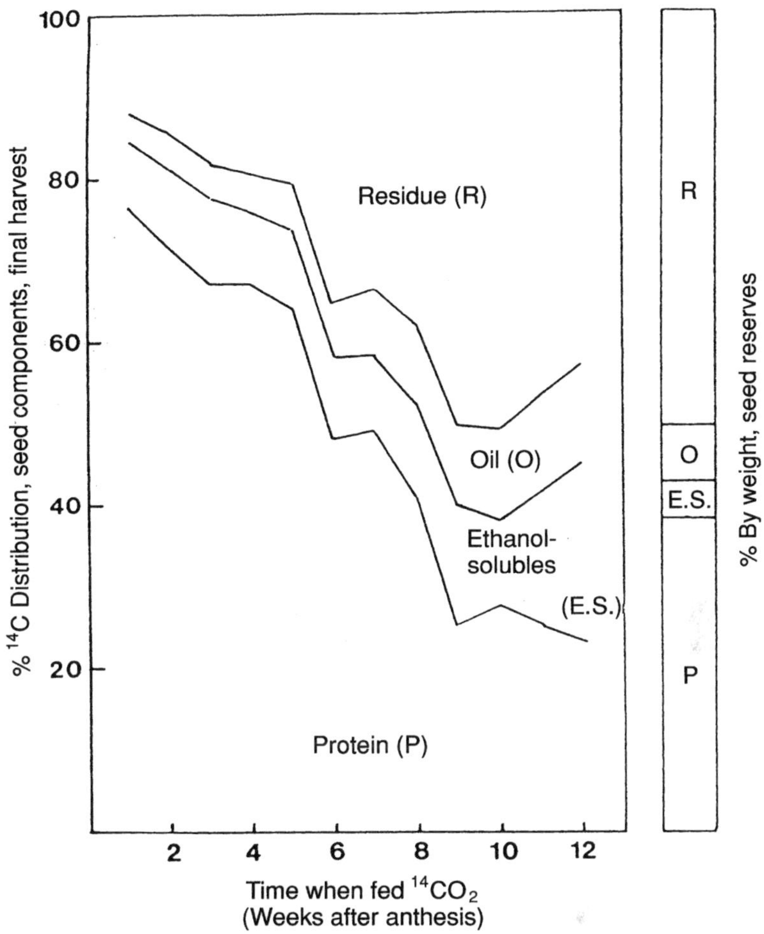

Fig. 7.12. Proportional (%) distribution of ^{14}C amongst reserve components of mature seeds of *L. angustifolius* after feeding $^{14}CO_2$ to parent plants (but not fruits) at different times after anthesis. The bar diagram on the right indicates proportions of seed total weight devoted to the various components. Residue (R) refers to wall reserves and fibres of the seed. Starch was not present in detectable amount in the mature seed. (From Pate *et al.*, 1980a.)

location and season, as dictated by any one or a combination of a number of critical climatic and edaphic factors experienced by the crop during its growth. Understanding the interaction of these many interacting elements poses daunting challenges to the crop physiologist attempting to understand the basis of variations in crop yield within and between genotypes and between sites and seasons.

Transport of Growth Substances and their Significance in Partitioning

Just as there are well-defined source–sink relations for solutes of C and N, water and mineral ions, so too can similar relations be expected for the growth-regulating substances translocated in xylem and phloem. Physiologists have gathered a vast body of data which has been interpreted in hypotheses that incorporate a critical role for these substances in determining the overall source : sink relations of the plant. In ways that are only very recently becoming accessible to investigation, these substances 'signal' changes in the environment inside and outside the plant and elicit responses at the cellular level in assimilate source or sink organs as a consequence of their changed concentration in the translocation channels. By definition, transport is not an essential role of plant hormones, and they may act directly on the tissue in which they are produced (Davies, 1995). However, coordination of growth during development or response to environmental change through inter-organ communication requires sophisticated regulation of transport processes.

Growth regulators such as abscisic acid (ABA), cytokinins (CKs) and the gibberellins (GAs) are all thought to affect plant growth in processes involving long-distance transport and have been implicated in numerous roles involving control of assimilate partitioning and growth architecture. To list a few, ABA is well established as an inhibitor of stomatal aperture (Jackson, 1993) and is thought to promote phloem unloading of assimilates into developing seeds (Brenner and Cheikh, 1995). Cytokinins are known as potent anti-senescence agents, and are involved in increasing stomatal opening (Badenoch-Jones *et al.*, 1996), promotion of lateral bud growth (Tamas, 1995), control of photosynthate allocation (Fetene and Beck, 1993) and increasing fruit set (Atkins and Pigeaire, 1993). Gibberellins are well-known promoters of cell and stem elongation (Phinney, 1984) and are thought to be involved in phloem loading (Daie *et al.*, 1986) and fruit set and development (Gianfagna, 1995). These few examples illustrate the great potential role that growth regulators have in controlling the accumulation and partitioning of assimilates. Most of the listed effects (and many others) are inferred from a broad range of species and there is only a weak understanding about the integration of the many roles any given hormone may concurrently play. The basis of this weak understanding is in large part due to a lack of knowledge of where growth regulators are synthesized and translocated.

In general, hormonal movement in vascular tissues is a neglected aspect of research (Hoad 1995), stemming from difficulty in obtaining sufficient quantities of transport fluids for analysis. Hormones operate at extremely low concentrations, usually less than 1 μM. Techniques for identification and quantification of hormones such as gas chromatography–mass spectrometry (GCMS), enzyme-linked immunosorbent assay (ELISA) and radioimmunoassay (RIA) are extremely sensitive and are able to detect down to femtomol levels of growth regulators, yet in many cases one still needs from the hundreds of microlitres to millilitre volumes of transport fluid for reliable detection.

Abscisic acid

ABA represents an excellent system for studying hormone transport since it has a relatively simple metabolism and only one active form. By contrast, the chemistry of the gibberellins and cytokinins is more complex, involving multiple active or reversibly inactivated forms, making it much more difficult to identify, quantify or model their transport. ABA is also present in greater abundance than other hormones and a highly sensitive immunoassay is widely available (Weiler and Zeigler, 1981). As little as 10 μl phloem or 100 μl xylem can be enough to generate a reliable signal for the hormone.

It is not surprising, then, that the most complete picture of hormone transport in any plant system has been drawn for ABA in *L. albus*, whose transport has been described according to the modelling approach outlined above for inorganic cations. Using flows of C and the C : ABA ratios in xylem and phloem saps, the flows of ABA were estimated under control and saline (40 mol NaCl m^{-3}) conditions (Wolf *et al.*, 1990). Accompanying data of C, N and ABA contents of individual organs were collected, enabling prediction of rates and sites of ABA biosynthesis and degradation. Like that described for K^+, much of the xylem-borne ABA exiting the root, originated from recirculation from the shoot after phloem-to-xylem transfer in the root (Fig. 7.13). Under control conditions ABA synthesis in the root took place at a rate such that xylem transport of ABA from root to shoot balanced the opposing flow of ABA in the phloem. Accordingly, only about 28% of ABA in xylem entering the aerial shoot was newly synthesized in the root.

The scenario changed considerably when plants were treated with NaCl (Fig. 7.13). From salt-stressed roots, xylem export of ABA exceeded the import via phloem by a factor of almost two, indicating that roots strongly increased ABA synthesis. In fact, root net biosynthesis of ABA increased 20-fold, while that of the basal leaf strata more than doubled. Overall, whilst the percentage of ABA exiting the roots in the xylem as recirculated from phloem decreased in salt-treated plants (from 72 to 45%), total ABA recirculation actually increased by a factor of 5.5.

The model provided firm answers for a longstanding debate regarding the relative contributions of the root versus the shoots in elevating ABA delivery to leaves for reducing stomatal conductance under stress conditions (Jackson, 1993). Empirically it was shown that under control conditions some ABA is synthesized in the root, but that considerably greater quantities are produced in the shoot. By contrast, when grown under NaCl stress the proportion of ABA synthesis is shifted from the shoot to the root, emphasizing the role of the root in the signalling of stomates in stressed plants. Thus both shoot and root production of ABA is important in lupin, with different emphasis depending on environmental situation. A recent model (Slovik *et al.*, 1995) indicated in accordance that ABA may function effectively as either a root-to-shoot or shoot-to-root signal.

Cytokinins

After ABA, the most intensively studied growth regulators in terms of transport are the cytokinins. In lupin there is in fact a greater number of studies involving translocated cytokinin than for ABA, but the picture is less holistic. This is probably because the CKs are less amenable to the type of translocation model outlined for ABA. Firstly, CKs are usually present at much lower concentrations than ABA and tenfold more xylem and phloem needs to be assayed for reliable

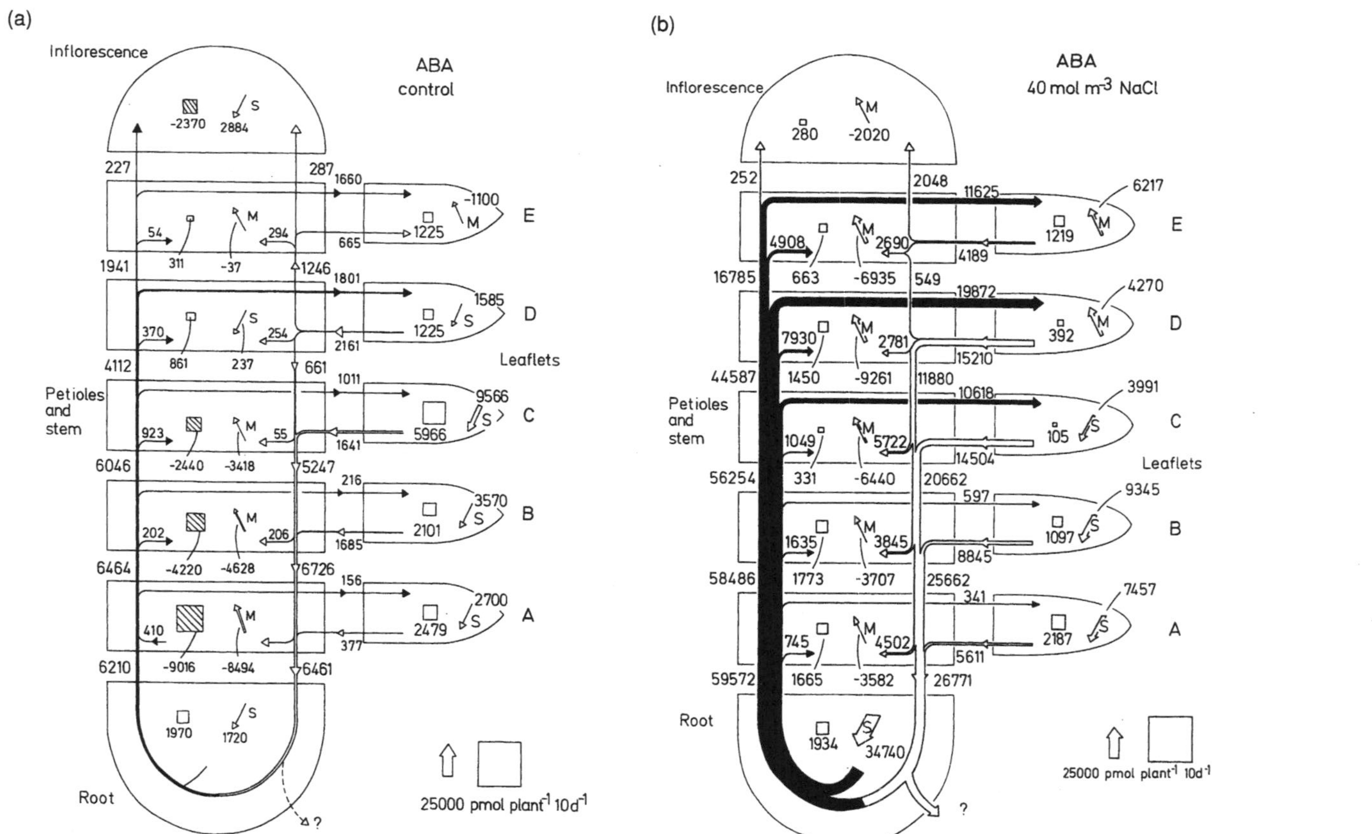

Fig. 7.13. Flows and utilization of abscisic acid in *L. albus* 41–51 days after germination. (a) Control; (b) with 40 mM NaCl. Open arrows indicate transport in phloem; closed arrows indicate transport in xylem. Squares indicate changes of ABA content within the study period. Diameters of arrows and areas of squares are drawn in proportion to rates of flow or utilization. Hatched squares indicate net loss of ABA. Rates of ABA synthesis or metabolic degradation are symbolized by an arrow marked with an S or M. (From Wolf *et al.*, 1990.)

measurements (R.J.N. Emery, Perth, 1997, personal observation). For xylem this entails collecting millilitre-range volumes. Second, the CKs are chemically complex, being represented by at least four active forms: zeatin (Z), zeatin riboside (ZR) and the corresponding dihydro derivatives of each (DZ and DZR). In addition certain CK conjugates such as nucleotides (CK-NTs) or *O*-glucosides (CK-OGs), even if they are not active *per se*, might serve as reversible storage or transport forms (Letham, 1994). Not only does this mean that a model of CK translocation must follow many molecular species, but it also entails more difficulty in quantification. Immunoassay measurements, although potentially rapid and very sensitive, are limited by cross-reactivities among antibodies and the different CK forms, necessitating complex chromatographic separation before assay of each form. Reliable quantification of individual CKs requires multiple chromatographic steps and GCMS analysis using isotopically labelled standards.

While a complete descriptive model of CK translocation is not yet available, several studies have nevertheless identified the forms of CKs present in sap and indicated some principles about their translocation. Early bioassay work (Davey and Van Staden, 1978a,b), and bioassay combined with GCMS identification (Taylor *et al.*, 1990) indicate that the flows of CKs share similarities with the movement of ABA. CK-like activity was highest during the onset of main-stem fruit development, and during this time the CK activity in phloem collected either from the base of the inflorescence or from petioles was relatively high compared with phloem or xylem obtained from the basal part of the stem. This is consistent with the idea that CKs move into developing inflorescence via phloem from the leaves rather than from basal stem phloem or directly from the xylem. Feeding lupins with radiolabelled CK reinforces this hypothesis. *L. angustifolius* fed with [^{3}H]ZR into the xylem via a cotton wick showed that the majority of label moved directly from the xylem laterally into the bark (Jameson *et al.*, 1987). In fact 60 to 80% of radioactivity was retained in the stem; and of that which moved to the leaves, much more was recovered from petioles than laminae. Negligible amounts of label entered the seeds though small amounts were recovered from the pod walls. The latter effect of low xylem contribution of CK to fruit has subsequently been confirmed by feeding labelled ZR and DZR through the xylem of first-order lateral shoots (Zhang and Letham, 1990).

Thus locational indicators imply that xylem-to-phloem exchange of CKs is commonplace, and that the extent of entry of CK into fruits directly from xylem is negligible. Further evidence of xylem-to-phloem transfer came from the recovery of [^{3}H]DZ in pod phloem sap after a feed of [^{3}H]ZR to roots of *L. albus* (Summons *et al.*, 1981). The robustness of the conclusions in at least one study (Jameson *et al.*, 1987) is backed up by the fact that naturally occurring xylem CKs were used (discussed below) and in some instances the radiolabelled metabolites were confirmed to be either active or storage form CKs.

A pivotal question that remains to be determined is the relative contribution of the different organs towards CK biosynthesis. Currently it is thought that the major site of synthesis is in the roots, and to a lesser extent in the developing seeds (Letham, 1994). On the other hand, shoot organs such as leaves, cambial zones and apices are thought to synthesize only minor amounts, if any. Before this can be thoroughly evaluated, a translocation model incorporating how much CK may be recycled via phloem-to-xylem transfer in the root must be undertaken. Before such a model can be attempted, the relevant CK present in both the xylem and phloem must be determined. Xylem CKs have been unambiguously identified by GCMS in *L. angustifolius*, showing that the ribosides, ZR and DZR, predominate during at least one stage of growth. Indications that a suite of other CKs, including, Z, DZ and various glucosides and nucleotide conjugates, were also present in lesser amounts was provided by radioimmunoassay.

Identification of phloem CKs has been surprisingly elusive for lupin, especially given the number of studies undertaken and the propensity of L. *angustifolius* and *L. albus* for phloem bleeding. Initial work relied on a combination of bioassay and enzyme hydrolysis (Hoad, 1995), an approach that does not have the resolution for rigorous identification. ZR and DZ have been tentatively identified as constituents of phloem (Taylor *et al.*, 1990), but unfortunately the identification was based on mass spectra derived from phloem mixed with xylem. In many cases the phloem concentration of solutes is far greater than that of xylem, but since no other study has provided unambiguous identification of CKs in lupin phloem, this early report should be regarded with caution. This is especially true since phloem CK identification across several species has been anomalous according to current thought. For example, isopentenyladenine types (which include isopentenyladenine, iP, and isopentenyladenosine, iPA) have been identified as the predominant CK forms in *Ricinus* and *Sinapsis alba* (Komor *et al.*, 1993; Lejeune *et al.*, 1994). Yet these are the precursors to the active Z-type forms, and one would not expect them to predominate in phloem, given that they are minor forms in xylem, and the shoot organs are not thought of as major sites of CK synthesis.

Clearly a modelling approach incorporating CK transport in relation to C and N flows could establish the source–sink relations for this group of regulators and contribute towards answering basic questions of CK physiology. Where does the primary synthesis of CK occur? Do these sites change with developmental progression or in response to stress? Are seed-produced CKs contributing to the CK status of other organs; and as a corollary, to what extent is the seed autonomous with respect to CK supply?

Gibberellins

Even more chemically complex than the CKs are the gibberellins (GAs). With over 100 documented naturally occurring GAs, the task of quantitatively describing synthesis and tracking their translocation in any plant system may seem daunting. However, upon the completion of initial exploratory work of describing which GAs are present, the process may be made more manageable. Transport fluid GAs have been rigorously identified in two studies of *L. albus*, which have shown GA metabolism in the species to be quite complicated. In phloem exudate collected from pod tips (Hoad *et al.*, 1993), 16 GAs representing four major pathways – (1) the non-hydroxylation pathway, (2) the early C-13-hydroxylation pathway, (3) the C-3 hydroxylation pathway and (4) the C-3, C-13 combined hydroxylation pathway – were confirmed by GCMS. It was noted that

GAs were present in much lower amounts than ABA which was measured in the same samples. Of the GAs that could be quantified, the concentrations ranged from 0.2 to 2.2% of those for ABA.

An earlier study (Thompson *et al.*, 1988) also identified GAs in pod-tip phloem, confirming the presence of five of the same GAs and two of the same pathways as the study by Hoad *et al.* (1993). In other saps, seven GAs were found in the stem phloem and two in the xylem. The spatial differences proved interesting since, firstly, three of those GAs reported for stem phloem were not detected in the pod phloem of either study. Secondly, no GA_1 could be found in the fruit phloem, only low amounts of GA_1 were detected in the stem phloem, and relatively large amounts of GA_1 occurred in the xylem sap. Hence it is possible either that early pathway (C-20) GAs are synthesized in the fruit before being converted into the highly biologically active GA_1 in the xylem or roots, or that a unique pathway leading to GA_1 occurs in the xylem.

Further work is needed to check for other phloem-borne GAs at different locations within the plant, and to confirm whether there is a broader range of xylem-borne GAs. The concern of which GAs should be modelled is largely a subjective one since the relative significance of the different forms is uncertain. GA_1 is generally accepted as the only active form for promoting stem elongation (Phinney, 1984), but there remains much to learn regarding the role of GAs and the relative importance of their molecular structure in processes such as flowering, fruit setting, seed filling and others.

Much of the distribution and transport of GAs is yet to be determined in any species, let alone lupin. For example, GAs are known to be synthesized in young tissues of the shoot and developing seeds, though the exact location is uncertain. There is particular uncertainty whether GAs may be synthesized in roots (Stoddart, 1983), and although it is suspected that GAs are transported in the xylem and phloem, this has not been demonstrated satisfactorily. Only through detailed measurements of GAs in transport fluids and subsequent account of their transport will many aspects of these questions be answered.

Other growth regulators

No consideration of growth regulators in higher plants would be complete without some mention of auxin and the gaseous senescence promoter, ethylene. There has been no study of the synthesis and translocation of auxin in lupin despite the fact that apical dominance and branching habit have been central themes in breeding strategies in *L. angustifolius* (Chapter 4). Clements *et al.* (1994) have shown that ethylene is involved in the regulation of pod set in *L. angustifolius* and in determining the rate of organ abscission and response to fungal infection, but there is no information as to the sites of ethylene synthesis or of translocation of its stable precursor, aminocyclopropane carboxylic acid (ACC). Although other regulatory molecules – such as salicylic acid (SA), methyl jasmonate or jasmonic acid (JA), the 18-amino acid polypeptide, systemin, chitosan and polygalacturonide oligosaccharides, which have recently been described in higher plants (Vick and Zimmerman, 1984; Doares *et al.*, 1995; Shulaev *et al.*, 1995; Schaller and Ryan, 1996) – are also likely to function in lupins, their precise roles and significance in the genus await further study. It is worth noting in this regard that because some lupins bleed from phloem they might offer a useful model species to study the kinetics and biochemistry of the SA-mediated systemic acquired resistance (SAR) response to wounding or infection (Jenns and Kuc, 1979: Shulaev *et al.*, 1995; Molders *et al.*, 1996) in crop legumes.

Molecular Basis for Source–Sink Relations of C and N

While the literature abounds with descriptions of C and N partitioning in plants, many of which are based on the long-distance transport of isotopically labelled molecules (e.g. Pate and Farrington, 1981; Pate *et al.*, 1983), there is a limited knowledge about mechanisms that fashion these patterns of flow. Furthermore, there is essentially no information as to the molecular basis of such source–sink relations or the ways in which plant growth regulators (known or unknown) might be directly involved in their determination. Clearly strategies that aim to modify source–sink relations and increase harvestable yield, whether by conventional means of selection and plant breeding or by exploiting the tools of recombinant DNA technology, would benefit greatly from application of such knowledge.

The source–sink relations of lupin for C and N described above are almost completely accounted for by the translocation of only three organic solutes: sucrose and the amides asparagine and glutamine. Furthermore, at sites where these solutes are loaded on to or unloaded from translocation channels, or where specific solute exchanges occur between vascular components (e.g. xylem : xylem, xylem : phloem or phloem : xylem; Figs 7.10 and 7.11), strategically placed groups of transfer cells have been identified (Fig. 7.2). The highly amplified plasma membrane surface of these cells could thus provide a means for enhanced rates of transport at the vascular 'bottlenecks' as well as exercising specificity of exchange through the activity of transport proteins specific for sucrose and the amides. Given this scenario, adjustments in the activity of these proteins through transcriptional, translational or post-translational regulation could reasonably be expected to influence both the development and day-to-day control of source–sink relations for C and N (Atkins, 1993). The transfer cell plasma membrane may also be the regulatory site for the long-distance source–sink relations of minor constituents in xylem and phloem, including inorganic cations (see Exchanges of C, N and inorganic cations between shoot and root, p.197) and the growth regulators (see Transport of Growth Substances and their Significance in Partitioning, p.207), and may in turn have its transport activities modulated by components of the transport channels. Clearly the discrimination between K^+ and Na^+ or other ions relies on differential short-distance transport to sites of loading in xylem and phloem. The transfer cell plasma membrane offers the obvious site for the exercise of such specificity.

Sucrose transport

Although transfer cells have yet to be isolated and characterized from lupins, studies have been initiated using other legume species (*Vicia faba, Phaseolus vulgaris, Glycine max*) as well as non-legumes. The most relevant to date have been those of the transport of sucrose within the developing seed of *V. faba* (Offler and Patrick,1993; McDonald *et al.*, 1995, 1996a,b; Patrick *et al.*, 1995; Wang *et al.*, 1995b) which have used intact tissues as well as isolated protoplasts of the dermal transfer cells of the cotyledons. These studies offer a useful model for what might be the situation in developing lupin seed.

Phloem-borne sucrose entering the symplast of the testa passes through the innermost cell layer, which is differentiated into thin-walled parenchyma transfer cells, and is exchanged through a putative sucrose/proton antiporter (Fieuw and Patrick, 1993) into the apoplastic cell discontinuity between the maternal tissue and embryo. The dermal transfer cells of the developing cotyledons take up sucrose from the apoplast (McDonald *et al.*, 1995, 1996a). McDonald *et al.* (1996b) have demonstrated that this activity has biochemical features consistent with a sucrose/proton symport mechanism driven by a proton motive force generated by a closely coupled H^+-ATPase in the plasma membrane (Bouche-Pillon *et al.*, 1994a,b; Walker *et al.*, 1995). Using immunocytochemical techniques BouchePillon *et al.* (1994a) showed that this ATPase was localized with a relatively high frequency in the wall ingrowth regions of the dermal transfer cells. A similar scenario has been indicated for the developing cotyledons of *P. vulgaris* (Wang *et al.*, 1995b) and, given a comparable mode of seed development in the grain legumes, is likely to apply also to lupin.

Sucrose transport proteins have been isolated from a range of species and their biochemical features characterized (reviewed in Bush, 1993) and, more recently, their encoding genes have been cloned (Sauer *et al.*, 1994; Bush *et al.*, 1996). Most are members of the major facilitator superfamily (MFS) of sugar transporters (Sauer *et al.*, 1994) found in a wide range of both eukaryotic and prokaryotic organisms, and similar proteins are likely to account for the majority of carrier-mediated sucrose transport in lupins.

Amino acid transport

Much less attention has been paid to the molecular mechanisms of amino acid transport by transfer cells. That they exercise specificity in transport of nitrogenous solutes may be inferred from the exchange processes, particularly those involving asparagine, detailed above for lupin (see Transport of Nitrogen in Xylem and Phloem, p. 185). Similar features have been ascribed to transfer cells of the sub-aleurone layer which lies between the endosperm cavity and the starchy endosperm of wheat (Wang *et al.*, 1994, 1995a). In developing maize kernels the equivalent layer of transfer cells is located strategically above a group of basal pedicel cells which express a unique form of glutamine synthetase (GS), and Muhitch *et al.* (1995) have speculated that this arrangement relates to a requirement in the specificity of N transfer to the developing embryo. A further example of the apparent specificity for amides in the transfer of N to metabolic sinks comes from nodulated soybean or cowpea, both of which translocate most of their N in xylem as the ureides, allantoin and allantoic acid (Atkins, 1991). These ureides are largely metabolized in leaves and stems and the N is reassimilated to form glutamine and asparagine which is loaded on to phloem (Pate *et al.*,

1984b; Atkins and Pate, 1987). In fact in the fruits of these species all incoming ureide-N is converted to glutamine before transfer to the growing embryo (Peoples *et al.*, 1985).

Bush (1993) has reviewed the now convincing evidence for plasma membrane proton-coupled symport as the most important contributor to the transport of amino acids in higher plants. cDNA expression libraries from *Arabidopsis* and sugar beet have been used to complement amino acid-transport deficient yeast mutants (Frommer *et al.*, 1993; Hsu *et al.*, 1993; Bush *et al.*, 1996) yielding five closely related amino acid symporters (AAP1–5) and NTR1 (Frommer *et al.*, 1994). It is now clear that there is a large family of similar transporters in plants which differ slightly in substrate specificity and pattern of expression (Kwart *et al.*, 1993; Fischer *et al.*, 1995) and, like the family of sucrose transporters, offer consensus sequences as a starting point for isolation of similar genes from lupins. However, these proteins preferentially translocate neutral amino acids and there has been no report of an amide-specific transporter. Furthermore, the current collection of yeast transport mutants does not include lesions for glutamine or asparagine transport. Zheng and Haselkorn (1996) have recently identified an operon (BztABCD) in *Rhodobacter capsulatus* encoding a binding-protein-dependent transporter with affinity for glutamate, glutamine, aspartate and asparagine. The four substrates share at least one component of this transport system which also includes genes for integral membrane proteins and ATP-binding proteins. Zheng and Haselkorn (1996) generated a Tn5 transposon mutant which could be rescued by a wildtype copy of the operon, suggesting that such a mutant may also be complemented by heterologous putative amide transporters. Williams *et al.* (1996) have used plasma membrane vesicles prepared from *Ricinus communis* seedling root tissue and presented evidence for a glutamine transport protein which has a higher affinity for the amide compared with the dicarboxylic amino acids, glutamate and aspartate. A specific asparagine transporter gene has been isolated from *Salmonella enterica* (Jennings *et al.*, 1995). The deduced amino acid sequence of this permease showed significant homology with a family of prokaryotic amino acid permeases, suggesting that the gene (*ansP*) may be a useful heterologous probe for asparagine permease from lupin transfer cells.

Proton pump ATPases

The amino acid transporters, like those for sucrose, are likely to be driven by a proton gradient generated through a plasma membrane H^+-ATPase. Indeed this conclusion has been suggested for amino acid uptake by the haustorial transfer cells of *Polytrichum formosum* sporophytes (Renault *et al.*, 1989) and the demonstration of a membrane localized H^+-ATPase in the minor leaf veins of *V. faba* (Bouche-Pillon *et al.*, 1994a,b) adds support to the idea.

Again the proton pump ATPases which appear to constitute the primary plasmalemma transport system in plants are coded for by a large multigene family (Harper *et al.*, 1994) with isoforms expressing different transport specificity in many cell types (Samuels *et al.*, 1992). Although the specific forms located in the plasma membrane of transfer cells have not been identified, Vancaeseele *et al.*

(1996) used antibodies raised against a plasma membrane H^+-ATPase from corn to demonstrate immunogold labelling of the inner surfaces of the wall projections of transfer cells of the embryo sac of unfertilized mature ovules of *Brassica napus*. Thus the growing base of genetic information is likely to provide consensus sequences which can be used to identify those that are functioning in lupin. Interestingly, DeWitt *et al.* (1991) found that the promoter for one isoform from *Arabidopsis* (AHA3) drove expression only in the phloem and suggested that this proton pump was associated with the loading of solutes to the assimilate stream. Using an epitope immunotag derived from the animal oncogene, *c-Myc*, De Witt and Sussman (1995) subsequently demonstrated that expression of AHA3 was restricted to the plasmalemma of phloem companion (transfer) cells.

Prospects for molecular studies of transport in lupins

McDonald *et al.* (1996a) have shown that cotyledon dermal transfer cells in *V. faba* can be isolated as protoplasts and materials such as this will provide a useful starting point from which to prepare cDNA libraries. Methods of mRNA and library preparation using PCR which require as few as 100 cells are now available (Dresselhaus *et al.*, 1994) so that very small samples of transfer cells prepared from leaf minor veins or stem internodes can also be used. As transfer cells develop in internodes only when N is supplied through the transpiration stream (Kuo *et al.*, 1980) it is likely that subtractive hybridization or differential display (Liang and Pardee, 1992) techniques could reveal transfer cell-specific genes including those responsible for solute exchange in these tissues. Thus many of the tools required for the application of a molecular approach to transfer cell function and more particularly to a determination of the role these cells play in regulation of long-distance translocation in plants are at hand. Importantly for lupin there are effective genetic engineering protocols available (Chapter 3) providing the means to generate transgenic plants which overexpress transfer cell transport proteins or in which antisense constructs for their genes have been inserted. Perhaps a first example of the impact of such an approach has already been generated. Geigenberger *et al.* (1996) have reported that phloem-specific expression of *E. coli* pyrophosphatase in transgenic tobacco inhibits long-distance transport of carbohydrates and amino acids, demonstrating the importance of pyrophosphate-dependent reactions in phloem energy metabolism.

The very detailed and substantial description of the physiological and biochemical features of long-distance transport which has been accumulated for *L. albus* and *L. angustifolius* over the past 20 or so years in Perth provides a sound framework in which the new information can be incorporated and its significance assessed. The challenge for the physiologist is to exploit the superb knowledge base afforded by the genus by making the connection between the genetic regulation of solute transport at the molecular level and the cellular mechanisms which transduce the 'signals' borne in phloem and xylem. By relating these in turn to partitioning in the whole plant a reliable judgement can be made as to the likelihood of increasing yield potential of crop plants through manipulating their transport properties.

References

Atkins, C.A. (1991) Ammonia assimilation and export of nitrogen from the legume nodule. In: Dilworth, M.J. and Glenn, A.R. (eds) *Biology and Biochemistry of Nitrogen Fixation*. Elsevier, Amsterdam, pp. 293–319.

Atkins, C.A. (1993) Nitrogen assimilation and translocation in relation to plant growth. In: Buxton, D.R., Shibles, R., Forsberg, R.A., Blad, B.L., Asay, K.H., Paulsen, G.M. and Wilson, R.F. (eds) *International Crop Science. I.* Crop Science Society of America, Madison, USA, pp. 807–811.

Atkins, C.A. and Flinn, A.M. (1978) CO_2 fixation in the carbon economy of developing seeds of *Lupinus albus* L. *Plant Physiology* 62, 486–490.

Atkins, C.A. and Pate, J.S. (1987) Integration of carbon and nitrogen metabolism. In: Biggins, J. (ed.) *Progress in Photosynthesis Research. Vol. 3.* Martinus Nijhoff Publishers, The Netherlands, pp. 535–544.

Atkins, C.A. and Pigeaire, A. (1993) Applications of cytokinins to flowers to increase pod set in *Lupinus angustifolius* L. *Australian Journal of Agricultural Research* 44, 1799–1819.

Atkins, C.A., Pate, J.S. and Sharkey, P.J. (1975) Asparagine metabolism – key to the nitrogen nutrition of developing legume seeds. *Plant Physiology* 56, 807–812.

Atkins, C.A., Kuo, J., Pate, J.S., Flinn, A.M. and Steele, T.W. (1977) The photosynthetic pod wall of pea (*Pisum sativum* L.): distribution of CO_2 fixing enzymes in relation to pod structure. *Plant Physiology* 60, 779–786.

Atkins, C.A., Pate, J.S. and Layzell, D.B. (1979) Assimilation and transport of nitrogen in non-nodulated (NO_3-grown) *Lupinus albus* L. *Plant Physiology* 64, 1078–1082.

Atkins, C.A., Pate, J.S., Griffiths, G.J. and White, S.T. (1980a) Economy of carbon and nitrogen in nodulated and non-nodulated (NO_3-grown) cowpea (*Vigna unguiculata* (L.) Walp). *Plant Physiology* 66, 978–983.

Atkins, C.A., Pate, J.S. and McNeil, D.L. (1980b) Phloem loading and metabolism of xylem-borne amino compounds in fruiting shoots of a legume. *Journal of Experimental Botany* 31, 1509–1520.

Atkins, C.A., Pate, J.S., Peoples, M.B. and Joy, K.W. (1983) Amino acid transport and metabolism in relation to the nitrogen economy of a legume leaf. *Plant Physiology* 71, 841–848.

Atkins, C.A., Pate, J.S. and Shelp, B.J. (1984) Effects of short-term N_2 deficiency on N metabolism of legume nodules. *Plant Physiology* 76, 705–710.

Atkins, C.A., Pate, J.S. and Peoples, M.B. (1986) Water relations of cowpea fruits during development. In: Lambers, H., Neeteson, J.J. and Stulen, I. (eds) *Fundamental, Ecological and Agricultural Aspects of Nitrogen Metabolism in Higher Plants.* Martinus Nijhoff Publishers, The Netherlands, pp. 235–238.

Badenoch-Jones, J., Parker, C.W., Letham, D.S. and Singh, S. (1996) Effect of cytokinins supplied via the xylem at multiples of endogenous concentrations on transpiration and senescence in derooted seedlings of oat and wheat. *Plant, Cell and Environment* 19, 504–516.

Bouche-Pillon, S., Fleurat-Lessard, P., Fromont, J.C., Serrano, R. and Bonnemain, J.-L. (1994a) Immunolocalization of the plasma membrane H^+-ATPase in minor veins of *Vicia faba* in relation to phloem loading. *Plant Physiology* 105, 691–697.

Bouche-Pillon, S., Fleurat-Lessard, P., Serrano, R. and Bonnemain, J.-L. (1994b) Asymmetric distribution of the plasma-membrane H^+-ATPase in embryos of *Vicia faba* with special reference to transfer cells. *Planta* 193, 392–397.

Brenner, M.L. and Cheikh, N. (1995) The role of hormones in photosynthate partitioning and seed filling. In: Davies, P.J. (ed.) *Plant Hormones, Physiology, Biochemistry and Molecular Biology*, 2nd edn. Kluwer Academic, Dordrecht, pp. 649–670.

Bush, D.R. (1993) Proton-coupled sugar and amino acid transporters in plants. *Annual Review of Plant Physiology and Plant Molecular Biology* 44, 513–542.

Bush, D.R., Chiou, T.-J. and Chen, L. (1996) Molecular analysis of plant sugar and amino acid transporters. *Journal of Experimental Botany* 47, 1205–1210.

Clements, J.C., Atkins, C.A. and Fletcher, N. (1994) Ethylene and pod set in lupins. In: Dracup, M. and Palta, J. (eds) *Proceedings of the First Australian Lupin Technical Symposium*. Department of Agriculture Western Australia, Perth, p. 261.

Daie, J., Watts, M., Aloni, B. and Wyse, R.E. (1986) *In vitro* and *in vivo* modification of sugar transport and translocation in celery by phytohormones. *Plant Sciences* 46, 35–41.

Davey, J.E. and Van Staden, J. (1978a) Cytokinin activity in *Lupinus albus*. I. Distribution in vegetative and flowering plants. *Physiologia Plantarum* 43, 77–81.

Davey, J.E. and Van Staden, J. (1978b) Cytokinin activity in *Lupinus albus*. II. Distribution in fruiting plants. *Physiologia Plantarum* 43, 82–86.

Davies, P.J. (1995) The plant hormone concept: concentration, sensitivity and transport. In: Davies, P.J. (ed.) *Plant Hormones, Physiology, Biochemistry and Molecular Biology*, 2nd edn. Kluwer Academic, Dordrecht, pp. 13–38.

DeWitt, N.D. and Sussman, M.R. (1995) Immunocytochemical localization of an epitope-tagged plasma membrane proton pump (H^+-ATPase) in phloem companion cells. *Plant Cell* 7, 2053–2067.

DeWitt, N.D., Harper, J.F. and Sussman, M.R. (1991) Evidence for a plasma membrane proton pump in higher plants. *Plant Journal* 1, 121–128.

Doares, S.H., Syrovets, T., Weiler, E.W. and Ryan, C.A. (1995) Oligosaccharides and chitosan activate plant defensive genes through the octadecanoid pathway. *Proceedings of the National Academy of Sciences (USA)* 92, 4095–4098.

Dresselhaus, T., Lorz, H. and Kranz, E. (1994) Representative cDNA libraries from few plant cells. *Plant Journal* 5, 605–610.

Fetene, M. and Beck, E. (1993) Reversal of the direction of photosynthate allocation in *Urtica dioica* L. plants by increasing cytokinin import into the shoot. *Botanica Acta* 106, 235–240.

Fieuw, S. and Patrick, J.W. (1993) Mechanism of photosynthate efflux from *Vicia faba* L. seed coats. I. Tissue studies. *Journal of Experimental Botany* 44, 63–74.

Fischer, W.N., Kwart, M., Hummel, S. and Frommer, W.B. (1995) Substrate specificity and expression profile of amino acid transporters (AAPs) in *Arabidopsis*. *Journal of Biological Chemistry* 270, 16315–16320.

Flinn, A.M., Atkins, C.A. and Pate, J.S. (1977) The significance of photosynthetic and respiratory exchanges in the carbon economy of the developing pea fruit. *Plant Physiology* 60, 412–418.

Frommer, W.B., Hummel, S. and Riesmeier, J.W. (1993) Expression cloning in yeast of a cDNA encoding a broad specificity amino acid permease from *Arabidopsis thaliana*. *Proceedings of the National Academy of Sciences (USA)* 90, 5944–5948.

Frommer, W.B., Hummel, S. and Rentsch, D. (1994) Cloning of an *Arabidopsis* histidine transporting protein related to nitrate and peptide transporters. *FEBS Letters* 347, 185–189.

Geigenberger, P., Lerchl, J., Stitt, M. and Sonnewald, U. (1996) Phloem-specific expression of pyrophosphatase inhibits long-distance transport of carbohydrates and amino acids in tobacco plants. *Plant Cell and Environment* 19, 43–55.

Gianfagna, T. (1995) Natural and synthetic growth regulators and their use in horticultural and agronomic crops. In: Davies, P.J. (ed.) *Plant Hormones, Physiology, Biochemistry and Molecular Biology*, 2nd edn. Kluwer Academic, Dordrecht, pp. 751–773.

Gunning, B.E.S. and Pate, J.S. (1969) 'Transfer cells'. Plant cells with wall ingrowths, specialized in relation to short distance transport of solutes – their occurrence, structure and development. *Protoplasm* 68, 107–133.

Gunning, B.E.S. and Pate, J.S. (1974) Transfer cells. In: Robards, A.W. (ed.) *Dynamic Aspects of Plant Ultrastructure*. McGraw Hill, New York, pp. 441–480.

Gunning, B.E.S., Pate, J.S. and Green, L.W. (1970) Transfer cells in the vascular system of stems: taxonomy, association with nodes, and structure. *Protoplasma* 71, 147–171.

Gunning, B.E.S., Pate, J.S., Minchin, F.R. and Marks, I. (1974) Quantitative aspects of transfer cell structure in relation to vein loading in leaves and solute transport in legume nodules. *Symposia of the Society for Experimental Biology* 28, 87–126.

Harper, J.F., Manney, L. and Sussman, M.R. (1994) Evidence for at least ten plasma membrane H^+-ATPase genes in *Arabidopsis* and complete genomic sequence and expression of a third isoform. *Molecular and General Genetics* 244, 572–587.

Hoad, G.V. (1995) Transport of hormones in the phloem of higher plants. *Plant Growth Regulation* 16, 173–182.

Hoad, G.V., Retamales, J.A., Whiteside, R.J. and Lewis, M. (1993) Phloem translocation of gibberellins in three species of higher plants. *Plant Growth Regulation* 13, 85–88.

Hocking, P.J. and Pate, J.S. (1977) Mobilization of minerals to developing seeds of legumes. *Annals of Botany* 41, 1259–1278.

Hocking, P.J. and Pate, J.S. (1978) Accumulation and distribution of mineral elements in the annual lupins *Lupinus albus* L. and *Lupinus angustifolius* L. *Australian Journal of Agricultural Research* 29, 267–280.

Hocking, P.J., Pate, J.S., Wee, S.C. and McComb, A.J. (1977) Manganese nutrition of *Lupinus* spp. especially in relation to developing seeds. *Annals of Botany* 41, 677–688.

Hocking, P.J., Pate, J.S., Atkins, C.A. and Sharkey, P.J. (1978) Diurnal patterns of transport and accumulation of minerals in fruiting plants of *Lupinus angustifolius* L. *Annals of Botany* 42, 1277–1290.

Hsu, L.C., Chiou, T.J., Chen, L.S. and Bush, D.R. (1993) Cloning a plant amino acid transporter by functional complementation of a yeast amino acid transport mutant. *Proceedings of the National Academy of Sciences, USA* 90, 7441–7445.

Jackson, M.B. (1993) Are plant hormones involved in root to shoot communication? *Advances in Botanical Research* 19, 104–187.

Jameson, P.E., Letham, D.S., Zhang, R., Parker, C.W. and Badenoch-Jones, J. (1987) Cytokinin translocation and metabolism in lupin species. I. Zeatin riboside introduced into the xylem at the base of *Lupinus angustifolius* stems. *Australian Journal of Plant Physiology* 14, 695–718.

Jennings, M.P., Anderson, J.K. and Beacham, I.R. (1995) Cloning and molecular analysis of the *Salmonella enterica ansP* gene, encoding an L-asparagine permease. *Microbiology* 141, 141–146.

Jenns, A. and Kuc, J. (1979) Graft transmission of systemic resistance of cucumber to anthracnose induced by *Colletotrichum lagenarium* and tobacco necrosis virus. *Phytopathology* 69, 753–756.

Jeschke, W.D. and Pate, J.S. (1995) Mineral nutrition and transport in xylem and phloem of *Banksia prionotes* and proteaceous trees of dimorphic root morphology. *Journal of Experimental Botany* 46, 895–905.

Jeschke, W.D., Atkins, C.A. and Pate, J.S. (1985) Ion circulation via phloem and xylem between root and shoot of nodulated white lupin. *Journal of Plant Physiology* 117, 319–330.

Jeschke, W.D., Pate, J.S. and Atkins, C.A. (1987) Partitioning of K^+, Na^+, Mg^{++}, and Ca^{++} through xylem and phloem to component organs of nodulated white lupin under mild salinity. *Journal of Plant Physiology* 128, 77–94.

Komor, E., Liegl, I. and Schobert, C. (1993) Loading and translocation of various cytokinins in phloem and xylem of the seedlings of *Ricinus communis* L. *Planta* 191, 252–255.

Kuo, J., Pate, J.S., Rainbird, R.M. and Atkins, C.A. (1980) Internodes of grain legumes – new location for xylem parenchyma transfer cells. *Protoplasma* 104, 181–185.

Kwart, M., Hirner, B., Hummel, S. and Frommer, W.B (1993) Differential expression of two related amino acid transporters with differing substrate specificity in *A. thaliana*. *Plant Journal*. 4, 993–1002.

Layzell, D.B., Pate, J.S., Atkins, C.A. and Canvin, D.T. (1981) Partitioning of carbon and nitrogen and the nutrition of root and shoot apex in a nodulated legume. *Plant Physiology* 67, 30–36.

Layzell, D.B., Rainbird, R.M., Atkins, C.A. and Pate, J.S. (1979) Economy of photosynthate use in nitrogen-fixating legume nodules: observations on two contrasting symbioses. *Plant Physiology* 64, 888–891.

Lejeune, P., Bernier, G., Requier, M.-C. and Kinet, J.-M. (1994) Cytokinins in phloem and xylem saps of *Sinapsis alba* during floral induction. *Physiologia Plantarum* 90, 522–528.

Letham, D.S. (1994) Cytokinins as phytohormones – sites of biosynthesis, translocation, and function of translocated cytokinin. In: Mok, D.S. and Mok, M.C. (eds) *Cytokinins: Chemistry, Activity and Function*. CRC Press, Boca Raton, pp. 57–80.

Liang, P. and Pardee, A.B. (1992) Differential display of eukaryotic messenger RNA by means of the polymerase chain reaction. *Science* 257, 967–971.

McDonald, R., Wang, H.L., Patrick, J.W. and Offler, C.E. (1995) Cellular pathway of sucrose transport in developing cotyledons of *Vicia faba* L. and *Phaseolus vulgaris* L. – a physiological assessment. *Planta* 196, 659–667.

McDonald, R., Fieuw, S. and Patrick, J.W. (1996a) Sugar uptake by the dermal transfer cells of developing cotyledons of *Vicia faba* L. Experimental systems and general transport properties. *Planta* 198, 54–63.

McDonald, R., Fieuw, S. and Patrick, J.W. (1996b) Sugar uptake by the dermal transfer cells of developing cotyledons of *Vicia faba* L. Mechanism of energy coupling. *Planta* 198, 502–509.

McNeil, D.L., Atkins, C.A. and Pate, J.S. (1979) Uptake and utilization of xylem-borne amino compounds by shoot organs of a legume. *Plant Physiology* 63, 1076–1081.

Molders, W., Buchala, A. and Metraux, J.-P. (1996) Transport of salicylic acid in tobacco necrosis virus-infected cucumber plants. *Plant Physiology* 112, 787–792.

Muhitch, M.J., Felker, F.C., Taliercio, E.W. and Chourey, P.S. (1995) Immunolocalization of a unique form of maize kernel glutamine synthetase using a monoclonal antibody. *Plant Physiology* 107, 757–763.

Offler, C.E and Patrick, J.W. (1993) Pathway of photosynthate transfer in the developing seed of *Vicia faba* L. A structural assessment of the role of transfer cells in unloading from the seed coat. *Journal of Experimental Botany* 44, 710–724.

Oghoghorie, C.G.O. and Pate, J.S. (1972) Exploration of the nitrogen transport system of a nodulated legume using ^{15}N. *Planta* 104, 35–49.

O'Neill, T.B (1961) Primary organization of *Lupinus* shoot. *Botany Gazette* 123, 1–9.

Pate, J.S. (1976) Nutrient mobilization and cycling: case studies for carbon and nitrogen in organs of a legume. In: Wardlaw, I.F. and Passioura, J.B. (eds) *Transport and Transfer Processes in Plants*. Academic Press, London, pp. 447–462.

Pate, J.S. (1979) Distribution of metabolites. In: Sutcliffe, J.F., Bidwell, R.G.S. and Krikorian, A.D. (eds) *Plant Physiology. A Treatise*, Vol. 8. Academic Press, London, pp. 335–401.

Pate, J.S. (1980) Transport and partitioning of nitrogenous solutes. *Annual Review of Plant Physiology* 31, 313–340.

Pate, J.S. (1984a) Nitrogen metabolism of lupins. In: *Proceedings of the 3rd International Lupin Congress, La Rochelle, France, June 1984*, pp. 290–309.

Pate, J.S. (1984b) The carbon and nitrogen nutrition of fruit and seed – case studies of selected grain legumes. In: Murray, D.R. (ed.) *Seed Physiology. Vol. 1 – Development*. Academic Press, London, pp. 41–81.

Pate, J.S. (1985a) Partitioning of carbon and nitrogen in N_2-fixing grain legumes. In: Shibles, R. (ed.) *World Soybean Research Conference III: Proceedings*. Westview Press, Boulder, Colorado, pp. 715–727.

Pate, J.S. (1985b) Physiology of pea – comparison with other legumes in terms of economy of carbon and nitrogen in whole-plant and organ functioning. In: Hebblethwaite, P.D., Heath, M.C. and Dawkins, T.C.K. (eds) *The Pea Crop. A Basis for Improvement*. Butterworths, London, pp. 279–296.

Pate, J.S. (1986a) Economy of symbiotic nitrogen fixation. In: Givnish, T.J. (ed.) *On the Economy of Plant Form and Function*. Cambridge University Press, Cambridge, pp. 299–325.

Pate, J.S. (1986b) Xylem-to-phloem transfer – vital component of the nitrogen-partitioning system of a nodulated legume. In: Cronshaw, J., Lucas, W.J. and Giaquinta, R.T. (eds) *Phloem Transport*. A.R. Liss, New York, pp. 445–462.

Pate, J.S. (1988) Water economy of fruits and fruiting plants: case studies of grain legumes. In: Nooden, L.D. and Leopold, A.C. (ed.) *Senescence and Ageing in Plants*, Academic Press, New York, pp. 219–239.

Pate, J.S. (1989a) Legume fruits – the structure : function equation. In: Stirton, C.H. and Zarucchi, J.L. (eds) *Advances in Legume Biology*. Monographs in Systematic Botany, 29, Missouri Botanical Garden, pp. 399–415.

Pate, J.S. (1989b) Origin, destination and fate of phloem solutes in relation to organ and whole plant functioning. In: Baker, D.A. and Milburn, J.A. (eds) *Transport of Photoassimilates*. Longmans, London, pp. 138–166.

Pate, J.S. (1989c) Synthesis, transport, and utilization of products of symbiotic nitrogen fixation. In: Poulton, J.E., Romeo, J.T. and Conn, E.E. (eds) *Plant Nitrogen Metabolism*. Plenum Publishing, New York, pp. 65–115.

Pate, J.S. (1996) Photoassimilate partitioning and consumption in nitrogen-fixing crop legumes. In: Zamski, E. and Schaffer, A.A. (eds) *Photoassimilate Partitioning in Plants and Crops: Source : Sink Relationships*. Marcel Dekker, New York, pp. 467–477.

Pate, J.S. and Atkins, C.A. (1983a) Nitrogen uptake, transport and utilization. In: Broughton, W.J. (ed.) *Nitrogen Fixation, Volume 3: Legumes*. Oxford University Press, New York, pp. 245–298.

Pate, J.S. and Atkins, C.A. (1983b) Xylem and phloem transport and the functional economy of carbon and nitrogen of a legume leaf. *Plant Physiology* 71, 835–840.

Pate, J.S. and Farrington, P. (1981) Fruit set in *Lupinus angustifolius* cv. Unicrop. II. Assimilate flow during flowering and early fruiting. *Australian Journal of Plant Physiology* 8, 307–318.

Pate, J.S. and Flinn, A.M. (1973) Carbon and nitrogen transfer from vegetative organs to ripening seeds of field pea (*Pisum arvense* L.). *Journal of Experimental Botany* 24, 1090–1099.

Pate, J.S. and Gunning, B.E.S. (1972) Transfer cells. *Annual Review of Plant Physiology* 23, 173–196.

Pate, J.S. and Hocking, P.J. (1978) Phloem and xylem transport in the supply of minerals to a developing legume (*Lupinus albus* L.) fruit. *Annals of Botany* 42, 911–921.

Pate, J.S. and Jeschke, W.D. (1995) Role of stems in transport, storage and circulation of ions and metabolites by the whole plant. In: Gartner, B. (ed.) *Stems and Trunks: Their Roles in Plant Form and Function*. Academic Press, New York, pp. 177–204.

Pate, J.S. and Kuo, J. (1981) Anatomical studies of legume pods – a possible tool in taxonomic research. In: Polhill, R.M. and Raven, P.H. (eds) *Advances in Legume Systematics*. Royal Botanic Gardens, Kew, pp. 903–912.

Pate, J.S. and Layzell, D.B. (1981) Carbon and nitrogen partitioning in the whole plant – a thesis based on empirical modelling. In: Bewley, J.D. (ed.) *Nitrogen and Carbon Metabolism*. Martinus Nijhoff/Junk, The Hague, pp. 94–134.

Pate, J.S. and Layzell, D.B. (1990) Energetics and biological costs of nitrogen assimilation. In: Stumpf, P.K. and Conn, E.E. (eds) *The Biochemistry of Plants,* Vol. 16. Academic Press, San Francisco, California, pp. 1–42.

Pate, J.S. and Minchin, F.R. (1980) Comparative studies of carbon and nitrogen nutrition of selected grain legumes. In: Summerfield, R.J. and Bunting, A.H. (eds) *Advances in Legume Science*. Royal Botanic Gardens, Kew, pp. 105–114.

Pate, J.S., Gunning, B.E.S. and Briarty, L.G. (1969) Ultrastructure and functioning of the transport system of the leguminous root nodule. *Planta* 85, 11–34.

Pate, J.S., Gunning, B.E.S. and Milliken, F.F. (1970) Function of transfer cells in the nodal regions of stems, particularly in relation to the nutrition of young seeds. *Protoplasma* 71, 313–334.

Pate, J.S., Sharkey, P.J. and Lewis, O.A.M. (1974) Phloem bleeding from legume fruits. A technique for study of fruit nutrition. *Planta* 120, 229–243.

Pate, J.S., Sharkey, P.J. and Lewis, O.A.M. (1975) Xylem to phloem transport of solutes in fruiting shoots of legume, studied by a phloem bleeding technique. *Planta* 122, 11–26.

Pate, J.S., Sharkey, P.J. and Atkins, C.A. (1977) Nutrition of a developing legume fruit. Functional economy in terms of carbon, nitrogen, water. *Plant Physiology* 59, 506–510.

Pate, J.S., Kuo, J. and Hocking, P.J. (1978) Functioning of conducting elements of phloem and xylem in the stalk of the developing fruit of *Lupinus albus* L. *Australian Journal of Plant Physiology* 5, 321–326.

Pate, J.S., Atkins, C.A., Hamel, K., McNeil, D.L. and Layzell, D.B. (1979a) Transport of organic solutes in phloem and xylem of a nodulated legume. *Plant Physiology* 63, 1082–1088.

Pate, J.S., Layzell, D.B. and Atkins, C.A. (1979b) Economy of carbon and nitrogen in a nodulated and non-nodulated (NO_3-grown) legume. *Plant Physiology* 64, 1083–1088.

Pate, J.S., Layzell, D.B. and McNeil, D.L. (1979c) Modelling the transport and utilization of carbon and nitrogen in a nodulated legume. *Plant Physiology* 63, 730–737.

Pate, J.S., Atkins, C.A. and Perry, M.W. (1980a) Significance of photosynthate produced at different stages of growth as carbon source for fruit filling and seed reserve accumulation in *Lupinus angustifolius* L. *Australian Journal of Plant Physiology* 7, 283–297.

Pate, J.S., Atkins, C.A., White, S.T., Rainbird, R.M. and Woo, K.C. (1980b) Nitrogen nutrition and xylem transport of nitrogen in ureide-producing grain legumes. *Plant Physiology* 65, 961–965.

Pate, J.S., Layzell, D.B. and Atkins, C.A. (1980c) Transport exchange of carbon, nitrogen and water in the context of whole plant growth and functioning – case history of a nodulated annual legume. *Bericht Deutschen Botanishchen Gesellschaft* 93, 243–255.

Pate, J.S., Atkins, C.A. and Rainbird, R.M. (1981) Theoretical and experimental costing of nitrogen fixation and related processes in nodules of legumes. In: Gibson, A.H. and Newton, W.E. (eds) *Current Perspectives in Nitrogen Fixation*. Australian Academy of Science, Canberra. Griffin Press, South Australia, pp. 105–116.

Pate, J.S., Peoples, M.B. and Atkins, C.A. (1983) Post-anthesis economy of carbon in a cultivar of cowpea. *Journal of Experimental Botany* 34, 544–562.

Pate, J.S., Atkins, C.A., Layzell, D.B. and Shelp, B.J. (1984a) Effects of N_2 deficiency on transport and partitioning of C and N in a nodulated legume. *Plant Physiology* 76, 59–64.

Pate, J.S., Peoples, M.B. and Atkins, C.A. (1984b) Spontaneous phloem bleeding from cryopunctured fruits of a ureide-producing legume. *Plant Physiology* 74, 499–505.

Pate, J.S., Peoples, M.B., Van Bel, A.J.E., Kuo, J. and Atkins, C.A. (1985) Diurnal water balance of the cowpea fruit. *Plant Physiology* 77, 148–156.

Pate, J.S., Atkins, C.A., Peoples, M.B. and Herridge, D.F. (1988) Partitioning of carbon and nitrogen in the nodulated grain legume: principles, processes and regulation. In: Summerfield, R.J. (ed.) *World Crops: Cool Season Food Legumes*. Kluwer Academic Publishers, Dordrecht, pp. 751–765.

Patrick, J.W., Offler, C.E. and Wang, X.-D. (1995) Cellular pathway of photosynthate transfer in the developing seed of *Vicia faba* L. and *Phaseolus vulgaris* L. I. Extent of transport through the coat symplast. *Journal of Experimental Botany* 46, 39–47.

Peoples, M.B., Pate, J.S. and Atkins, C.A. (1983) Mobilization of nitrogen in fruiting plants of a cultivar of cowpea. *Journal of Experimental Botany* 34, 563–578.

Peoples, M.B., Atkins, C.A., Pate, J.S. and Murray, D.R. (1985) Economy of water, carbon and nitrogen in the developing cowpea fruit. *Plant Physiology* 77, 142–147.

Phinney, B.O. (1984) Gibberellin A_1, dwarfism and the control of shoot elongation in higher plants. In: Crozier, A. and Hillman, J.R. (eds) *The Biosynthesis and Metabolism of Plant Hormones*. Cambridge University Press, Cambridge, pp. 17–41.

Rainbird, R.M., Atkins, C.A. and Pate, J.S. (1983a) Diurnal variation in the functioning of cowpea nodules. *Plant Physiology* 72, 308–312.

Rainbird, R.M., Atkins, C.A., Pate, J.S. and Sanford, P. (1983b) Significance of hydrogen evolution in the carbon and nitrogen economy of nodulated cowpea. *Plant Physiology* 71, 122–127.

Renault, S., Caussin, C., Bonnemain, J. and Delrot, S. (1989) The proton electrochemical transmembrane gradients generated by the transfer cells of the haustorium of *Polytrichum formosum* and their use in the uptake of amino acids. *Plant Physiology* 90, 913–920.

Samuels, A.L., Fernando, M. and Glass, D.M. (1992) Immunoflourescent localization of plasma membrane H^+-ATPase in barley roots and effects of K^+ nutrition. *Plant Physiology* 99, 1509–1514.

Sauer, N., Baier, K., Gahrtz, M., Stadler, R., Stolz, J. and Truernit, E. (1994) Sugar transport across the plasma membranes of higher plants. *Plant Molecular Biology* 26, 1671–1679.

Schaller, A. and Ryan, C.A. (1996) Systemin – a polypeptide defense signal in plants. *Bioessays* 18, 27–33.

Sharkey, P.J. and Pate, J.S. (1975) Selectivity in xylem to phloem transfer of amino acids in fruiting shoots of white lupin (*Lupinus albus* L.). *Planta* 127, 251–262.

Sharkey, P.J. and Pate, J.S. (1976) Translocation from leaves to fruits of a legume studied by a phloem bleeding technique. Diurnal changes and effects of continuous darkness. *Planta* 128, 63–72.

Shulaev, V., Leon, J. and Raskin, I. (1995) Is salicylic acid a translocated signal of systemic acquired resistance in tobacco? *Plant Cell* 7, 1691–1701.

Slovik, S., Daeter, W. and Hartung, W. (1995) Compartmental redistribution and long-distance transport of abscisic acid (ABA) in plants as influenced by environmental changes in the rhizosphere – a biomathematical model. *Journal of Experimental Botany* 46, 881–894.

Stoddart, J.L. (1983) Sites of gibberellin biosynthesis and action. In: Crozier, A. (ed.) *The Biochemistry and Physiology of Gibberellins*, Vol 2. Praeger, New York, pp. 1–55.

Summons, R.E., Letham, D.S., Gollnow, B.I., Parker, C.W., Entsch, B., Johnson, L.P., Macleod, J.K. and Rolfe, B.G. (1981) Cytokinin translocation and metabolism in species of the leguminoseae: studies in relation to shoot and nodule development. In: Guern, J. and Peaud-Lenoel, C. (eds) *Metabolism and Molecular Activities of Cytokinins.* Springer-Verlag, New York, pp. 69–79.

Tamas, I.A. (1995) Hormonal regulation of apical dominance. In: Davies, P.J. (ed.) *Plant Hormones, Physiology, Biochemistry and Molecular Biology*, 2nd edn. Kluwer Academic, Dordrecht, pp. 572–597.

Taylor, J.S., Thompson, B., Pate, J.S., Atkins, C.A. and Pharis, R.P. (1990) Cytokinins in the phloem sap of white lupin (*Lupinus albus* L.) *Plant Physiology* 94, 1714–1720.

Thompson, B.E., Pharis, R.P., Taylor, J.S. and Pate, J.S. (1988) Endogenous gibberellins in *Lupinus albus* phloem sap and germinant seedlings. *Proceedings of the 13th International Conference on Plant Growth Substances, University of Calgary, Calgary*, Abstract 371.

Unkovich, M.J., Pate, J.S. and Hamblin, J. (1994) The nitrogen economy of broadacre lupin in southwest Australia. *Australian Journal of Agricultural Research* 45, 149–164.

Vancaeseele, L., Klingner, B. and Sumner, M.J. (1996) The immunolocalization of plasma membrane H^+-ATPase in the transfer cell region of *Brassica napus* (Brassicaseae) ovules. *American Journal of Botany* 83, 1386–1390.

Vick, B.A. and Zimmerman, D.C. (1984) Biosynthesis of jasmonic acid by several plant species. *Plant Physiology* 75, 458–461.

Walker, N.A., Patrick, J.W., Zhang, W. and Fieuw, S. (1995) Mechanism of photosynthate efflux from seed coats of *Phaseolus vulgaris*: a chemiosmotic analysis. *Journal of Experimental Botany* 46, 539–549.

Wang, H.L., Offler, C.E. and Patrick, J.W. (1994) Nucellar projection transfer cells in the developing wheat grain. *Protoplasma* 182, 39–52.

Wang, H.L., Patrick, J,W., Offler, C.E. and Wang, X.D. (1995a) The cellular pathway of photosynthate transfer in the developing wheat grain. 3. A structural analysis and physiological studies of the pathway from the endosperm cavity to the starchy endosperm. *Plant Cell and Environment* 18, 389–407.

Wang, X.D., Harrington, G., Patrick, J.W., Offler, C.E. and Fieuw, S. (1995b) Cellular pathway of photosynthate transport in coats of developing seed of *Vicia faba* L. and *Phaseolus vulgaris* L. 2. Principal cellular sites of efflux. *Journal of Experimental Botany* 46, 49–63.

Weiler, E.W. and Zeigler, H. (1981) Determination of phytohormones in phloem exudate from tree species by radioimmunoassay. *Planta* 152, 168–170.

Williams, L.E., Bick, J.A., Neelam, A., Weston, K.N. and Hall, J. (1996) Biochemical and molecular characterization of sucrose and amino acid carriers in *Ricinus communis*. *Journal of Experimental Botany* 47, 1211–1216.

Wolf, O., Jeschke, W.D. and Hartung, W. (1990) Long distance transport of abscisic acid in salt stressed *Lupinus albus* plants. *Journal of Experimental Botany* 41, 593–600.

Zhang, R. and Letham, D.S. (1990) Cytokinin translocation and metabolism in lupin species III. Translocation of xylem cytokinin into the seeds of lateral shoots of *Lupinus angustifolius*. *Plant Science* 70, 65–71.

Zheng, S. and Haselkorn, R. (1996) A glutamate/glutamine/aspartate/asparagine transport operon in *Rhodobacter capsulatus*. *Molecular Microbiology* 20, 1001–1011.

8 Responses to Abiotic Stresses

MILES DRACUP[1,2], NEIL C. TURNER[1,3], CAIXIAN TANG[1], MARK READER[1,2] AND JAIRO PALTA[1,3]

[1]Centre for Legumes in Mediterranean Agriculture (CLIMA), University of Western Australia, Nedlands, Western Australia 6907; [2]Agriculture Western Australia, 3 Baron-Hay Court, South Perth, Western Australia 6151; [3]CSIRO Plant Industry, Centre for Mediterranean Agricultural Research, Private Bag, PO Wembley, Western Australia 6014

Introduction

Much information has accumulated on responses of lupins to soil and climatic stresses. With the growing interest in the cultivation of lupins there is a need to integrate the information to help to develop principles for the adaptation and selection of the various lupin species. This need is amplified by the growing importance of stresses arising from human activities; salinity, compaction and acidity are prime examples. It is also increasingly important to maximize productivity with the ever increasing population and pressure on available land.

Lupins characteristically grow on well-drained acidic to neutral soils and are generally intolerant of waterlogged or extremely alkaline or saline soils. However, there is distinct variation among, and even within, the species. This is hardly surprising given the range of environments in which lupins are distributed and the range of soils from which they have been collected (Cowling and Clements, 1993; Cowling, 1994; Gladstones, 1994), even within one species such as *L. angustifolius*.

In this chapter we discuss responses of lupins to the major abiotic stresses of drought, high and low temperature and pH, salinity, waterlogging and compaction. Responses to nutrient availability are discussed in Chapter 5 and specific aspects of N fixation can be found in Chapter 6. Within the confines of space we have endeavoured to give a comprehensive synthesis and interpretation of the information on responses to abiotic stresses, but, in line with research activity, have concentrated on *L. angustifolius*, *L. albus* and *L. luteus*.

Drought

In their centre of origin around the Mediterranean Basin (Chapter 1), annual lupins encounter a range of environments in which the availability of water is uncertain. A feature of mediterranean climates is the high probability of rainfall in the cool winter months with increasing water shortage during spring and early summer, when seed filling occurs (Turner, 1992; Loss and Siddique, 1994). While lupins are still grown in mediterranean climatic regions of Australia, the Mediterranean Basin and South America, they are also grown in more temperate regions in which droughts can occur at any stage of growth.

Tissue water relations

Lupins have a high turgid weight : dry weight ratio, and cells with high elasticity and low osmotic pressures at full turgor. Studies using the pressure/volume technique (Scholander *et al.*, 1965; Turner, 1988) have shown that the turgid weight : dry weight ratio varies from 7 in *L. angustifolius* to over 11 in *L. atlanticus*, the bulk modulus of elasticity varies from 7 MPa (most elastic) in *L. luteus* to over 20 MPa (least elastic) in *L. cosentinii*, and the osmotic pressure at full turgor varies from 0.6 MPa in *L. cosentinii* to 0.9 MPa in *L. angustifolius* (Table 8.1). These values are typical of mesophytic species, but the cells are more elastic and the osmotic pressure at full turgor is lower than in cereals (Turner and Henson, 1989). The moisture release curves for *L. angustifolius* and *L. cosentinii*, given in Fig. 8.1, demonstrate the higher elasticity of the leaves in *L. angustifolius* than in *L. cosentinii*. The apoplastic water content in *L. angustifolius* leaves has been estimated to be 8% (Turner and Henson, 1989).

There have been very limited studies on the tissue water relations of roots of lupins. Turner (1987) showed that root cells were much more elastic than leaf cells in *L. angustifolius* (Fig. 8.1), resulting in much lower relative water contents in roots compared with leaves at modest water deficits.

Development of water deficits and osmotic adjustment

On clear sunny days, the leaf water potential near midday of adequately watered plants in the field and glasshouse is –0.6 to –0.8 MPa in *L. angustifolius*, *L. cosentinii* and *L. albus* (Henson *et al.*, 1989a,b; Jensen *et al.*, 1989; Turner and Henson, 1989; Leport *et al.*, 1998), rising to these values from near zero at dawn (Henson *et al.*, 1989a,b). These midday values are higher (less negative) than those in cereals, reflecting the higher hydraulic conductance of the roots and shoots in lupins than in cereals (Passioura and Munns, 1984; Gallardo *et al.*, 1996). As soil matric potential decreased below only –20 kPa, the leaf water potential decreased to values of about –2 MPa (Henson *et al.*, 1989a; Turner and

Table 8.1. Turgid weight : dry weight ratio (TW/DW), bulk modulus of elasticity (ε) and osmotic pressure at full turgor (π_{100}) in adequately watered plants of five lupin species (data from Turner *et al.*, 1987; Turner and Henson, 1989; Jensen and Henson, 1990).

Species	TW/DW	ε (MPa)	π_{100} (MPa)
L. angustifolius	7–9	11–18	0.5–0.9
L. cosentinii	7–9	20–30	0.6–0.8
L. atlanticus	11	10	0.6
L. luteus	9	7–8	0.7–0.8
L. pilosus	9	10	0.7

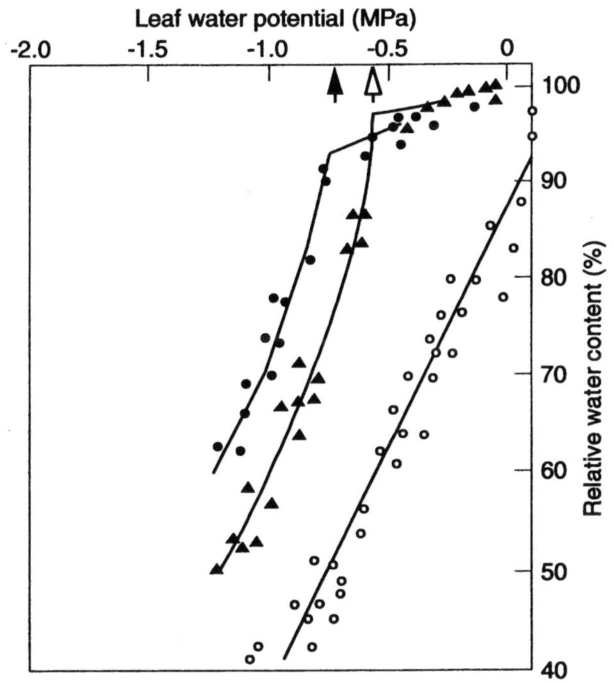

Fig. 8.1. Moisture release curves for leaves (closed symbols) and roots (open symbols) of *L. angustifolius* (●,○) and *L. cosentinii* (▲). The arrows denote the osmotic potential at full turgor for the leaves of *L. angustifolius* (closed arrow) and *L. cosentinii* (open arrow). Adapted from Turner *et al.* (1987) and Turner (1987).

Henson, 1989; Leport *et al.*, 1998), at which stage the leaves senesced and abscised.

Many species have the ability to increase the solute concentration of their cells as water deficits develop so that cell turgor is maintained (Turner and Jones, 1980). The ability to adjust osmotically in response to developing water deficits has been studied in a range of lupin species. In leaves, the degree of osmotic adjustment is about 0.2 MPa in *L. angustifolius*, *L. cosentinii*, *L. albus* and *L. luteus* and 0.4 to 0.5 MPa in *L. atlanticus* and *L. pilosus* (Turner *et al.*, 1987; Turner and Henson, 1989; Rodriguez-Maribona *et al.*, 1989; Jensen and Henson, 1990; Kuang *et al.*, 1990; Leport *et al.*, 1998). Roots have little capacity to adjust osmotically in response to water deficits (Turner *et al.*, 1987), although diurnal changes in osmotic pressures have been observed in adequately watered *L. cosentinii* roots (Jensen *et al.*, 1989), presumably due to the concentration of solutes as water is removed from the highly elastic root cells (Turner, 1987).

Influence on leaf growth and leaf gas exchange

Leaf expansion, the rate of net photosynthesis and the stomatal conductance of all lupin species that have been studied decrease markedly at a midday leaf water potential of about –0.9 MPa (Fig. 8.2) and a pre-dawn leaf water potential of –0.2 MPa (Rodriguez-Maribona *et al.*, 1989). This response is very different from the response observed in cereals, in which the conductance decreases linearly with midday leaf water potential when the soil is slowly dried (e.g. Turner and Henson, 1989), but is similar to the response in a range of other cool-season

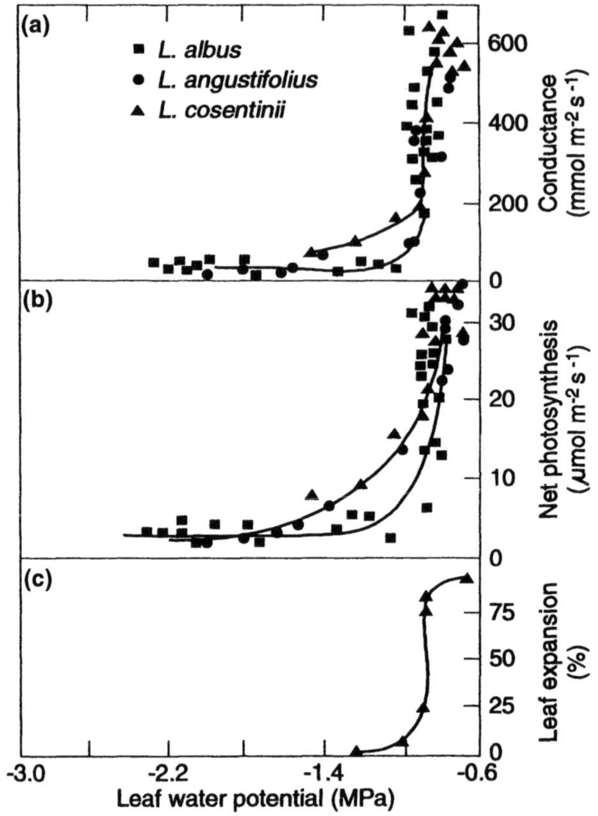

Fig. 8.2. Relationship between leaf water potential and (a) leaf conductance, (b) rate of net photosynthesis and (c) relative leaf expansion rate in *L. albus* (■), *L. angustifolius* (●) and *L. cosentinii* (▲). Adapted from Henson *et al.* (1989a), Turner and Henson (1989) and Leport *et al.* (1998).

pulses (Turner *et al.*, 1996; Leport *et al.*, 1998). The rapid decrease in leaf expansion, photosynthesis and stomatal conductance at only –0.9 MPa may, in part, be due to the small degree of osmotic adjustment in some lupin species, but is more likely to be a result of phytohormone production by the plants in drying soil. Studies by Henson *et al.* (1989b) and Turner and Henson (1989) showed that the decrease in stomatal conductance, photosynthesis and leaf expansion in *L. angustifolius* and *L. cosentinii* occurred when the concentration of abscisic acid (ABA) in the leaf increased dramatically. Subsequent work has shown that leaf expansion decreases as soil water is depleted even when high leaf turgor is maintained (Kuang *et al.*, 1990), suggesting that soil water content, not leaf water relations, is affecting leaf expansion. Work with other species has shown that, as the soil dries, ABA is produced in the roots and transferred in the xylem to the leaves and meristems, where it directly affects leaf expansion, stomatal conductance and leaf photosynthesis (Davies and Zhang, 1991; see also Chapter 7, this volume). In lupins, differences among species in the inherent levels of ABA in root tips at full turgor (100% relative water content) were observed (Table 8.2). When the roots were dried to a water content of 50% the various lupin species produced different amounts of ABA. *L. digitatis*, a species from the desert margins

Table 8.2. Concentration of free abscisic acid (nmol g dw^{-1}) in roots of 3- to 5-day-old lupin seedlings at 100% and 50% water content. Data are means ± one standard error of the mean of four replicates. (Adapted from Hartung and Turner, 1998.)

	Root water content	
Species	100%	50%
L. albus	4.63 ± 1.34	9.70 ± 1.07
L. angustifolius	2.69 ± 1.34	10.11 ± 1.78
L. atlanticus	2.63 ± 1.32	4.92 ± 1.19
L. cosentinii	8.47 ± 1.44	14.91 ± 1.52
L. digitatus	3.55 ± 1.59	22.12 ± 1.59
L. luteus	2.43 ± 0.27	3.57 ± 0.56
L. pilosus	1.55 ± 0.48	8.66 ± 0.75

of Sudan and Egypt (Gladstones, 1974), accumulated the highest amount of ABA in the root tips, while *L. luteus*, which is found in cooler maritime coastal regions of the Iberian Peninsula, coastal Morocco and coastal regions of the Mediterranean Basin (Gladstones, 1974), accumulated the least amount of ABA (Table 8.2). Thus, the accumulation of ABA in response to water deficits appears to be greater in lupin species originating from dry than from wet regions.

While there are some unanswered questions about ABA as the stress signal in other species (Munns and King, 1988), both endogenous and exogenous ABA have been shown to close stomata and reduce photosynthesis and transpiration in *L. cosentinii* (Henson *et al.*, 1989b). Stomatal conductance decreases as ABA concentration in the leaf rises, consistent with ABA being active in stomatal closure of lupins. In a few tested species of lupin, there was a logarithmic relationship between leaf conductance and both leaf ABA concentration and the concentration of exogenous ABA supplied to the xylem stream (Fig. 8.3) (Henson *et al.*, 1989b; Turner and Henson, 1989; Henson and Turner, 1991; Gallardo *et al.*, 1994). The results obtained by Henson and Turner (1991) with *L. luteus* contrast with earlier studies by Lancaster *et al.* (1977) in which the response of leaf conductance to exogenous ABA was smaller than reported in the studies by Henson and Turner (1991); these differences may reflect the more sensitive methods for measuring leaf conductance available to the later studies.

Influence on dry matter partitioning and yield

Water deficits during flowering, pod set and pod filling can have major effects on yield. Studies with a range of lupin species over two years at a low-rainfall site in Western Australia demonstrated that *L. angustifolius* and *L. albus* yielded 2 to 4 t ha^{-1} when irrigated, compared with yields of about 1.5 t ha^{-1} in *L. atlanticus* and *L. pilosus*. Yields of all four species were only 0.25 to 0.45 t ha^{-1} in the rainfed plots in a year with 170 mm of growing-season rainfall, whereas *L. albus* and *L. angustifolius* yielded 2 t ha^{-1} and *L. atlanticus* and *L. pilosus* yielded only 0.6 to 0.9 t ha^{-1} in a year with 270 mm growing-season rainfall (Thompson *et al.*, 1997).

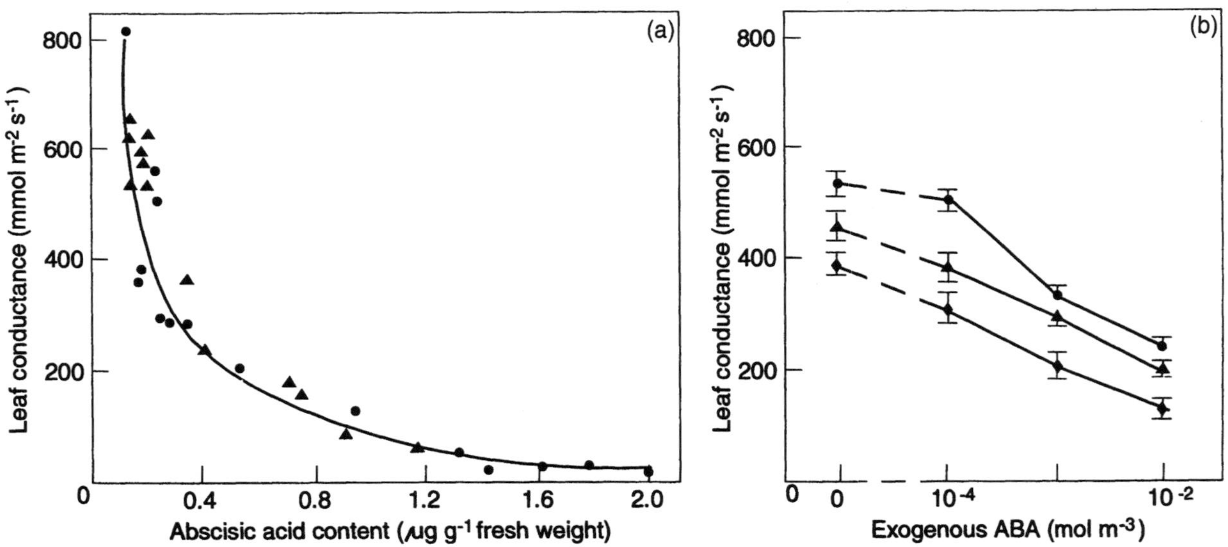

Fig. 8.3. Relationship between (a) leaf conductance and abscisic acid content of the leaves or (b) leaf conductance and exogenous abscisic acid concentration fed into the xylem stream of a transpiring leaf in *L. angustifolius* (●), *L. cosentinii* (▲) and *L. luteus* (◆). Adapted from Henson *et al.* (1989b), Turner and Henson (1989) and Henson and Turner (1991).

As lupin has an indeterminate flowering habit (Dracup and Kirby, 1996a), the effect of terminal stress is greater on the later-formed pods (Dracup and Kirby, 1996b). Biddiscombe (1975) observed a 47% reduction in yield of *L. angustifolius* due to water deficit during flowering. This arose from a 15% increase in flower drop and a reduction in flowering nodes and number of pods as a result of a reduction in lateral branch development. Similar results were obtained by French and Turner (1991) in a field study in south-western Australia. In a season with 185 mm of growing-season rainfall, irrigation of *L. angustifolius* that kept the soil water content near field capacity doubled dry matter production and seed yield. Seed yield on the main stem was actually higher at 3.2 g per plant in the rainfed lupins compared with 2.7 g per plant in the irrigated plants. However, the seed yield of plants on the rainfed plots was only 70% and 20% of the seed yield in the irrigated plots on the first- and second-order branches, respectively. Likewise, water deficits during pod set and pod filling in *L. angustifolius* and *L. albus* had little effect on main-stem dry matter, but progressively reduced the amount of dry matter accumulated in higher-order branches (Withers, 1979a; French and Turner, 1991). Dracup and Kirby (1996b) found that water deficit during seed filling hastened seed development and caused seed and pod abortion. Subsequent work (Dracup *et al.*, 1998) has found that water deficit at this time reduced harvest index from 30 to 18% in *L. angustifolius.*

Transient periods of water deficit during flowering also have a marked and lasting influence on dry matter partitioning. Withers (1979a,b) applied a water deficit at five different stages of development that reduced the leaf relative water content of *L. albus* to about 80% compared with a leaf relative water content of 90% in the controls. A water deficit in the vegetative phase did not reduce seed yields, but a deficit, particularly during flowering on the main stem and first-order laterals, reduced the total seed yield by reducing the yield of the first-order laterals, but not the seed yield of the main-stem pods. Similarly, a transient mild deficit, which reduced stomatal conductance of *L. angustifolius* by 60% and reduced the midday leaf water potential from –0.7 to –1.1 MPa over a period of 10 days, increased the rate of seed filling in the main-stem pods and increased seed yield by 10% over that in the frequently irrigated plants (French and Turner, 1991). A mild water deficit imposed over 5 days reduced yield of the pods on the first- and second order-branches without decreasing yield on the main stem (Fig. 8.4). The transient stress induced a shift in allocation from vegetative to reproductive growth that was similar to that in unirrigated lupins, in which the stress was progressively more severe (French and Turner, 1991).

Lupins have high rates of photosynthesis and use water rapidly when soil water is readily available (Henson *et al.*, 1989a; Turner and Henson, 1989; Greenwood *et al.*, 1992; Leport *et al.*, 1998), but stomata close rapidly with the development of mild water deficits. Nevertheless, photosynthesis remains positive, but low, in the remaining leaves and this may enable grain filling to proceed even at severe soil water shortage. This opportunistic strategy to water shortage in lupins appears to be common to a range of other pulses (Leport *et al.*, 1998; Turner *et al.*, 1996), but is different from that in cereals, which are much more conservative when water is plentiful but are less sensitive to increasing water deficits (Turner and Henson, 1989).

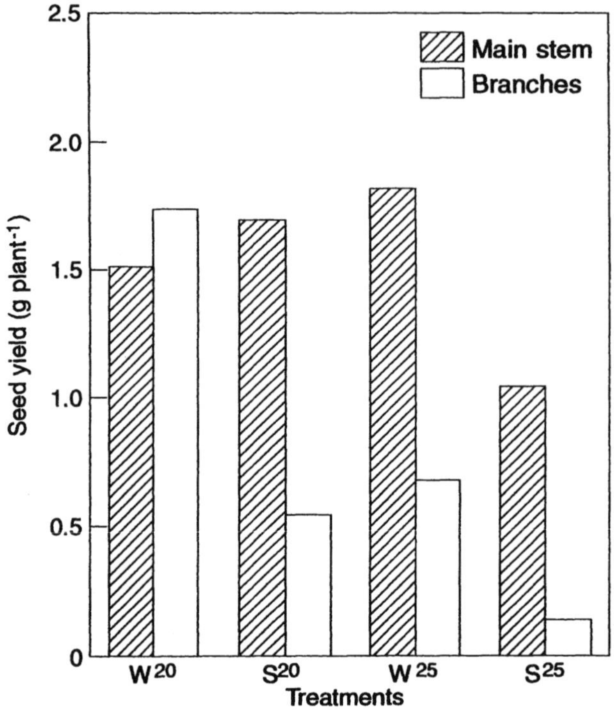

Fig. 8.4. Seed yield on the main stem (hatched columns) and branches (open columns) of *L. angustifolius* plants that were watered adequately (W) or given a mild water stress for 5 days (S) at 20°C throughout (^{20}) and at an elevated temperature of 25°C for 5 days (^{25}) in early pod filling. Data from Palta and Plaut (1998).

Temperature

It is generally recognized that plants grow faster as temperatures rise. However, above an optimum temperature, growth slows and the plant may be injured, so that plant size never fully recovers even if optimal temperatures are restored. Similarly, at temperatures below freezing, growth stops and, depending on the plant's degree of freezing tolerance, the plant may also be injured. Plants may also suffer chilling injury without experiencing freezing temperatures. The temperatures at which injury occurs vary with the species, the organ concerned and the degree of prior acclimatization.

Low temperature

The stage of growth when lupins experience low temperatures can be divided into two categories, depending on location and the time of planting. In cool temperate climates such as in northern Europe and North America, crops sown in the autumn generally experience freezing temperatures during early vegetative growth in winter. In mediterranean climates, which prevail in the main lupin growing regions of southern Australia and in the Mediterranean Basin, lupins are sown in autumn, and may be injured by low temperatures during early vegetative growth. However, in mediterranean climates, lupins are more likely to be injured by low temperatures during early reproductive growth in late winter or early spring because reproductive tissues tend to be more sensitive than vegetative tissues.

Some lupins tolerate quite low temperatures; for example, Huyghe and Papineau (1990) found that roots of *L. albus* 10 cm below the soil surface could survive temperatures as low as −3.5°C. They also found that survival of roots was increased by sowing earlier in autumn so that roots were older and thicker when freezing occurred. This is supported by results of Milford and co-workers who found that survival of soil freezing was associated with lignification of the central root stele and development of secondary root structure (Shields *et al.*, 1996). Early sowing, however, resulted in greater freezing damage to the shoot apex when stem elongation (and presumably floral initiation) had occurred. By comparison, later sown plants, which presumably had not reached floral initiation when freezing occurred, lost their apices less often (Papineau, 1987). Villais *et al.* (1991) found that acclimatization to low temperatures in *L. albus* involves changes in protein expression.

The degree of winter hardiness of the different lupin species varies greatly, depending on their origin. *L. mutabilis* is considered the most cold sensitive of the cultivated lupin species (C. Huyghe, Lusignan, 1996, personal communication), while it is purported that *L. cosentinii* is highly susceptible to frost and that *L. luteus* tolerates mild frosts (Gladstones, 1970). There are lines of both *L. angustifolius* and *L. albus* which tolerate frost better than others (C. Huyghe, Lusignan, 1996, personal communication).

In the case of low temperatures during flowering and seed filling, reproductive tissues can be injured by temperatures above freezing causing abortion of flowers. Differentiating tissues in seeds and the growing pollen tubes appear to be more sensitive to low temperatures than vegetative tissues; the leaves enclosing the meristem tend to protect vegetative apices from freezing damage. Actively growing tissues tend to be more sensitive to low temperatures than dormant tissues (Bidwell, 1979). Most lupins produce one or more orders of branches and do not flower synchronously; thus if, say, some or all of the flowers on the main shoot inflorescence are lost due to frost, the plant has the potential to compensate to some extent with extra seed production on the branches. For example, in part of a crop affected by frost, M. Dracup (Western Australia, 1995, unpublished data) found a 40% reduction in pods on the main shoot and first-order branches but a 100% increase in pods on basal, second- and third-order branches. Furthermore, although there were 25% fewer seeds on the frosted plants, the seeds were 30% heavier.

In addition to restricted seedling growth at temperatures down to 2–5°C (Rahman and Gladstones, 1993; Dracup *et al.*, 1993a), low temperature similarly restricts nodulation and nodule development (Peltzer, 1997) in *L. angustifolius*. The inhibition appears to be the result of a markedly reduced level of synthesis of *nod* gene inducing furtons (Chapter 6) produced by lupin roots under conditions of 7°C or below (Peltzer, 1997).

High temperature

High temperatures may limit productivity of lupins in both mediterranean (where lupins fill their seeds in spring) and cool temperate climates (where lupins flower and fill seeds in summer). The effects are often exacerbated by water deficits, which are common when temperatures are high. For example, in one experiment, *L. angustifolius* plants were grown at 20°C with adequate water, but exposed to 25°C and/or a mild water deficit for 5 days at early pod filling. Yield declined by about 25% when the stresses were applied separately or by 60% when they occurred together (Fig. 8.4). The following discussion therefore focuses on studies where plants were watered throughout experiments, although this does not guarantee that their water status remained high. Vapour pressure deficit would also have usually increased with temperature.

High temperatures induce the synthesis of specific proteins and suppress the expression of others. This has been demonstrated in *L. luteus* roots exposed to temperatures around 39°C (Porankiewicz and Gwozdz, 1995). Such specifically expressed proteins are known as heat shock proteins and are thought to be associated with the acquisition of increased ability to survive high temperatures, though their exact functions remain unknown (Vierling, 1990).

Germination and emergence of *L. angustifolius* were reduced by soil temperatures greater than 20°C and dropped to almost zero at 30°C (Dracup *et al.*, 1993a). Thus high temperatures shortly after sowing may reduce establishment. The inability to germinate at high temperatures is probably due to the destabilizing effect of high temperatures on membranes during their reconstitution in imbibing seeds; membrane reconstitution during imbibition is probably one of the most heat-sensitive stages of plant development. This is supported by work that showed if seeds of *L. angustifolius* were imbibed at 20°C for 18 h they would germinate and grow well in the dark at temperatures as high as 34°C with close to 100% survival over 8 days (M.A. Reader, Perth, 1995, unpublished data).

Reproductive tissues appear to be particularly sensitive to high temperatures. High temperature (33/13°C day/night) around flowering causes abortion of flowers and pods (Downes and Gladstones, 1984). However, attempts to reproduce these results have not been successful (C.A. Atkins and P. Storer, Perth, 1996, personal communication) and might be related to plants in the latter experiment having a higher water status. Other likely consequences of high temperatures around flowering are: male sterility, reduced pollen tube growth and lowered pod and seed set (*Phaseolus vulgaris*; Gross and Kigel, 1994); and reduced seed growth rate (soybean; Egli and Wardlaw, 1980).

High temperatures during seed filling can also affect seed growth and the effects of high temperatures are exacerbated by the pods heating up relative to the surrounding air. M.A. Reader and M. Dracup (Perth, 1997, unpublished data) found pods of field-grown *L. angustifolius* reached 3–5°C higher daily temperatures than the air. In *L. angustifolius*, a total of 6 h of 36°C inside pods (two consecutive days with a diurnal cycle stepping up from 15°C at night to 36°C over midday) during seed filling was sufficient to reduce weight per seed by 8% (M.A. Reader and M. Dracup, Perth, 1996, unpublished data), without increasing seed abortion. Elsewhere, high temperatures (about 32/28°Cd per night) during seed filling increased the proportion of small and unfilled seeds in common bean (Abdus Siddique and Goodwin, 1980) and in soybean, where seed viability and seedling vigour were also reduced (Keigley and Mullen, 1986). Although seeds from heat-treated *L. angustifolius* (38°C for two consecutive days with a diurnal cycle stepping up from 20°C night to 38°C over midday) were 12% smaller, they were equally viable and seedlings emerged at the same rate as seedlings from larger seeds from control plants (M.A. Reader and M. Dracup, Perth, 1996, unpublished data).

Soil Alkalinity

Lupin species respond differently to soil alkalinity (Gladstones, 1970; White, 1990). *L. angustifolius, L. luteus* and *L. albus* generally grow worse on alkaline/calcareous soils than *L. pilosus, L. atlanticus* and *L. cosentinii* (Tang *et al.*, 1995b). The poor growth of sensitive lupin species has been blamed variously on high pH, high HCO_3^-, Fe deficiency and/or Ca toxicity. These factors are addressed below, although since high pH is usually associated with high HCO_3^- and effects of the two are difficult to separate, they are discussed together.

High pH/HCO_3^-

Root growth of *L. angustifolius* is more sensitive to high pH than, in order, root growth of *L. luteus, L. atlanticus, L. albus, L. cosentenii* and *L. pilosus* (Fig. 8.5). When the nutrient solution was buffered with organic buffers and bubbled with CO_2-free air to avoid high HCO_3^- concentrations, root elongation of *L. angustifolius* also decreased markedly as the solution pH rose from 5.5 to 6.0 (Tang *et al.*, 1992), indicating that high HCO_3^- does not necessarily cause the growth reduction. Root elongation of *Pisum sativum* and *L. pilosus* was not affected significantly by pH in the range of 4.5 to 8.0 (Tang *et al.*, 1992, 1993, 1996).

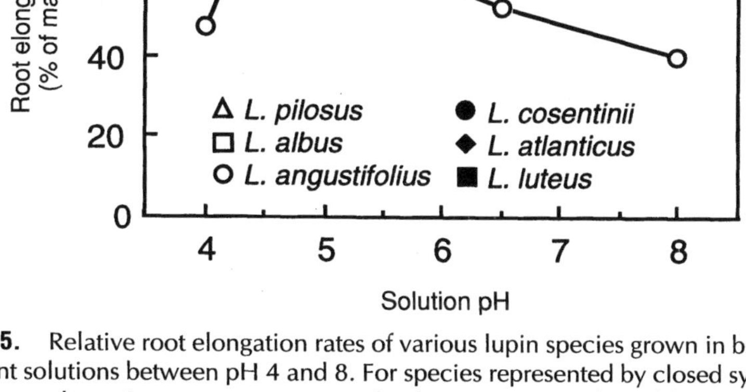

Fig. 8.5. Relative root elongation rates of various lupin species grown in buffered nutrient solutions between pH 4 and 8. For species represented by closed symbols maximum elongation rates were assumed to be at pH 5.2. (Adapted from Tang *et al.*, 1992, 1993, 1996.)

Decreased root elongation of *L. angustifolius* above pH 6 was due to decreased cell elongation, rather than cell division, and occurred within one hour (Tang *et al.*, 1992). Roots exposed to pH above 6 showed physical damage to the root surface, with epidermal cells peeling off and reduced root hair formation (Fig. 8.6). However, the membrane integrity of root cortical cells appears to be unaffected by high pH (Tang *et al.*, 1993, 1996).

Shoot growth of *L. angustifolius, L. luteus* and *L. albus* also decreased at high pH but was more tolerant than root growth. Shoot weight of N_2-fixing *L. angustifolius* grown in nutrient solution for 26 d was 20% lower at pH 6 and 60% lower at pH 7.5 than at pH 5, whereas shoot weight of *L. pilosus* increased slightly with pH (Tang and Robson, 1993a). Independent effects of high pH on N_2 fixation cannot be excluded.

The symbiosis of lupin–bradyrhizobium appears to be impaired by high pH. Decreased nodule number at high pH has been observed on *L. angustifolius* grown on limed soil (Jessop and Mahoney, 1982; Tang *et al.*, 1995a), on *L. albus*, *L. angustifolius* and *L. pilosus* grown in high pH solution (Subramaniam *et al.*, 1982; Bertoni *et al.*, 1992; Tang and Robson, 1993a) and on *L. angustifolius* and *L. pilosus* grown on an alkaline soil (Tang and Robson, 1995). The cause of poor nodulation at high pH is not understood. In solution culture, nodule initiation appears to be reduced independently of host plant growth (Tang and Robson, 1993a), and Cobley (1993) postulated that high pH impairs the bradyrhizobial root-recognition and attachment process.

While *Rhizobium* is generally sensitive to low pH, *Bradyrhizobium* appears to be sensitive to high pH. The growth of bradyrhizobia WU425 at pH 7 or 8 is markedly less than at pH 6 (Cobley, 1993). In a survey of over 100 sites in France, Amarger *et al.* (1984) showed that soils with pH < 6 had over 100 lupin bradyrhizobia per gram of soil, whereas those of pH > 7 had fewer than 1. In contrast, Mock and Gibson (1994) found that *Bradyrhizobium* strain WU425 persisted in alkaline soils for at least 17 years at populations exceeding 1000 g^{-1} of soil.

Iron deficiency

Plants grown on alkaline and calcareous soils often suffer from iron deficiency and lupin species differ in their sensitivity to this. *L. angustifolius* and *L. luteus* are more sensitive than *L. albus* which, in turn, is more sensitive than *L. pilosus* and *L. cosentinii* (Gladstones, 1970; White, 1990; Cowling and Clements, 1993; Tang and Robson, 1993b; Tang *et al.*, 1995b). Rankings of iron chlorosis among *L. angustifolius, L. luteus* and *L. pilosus* on an alkaline soil in the field (Tang *et al.*, 1995b) were consistent with their sensitivity to iron deficiency in nutrient solution (Tang and Robson, 1993b). In nutrient solution, the higher sensitivity of *L. luteus* and *L. angustifolius* than *L. pilosus* to iron deficiency was ascribed to their higher internal and external iron requirements and lower ability to translocate iron from roots to shoots (Tang and Robson, 1993b). However, the susceptibility to iron deficiency of *L. angustifolius* does not appear to be due to poor ability to reduce Fe^{3+} or absorb or retranslocate iron (White, 1990). The severity of iron chlorosis correlates poorly with shoot growth and seed yield in the field (Tang *et al.*, 1995b)

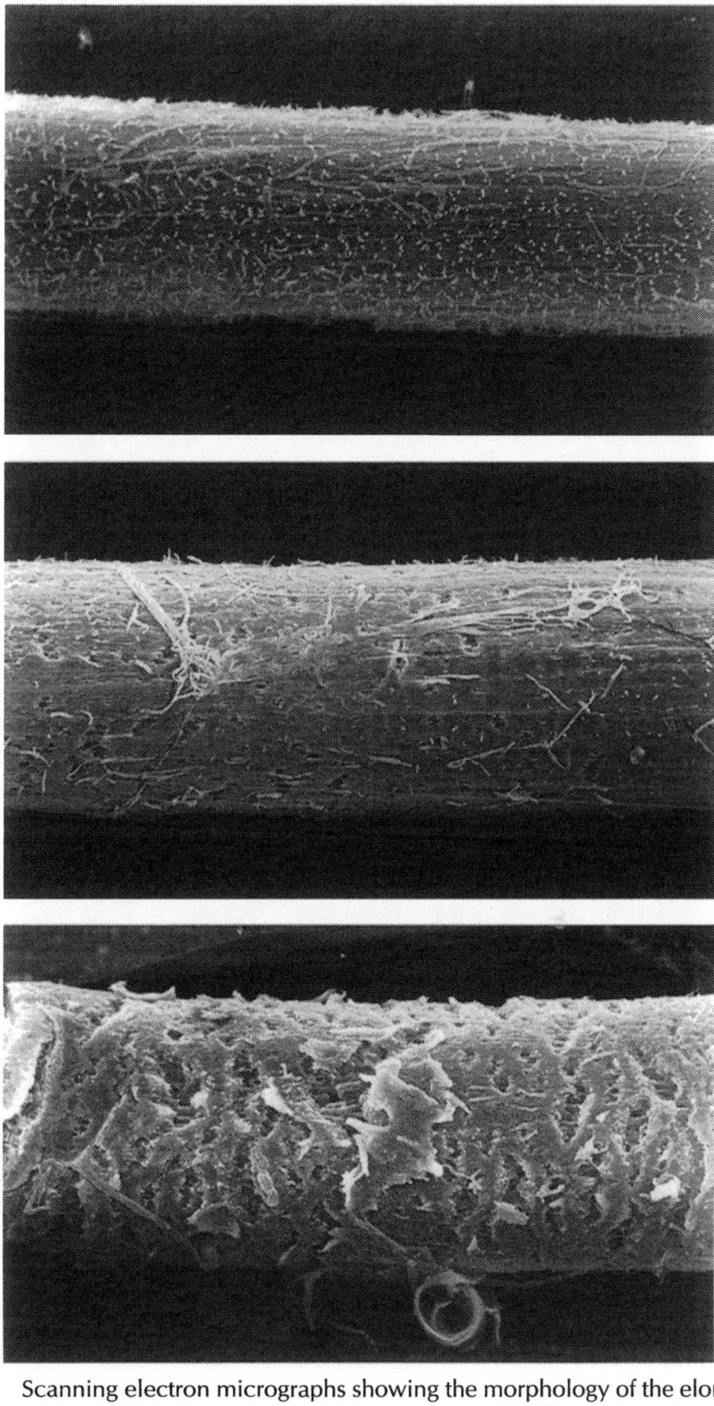

Fig. 8.6. Scanning electron micrographs showing the morphology of the elongating zone of *L. angustifolius* roots grown at pH 5.2 (top), 6.0 (middle) and 7.5 (bottom) in buffered solutions. (C. Tang, Perth, 1996, unpublished data.)

and plants showing iron chlorosis at early growth stages can grow out of the chlorosis (White and Robson, 1989b; Tang *et al.*, 1995b).

Iron is required for the lupin–bradyrhizobium symbiosis and plants reliant on N_2 fixation have higher requirements for iron than plants supplied with NH_4NO_3 (Tang, 1995). Iron deficiency decreases nodule formation and leghaemoglobin production in *L. angustifolius* and nodule initiation is more sensitive to iron deficiency than other phases of the symbiosis (Tang, 1995). Iron deficiency limits division of cortical cells to form nodule initials and inhibits bradyrhizobial proliferation in roots (Tang, 1995). However, the importance of iron deficiency in causing poor nodulation of lupins in some alkaline soils is unknown.

Calcium toxicity

High calcium concentrations in alkaline soils may be toxic to some lupin species. Increasing the Ca concentration – as $Ca(NO_3)_2$ – from 1 to 15 mol m^{-3} decreased leaf extension, leaf transpiration and net assimilation rate in *L. luteus* (De Silva *et al.*, 1994). Also, growing *L. albus* on 44 different soils where the free $CaCO_3$ level ranged from 0 to 26%, Duthion (1992) showed that poor plant growth was related to $CaCO_3$ level but not extractable Fe or soil pH. In another study where *L. angustifolius* was grown on mixtures of two clay soils with $CaCO_3$ concentrations varying from 1.5 to 6.0%, with pH between 6.5 and 7.8 (in $CaCl_2$), Jessop *et al.* (1990) found that the weight of shoots and roots decreased as lime concentration in soil increased. In contrast, Tang *et al.* (1995a) found that adding 0.33 g of $CaSO_4$ kg^{-1} soil containing 0–10 g of $CaCO_3$ kg^{-1} soil (soil solution pH ranging from 4.8 to 7.3) increased calcium concentrations in soil solution by 1.7- to 6.6-fold and in leaves by 10 to 30%, but did not decrease shoot growth of *L. angustifolius* accordingly. Further, increasing the $CaCl_2$ to 20 mol m^{-3} in nutrient solution at pH 5.2 and 7 had a similar effect on the growth of *L. angustifolius* as did equivalent ionic strengths of KCl (Tang *et al.*, 1995a). The causes of such variable effects of high Ca have not been identified.

The reason for inconsistencies between studies is not clear. Further work is required to understand why lupins are sometimes adversely affected by high soil Ca levels and how (and whether) the general intolerance of lupins to alkaline soils relates to the Ca/lime concentration in those soils.

Soil Acidity

Plant growth can be affected by soil acidity through effects of low pH, Al and Mn toxicities, and Mg, Ca and other nutrient deficiencies (Marschner, 1991). Lupins are generally considered tolerant to soil acidity compared with other legumes.

Aluminium toxicity

Lupins are considered comparatively tolerant to Al toxicity, though the tolerance varies between species. In nutrient solution containing 1 mol m^{-3} Ca^{2+}, root elongation of barley and faba bean decreased to about 50% of the control at 1.8

and 9.3 mmol m^{-3} Al, respectively, while *L. luteus* only showed a comparable inhibition of root elongation at 222 mmol m^{-3} Al (Grauer and Horst, 1990). Carr and Sweetingham (1994) found that root growth of *L. angustifolius* was reduced to 40% of the control in acidic subsoils containing 26 mmol m^{-3} KCl-extractable Al, whereas *L. luteus* was unaffected, even at 105 mmol m^{-3}.

Mechanisms of Al tolerance in lupins are not understood. However, some lupins extrude large amounts of organic compounds from roots (Guerrier *et al.*, 1977; Gardner *et al.*, 1981) which can bind Al and decrease its toxicity. Certain lupin species, such as *L. luteus,* may exclude Al from sensitive root sites (Horst and Göppel, 1986) and have low Al intake (Guerrier and Morard, 1978).

High manganese

Lupins appear to be comparatively tolerant of high Mn levels. Growth of most legumes is reduced when Mn concentrations reach 140–1600 ppm in the shoots (Helyar, 1978). In *L. angustifolius*, toxicity occurred at 2000 ppm Mn in shoots (Reuter and Robinson, 1986) but threshold concentrations were not identified. *L. albus* is known to accumulate high concentrations of Mn (Gladstones and Drover, 1962; see Chapter 5, this volume).

Low pH

In nutrient solution, root elongation of *L. angustifolius* started to decrease at pH < 5 and decreased to 50% of the maximum at pH 4 (Fig. 8.5). In comparison, at pH 4 root growth was 90% of maximum in *L. albus* (Fig. 8.5) and 70% of maximum in *Pisum sativum* (Tang *et al.*, 1992).

Symbiotic N_2 fixation in lupin may be less affected by acidic soils than in other legumes (Hartmann and Aldag, 1989) but lupin nodulation seems to be impaired by low pH and may be as sensitive to low pH as that of other legumes. For example, *L. angustifolius* nodulated poorly at pH 4.3 independently of Al concentrations, while other legumes, such as soybean, mungbean, cowpea and peanut, nodulated well at the same pH (Alva and Edwards, 1990). In Al-free solution, early nodule formation was inhibited in *L. angustifolius* at pH 4 (Cobley, 1993) and 4.8 (Munns, 1977), in *L. albus* and *L. pilosus* at pH 4 (C. Tang, Perth, unpublished data) and in *L. luteus* at pH 4.3 (Munns, 1977). In soil, nodule number decreased in *L. angustifolius* grown at pH 5.4 (measured in water) (Jessop and Mahoney, 1982) and in *L. albus* at pH 4 (measured in $CaCl_2$; B. Thomson and C. Tang, 1996, Perth, unpublished data). The decreased nodule formation at low pH is independent of host plant growth in *L. angustifolius* and results in low N concentration in shoots and decreased plant growth in *L. albus*.

Bradyrhizobium (*Lupinus*) has an optimal pH for growth lower than most (brady)rhizobium species (Graham and Parker, 1964) and is more tolerant to low pH than *Rhizobium* in general (Rerkasem, 1977). Al is more toxic to *Rhizobium* than *Bradyrhizobium,* including lupin bradyrhizobium (Rerkasem, 1977). However, growth of *Bradyrhizobium* WU425 in broth culture decreased markedly at pH 4 compared with pH 5 and 6 (Cobley, 1993) and its population in acidic soils in the field generally declines with time (Slattery and Coventry, 1989).

Salinity

Soil salinity reduces plant growth when specific ions in the tissue reach toxic levels. It also interferes with nutrient uptake and decreases the water potential outside the plant. The concentrations at which ion toxicity occurs differ with the genetic capacity of the species to cope with high ion concentrations, growth stage, environmental interactions and ion species involved. Lupins do not tolerate high levels of salinity; they are considered non-halophytes (or glycophytes – see Chapter 7, this volume).

Symptoms of salt toxicity

Symptoms of salt toxicity in lupin *(L. albus)* have been described by Van Steveninck *et al.* (1982), Jeschke *et al.* (1986), Munns *et al.* (1988) and Treeby and Van Steveninck (1988). Initially there is a grey blotching of leaflets; then marginal to complete leaf necrosis develops, followed by wilting and abscission of leaflets. It appears that, in lupin, wilting does not occur until salts reach comparatively high levels in leaves (B.J. Atwell, Sydney, 1996, personal communication). Initial necrosis at the leaf margins may reflect the concentrating of salts in the transpiration stream as it moves toward the margins (Munns, 1988). Toxicity symptoms occur first on the oldest main-stem leaves and then on progressively younger leaves, and eventually, at high salt levels, the plants lose all leaves. The yellowing that generally accompanies leaf senescence does not occur.

Growth

In most plant species, root growth is less affected by salinity than shoot growth (Munns and Termaat, 1986). However, in lupins, all known studies which have presented growth data show that root growth is most affected (*L. luteus*: Van Steveninck *et al.*, 1982; *L. albus*: Jeschke *et al.*, 1986; *L. pilosus* and *L. angustifolius*: B.J. Atwell, Sydney, 1996, personal communication).

Lupin species vary in their salt tolerance but can generally be classified as moderately sensitive to salinity (Fig. 8.7). Notwithstanding problems with comparing responses at different ages, lengths of salinization and experimental conditions, *L. pilosus* appears more tolerant than *L. angustifolius*, *L. arboreus* and *L. albus*, while data from Van Steveninck *et al.* (1982) suggest that *L. luteus* is also more tolerant than *L. angustifolius*. For comparison, other moderately sensitive legumes are soybean (Maas and Hoffman, 1977; Lauchli, 1984) and faba bean (Subbarao and Johansen, 1994), while chickpea is sensitive (Subbarao and Johansen, 1994).

Seed germination and seedling growth of lupin appear considerably more tolerant of salinity than plant growth. In *L. albus* (termed *L. termis* in the article but synonomous; Gladstones, 1974), final germination was not reduced until NaCl exceeded 160 mol m^{-3}, while seedling growth declined at NaCl concentrations above 80 mol m^{-3} (Shaddad *et al.*, 1990).

There is only one known study on the response of lupin to aerial rather than soil salinity. In this study, Sykes and Wilson (1988) showed that *L. arboreus* is

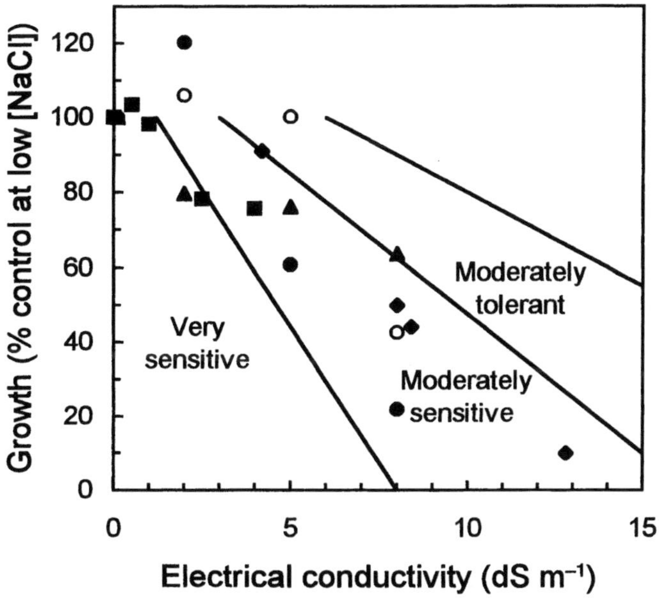

Fig. 8.7. Effects of salinity on relative growth rates of whole lupin plants compared with non-salinized controls: *L. albus* (■) (Jeschke *et al.*, 1986); *L. arboreus* (◆) (Sykes and Wilson, 1988, 1989); *L. angustifolius* cv. Yandee (o) and cv. 75A258 (●) and *L. pilosus* (▲) (B.J. Atwell, Sydney, 1996, personal communication). Classifications of salt tolerance are a synthesis of those used by Greenway and Munns (1980) and Maas and Hoffman (1977). Electrical conductivities were calculated from salt concentrations in the nutrient solutions.

tolerant to salt spray compared with many other dune scrub species. This contrasts with its sensitivity to soil salinity.

Physiology

Physiological responses of lupins to salinity have been studied in most detail with *L. albus*, due to its ability to bleed from xylem and phloem, which enables the movement of ions within the plant to be investigated.

The earliest response of non-halophytes to salinity is slowed leaf growth (Munns and Termaat, 1986). In *L. albus* the hydraulic resistance of the plant increases substantially shortly after salinization and must greatly exacerbate the effect of reduced soil water potential in lowering the leaf water potential (Munns and Passioura, 1984). However, this lowered leaf water potential is not directly responsible for reduced growth because even if leaf water potential is artificially elevated, leaf expansion remains reduced (Munns and Termaat, 1986). This continued slower leaf expansion suggests that a signal, probably a growth regulator, comes from the roots in response to low external water potential (Munns and Termaat, 1986). As in the case of drought, the stress hormone ABA may perform this function as its transport in xylem of *L. albus* increased tenfold after applying 40 mol NaCl m^{-3} to the growing medium, with half of this ABA synthesized in the roots (Wolf *et al.*, 1990).

Long-term salt tolerance of non-halophytes is largely related to their ability to exclude salt from the shoot and how they deal with the inevitable arrival of some salt to the shoot. Lupins are poor at excluding salt from the xylem. Munns (1988) found that at between 0 and 100 mol NaCl m^{-3} outside the root, the concentrations of Na^+ and Cl^- in xylem of *L. albus* rose linearly and averaged 10% of the external concentration, compared to only 4% in barley (Munns, 1985). At about 150 mol m^{-3}, salt concentrations in the xylem rose markedly, suggesting the mechanism(s) in the root for removing salt from the transpiration stream then became overloaded (Munns, 1988). Jeschke *et al.* (1986) also found linear increases in Na^+ and Cl^- concentrations in xylem sap, but under their conditions of growth/xylem collection, xylem concentrations were far higher, reaching about 35 mol m^{-3} when the external concentration was 50 mol m^{-3}.

While high concentrations of Na^+ and Cl^- reach the leaves of *L. albus* via the xylem, large concentrations also exit the leaves in phloem; concentrations in the phloem are nearly as high (Munns *et al.*, 1988), if not higher, than in the xylem (Jeschke *et al.*, 1986). However, uptake of salts by leaf cells still falls behind the rate of delivery (Munns, 1988), eventually damaging the leaves through either build-up in the cell walls, which causes desiccation, or build-up in the cytoplasm, which interferes with metabolism (Flowers and Yeo, 1986, 1989; Munns and Termaat, 1986; Yeo and Flowers, 1986; Munns, 1988).

Meanwhile, the high concentrations of Na^+ and Cl^- translocated in phloem mean that the youngest leaves are not protected from the salt; and the dividing cells, which have little vacuolation to sequester salts, are susceptible to metabolic damage. This is unlikely to explain the lower final leaf areas of plants grown with high NaCl (Munns *et al.*, 1988) because the young leaves are the last to die.

The large concentrations of Na^+ and Cl^- in the phloem also mean that large amounts return to the root (33% at 10 mol NaCl m^{-3}; Jeschke and Pate, 1991) and subsequent excretion into the external solution might help the plant to reduce its salt load.

Investigations into mechanisms behind differences in salt tolerance between lupin species have been less detailed. Surprisingly, Van Steveninck *et al.* (1982) and B.J. Atwell (Sydney, 1996, personal communication) found that at below 50 mol NaCl m^{-3} in the root medium, concentrations (on a fresh weight basis) of Na^+ and Cl^- in shoots of *L. angustifolius* were well below those of the more salt-tolerant species, *L. luteus* and *L. pilosus*. However, at higher external salt concentrations, *L. angustifolius* had higher Na^+ concentrations in the shoots and much reduced growth. Differences in tolerance might therefore be related to differences in the ability of roots to exclude salts, at least from shoots, above certain levels of external NaCl. Alternatively, the more tolerant species might be more successful at sequestering Na^+ and Cl^- within the vacuole. Markedly greater substitution of Na^+ for K^+ in *L. pilosus* than in *L. angustifolius* (B.J. Atwell, Sydney, 1996, personal communication) supports selective sequestering of Na^+ in the vacuole. The success of sequestering Na^+ in vacuoles would be promoted by the succulence (hence dilution of salts) of *L. luteus* foliage; the fresh weight to dry weight ratio of *L. luteus* was 35% higher than *L. angustifolius* at 20 mol NaCl m^{-3} (calculated from Van Steveninck *et al.*, 1982).

Nitrogen fixation

We are not aware of any research on the sensitivity of nitrogen fixation in lupin to salinity. However, rhizobia generally appear more tolerant than the host plant (Lauchli, 1984; Subbarao and Johansen, 1994) and, once formed, the development and function of nodules are generally not directly affected by salinity (Subbarao and Johansen, 1994); in pea, for example, N fixation was reduced as much by salinity as was plant growth (Minchin and Pate, 1975). The infection process appears to be the most likely sensitive element of the symbiosis (Subbarao and Johansen, 1994).

Environmental interactions

Comparisons of salt tolerance and plant functioning between experiments can be misleading because of the influences of environmental conditions. Any factor that reduces the amount of transpiration per unit of C fixed will also reduce the salt concentration in leaves (e.g. interaction with humidity in *L. albus*; Shennan, 1986) and therefore prolong their effective life. This would occur if the plant had an inherently high water-use efficiency, or was grown under conditions that increased the ratio of photosynthesis to transpiration such as high humidity, high CO_2 or high light (Munns and Termaat, 1986). Similarly, legumes are generally less sensitive to salinity if their nitrogen requirements are supplied non-symbiotically (Lauchli, 1984) and N supplied as NO_3^- might also help reduce Cl^- uptake.

Soil conditions are also important to salt responses. Low supply of nutrients such as K, Ca and P can exacerbate salinity effects (e.g. Treeby and Van Stevininck, 1988; Jeschke *et al.*, 1992; Subbarao and Johansen, 1994). For example, B.J. Atwell (Sydney, 1996, personal communication) found that at 80 mol NaCl m^{-3} the relative growth rate of *L. angustifolius* doubled if the external Ca^{2+} concentration was raised from 0.5 to 2.5 mol m^{-3}. Salinity is also frequently accompanied by waterlogging, which decreases the availability of metabolic energy in the root, reducing the root's ability to exclude salt (Barrett-Lennard, 1986).

Waterlogging

The concentration of rainfall during winter in mediterranean environments makes soils prone to waterlogging, particularly where internal drainage is low and/or there is little slope. Waterlogging, or saturation of soil by water, reduces the rate of diffusion of gases into and out of the soil. Thus gases such as CO_2 and ethylene build up while O_2 decreases. As a consequence of low O_2 concentrations the solubility of manganese and iron and other heavy metals can increase and denitrification can occur (Jackson and Drew, 1984; Drew and Stolzy, 1991; Pezeshki, 1994). O_2 supply is considered the most important limitation to plant growth during waterlogging (Armstrong, 1979), and O_2 deficiency is considered likely when the air-filled porosity of soil is below 10%, (Hillel, 1980). Plants vary in their responses to low O_2 and lupins are considered relatively intolerant, though there is a range of tolerance within the genus.

Symptoms of waterlogging damage

Wilting, chlorosis (yellowing) and pigmentation of the oldest leaves and cotyledons are the earliest visible symptoms of waterlogging in lupins. Which of these responses occurs first depends on the species and experimental conditions; for example, wilting occurred in *L. albus* and chlorosis in *L. cosentenii* (Phuphak, 1989), while older leaves and the stem of *L. luteus* become pigmented. Cotyledon abscission is hastened by waterlogging (Phuphak, 1989) and leaflets of old leaves abscise. Meanwhile, root tips lose turgor and their coating of mucilage (Hanbury, 1988), the root cortex may degenerate (Atwell, 1991a), roots die and air spaces may develop within roots close to the surface (Hanbury, 1988; Kuo, 1992).

Growth and survival

Seeds, seedlings and mature plants respond to waterlogging in different ways and the basis for their responses and relative tolerances will vary. Likewise, the stage at which waterlogging occurs and its duration are important in determining how plant functioning overall is affected.

Germination is accompanied by high rates of respiration, so when seeds are surrounded by a film of water in which gases diffuse only very slowly, O_2 supply may become limiting. Lupin seeds are particularly sensitive to waterlogging. Sarlistyaningsih *et al.* (1995) found no germination of *L. angustifolius* seeds in waterlogged soil and the seeds were dead within four days. Seeds of clover, mungbean and faba bean coped with waterlogging better, possibly because of their thinner seed coats (Sarlistyaningsih, 1993), and, in the case of clover and mung bean, the smaller seeds. Sarlistyaningsih *et al.* (1995, 1996) showed that O_2 depletion from the soil, resulting from microbial respiration, is the major factor causing reduced survival of lupin seeds in waterlogged soil.

After the seedling stage, tolerance of lupins to waterlogging decreases – at least until late in vegetative growth (Broué *et al.*, 1976). Waterlogging generally decreases growth of roots before shoots, and root extension is particularly sensitive (Jackson and Drew, 1984). Hanbury (1988) found that the dry weight of *L. albus* roots decreased within five days of imposing hypoxia (low O_2) but shoot dry weight did not decrease significantly for ten days. Similarly, measurements made about two weeks after imposing waterlogging have shown that root growth of lupins is reduced far more than shoot growth (Fig. 8.8) (Davies *et al.*, 1994). For example, C.L. Davies (Perth, 1996, personal communication) found that the root : shoot ratio of *L. angustifolius* after waterlogging (between 8 and 10 weeks) was 0.6, compared with 1.0 in control plants.

With prolonged waterlogging, root apices die (Jackson and Drew, 1984) and new roots, particularly adventitious roots, may develop (for lupins: Hanbury, 1988; Davies *et al.*, 1996) while others are dying (Jackson and Drew, 1984). For example, two weeks after waterlogging (between eight and ten weeks),

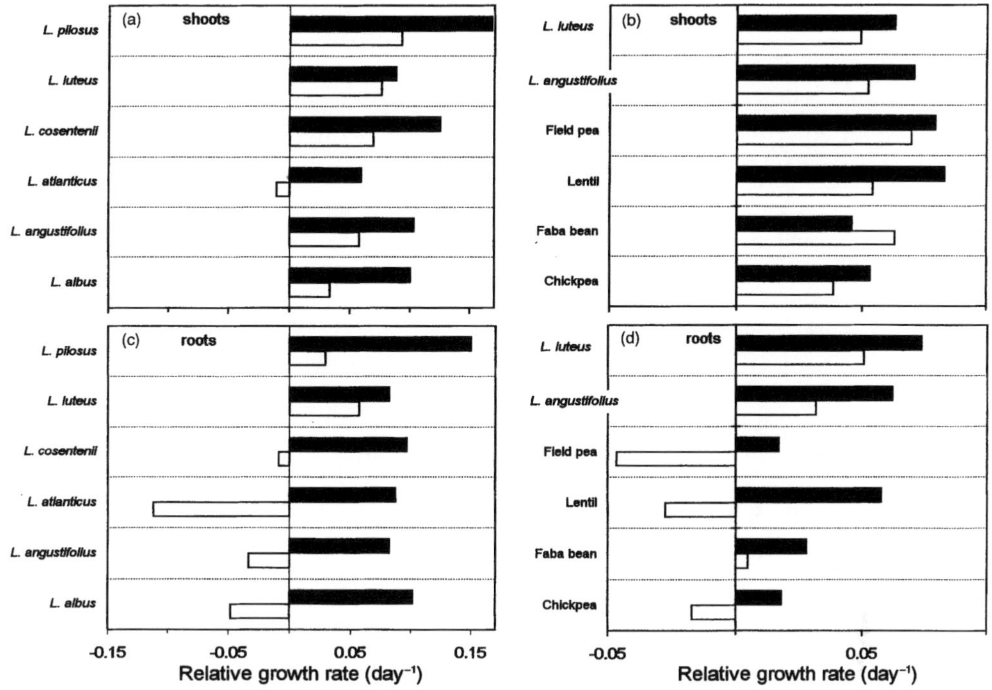

Fig. 8.8. Effects of waterlogging on relative growth rates of shoots and roots. (a, c). Comparison of lupin species waterlogged between 17 and 29 days after sowing (DAS) (Phuphak, 1989). (b, d) Comparison of two lupin species and four other legumes waterlogged between 35 and 42 DAS (B.D. Thomson, Perth, 1996, personal communication). Solid bars represent non-waterlogged controls and open bars represent the waterlogged treatments.

adventitious (from the hypocotyl) root weight increased by 20% in *L. angustifolius* and by 170% in *L. luteus*, compared with 70 and 95% reductions, respectively, in the rest of the root systems (C.L. Davies, Perth, 1996, personal communication). This is consistent with observations made by Dracup *et al.* (1992) after natural waterlogging of lupins in the field.

Effects of waterlogging on shoots are secondary to those on roots and, among other things, stem elongation, leaf orientation, leaf expansion and dry matter accumulation can be modified (Drew, 1983; Jackson and Drew, 1984). In *L. angustifolius* and *L. luteus*, leaf expansion decreased more than the accumulation of dry matter during waterlogging (between eight and ten weeks), causing specific leaf area to decline by 60 and 40%, respectively (C.L. Davies, Perth, 1996, personal communication).

L. luteus and *L. pilosus* are the most tolerant to waterlogging, followed by *L. cosentenii*, *L. angustifolius*, *L. albus* and *L. atlanticus* (Fig. 8.8). This order of tolerance is supported by Broué *et al.* (1976) (based on dry weight at the end of waterlogging), who waterlogged several lupin species at 20–34 days. Their data also indicate that the tolerance of *L. mutabilis* is comparable to that of *L. albus* and *L. angustifolius*. The superior tolerance of *L. luteus* over *L. angustifolius* has been confirmed in several studies (Broué *et al.*, 1976: 1–2 weeks of waterlogging imposed 2–4 weeks after emergence; Davies *et al.*, 1994, 1996: 2 weeks of waterlogging imposed at 4 or 8 weeks: and B.D. Thomson, Perth, 1996, personal communication: 1 week of waterlogging imposed at 5 weeks) (see also Fig. 8.8). The differences in tolerance between these species thus hold true across a range of ages and waterlogging regimes, while the differences between species are larger than differences within species (Broué *et al.*, 1976; Phuphak, 1989). Growth subsequent to waterlogging has rarely been examined. Although Davies *et al.* (1994) found that *L. luteus* still performed better than *L. angustifolius*, the growth reductions were far greater in the two weeks after waterlogging than during the two weeks of waterlogging. It would be interesting to investigate whether this is due to a flush of aminocyclopropane-1-carboxylic acid (ACC) transport from roots to shoots, followed by oxidation to ethylene, as found during recovery of mandarin seedlings from drought (Tudela and Primo-Millo, 1992).

In a field experiment in Western Australia with two weeks of controlled waterlogging starting eight weeks after sowing, Davies *et al.* (1996) found the superior tolerance of *L. luteus* over *L. angustifolius* was reflected in root distributions during the season and shoot dry matter and seed yield at maturity. Root length densities of *L. luteus* decreased less with waterlogging than in *L. angustifolius*, especially at depth. At maturity, shoot dry matter was reduced by 20% in *L. luteus* compared with 55% in *L. angustifolius*; there was negligible effect on seed yield in *L. luteus*, but a 60% reduction in *L. angustifolius*.

Understandably, waterlogging injury increases with the duration of waterlogging and decreases if waterlogging is only intermittent (Broué *et al.*, 1976). Thus the greater the proportion of root system in saturated soil and the longer this occurs, the greater the damaging effect.

Among other cool-season grain legumes, faba bean appears to have comparable tolerance of waterlogging to *L. luteus*: although shoots of faba bean grew better during waterlogging, the reverse was true of root growth (Fig. 8.8). Chickpea, lentil and field pea appear to be even less tolerant than *L. angustifolius* (Fig. 8.8).

Physiology

Root growth and functioning

There are large differences in root thickness between orders of branching and species which will affect the O_2 fluxes necessary for growth and survival. When waterlogging begins, roots start to degenerate – particularly at the tips, which have high energy requirements relative to fully expanded tissues (Jackson and Drew, 1984). In *L. albus*, tips appear to be more susceptible to damage as distance from the shoot increases (Hanbury, 1988), consistent with the importance of the length of the diffusion pathway for O_2. Recovery of the root system after death of apices therefore relies on root primordia, which are more tolerant to waterlogging (Jackson and Drew, 1984). These new roots, which are often adventitious, tend to develop close to the surface, leading to a shallower root system than in non-waterlogged plants (e.g. Dracup *et al.*, 1992; Davies *et al.*, 1996). The development of adventitious roots may be related to better O_2 supply through the soil or stem tissue (Jackson and Drew, 1984). The adventitious roots may then take over the functions of the original root system, though they will be vulnerable to surface drying, and the profile for water extraction will be shallower.

The functioning of roots during waterlogging is often aided by development of aerenchyma for transporting gases. We are not aware of any reports of aerenchyma in lupin, but, during waterlogging of *L. albus*, Hanbury (1988) reported production of roots close to the surface which had enhanced air spaces.

Nitrogen nutrition

Ion uptake by roots slows markedly during waterlogging since there is less energy available, root growth is slowed and availability of some nutrients decreases. For example, after two weeks of waterlogging of four-week-old plants supplied with inorganic N, the N content of *L. angustifolius* was 60% less than in controls and 30% less in the more waterlogging-tolerant *L. luteus* (C.L. Davies, Perth, 1996, personal communication); the corresponding dry matter reductions were only 45% and 5%, respectively. This deficiency of N could be at least partly responsible for the premature chlorosis and senescence of leaves which occurred earlier in *L. angustifolius* than in *L. luteus* (C.L. Davies, Perth, 1996, personal communication). The chlorosis is associated with decreased chlorophyll concentrations: in the lowest leaf of several species of lupin, Phuphak (1989) found chlorophyll concentrations declined by 30–65% during 12 days of waterlogging, with the greatest reductions occurring in the most intolerant species and the lowest reductions in the most tolerant species. The basis for these differences in N uptake during waterlogging warrants research.

There will be additional limitations from waterlogging in plants reliant on symbiotically fixed N, since nodulation and nitrogen fixation are very sensitive to reduced O_2 supply (Minchin and Pate, 1975; Jackson and Drew, 1984; Chapter 6, this volume). Acetylene reduction by lupin nodules declines as the external O_2 concentration declines (Trinick *et al.*, 1976) and, in the field, stopped within a week of 12-week-old *L. angustifolius* plants being waterlogged (Farrington *et al.*, 1977). Waterlogging resulted in senescence and breakdown of the nodules and so recovery in acetylene reduction relied on formation of new nodules after the waterlogging was relieved (Farrington *et al.*, 1977).

Leaf gas exchange

Waterlogging generally causes stomatal closure, which may be accompanied by wilting (Drew, 1983; Pezeshki, 1994). The causes of stomatal closure during waterlogging are not clear, but in the short term could be associated with increased resistance to water flow within the plant (Pezeshki, 1994; for *L. albus*: Hanbury, 1988). In the longer term, hormonal imbalances or a feedback response to accumulation of photosynthate (supported by increased specific leaf area in *L. luteus* and *L. angustifolius*; C.L. Davies, Perth, 1996, personal communication) may be responsible. Net C fixation (as net assimilation rate) also decreases and the extent to which this reflects stomatal conductance is not clear.

Environmental interactions with waterlogging

Temperature is a key factor in the impact of waterlogging on plants; as temperatures rise, the solubility of O_2 declines and microbial respiration increases while respiratory demands from plant roots increase (Drew, 1983; Jackson and Drew, 1984; McFarlane *et al.*, 1989; Drew and Stolzy, 1991). Thus high temperatures exacerbate the problems of O_2 supply to roots during waterlogging. More rapid transpiration at high temperatures (or low humidity) also makes plants more susceptible to wilting during waterlogging (Jackson and Drew, 1984).

Soil organic matter is also important since microbial respiration, and thus O_2 depletion, is quicker where organic matter is high. Anaerobic bacteria can then proliferate, leading to increased solubility of microelements, which may then reach toxic levels. Breakdown of organic matter by microorganisms may also result in increased concentrations of toxic organic acids.

In contrast to the major nutrient ions, sodium concentrations in shoots rise with waterlogging since maintenance of low sodium levels in plants depends on active efflux of sodium out of roots (Barrett-Lennard, 1986). Waterlogging can therefore exacerbate the effects of even mild salinity and can be very important since both stresses often occur together (Barrett-Lennard and Nulsen, 1989).

Compaction

Compaction is probably the most ubiquitous constraint to root growth. Wheels of agricultural traffic compress soil, reducing the average size of pores in the soil. Zones of compaction are frequently evident at 15–30 cm and are called, among other terms, tillage/traffic/plough pans or soles. Compaction occurs most readily with moist, coarse-textured soils (Hamblin, 1985) and restricts root growth because of increased mechanical resistance, waterlogging and/or poor aeration. The restricted root growth can, in turn, prevent the plant from obtaining adequate water or nutrients from the soil. Hormonal signals from compacted roots might also cause reduced shoot growth (Masle and Passioura, 1987). In one field trial, Henderson (1991) found that subsoil compaction reduced the yield of *L. angustifolius* by 26%. Although poor aeration could limit nodule functioning, in fact nodules of a number of legumes show distinct morphological adaptation which permits N_2 fixation to be maintained at sub-ambient pO_2 (Chapter 6, this volume).

Root growth

In compact soils there is an inevitable decrease in root elongation (Table 8.3). This is associated with compression of the zone of cell elongation in the root tip which, for example, extended 11 mm back from the apex for *L. angustifolius* roots grown at a bulk density of 1.6 Mg m^{-3} but only 8 mm back when grown at 1.8 Mg m^{-3} (Atwell and Newsome, 1990). Such compression also leads to lateral roots appearing closer to the root apex and distorted root tips (e.g. Atwell, 1988; Dracup *et al.*, 1993b) (Fig 8.9).

Patterson *et al.* (1995) found that extension of the tap root of *L. angustifolius* decreased by 75% when soil bulk density increased from 1.3 to 1.7 Mg m^{-3} (Fig. 8.10). However, the weight of roots decreases little, if at all (Atwell, 1992;

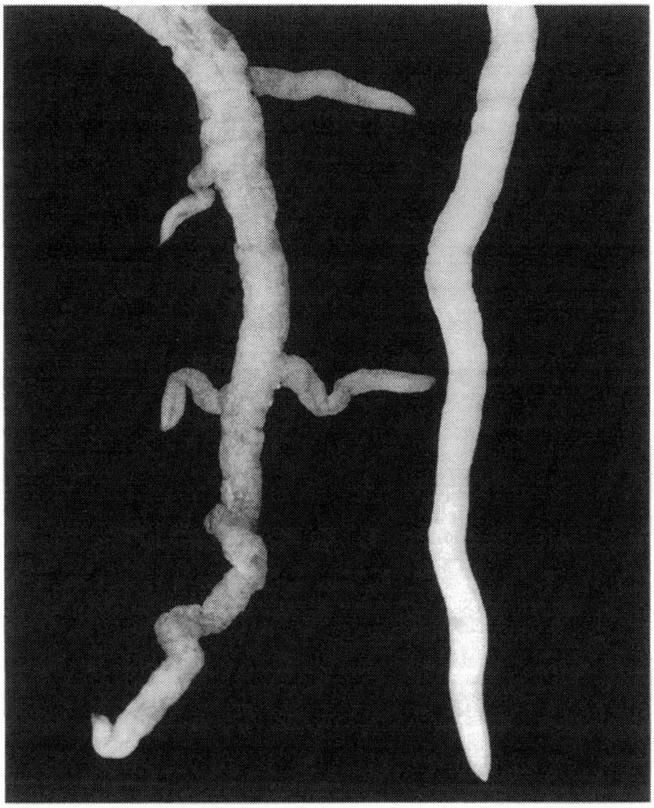

Fig. 8.9. Tap roots of *Lupinus angustifolius* from a compact (left) and loose (right) soil. Note distortions and irregular swelling of the root from compact soil, reflecting spatial variation in soil strength. Note also the lateral roots emerging closer to the tip of the tap root from compact soil.

Dracup *et al.*, 1993b), due to accompanying root swelling (Atwell, 1993; Patterson *et al.*, 1995) (Fig. 8.10; Table 8.3). It is somewhat paradoxical that an impeded root should swell, but the swelling seems to aid passage through tight soils; as the root swells, it exerts pressure against the surrounding soil to cause a reduction in the soil's mechanical impedance at the root tip (Atwell, 1993). If the reduction in impedance is sufficient, the tip can advance. Because of root swelling, reduced root extension also does not necessarily translate into reduced volume expansion in compact soils. In one experiment with *L. angustifolius*, Atwell and Newsome (1990) found that despite a 34% decrease in elongation rate between bulk densities of 1.6 and 1.8 Mg m^{-3} the calculated change in rate of volume increase was negligible. In another experiment with *L. angustifolius*, Atwell (1988) found a 38% decrease in elongation rate and calculated a 17% decrease in expansion rate of the tap root.

Root swelling is mainly confined to the cortex, rather than the stele, and is caused by wider cells rather than more cells (Atwell, 1988, 1993). The swelling is relatively permanent and so reflects changes in the impedance encountered by the tip as it moves through the soil. The root diameter can therefore vary markedly along the root (Fig. 8.9).

The causes of the radial expansion of cortical cells under compact conditions are unclear and results from experiments investigating the role of growth

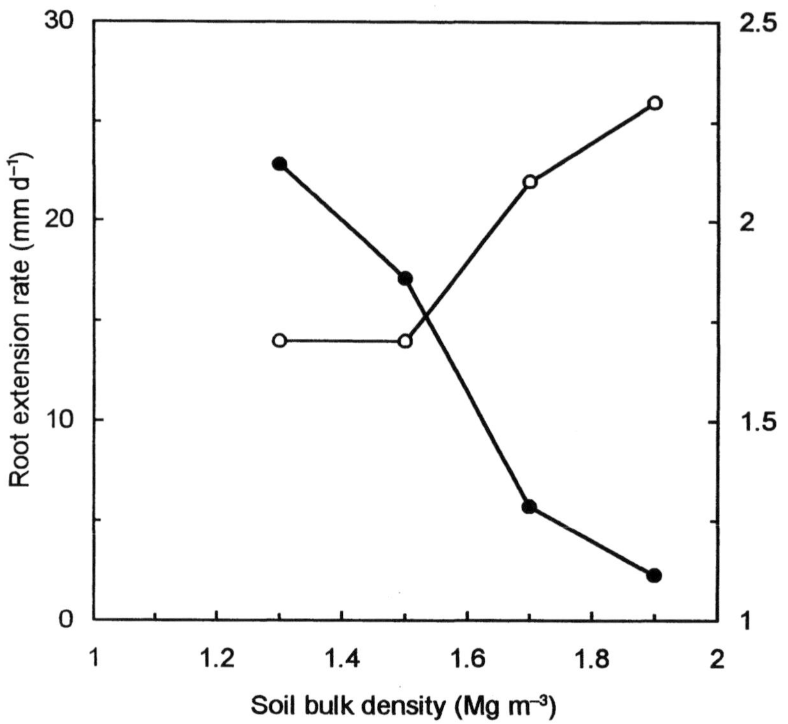

Fig. 8.10. Effects of soil bulk density on root extension (●) and root diameter (○) of *L. angustifolius*, calculated from Patterson *et al.* (1995).

regulators have been inconsistent (Atwell, 1993), though ABA is unlikely to play a role – at least in lupins (Atwell, 1993). To exert more pressure on the surrounding soil the osmotic pressure inside the root must be raised. Atwell (1988) found that the osmotic pressure of compacted apices of *L. angustifolius*, albeit after removal from the soil, was 1.47 MPa at high soil strength compared with 1.16 MPa at low strength (i.e. an increase of 27%). This increase in solutes could at least partly result from decreased volume expansion, calculated as 17%, and is supported by the relative proportions of the major osmotic solutes being unchanged.

The thickness of compacted roots and the tendency of roots to swell in response to mechanical impedance appears to explain a large amount of the differences between species in their ability to penetrate compact soils (Atwell, 1993; Materechera *et al.*, 1991, 1992, 1993). Materechera *et al.* (1993) found that a higher proportion of the roots of a selection of dicotyledonous species (with thick roots) penetrated a tillage pan than from a selection of monocotyledonous species (with thinner roots). Compared with root growth in deeply tilled soil profiles, root proliferation within and beneath a compact layer was reduced least in *L. angustifolius* and safflower (followed by peas, oats, wheat and faba bean). In separate experiments, using artificially compacted soil columns, root elongation of *L. angustifolius* was similar to pea (Atwell, 1992) and least affected by compaction out of 22 species (Materechera *et al.*, 1991).

Table 8.3. Effect of soil bulk density on roots of *Lupinus angustifolius*.

Reference	Soil bulk density (Mg m^{-3})	Elongation rate (mm day^{-1})	Root diameter (mm)
Atwell (1988)	1.23	45.5	1.56
	1.42	33.2	1.68
	1.64	28.1	1.81
Atwell and Newsome (1990)	1.6	15.6	1.69
	1.8	10.3	2.08
Materechera *et al.* (1991)	Low*	6.9	0.98
	1.35	0.7	1.81
Materechera *et al.* (1992)	1.25	nd	0.83
	1.57	nd	1.47
Patterson *et al.* (1995)	1.3	50	1.7
	1.5	40	1.7
	1.7	20	2.1
	1.9	14	2.3

*Roots grown in vermiculite.

The effects of compaction can be exacerbated by accompanying effects of reduced gaseous diffusion in a compact soil. For example, Asady and Smucker (1989) found that the oxygen diffusion rate decreased by more than 50% when the soil bulk density increased from 1.1 to 1.4 Mg m^{-3}. We are not aware of any investigations on this interaction with lupins, but effects of poor aeration have been discussed in the waterlogging section of this chapter.

Soil crusting

Surface soil crusting is another form of compaction that can impede or even reduce seedling emergence and thus limit crop establishment. Since sowing will disturb a surface crust, crusting will only be a problem where it can re-establish itself before seedling emergence. Therefore crops will be more prone to the effects of surface crusting if seeds are sown deeply, if the genotype emerges slowly (e.g. *L. angustifolius* is faster than *L. luteus*; Dracup and Kirby, 1996a) or if poor quality seed is used. Lupins are particularly sensitive to surface crusts (White and Robson, 1989a) because of their epigeal pattern of emergence, where hypocotyl elongation brings the cotyledons above the soil surface. The size of the cotyledons means that comparatively large forces must be generated to force them through a crusted soil and, like roots, the hypocotyl becomes swollen. Seedlings emerging through a crusted soil will therefore be weakened and thus more prone to predation, diseases and environmental stresses.

Texture

L. angustifolius is generally regarded as poorly adapted to fine-textured soils (e.g. White, 1990), particularly relative to pea. However, the greater tolerance to compaction by *L. angustifolius* compared with pea (Atwell, 1992) was irrespective of whether the soil was a sand or fine-textured clay loam. Moreover, Patterson *et al.* (1995) showed that the relationship between elongation of *L. angustifolius* tap roots and bulk density was not affected by the soil texture. Thus texture alone cannot explain the inferior growth of *L. angustifolius* on fine-textured soils. Although fine-textured soils are more prone to waterlogging, it would not explain the better adaptation of peas since they are less tolerant of waterlogging than is *L. angustifolius* (see section on waterlogging). However, fine-textured soils often have a high pH which is likely to directly reduce root growth of *L. angustifolius* more than that of peas, particularly where there are high levels of free lime (see section on soil pH). This agrees with deductions made by Atwell (1991a,b) for causes of poor growth of *L. angustifolius* on an alkaline, fine-textured soil at Merredin in Western Australia.

Acknowledgements

George Milford, Rana Munns, Tim Setter and Zed Rengel provided valuable comments. Brian Atwell, Christine Davies, Christian Huyghe, George Milford and Brett Thomson allowed us to cite unpublished data.

References

Abdus Siddique, M.D. and Goodwin, P.B. (1980) Seed vigour in bean (*Phaseolus vulgaris* L. cv. Apollo) as influenced by temperature and water regime during development and maturation. *Journal of Experimental Botany* 31, 313–323.

Alva, A.K. and Edwards, D.G. (1990) Responses of lupin cultivars to concentration of calcium and activity of aluminium in dilute nutrient solutions. *Journal of Plant Nutrition* 13, 57–76.

Amarger, N., Sommer, G., Giraud, J.-J. and Lagacherie, B. (1984) *Rhizobium lupini*: Presence et survie dans les sols, selection des souches et inoculation des semences de lupin. In: *Proceedings of 3rd International Lupin Conference*. International Lupin Association, La Rochelle, France, pp. 599–600.

Armstrong, W. (1979) Aeration in higher plants. *Advances in Botanical Research* 7, 225–331.

Asady, G.H. and Smucker, A.J.M. (1989) Compaction and root modifications of soil aeration. *Soil Society of America Journal* 53, 251–254.

Atwell, B.J. (1988) Physiological responses of lupin roots to soil compaction. *Plant and Soil* 111, 277–282.

Atwell, B.J. (1991a) Factors which affect the growth of grain legumes on a solonized brown soil. I. Genotypic responses to soil physical factors. *Australian Journal of Agricultural Research* 42, 95–105.

Atwell, B.J. (1991b) Factors which affect the growth of grain legumes on a solonized brown soil. II. Genotypic responses to soil chemical factors. *Australian Journal of Agricultural Research* 42, 107–119.

Atwell, B.J. (1992) The effect of mechanical impedance on lupin (*L. angustifolius*) and pea (*P. sativum*). In: Kutschera, L., Hübl, E., Lichtenegger, E., Persson, H. and Sobotik, M. (eds) *Root Ecology and its Practical Application. Third International Society of Root Research Symposium, Vienna.* Verein für Wurzelforschung, Klagenfurt, pp. 125–128.

Atwell, B.J. (1993) Response of roots to mechanical impedance. *Environmental and Experimental Botany* 33, 27–40.

Atwell, B.J. and Newsome, J.C. (1990) Turgor pressure in mechanically impeded lupin roots. *Australian Journal of Plant Physiology* 17, 49–56.

Barrett-Lennard, E.G. (1986) Effects of waterlogging on the growth and NaCl uptake by vascular plants under saline conditions. *Reclamation and Revegetation Research* 5, 245–261.

Barrett-Lennard, E.G. and Nulsen, R.A. (1989) Dryland soil salinity – cure, containment or catastrophe? In: *Proceedings of the 5th Australian Agronomy Conference*. The Australian Society of Agronomy, Victoria, pp. 212–220.

Barrett-Lennard, E.G., Leighton, P.D., McPharlin, I.R., Setter, T. and Greenway, H. (1986) Methods to experimentally control waterlogging and measure soil oxygen in field trials. *Australian Journal of Soil Research* 24, 477–483.

Bertoni, G.M., Pissaloux, A., Morard, P. and Sayag, D.R. (1992) Bicarbonate-pH relationship with iron chlorosis in white lupine. *Journal of Plant Nutrition* 15, 1509–1518.

Biddiscombe, E.F. (1975) Effect of moisture stress on flower drop and seed yield of narrow-leafed lupin (*Lupinus angustifolius* L. cv. Unicrop). *Journal of the Australian Institute for Agricultural Science* 41, 70–72.

Bidwell, R.G.S. (1979) *Plant Physiology*, 2nd edn. Macmillan, New York, 762 pp.

Broué, P., Marshall, D.R. and Munday, J. (1976) The response of lupins to waterlogging. *Australian Journal of Experimental Agriculture and Animal Husbandry* 16, 549–554.

Carr, S.J. and Sweetingham, M.W. (1994) Root elongation of *Lupinus angustifolius* and *L. luteus* differs in the presence of phytotoxic aluminium. In: Dracup, M. and Palta, J. (eds) *Proceedings of First Australian Lupin Technical Symposium*. Department of Agriculture, Western Australia, p. 260.

Cobley, B.T. (1993) The mechanisms by which high pH impairs the nodulation of *Lupinus angustifolius* L. Honours dissertation, School of Agriculture, The University of Western Australia, Perth, Australia.

Cowling, W.A. (1994) Breeding for yield and adaptation in *Lupinus* species. In: Dracup, M. and Palta, J. (eds) *Proceedings of First Australian Lupin Technical Symposium*. Department of Agriculture, Western Australia, pp. 129–139

Cowling, W.A. and Clements, J.C. (1993) Association between collection site soil pH and chlorosis in *Lupinus angustifolius* induced by a fine-textured, alkaline soil. *Australian Journal of Agricultural Research* 44, 1821–1836.

Davies, W.J. and Zhang, J. (1991) Root signals and the regulation of growth and development of plants in drying soil. *Annual Review of Plant Physiology and Plant Molecular Biology* 42, 55–76.

Davies, C.L., Dracup, M. and Turner, D.W. (1994) Does *Lupinus luteus* tolerate waterlogging better than *L. angustifolius*? In: Dracup, M. and Palta, J. (eds) *Proceedings of the First Australian Lupin Technical Symposium*. Department of Agriculture, Western Australia, p. 263.

Davies, C.L., Turner, D.W. and Dracup, M. (1998) Waterlogging tolerance of yellow lupin (*L. luteus*). In: Hill, G. (ed.) *Proceedings of the 8th International Lupin Conference, Monterey, California* (in press).

De Silva, D.L.R., Ruiz, L.P., Atkinson, C.J. and Mansfield, T.A. (1994) Physiological disturbances caused by high rhizospheric calcium in the calcifuge *Lupinus luteus*. *Journal of Experimental Botany* 45, 585–590.

Downes, R.W. and Gladstones, J.S. (1984) Physiology of growth and seed production in *Lupinus angustifolius* L. I. Effects on pod and seed set of controlled short duration high temperatures at flowering. *Australian Journal of Agricultural Research* 35, 493–499.

Dracup, M. and Kirby, E.J.M. (1996a) *Lupin Development Guide.* The University of Western Australia Press, Perth, 112 pp.

Dracup, M. and Kirby, E.J.M. (1996b) Pod and seed growth and development of narrow-leafed lupin in a water limited mediterranean type environment. *Field Crops Research* 48, 209–222.

Dracup, M., Belford, R.K. and Gregory, P.J. (1992) Constraints to root growth of wheat and lupin crops in duplex soils. *Australian Journal of Experimental Agriculture* 32, 947–961.

Dracup, M., Davies, C. and Tapscott, H. (1993a) Temperature and water requirements for germination and emergence of lupin. *Australian Journal of Experimental Agriculture* 33, 759–766.

Dracup, M., Gregory, P.J. and Belford, R.K. (1993b) Restricted growth of lupin and wheat roots in the sandy A horizon of a yellow duplex soil. *Australian Journal of Agricultural Research* 44, 1273–1290.

Dracup, M., Reader, M.A. and Palta, J.A. (1998) Variation in yield of narrow-leafed lupin caused by terminal drought. *Australian Journal of Agricultural Research* 49 (in press).

Drew, M.C. (1983) Plant injury and adaptation to oxygen deficiency in the root environment: a review. *Plant and Soil* 75, 179–199.

Drew, M.C. and Stolzy, L.H. (1991) Growth under oxygen stress. In: Waisel, Y. and Eshel, A. (eds) *Plant Roots – The Hidden Half.* Marcel Dekker, New York, pp. 331–350.

Duthion, C. (1992) Comportement du lupin blanc, *Lupinus albus* L, cv. Lublanc, en sols calcaires seuils de tolerance à la chlorose. *Agronomie* 12, 439–445.

Egli, D.B. and Wardlaw, I.F. (1980) Temperature response of seed growth characteristics of soybeans. *Agronomy Journal* 72, 560–564.

Farrington, P. and Pate, J.S. (1981) Fruit set in *Lupinus angustifolius* cv. Unicrop. I Phenology and growth during flowering and early fruiting. *Australian Journal of Plant Physiology* 8, 293–305.

Farrington, P., Greenwood, E.A.N., Titmanis, Z.V., Trinick, M.J. and Smith, D.W. (1977) Fixation, accumulation, and distribution of nitrogen in a crop of *Lupinus angustifolius* cv. Unicrop. *Australian Journal of Agricultural Research* 28, 237–248.

Flowers, T.J. and Yeo, A.R. (1986) Ion relations of plants under drought and salinity. *Australian Journal of Plant Physiology* 13, 75–91.

Flowers, T.J. and Yeo, A.R. (1989) Effects of salinity on plant growth and crop yields. In: Cherry, J.H. (ed.) *Environmental Stress in Plants: Biochemical and Physiological Mechanisms.* Springer-Verlag, New York, pp. 101–119.

French, R.J. and Turner, N.C. (1991) Water deficits change dry matter partitioning and seed yield in narrow-leafed lupins (*Lupinus angustifolius* L.). *Australian Journal of Agricultural Research* 42, 471–484.

Gallardo, M., Turner, N.C. and Ludwig, C. (1994) Water relations, gas exchange and abscisic acid content of *Lupinus cosentinii* leaves in response to drying different proportions of the root system. *Journal of Experimental Botany* 45, 909–918.

Gallardo, M., Eastham, J., Gregory, P.J. and Turner, N.C. (1996) A comparison of plant hydraulic conductances in wheat and lupins. *Journal of Experimental Botany* 47, 233–239.

Gardner, W.K., Parbery, D.G. and Barder, D.A. (1981) Proteoid root morphology and function in *Lupinus albus. Plant and Soil* 60, 143–147.

Gladstones, J.S. (1970) Lupins as crop plants. *Field Crop Abstracts* 23, 123–148.

Gladstones, J.S. (1974) Lupins of the Mediterranean region and Africa. *Western Australian Department of Agriculture Technical Bulletin*, 26, 3–48.

Gladstones, J.S. (1994) An historical review of lupins in Australia. In: Dracup, M. and Palta, J. (eds) *Proceedings of First Australian Lupin Technical Symposium.* Department of Agriculture, Western Australia, pp. 1–39.

Gladstones, J.S. and Drover, D.P. (1962) The mineral composition of lupins. *Australian Journal of Experimental Agriculture and Animal Husbandry* 2, 46–53.

Graham, P.H. and Parker, C.A. (1964) Diagnostic features in the characterisation of the root-nodule bacteria of legumes. *Plant and Soil* 20, 383–396.

Grauer, U.E. and Horst, W.J. (1990) Effect of pH and nitrogen source on aluminium tolerance of rye (*Secale cereale* L.) and yellow lupin (*Lupinus luteus* L.). *Plant and Soil* 127, 13–21.

Greenway, H. and Munns, R. (1980) Mechanisms of salt tolerance in nonhalophytes. *Annual Review of Plant Physiology* 31, 149–190.

Greenwood, E.A.N., Turner, N.C., Schulze, E.-D., Watson, G.D. and Venn, N.R. (1992) Groundwater management through increased water use by lupin crops. *Journal of Hydrology* 134, 1–11.

Gross, Y. and Kigel, J. (1994) Differential sensitivity to high temperature of stages in the reproductive development of common bean (*Phaseolus vulgaris* L). *Field Crops Research* 36, 201–212.

Guerrier, G. and Morard, P. (1978) Influence de l'aluminium sur la nutrition minerale de la feverole (calcicole) et du lupin (calcifuge). *Soil Science and Plant Nutrition* 24, 441–448.

Guerrier, G., Morard, P., Cabassy, J.-P. and Berducou, J. (1977) Importance des acides organiques dans l'absorption de l'aluminium chez le lupin (*Lupinus luteus*) et chez la feverole (*Vicia faba* minor). *Academie d'Agriculture de France* 63, 464–469.

Hamblin, A.P. (1985) The influence of soil structure on water movement, crop growth and water uptake. *Advances in Agronomy* 38, 95–158.

Hanbury, C. (1988) Root hypoxia and effects on growth and water relations of *Lupinus albus*. Honours dissertation, School of Agriculture, The University of Western Australia, Perth, Australia.

Hartmann, C. and Aldag, R. (1989) N_2-fixierung und ertragsstruktur der weiben lupine (*Lupinus albus* L.) im vergleich zu *Vicia faba* L. und *Glycine max* (L.) Merr. auf verschiedenen standorten. *Journal of Agronomy and Crop Science* 163, 201–211.

Hartung, W. and Turner, N.C. (1998) Abscisic acid relations in stressed roots. In: Altman, A. and Waisel, Y. (eds) *Biology of Root Formation and Development.* Plenum Press, New York (in press).

Helyar, K.R. (1978) Effects of aluminium and manganese toxicity on legume growth. In: Kamprath, E.J. and Andrew, C.S. (eds) *Mineral Nutrition of Legumes in Tropical and Subtropical Soils.* CSIRO, Melbourne, pp. 207–231.

Henderson, C.W.L. (1991) Sensitivity of eight cereal and legume species to the compaction status of deep, sandy soils. *Australian Journal of Experimental Agriculture* 31, 347–355.

Henson, I.E. and Turner, N.C. (1991) Stomatal responses to abscisic acid in three lupin species. *New Phytologist* 117, 529–534.

Henson, I.E., Jensen, C.R. and Turner, N.C. (1989a) Leaf gas exchange and water relations of lupins and wheat. I. Shoot responses to soil water deficits. *Australian Journal of Plant Physiology* 16, 401–413.

Henson, I.E., Jensen, C.R. and Turner, N.C. (1989b) Leaf gas exchange and water relations of lupins and wheat. III. Abscisic acid and drought-induced stomatal closure. *Australian Journal of Plant Physiology* 16, 429–442.

Hillel, D. (1980) *Fundamentals of Soil Physics.* Academic Press, New York.

Holaday, A.S., Martindale, W., Alred, R., Brooks, A.L. and Leegood, R.C. (1992) Changes in activities of enzymes of carbon metabolism in leaves during exposure of plants to low temperature. *Plant Physiology* 98, 1105–1114.

Horst, W.J. and Göppel, H. (1986) Aluminium-toleranz von ackerbohne (*Vicia faba*), lupine (*Lupinus luteus*), gerste (*Hordeum vulgare*) und roggen (*Secale cereale*). II. Mineralstoffgehalte in sprob und wurzeln in abhangigkeit vom aluminium-angebot. *Zeitschrift für Pflanzenernahrung und Bodenkunde* 149, 94–109.

Huyghe, C. and Papineau, J. (1990) Winter development of autumn sown white lupin: agronomic and breeding consequences. *Agronomie* 10, 709–716.

Jackson, M.B. and Drew, M.C. (1984) Effects of flooding on growth and metabolism of herbaceous plants. In: Kozlowski, T.T. (ed.) *Flooding and Plant Growth*. Academic Press, New York, pp. 47–128.

Jensen, C.R. and Henson, I.E. (1990) Leaf water relations characteristics of *Lupinus angustifolius* and *L. cosentinii*. *Oecologia* 82, 114–121.

Jensen, C.R., Henson, I.E. and Turner, N.C. (1989) Leaf gas exchange and water relations of lupins and wheat. II. Root and shoot water relations of lupin during drought-induced stomatal closure. *Australian Journal of Plant Physiology* 16, 415–428.

Jeschke, W.D. and Pate, J.S. (1991) Cation and chloride partitioning through xylem and phloem within the whole plant of *Ricinus communis* L. under conditions of salt stress. *Journal of Experimental Botany* 42, 1105–1116.

Jeschke, W.D., Pate, J.S. and Atkins, C.A. (1986) Effects of NaCl salinity on growth, development, ion transport and ion storage in white lupin (*Lupinus albus* L. cv. Ultra). *Journal of Plant Physiology* 124, 257–274.

Jeschke, W.D., Wolf, O. and Hartung, W. (1992) Effect of NaCl salinity on flows and partitioning of C, N, and mineral ions in whole plants of white lupin, *Lupinus albus* L. *Journal of Experimental Botany* 43, 777–788.

Jessop, R.S. and Mahoney, J. (1982) Effects of lime on the growth and nodulation of four grain legumes. *Australian Journal of Soil Research* 20, 265–268.

Jessop, R.S., Roth, G. and Sale, P. (1990) Effects of increased levels of soil $CaCO_3$ on lupin (*Lupinus angustifolius*) growth and nutrition. *Australian Journal of Soil Research* 28, 955–962.

Keigley, P.J. and Mullen, R.E. (1986) Changes in soybean seed quality from high temperature during seed fill and maturation. *Crop Science* 26, 1212–1216.

Kuang, J., Turner, N.C. and Henson, I.E. (1990) Influence of xylem water potential on leaf elongation and osmotic adjustment of wheat and lupin. *Journal of Experimental Botany* 41, 217–221.

Kuo, J. (1992) Morphological and anatomical responses to waterlogging and submergence in selected plants. In: Kuo, C.G. (ed.) *Adaptation of Food Crops to Temperature and Water Stress.* Asian Vegetable Research and Development Centre, Taiwan, pp. 205–219.

Lancaster, J.E., Mann, J.D. and Porter, N.G. (1977) Ineffectiveness of abscisic acid in stomatal closure of yellow lupin, *Lupinus luteus* var. Weiko III. *Journal of Experimental Botany* 28, 184–191.

Lauchli, A. (1984) Salt exclusion: an adaptation of legumes for crops and pastures under saline conditions. In: Staples, R.C. and Toenniessen, G.H. (eds) *Salinity Tolerance in Plants – Strategies for Crop Improvement.* John Wiley and Sons, New York, pp. 171–187.

Leport, L., Turner, N.C., French, R.J., Tennant, D., Thomson, B.D. and Siddique, K.H.M. (1998) Water relations, gas exchange growth and water use of cool-season grain legumes in a mediterranean-type environment. *Crop Science* (in press).

Loss, S.P. and Siddique, K.H.M. (1994) Morphological and physiological traits associated with wheat yield increases in Mediterranean environments. *Advances in Agronomy* 52, 229–276.

Maas, E.V. and Hoffman, G.J. (1977) Crop salt tolerance – current assessment. *American Society of Civil Engineers – Journal of Irrigation and Drainage Engineering* 103, 115–134.

Marschner, H. (1991) Mechanisms of adaptation of plants to acid soils. *Plant and Soil* 134, 1–20.

Masle, J. and Passioura, J.B. (1987) The effect of soil strength on the growth of young wheat plants. *Australian Journal of Plant Physiology* 14, 643–656.

Materechera, S.A., Dexter, A.R. and Alston, A.M. (1991) Penetration of very strong soils by seedling roots of different species. *Plant and Soil* 135, 31–41.

Materechera, S.A., Alston, A.M., Kirby, J.M. and Dexter, A.R. (1992) Influence of root diameter on the penetration of seminal roots into a compacted subsoil. *Plant and Soil* 144, 297–303.

Materechera, S.A., Alston, A.M., Kirby, J.M. and Dexter, A.R. (1993) Field evaluation of laboratory techniques for predicting the ability of roots to penetrate strong soil and of the influence of roots on water sorptivity. *Plant and Soil* 149, 149–158.

McFarlane, D.J., Barrett-Lennard, E.G. and Setter, T.L. (1989) Waterlogging: a hidden constraint to crop and pasture production in southern regions of Australia. In: *Proceedings of the 5th Australian Agronomy Conference*. The Australian Society of Agronomy, Victoria, pp. 74–83.

Minchin, F.R. and Pate, J.S. (1975) Effects of water, aeration, and salt regime on nitrogen fixation in a nodulated legume – definition of an optimum root environment. *Journal of Experimental Botany* 26, 60–69.

Mock, I.T. and Gibson, A.H. (1994) Evaluation of *Lupinus* species and strains of *Bradyrhizobium lupini* for improved production potential on calcareous soils. In: Dracup, M. and Palta, J. (eds) *Proceedings of First Australian Lupin Technical Symposium, Perth, Western Australia*, pp. 182–186.

Munns, D.N. (1977) Soil acidity and related factors. In: Vincent, J.M., Whitney, A.S. and Bose, J. (eds) *Exploiting the Legume–Rhizobium symbiosis in Tropical Agriculture*. University of Hawaii, pp. 211–236.

Munns, R. (1985) Na^+, K^+ and Cl^- in xylem sap flowing to shoots of NaCl-treated barley. *Journal of Experimental Botany* 36, 1032–1042.

Munns, R. (1988) Effect of high external NaCl concentrations on ion transport within the shoot of *Lupinus albus*. I. Ions in xylem sap. *Plant, Cell and Environment* 11, 283–289.

Munns, R. and King, R.W. (1988) Abscisic acid is not the only stomatal inhibitor in the transpiration stream of wheat plants. *Plant Physiology* 88, 703–708.

Munns, R. and Passioura, J.B. (1984) Hydraulic resistance of plants. III. Effects of NaCl in barley and lupin. *Australian Journal of Plant Physiology* 11, 351–359.

Munns, R. and Termaat, A. (1986) Whole-plant responses to salinity. *Australian Journal of Plant Physiology* 13, 143–160.

Munns, R., Tonnet, M.L., Shennan, C. and Gardner, P.A. (1988) Effects of high external NaCl concentration on ion transport within the shoot of *Lupinus albus*. II. Ions in phloem sap. *Plant, Cell and Environment* 11, 291–300.

Palta, J.A. and Plaut, Z. (1998) Yield and components of seed yield of narrow-leafed lupin (*Lupinus angustifolius* L.) subjected to transient water deficit. *Journal of Experimental Botany* (in press).

Papineau, J. (1987) La resistance au froid des lupins blancs (*Lupinus albus* L.) variabilité génétique, séléction, création de matérial vegetal doux. Memoire pour l'obtention du titre d'Ingenieur Diplome Par L'Etat, École National Supérieure Agronomique de Montpellier, Montpellier, France.

Passioura, J.B. and Munns, R. (1984) Hydraulic resistance of plants. II. Effects of rooting medium, and time of day, in barley, and lupin. *Australian Journal of Plant Physiology* 11, 341–350.

Patterson, A., Gregory, P.J. and Gilkes, R.J. (1995) The effect of density of sandy soils on soil strength and root growth. In: So, H.B., Smith, G.D., Raine, S.R., Schafer, B.M. and Loch, R.J. (eds) *Sealing, Crusting and Hardsetting Soils: Productivity and Conservation. The Second International Symposium on Sealing, Crusting and Hardsetting Soils and Continuing Education*, pp. 441- 444.

Peltzer, S.E. (1997) The effect of low temperature on the growth and nodulation of *Lupinus angustifolius*. PhD thesis, The University of Western Australia, Perth, Australia.

Pezeshki, S.R. (1994) Plant response to flooding. In: Wilkinson, R.E. (ed.) *Plant–Environment Interactions*. Marcel Dekker, New York, pp. 289–321.

Phuphak, S. (1989) Adverse effects of waterlogging on growth of lupin (*Lupinus* spp.) and field pea (*Pisum sativum*) cultivars. Masters preliminary dissertation, School of Agriculture, The University of Western Australia, Perth, Australia.

Porankiewicz, J. and Gwozdz, E.A. (1995) Protein synthesis in lupin roots exposed to heat shock. *Acta Physiologiae Plantarum* 17, 47–54.

Rahman, M.S. and Gladstones, J.S. (1993) Temperature response of lupin species during germination and flower development. *Journal of the Australian Institute of Agricultural Science* 39, 259–262.

Rerkasem, B. (1977) Differential sensitivity to soil acidity of legume – *Rhizobium* symbiosis. PhD thesis, The University of Western Australia, Perth, Australia.

Reuter, D.J. and Robinson, J.B. (1986) *Plant Analysis – An Interpretation Manual*. Inkata Press, Melbourne, 218 pp.

Rodriguez-Maribona, B., Tenorio, J.L., Conde, J.R. and Ayerbe, L. (1992) Correlation between yield and osmotic adjustment of peas (*Pisum sativum* L.) under drought stress. *Field Crops Research* 29, 15–22.

Sarlistyaningsih, L. (1993) The relative tolerance of the germination and survival of lupins (*L. angustifolius* L. cv. Gungurru) to waterlogging. Master of Science thesis, University of Western Australia, Perth, Australia.

Sarlistyaningsih, L., Sivasithamparam, K. and Setter, T.L. (1995) Influence of waterlogging on germination and survival of lupin seeds (*Lupinus angustifolius* L. cv. Gungurru) coated with calcium peroxide and streptomycin. *Australian Journal of Experimental Agriculture* 35, 537–541.

Sarlistyaningsih, L., Sivasithamparam, K. and Setter, T.L. (1996) The role of limited oxygen supply and soil microorganisms on germination and survival of lupin seeds (*Lupinus angustifolius* L. cv. Gungurru) in waterlogged soil. *Australian Journal of Experimental Agriculture* 36, 323–329.

Scholander, P.F., Hammel, H.T., Bradstreet, E.D. and Hemmingsen, E.A. (1965) Sap pressure in vascular plants. *Science* 148, 339–346.

Shaddad, M.A., Radi, A.F., Abdel-Rahman, A.M. and Azooz, M.M. (1990) Response of seeds of *Lupinus termis* and *Vicia faba* to the interactive effect of salinity and ascorbic acid pyridoxine. *Plant and Soil* 122, 177–183.

Shennan, C. (1986) Growth and ion relations of *Lupinus albus* grown at low and moderate salinity. In: Beilby, M.J., Walker, N.A. and Smith, J.R. (eds) *Membrane Transport in Plants and Fungi*, pp. 457–462.

Shield, I., Stevenson, H.J., Leach, J.E., Scott, T., Day, J.M. and Milford, G.F.J. (1996) Effects of sowing date and planting density on the structure and yield of autumn-sown, florally-determinate white lupins (*Lupinus albus*) in the United Kingdom. *Journal of Agricultural Science, Cambridge* 127, 183–191.

Slattery, J.F. and Coventry, D.R. (1989) Populations of *Rhizobium lupini* in soils used for cereal–lupin rotations in north-east Victoria. *Soil Biology and Biochemistry* 21, 1009–1010.

Subbarao, G.V. and Johansen, C. (1994) Potential for genetic improvement in salinity tolerance in legumes: pigeon pea. In: Pessarakli, M. (ed.) *Handbook of Plant and Crop Stress*. Marcel Dekker, New York, pp. 581–595.

Subramaniam, P., Pradinas, R.M. and Rodriguez-Barrueco, C. (1982) Nodulation and growth of *Lupinus angustifolius* L. as influenced by varied pH levels of the rooting medium. In: *Proceedings of 2nd International Lupin Conference, Torremolinos (Espana)-Mayo*. Publicaciones Agronas, Ministry of Agriculture, Madrid, pp. 134–138.

Sykes, M.T. and Wilson, J.B. (1988) An experimental investigation into the response of some New Zealand sand dune species to salt spray. *Annals of Botany* 62, 159–166.

Sykes, M.T. and Wilson, J.B. (1989) The effect of salinity on the growth of some New Zealand sand dune species. *Acta Botanica Neerlandica* 38, 172–182.

Tang, C. (1995) Iron in symbiotic nitrogen fixation in legumes. In: Hemantaranjan, A. (ed.) *Advances in Iron Nutrition Research*. Scientific Publishers, Jodhpur, pp. 145–175.

Tang, C. and Robson, A.D. (1993a) pH above 6.0 reduces nodulation in *Lupinus* species. *Plant and Soil* 152, 269–276.

Tang, C. and Robson, A.D. (1993b) *Lupinus* species differ in their requirements for iron. *Plant and Soil* 157, 11–18.

Tang, C. and Robson, A.D. (1995) Nodulation failure is important in the poor growth of two lupin species on an alkaline soil. *Australian Journal of Experimental Agriculture* 35, 87–91.

Tang, C., Longnecker, N.E., Thomson, C.J., Greenway, H. and Robson, A.D. (1992) Lupin (*Lupinus angustifolius* L.) and pea (*Pisum sativum* L.) roots differ widely in their sensitivity to pH above 6.0. *Journal of Plant Physiology* 140, 715–719.

Tang, C., Robson, A.D., Longnecker, N.E. and Greenway, H. (1993) Physiological responses of lupin roots to high pH. *Plant and Soil* 156, 509–512.

Tang, C., Robson, A.D. and Adams, H. (1995a) High Ca is not the primary factor in poor growth of *Lupinus angustifolius* L. in high pH soil. *Australian Journal of Agricultural Research* 46, 1051–1062.

Tang, C., Robson, A.D., Longnecker, N.E. and Buirchell, B.J. (1995b) The growth of *Lupinus* species on alkaline soils. *Australian Journal of Agricultural Research* 46, 255–268.

Tang, C., Adams, H., Longnecker, N.E. and Robson, A.D. (1996) A method to identify lupin species tolerant of alkaline soils. *Australian Journal of Experimental Agriculture* 36, 595–601.

Thomson, B.D., Siddique, K.H.M., Barr, M.D. and Wilson, J.M. (1997) Grain legume species in low rainfall Mediterranean-type environments. I. Phenology and seed yield. *Field Crops Research* 54, 173–187.

Treeby, M.T. and Van Steveninck, R.F.M. (1988) Effects of salinity and phosphate on ion distribution in lupin leaflets. *Physiologia Plantarum* 73, 317–322.

Trinick, M.J., Dilworth M.J and Grounds, M. (1976) Factors affecting the reduction of acetylene by root nodules of *Lupinus* species. *New Phytologist* 77, 359–370.

Tudela, D. and Primo-Millo, E. (1992) 1-Aminocylopropane-1-carboxylic acid transported from roots to shoots promotes leaf abscission in Cleopatra mandarin (*Citrus reshni* Hort. ex Tan.) seedlings re-hydrated after water stress. *Plant Physiology* 100, 131–137.

Turner, N.C. (1987) The use of the pressure chamber in studies of plant water status. In: Hanks, R.J. and Brown, R.M. (eds) *Proceedings of the International Conference on Measurement of Soil and Plant Water States. Vol 2. Plants.* Utah State University, Logan, pp. 13–24.

Turner, N.C. (1988) Measurement of plant water status by the pressure chamber technique. *Irrigation Science* 9, 289–308.

Turner, N.C. (1992) Crop production on duplex soils: an introduction. *Australian Journal of Experimental Agriculture* 32, 797–800.

Turner, N.C. and Henson, I.E. (1989) Comparative water relations and gas exchange of wheat and lupins in the field. In: Kreeb, K.H., Richter, H. and Hinckley, T.M. (eds) *Structural and Functional Responses to Environmental Stresses: Water Shortage.* SPB Academic Publishing, The Hague, pp. 293–304.

Turner, N.C. and Jones, M.M. (1980) Turgor maintenance by osmotic adjustment: a review and evaluation. In: Turner N.C. and Kramer, P.J. (eds) *Adaptation of Plants to Water and High Temperature Stress.* John Wiley, New York, pp. 87–103.

Turner, N.C., Stern, W.R. and Evans, P. (1987) Water relations and osmotic adjustment of leaves and roots of lupins in response to water deficits. *Crop Science* 27, 977–983.

Turner, N.C., Leport, L., French, R.J., Thomson, B.D. and Siddique, K.H.M. (1996) Adaptation of cool-season pulses to water-limited environments. In: Asghar, M. (ed.) *Proceedings of the 8th Australian Agronomy Conference*, Toowoomba, Queensland, January/February 1996. Australian Society of Agronomy, Carlton, pp. 554–557.

Van Steveninck, R.F.M., Van Steveninck, M.E., Stelzer, R. and Lauchli, A. (1982) Studies on the distribution of Na and Cl in two species of lupin (*Lupinus luteus* and *Lupinus angustifolius*) differing in salt tolerance. *Physiologia Plantarum* 56, 465–473.

Vierling, E. (1990) Heat shock protein function and expression in plants. In: Alscher, R.G. and Cumming, J.R. (eds) *Stress Responses in Plants: Adaptation and Acclimation Mechanisms.* Wiley-Liss, New York, pp. 357–375.

Villais, P., Kersulec, A. and Citharel, J. (1991) Effects of low temperatures on the first development stages and protein composition of two genotypes of *Lupinus albus* L. *Agronomie* 11, 689–698.

White, P.F. (1990) Soil and plant factors relating to the poor growth of *Lupinus* species on fine-textured alkaline soils – a review. *Australian Journal of Agricultural Research* 41, 871–890.

White, P.F. and Robson, A.D. (1989a) Emergence of lupins from a hard setting soil compared with peas, wheat and medic. *Australian Journal of Agricultural Research* 40, 529–537.

White, P.F. and Robson, A.D. (1989b) Lupin species and peas vary widely in their sensitivity to Fe deficiency. *Australian Journal of Agricultural Research* 40, 539–547.

Withers, N.J. (1979a) Effects of water stress on *Lupinus albus*. I. Response of vegetative growth to water stress during a single growth stage at two humidity levels. *New Zealand Journal of Agricultural Research* 22, 445–454.

Withers, N.J. (1979b) Effects of water stress on *Lupinus albus*. II. Response of seed yield to water stress during a single growth stage at two humidity levels. *New Zealand Journal of Agricultural Research* 22, 455–461.

Wolf, O., Jeschke, W.D. and Hartung, W. (1990) Long distance transport of abscisic acid in NaCl-treated intact plants of *Lupinus albus*. *Journal of Experimental Botany* 41, 593- 600.

Yeo, A.R. and Flowers, T.J. (1986) Salinity resistance in rice (*Oryza sativa* L.) and a pyramiding approach to breeding varieties for saline soils. *Australian Journal of Plant Physiology* 13, 161–173.

9 Diseases and Pests

M.W. Sweetingham[1,2], R.A.C. Jones[1,2] and A.G.P. Brown[1]

[1]*Agriculture Western Australia, 3 Baron-Hay Court, South Perth, Western Australia 6151;* [2]*Centre for Legumes in Mediterranean Agriculture, University of Western Australia, Nedlands, Western Australia 6907*

Introduction

Although lupins have long been cultivated for forage and green manure, it was not until they were domesticated as a low alkaloid grain crop, mid-way through this century, that pests and diseases were recognized as economically important. Lupins are now cultivated in many contrasting environments around the world which influence the spectrum and severity of pest and disease challenge.

In Australia, Chile, South Africa and southern Europe lupins are grown predominantly in regions with a mediterranean climate. This environment favours pests and pathogens that are well adapted to survive the hot dry summer without herbaceous hosts in soil, crop residues or seed. To cause significant losses, they must be dispersed or vectored efficiently and have sufficiently rapid life cycles to generate an epidemic in the short winter–spring growing season.

In eastern Europe and parts of North America the crop is grown over summer, experiencing warmer temperatures and longer day-length during the seedling and vegetative phase than in mediterranean environments. Here, the greater growth rate of the crop reduces the impact of many seedling pests and diseases. This contrasts with western Europe, where lupins are now mostly autumn sown and make little vegetative growth during the cold winter. In the south-eastern United States lupins are being evaluated as a winter crop and the warm, humid environment favours some diseases not important in lupins elsewhere. In the Andean highlands of tropical South America, lupins are a rainy season crop grown under cool conditions at a high altitude.

Advances in pest and disease management have been achieved mostly using an integrated approach, combining various cultural practices with resistance breeding and chemical control. Development of cultural practices to reduce damage has required a detailed understanding of the biology of the pests and the epidemiology of the diseases in different environments. Fungicides, other than seed treatments, are seldom employed as the control achieved rarely provides an economic return. By contrast, insecticide application, which is

 Lupins as Crop Plants: Biology, Production and Utilization (eds J.S. Gladstones, C.A. Atkins and J. Hamblin)

relatively cheap and effective, is often employed when pest populations exceed economic thresholds.

Most diseases of lupins are caused by fungi or viruses. No important bacterial pathogens are known and there are few reports of nematodes causing significant damage. A broad range of insects are pests of lupins, while some mites and molluscs can also cause damage. This chapter does not discuss vertebrate pests such as birds and mammals.

Diseases

Leaf spots

Brown spot (*Pleiochaeta setosa* [Kirchn.] Hughes)

Brown spot of lupins was first attributed to the fungus *Pestalotia lupini* in 1898, but was later recognized to be synonymous with *Ceratophorum setosum* described from *Cytisus* in 1892. Hughes (1951) reviewed the taxonomy and proposed the currently accepted binomial *Pleiochaeta setosa*.

Brown spot has been recorded in all continents where lupins are cultivated. *P. setosa* is the most widespread and damaging pathogen of *L. angustifolius* and *L. albus* in Australia, where it also causes root rot (see section on Root diseases). It is of less importance in Europe but nonetheless capable of significant losses to autumn sowings of *L. albus* in France (Gondran *et al.*, 1994). Serious outbreaks in *L. angustifolius* have occurred in the winter rainfall regions of South Africa (Van Jaarsveld, 1980) and in Chile destruction of entire *L. albus* crops has been reported (Lara and Andrade, 1983). It has also caused losses in experimental plots of *L. albus* in Minnesota (Kalis-Kuznia *et al.*, 1991) and maritime Canada (Paulitz and Atlin, 1992). It is seen as a threat to the emerging lupin industry in south-eastern USA (Collins *et al.*, 1996).

Infection produces angular or net-like dark brown leaf lesions. Diseased leaflets are often distorted and twisted if they become infected before they are fully expanded. Stems and pods can develop large lesions of similar colour. When *L. albus* pods become infected, malformed and brown-stained seed often result. Seed infection is less common in *L. angustifolius*.

As well as being transmitted by seed (Germar, 1939), *P. setosa* can survive on stubble (Wood and McLean, 1982). Hogetop (1937), in Brazil, was the first to suggest that the fungus could survive over summer in soil as conidiospores. The central cells of these spores are thick-walled and pigmented, adapting them to long-term survival. Until recently, the predominant role of soil-borne spores in both foliar and root infection had not been appreciated (Sweetingham, 1990). Stubble carries relatively little inoculum and seed contributes mainly by moving disease to regions that have not previously grown lupins. Seed-borne infection may play a greater role in *L. albus*.

Soil-borne spores are rain-splashed and deposit predominantly on the lower leaf surface. Within two to three days necrotic lesions develop and after two weeks infected leaves begin to senesce and defoliate. The fungus sporulates on the defoliated leaves that accumulate on the soil surface under the crop. This secondary inoculum is in turn rain-splashed, which accelerates the epidemic (Yang *et al.*, 1996).

Early seedling infection can drastically set back the crop, even killing plants by complete defoliation and infection of the growing point (Sweetingham *et al.*, 1993). Epidemics developing later in the season are less dramatic but still reduce yields significantly. In Western Australia losses have been greatest in colder regions with a short growing season and on less favourable soils. Here lupins remain smaller and exposed to rain-splash for longer, and the short season provides plants with less opportunity to compensate for early reductions in growth. Under warmer conditions lupins rapidly grow out of the susceptible seedling stage and develop tall stands, reducing the incidence of rain-splash of spores on to the younger foliage. Poor phosphorus nutrition can also increase the impact of brown spot through reduced crop height and canopy development (Sweetingham, 1990).

P. setosa spore populations decline in soil in the absence of a lupin host and so crop rotation is fundamental for brown spot control. In Australia, the minimum break between lupin crops varies from one to three years, depending on climatic and soil type risk (Sweetingham, 1990). Dicarboximide fungicides are widely used as a seed treatment. They provide effective control of disease in young seedlings and can prevent large yield losses. Brown spot is more severe where cereal stubble residues are removed. A stubble mulch provides a physical barrier which reduces the rain-splash of spores from the soil surface to the lupin foliage. The benefits of retaining stubble and fungicide seed treatment are additive, but stubble offers the additional benefit of protection throughout the season (Sweetingham *et al.*, 1993). Similarly, pearl millet stubble has been shown to reduce brown spot in Alabama (Collins *et al.*, 1996).

No major resistance genes for brown spot have been found in *L. angustifolius*, but there are quantitative differences among breeding lines (Cowling, 1988). The cultivar Myallie has moderate resistance to defoliation. By contrast, greater variation in resistance exists in *L. albus*. Pod and seed infection is less in taller lines through escape from rainsplash of soil-borne spores. Papineau and Huyghe (1989) have reported *L. albus* ecotypes with resistance to *P. setosa* from the Azores Archipelago. The Ukrainian *L. luteus* cultivar Motiv 369 has a high degree of resistance to both infection and defoliation due to brown spot (Yang *et al.*, 1996). This also occurs in several other European *L. luteus* cultivars.

Grey leaf spot (*Stemphylium botryosum* Wallr., *Stemphylium solani* Weber)

Grey leaf spot was first reported in south-eastern USA in the 1950s, associated with the fungi *Stemphylium botryosum* and *St. solani.* Small circular lesions develop on leaves and sometimes larger lesions on stem and pods. They cause complete defoliation on susceptible *L. angustifolius* lines but, in inoculation tests, *L. albus* and *L. luteus* were not susceptible (Wells *et al.*, 1956). Resistance was found in *L. angustifolius* accessions from natural stands in Portugal and from a mutation in a breeding line from Georgia, USA (Forbes *et al.*, 1965).

In Western Australia in the 1970s, grey leaf spot periodically developed in late spring in tall late-flowering cultivars of *L. angustifolius* grown in high rainfall coastal environments. In 1976 cv. Marri and in 1979 cv. Illyarrie, both of which carried resistance from the Georgian breeding programme, were released (Gladstones, 1982). Whether this resistance is present in subsequent Australian cultivars is not known as screening was discontinued. However, grey leaf spot epidemics are now rare in Western Australia due to resistance or to escape because of the earlier maturity of newer cultivars.

Minor leaf spots

A *Phoma* sp. causes reddish-brown leaf and stem lesions on *L. albus* in the maritime provinces of eastern Canada (Faluyi *et al.,* 1993). On the basis of its specificity to lupins, Paulitz and Cote (1991) believe the fungus is probably *Phoma lupini* Ellis & Everh., which was also reported as a pathogen on *L. mutabilis* in Peru and on native lupins in western USA. It has not been reported on lupins in Europe or Australia.

Wood and McLean (1982) described a leaf spot caused by a *Cladosporium* sp. in Western Australia. It causes a dark grey spotting on flowers, young leaves and growing tips under warm and wet winter conditions. The disease does not appear to affect yield as plants compensate with the onset of warmer and drier spring weather. Occasionally it can infect seed, resulting in reduced emergence if used for sowing.

Serious outbreaks of *Ascochyta lupinicola* were described by Pape (1927) in Germany but there have been no subsequent records.

Obligate parasites

Powdery mildew (*Erisiphe polygoni* D.C. and *Microsphaera diffusa* Z.)

Powdery mildew can affect all lupin species but *L. polyphyllus* is particularly sensitive. *Erysiphe polygoni* has been subdivided and two species occur on lupins: *E. martii*, which is probably the main species on *L. polyphyllus*, and *E. pisi*, which can infect a range of legumes. The specificity of the *E. pisi* strain infecting lupin is not known. Leaves, stems and pods develop a typical white powdery growth composed of mycelia and abundant conidiospores. Severe infection leads to necrosis and defoliation.

In western and eastern Europe, powdery mildew infection usually develops too late in the growing season to cause significant plant damage. It is more important in regions with a mediterranean climate (Gondran *et al.*, 1994). Epidemics resulting in serious losses became common in the 1970s in the Western Cape province of South Africa, particularly in *L. angustifolius* (Van Jaarsveld, 1983). It has not been serious in any part of Australia.

In England, Smith and Wheeler (1969) showed that cleistothecia over-wintered on the stubble of *L. polyphyllus* and that these discharged wind-borne ascospores in the spring. Epidemics depend on the asexual conidiospores, which disperse in the crop to infect new tissue. These spores need high humidity but not free moisture on the plant surface to germinate and infect.

In south-eastern USA, *Microsphaera diffusa* causes powdery mildew on lupins. This fungus is most virulent on *L. angustifolius* and less damaging on *L. albus* and *L. luteus* (Luttrell and Samples, 1954).

Fungicides have been used on ornamental lupins but are unlikely to give economic control in the grain crops. Van Jaarsveld (1983) reported resistance to *E. polygoni* in wild *L. angustifolius* lines and in *L. albus* cv. Ultra.

Rust (*Uromyces lupinicolus* Bub., *Uromyces renovatus* Syd.)

The rusts *Uromyces lupinicolus* and *U. renovatus* were both recorded as causing identical symptoms on lupins in Germany by Pape (1927). Typical orange pustules are formed on leaves and stems, usually on plants approaching maturity. Serious yield loss only occurs if infections develop in early stages of crop development. The two rust species are distinguished on the basis of the uredospore wall thickness (Gondran *et al.*, 1994). Rust is of minor importance in western and eastern Europe, but it is quite common on *L. hirsutus*, *L. angustifolius* and *L. luteus* in Morocco and the Iberian Peninsula. It also occurs in South America, but not North America or Australia.

Control with foliar application of systemic triazole fungicides is effective but not economical. Blaszczak (1962) reported that *L. mutabalis* and *L. polyphylus* were the most resistant species. Resistance is also believed to exist in an *L. albus* ecotype from the Azores Archipelago (Papineau and Huyghe, 1989).

Vascular wilts

Fusarium wilt (*Fusarium oxysporum* Schl. f. sp. *lupini* Snyd. & Hans.)

A wilt disease of lupins was described in 1906 in Germany caused by *Fusarium vasinfectum* (Pape, 1927). The incidence of the disease increased and by the 1930s production of *L. luteus* in Germany was seriously endangered (Richter, 1941). In their taxonomic review of *Fusarium*, Snyder and Hansen (1940) proposed the name *F. oxysporum* formae specialis *lupini* to distinguish the lupin wilt pathogen from other root-infecting strains of the same species.

Fusarium wilt remains an important disease of lupins in eastern Europe (Truszkowska, 1977; Korniichuk, 1991). It also occurs in *L. albus* grown in the highveld summer rainfall areas of South Africa (Malan and Knox-Davies, 1990) and in Egypt (Osman *et al.*, 1986).

Infection first becomes apparent in the vegetative stage as leaves begin to darken. Obvious symptoms begin at budding or flowering time when leaves wilt, desiccate and rapidly defoliate. At this stage infected roots are almost symptomless except for a brown zone under the epidermis. Brown streaking of the vascular tissue is sometimes visible on the upper stem and, under wet conditions, pink spore masses develop from the vascular streaks (Korniichuk, 1985). Infected plants eventually die.

The existence of races of *F. oxysporum* f.sp. *lupini* specific to different lupin species emerged from the surveys of Richter (1941). Lamberts (1955) and Armstrong and Armstrong (1964) defined three races from European isolates. Race 1 was pathogenic to *L. luteus* and some cultivars of *L. albus* but non-pathogenic to *L. angustifolius*. Race 2 was pathogenic to *L. luteus* and possibly all cultivars of *L. albus* but non-pathogenic to *L. angustifolius*. Race 3 was pathogenic to *L. angustifolius* and *L. albus* but non-pathogenic to *L. luteus*. All three races of *F. oxysporum* f.sp. *lupini* in Europe were specific to lupins. Fusarium wilt pathogens from cowpea and cotton in the USA were pathogenic on some *L. luteus* cultivars (Armstrong and Armstrong 1964).

The specificity of the pathogen means that crop rotation is an effective management strategy and it is widely adopted in eastern Europe. Truszkowska

(1977) in Poland showed that a five-year break in the cultivation of *L. luteus*, with intervening crops such as rye, potato, oats and maize, was able to lower the soil population of *F. oxysporum* f.sp. *lupini* to insignificant levels. Infected crop debris is the main source of disease carry-over, but the fungus can also be carried on seed (Richter, 1941). It is important to avoid introducing the disease into clean areas by sowing infected seed. Soil solarization has been employed effectively under irrigated crop situations in Egypt (Osman *et al.*, 1986) but would be uneconomic for larger-scale production.

Lamberts (1955) reported resistance to *F. oxysporum* f.sp. *lupini* conferred by a single dominant gene in Portuguese *L. luteus* lines. Some lines were identified with resistance to all three races. Several European breeding programmes have successfully produced resistant *L. luteus* cultivars. However, Furgal-Wegrzycka (1984) described a new race of *F. oxysporum* f.sp. *lupini* from Poland which infects resistant *L. luteus* cvs Afus, Jantar and Refusa. In Ukraine, Korniichuk (1991) has produced resistant *L. albus* cultivars. Resistance also occurs in *L. angustifolius* and *L. cosentinii* (Kiselev *et al.*, 1988).

Verticillium wilt (*Verticillium* sp.)

Verticillium has been reported to cause vascular infections and wilt symptoms, similar to those caused by *Fusarium*, in several lupin species in Germany (Pape, 1927) and in Wales (Griffiths and Isaac, 1963). Strains of *Verticillium* from sunflower also attack lupin (Griffiths and Isaac, 1963). It is not a significant pathogen of lupin anywhere in the world.

Stem and pod blights

Anthracnose (*Colletotrichum gloeosporiodes* [Penz.] Penz. & Sacc. and *Colletotrichum acutatum* Simm. & Simmonds)

Anthracnose of lupins was first reported on *L. angustifolius* in south-eastern USA in 1939. Both acervuli of *Colletotrichum gloeosporiodes* and perithecia of the teleomorph *Glomerella cingulata* were described from stem lesions on infected plants (Weimer, 1952). However, the teleomorph has not been noted in subsequent descriptions of lupin anthracnose elsewhere.

There is no record of anthracnose of lupins in Europe until the early 1970s, when it was recorded on *L. mutabilis* and subsequently on *L. albus* (Gondran, 1984). It is now considered the most serious disease of *L. albus* in France and Chile (Gondran *et al.*, 1994). Its importance in Brazil (Reis *et al.*, 1983), Russia (Agaev, 1993) and Canada (Paulitz *et al.*, 1995) has also been noted.

C. gloeosporioides is a species complex of morphologically related forms which infect a diverse range of plants. Significant variation in spore morphology and colony appearance exists and confusion with *C. acutatum* has occurred. DNA sequencing of isolates identified as *C. gloeosporioides* from lupins in the UK indicated greater homology to *C. acutatum* than *C. gloeosporiodes* (Sreenivasapprasad *et al.*, 1994). There are reports of *C. acutatum* on *L. arboreus* in New Zealand (Dingley and Gilmour, 1972) and on *L. albus* and *L. mutabilis* in France (J. Gondran, personal communication). *C. acutatum* is commonly isolated from strawberry.

Dick (1994) described the rapid decline in *L. arboreus* stands in New Zealand pine forests caused by anthracnose since 1988. She suggested that the sudden appearance of the disease was indicative of the introduction of a new strain. The isolates were morphologically distinct from *C. acutatum* earlier reported by Dingley and Gilmour (1972) on *L. arboreus* and were attributed to *C. gloeosporiodes.* Two strains of *C. gloeosporioides* that can cause lupin anthracnose have been distinguished by vegetative compatibility grouping and by RAPD-PCR (Shivas *et al.*, 1998). VCG-1 and VCG-2 both occur in Europe. Circumstantial evidence suggests that VCG-2, which occurs in South America and attacked *L. arboreus* in New Zealand, may be more pathogenic than VCG-1.

In 1994 an outbreak of anthracnose occurred in Western Australia in breeders' plots and was eradicated. The source of infection was *L. albus* seed imported from Germany (Sweetingham *et al.*, 1995). A serious outbreak occurred in commercial crops in 1996. Infection was mainly in *L. albus* but also in *L. angustifolius* and *L. cosentinii* in the northern grainbelt of Western Australia. Several infected crops of *L. angustifolius* were also found on the Eyre Peninsula of South Australia. The spore morphology of isolates from the 1994 and 1996 outbreaks were consistent with *C. gloeosporiodes* and were both VCG-2.

A characteristic symptom of anthracnose in lupins is the bending-over of stems and petioles with a lesion inside the crook. These lesions can girdle and break off the stem or twist and deform the pods. Lesions tend to be purplish or brown with masses of pink conidiospores visible in the centre. Pods are often twisted and develop large sunken lesions with pink spore masses. Seed can become infected, developing a brown discoloration and crinkled appearance, and sometimes pink spores are visible. Symptoms are similar in all lupin species but *L. albus* and *L. mutabilis* are the most susceptible.

Lupin anthracnose was first shown to be seed-borne by Decker (1947). Seed transmission has clearly been responsible for its global movement. The longevity of the pathogen has not been evaluated under different conditions for lupins. However, bean anthracnose, caused by *Colletotrichum lindemuthianum,* does not survive the Canadian winter in residues buried in soil (Tu, 1983). Survival is more significant over summer in a mediterranean climate that does not favour residue breakdown. Conidiospores of *Colletotrichum* are dispersed relatively short distances (a few metres) in the crop by rain-splash. The combined action of wind and rain may spread some spores further. The fungus has a rapid infection cycle and secondary inoculum can be produced within two weeks (Fig. 9.1).

Lupin has been reported as a host for *Colletotrichum* species from peach (Weimer, 1952), alfalfa and strawberry (Welty, 1984). The specificity of the lupin anthracnose is still unclear. Reed *et al.* (1996) found that *C. acutatum* from ornamental lupin only developed typical symptoms on *Lupinus* species. *C. gloeosporiodes* VCG-2 from Western Australia is non-pathogenic to a range of legume species (M.W. Sweetingham, 1997, unpublished data).

Avoiding sources of inoculum by selecting healthy seed and through crop rotation is of fundamental importance. The length of rotational break required to reduce disease to uneconomic levels has not been determined but will vary with region and cultural practice. Fungicide seed treatments with carbendazim and iprodione have been reported to assist with control (Gondran *et al.*, 1994).

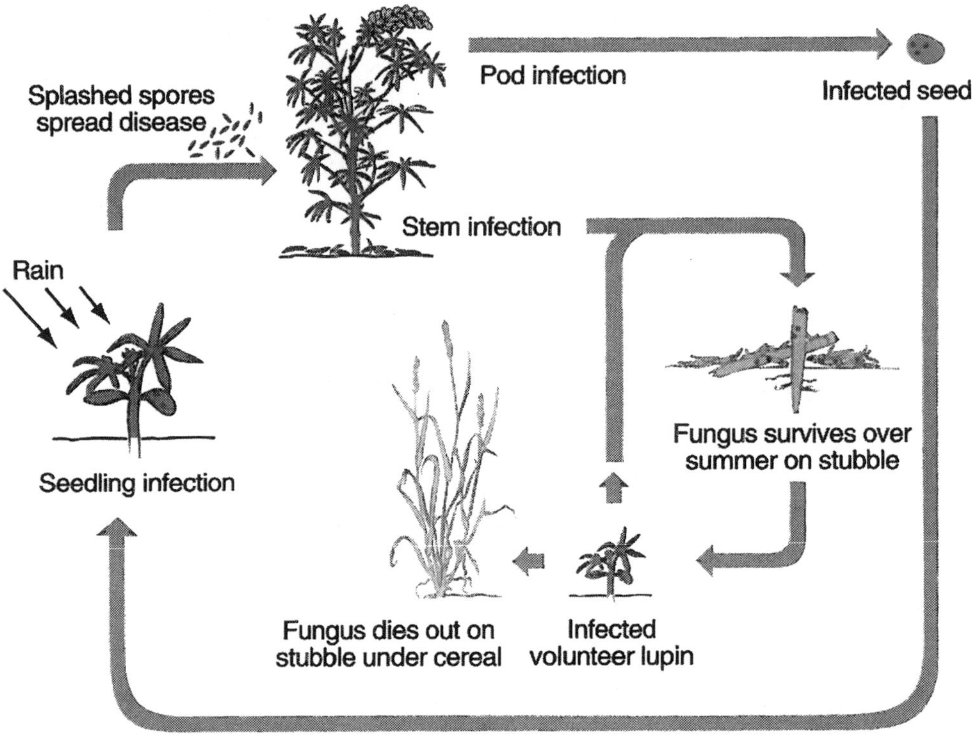

Fig. 9.1. The infection cycle of the lupin anthracnose fungus in a lupin-cereal rotation in a Mediterranean environment.

Resistance attributable to a single dominant gene has been reported in *L. angustifolius*. Action of the gene is temperature sensitive, becoming progressively less effective from 22 to 28°C (Wells and Forbes, 1967). Resistant cultivars, such as Rancher, were released from the Georgian breeding programme. The Western Australian cv. Unicrop was susceptible. Resistance from the Georgian programme was incorporated into cv. Illyarrie (Gladstones, 1982). Shivas *et al.* (1998) found that Illyarrie and Unicrop were both susceptible to VCG-2, which suggests that the resistance may only hold against VCG-1. No resistant cultivars of *L. luteus* are known. However, resistance is believed to exist in *L. albus* ecotypes collected from the Azores Archipelago (Papineau and Huyghe, 1989).

Phomopsis (*Diaporthe toxica* Williamson)

Phomopsis leptostromiformis (Kühn) Bubák infection of lupins can result in a mycotoxicosis that can kill livestock grazing stubbles. It has been a major problem for mixed crop and livestock farms in Australia and South Africa. The fungus produces the potent mycotoxin phomopsin during the saprophytic colonization of stems (Chapter 12, this volume). Based on distinct morphological and biochemical characteristics, Williamson *et al.* (1994) erected a new species name for the teleomorph (*Diaporthe toxica*) leaving the name *P. leptostromiformis* for the anamorph in doubt.

Latent infection occurs during the growing season as subcuticular coralloid mycelia (Williamson *et al.*, 1991). These develop into characteristic dark purplish lesions on infected stems and pods at crop maturity. If the plant is stressed by frost, drought or herbicide damage, stem and pod lesions can develop earlier and in such cases the main stem can collapse, reducing seed yields. Where pod lesions develop before harvest, seed may also become infected, showing a gold or brown discoloration. Infected grain may also carry low levels of the mycotoxin (Chapter 12).

Under Australian conditions, *D. toxica* over-summers on colonized stubble as both perithecia and pycnidia contained in dark spherical stromata (up to 3 mm in diameter). Ascospores are wind dispersed early in the winter along with rain-splashed conidiospores. The fungus remains viable in woody stem fragments for at least two years (Wood and McLean, 1982). Winter and spring infection can be accelerated by secondary conidiospore production on prematurely senescent and defoliated leaflets (Brown, 1984). The fungus has been reported as a stem pathogen in Europe (Kochman and Kubika, 1974) and as causing significant damage in green *L. luteus* in south-eastern USA (Ostazeski and Wells, 1960).

Resistance offers the only practical control for *D. toxica*. It reduces the risk of lupinosis and seed discoloration. Under Australian conditions, *L. albus* cvs Ultra and Kiev Mutant have a good level of resistance. The first resistant *L. angustifolius* was cv. Gungurru. It carried resistance from wild lupins collected from the western Mediterranean region (Gladstones, 1989).

Kochman and Kubika (1974) found that *L. luteus* was the most susceptible species in Poland. In South Africa a range of *L. albus* were more susceptible than *L. luteus*, while a *L. mutabilis* line was the most resistant species (Van Jaarsveld and Knox-Davies, 1974). These apparent discrepancies possibly relate to strains of *D. toxica* with species preference within *Lupinus* (Shankar *et al.*, 1995).

Sclerotinia (*Sclerotinia sclerotiorum* [Lib.] de Bary and *S. minor* Det.)

Sclerotinia sclerotiorum was first reported causing stem rot of lupins in Germany (Pape, 1927). It has since been reported on *L. angustifolius* and *L. albus* in Australia (Wood and McLean, 1982), in Europe (Gondran *et al.*, 1994) and in the USA on the ornamental *L. texensis*. Usually it is of minor importance except when lupins are grown following a susceptible host. *S. sclerotiorum* can infect over 300 plant species, including numerous field crops and vegetables, but notably not cereals. *S. minor* also has a wide host range.

Wilting plants usually become evident soon after flowering. Large brown water-soaked lesions are readily visible on stems from which a profusion of white mycelia and black sclerotia develop. *S. sclerotiorum* typically produces lesions 30–60 cm up the stem, and main-stem pods are often infected. Mycelia are often found ramifying inside the pith cavity with black irregularly shaped sclerotia, usually 4–8 mm in diameter. Sclerotia can form in infected pods and contaminate harvested grain. The lesion kills the stem above the lesion, but basal branches may survive. By contrast, infection with *S. minor* is confined to the stem base close to ground level. Sclerotia are smaller (typically 1–2 mm diameter) and more spherical. Lesions typically kill the whole plant (Sweetingham, 1986).

Under suitable temperature and moisture conditions, sclerotia of *S. sclerotiorum* germinate to produce cup-shaped apothecia (0.5–2 mm in diameter), from which ascospores are dispersed by wind to infect nearby plants. Sclerotia can also produce conidiospores, which are dispersed by rain-splash. Alternatively, hyphae emanating from sclerotia in the soil may directly infect the stem base. The latter appears most common in *S. minor*. Infection by both species is favoured by wet and warm conditions.

Crop rotation is the key to disease management, reducing sclerotial populations in soil. In Australia, lupins sown directly after canola or a pasture dominated by capeweed (*Arctotheca calendula*) are at much greater risk than after cereal. Resistance is not known in any lupin species. *L. albus* appears more susceptible than *L. angustifolius* to pod infection by *S. sclerotiorum*.

Southern blight (*Sclerotium rolfsii* Sacc.)

Southern blight, also known as white mould, has been recognized as a problem disease in south-eastern USA since the 1890s. It infects more than 500 species from 100 plant families, including a wide range of legumes. The causal organism, *Sclerotium rolfsii*, was first reported causing foot rot of *L. albus*, *L. angustifolius* and *L. luteus* in Florida by Weimer (1944). Development of southern blight is generally favoured by temperatures of 25–35°C and high soil moisture and is usually found in the tropics. It is still considered an important disease of *L. albus* in Alabama (Collins *et al.*, 1993). It has not been reported from Europe or in commercial crops in Australia but can occur in irrigated experimental plots in Australia.

A canker is formed near the surface of the soil and can girdle the entire stem. Diseased plants are stunted and subsequently wilt and die. In humid environments, white mycelia develop on the soil surface around the stem base. Characteristic spherical sclerotia (2–3 mm in diameter), initially white but turning shiny brown with maturity, are formed on the lesions or adjacent soil. Plants of all ages may be infected during spring and early summer. The disease tends to affect clusters of plants along a drill row. *S. rolfsii* survives from one season to the next as sclerotia on or near the soil surface. Sclerotia are dispersed in soil moved by water, wind or tillage equipment.

The disease is managed primarily by sanitation and crop rotation. It influences the viability of potential cropping sequences in south-eastern USA, including a lupin–cotton rotation. Soil solarization with clear plastic sheeting resulted in large reductions in disease incidence and yield increases in experimental plots of *L. albus* in Alabama (Collins *et al.*, 1993). Burying crop debris is often practised for control of southern blight. There have been no reports of resistance in lupins.

Botrytis (*Botrytis cinerea* Pers.)

In Europe, grey mould caused by *Botrytis cinerea* can cause significant losses in *L. albus* crops. It attacks flowers and pods, leading to abortion, but only under unusually humid spring weather (Gondran *et al.*, 1994). Infection may result in large, slightly sunken lesions that girdle and kill branches. A characteristic grey mould is usually associated with mature lesions in which large black sclerotia, similar to those of *S. sclerotorium*, may appear. It has been found in *L. angustifolius* in south-eastern USA (Weimer, 1943) and occasionally in Australia.

The fungus has a wide host range and is also a competent saprophyte, so crop rotation has little impact on incidence. It tends to be worse at high plant densities and on more poorly drained sites. There have been no reports of resistance in lupins.

Root diseases

Damping-off (*Pythium* spp., *Fusarium* spp., *Rhizoctonia* spp.)

Like other legume crops, lupins are susceptible to damping-off – the pre- or post-emergent death of seedlings due to root or hypocotyl infection. The fungi involved, typically *Pythium*, *Fusarium* and *Rhizoctonia* species, are common in all agricultural soils. When environmental conditions favour their development or are suboptimal for lupin growth, these fungi can cause economic losses. Damage from *Pythium* is commonly associated with waterlogged soil conditions. More than one fungus may be involved simultaneously.

Hypocotyls and roots of lupins can develop both dry and water-soaked lesions, ranging in colour from reddish to brown to black depending on the pathogen and soil conditions. Infected seedlings wilt and die or remain as stunted plants that often die later in the season.

Weimer (1944) investigated symptoms based on inoculation of fungi from lupin roots in south-eastern USA. He found *R. solani* and *P. graminicola* caused distinct hypocotyl rots and described *F. oxysporum* f. sp. *radicis lupini*, a rootrotting pathogen distinct from the vascular wilt pathogen. In Western Australia, *P. irregulare* and *P. acanthicum* cause mild root rot on *L. angustifolius* but *F. oxysporum* and other *Fusarium* species are extremely weak pathogens (Sweetingham, 1989).

Hypocotyl rot of *L. angustifolius* in Western Australia was first described by Wood and McLean (1982) as a problem on infertile sands with low organic matter. The dry, sunken lesions had a very distinctive red-brown colour. It is caused by a specific strain of *R. solani* (AG 11) (Carling *et al.*, 1994). This strain has only been found in the northern grainbelt of Western Australia and on cotton in Arkansas, USA. Two other strains of *R. solani* that are well distributed in agricultural soils around the world, (AG 2-2) and (AG 2-1), cause a similar hypocotyl rot but can also infect roots. In Minnesota, *R. solani* (AG 5) causes a very similar hypocotyl rot in *L. albus* (Leach and Clapham, 1992).

Deep sowing weakens seedlings, exposes more hypocotyl tissue to infection and increases seedling mortality due to *R. solani* (AG 11) (Sweetingham, 1990). Seed treatment with the fungicide iprodione gives some disease suppression. In Western Australia, infection by *R. solani* (AG 11), (AG 2-1) and (AG 2-2) may be a greater risk where lupins are sown after subterranean clover pasture, which is a good host for each strain.

In pot experiments, European cultivars of *L. luteus* are more susceptible than *L. angustifolius* to *R. solani* (AG 11), apparently associated with their slower germination and emergence. *L. albus* cv. Kiev Mutant is moderately resistant (M.W. Sweetingham, 1994, unpublished paper).

Pleiochaeta root rot (*Pleiochaeta setosa* [Kirchn.] Hughes)

The brown spot pathogen, *P. setosa*, causes root rot of seedling lupins. In contrast to the damping-off fungi, it is primarily a lupin pathogen. It is the most important root pathogen of lupins in Australia and has also been reported in south-eastern USA (Collins *et al.* 1996), Minnesota (Kuznia *et al.*, 1993) and Belarus (V. Kupschov, Minsk, 1996, personal communication).

Pulselli (1928) isolated *P. setosa* from below-ground tissues of *L. albus*, believing it to be secondary to attack by *Thielaviopsis*. Proof that *P. setosa* could reproduce typical root disease symptoms was obtained by Sweetingham (1984). Hogetop (1937) had suggested that conidiospores of the fungus survive in soil but appeared unaware of the connection with root rot. Infection commences soon after seed germination and dark brown root lesions develop within three to four weeks. The tap root can rot away completely, causing seedling death. The reduced stand densities lower yield potential of the crop and plants with a partially rotted root system are less vigorous.

Disease severity is mostly influenced by the spore population and its distribution in soil. Where lupins have been grown recently, *P. setosa* spores are most concentrated in the top 2 cm of the profile, rapidly declining with none below the depth of tillage. Consequently root rot is more severe with shallow sowing because the tap root has to grow through the zone of concentrated spores. As hypocotyl tissue is resistant to infection by *P. setosa*, a sowing depth of 5 cm

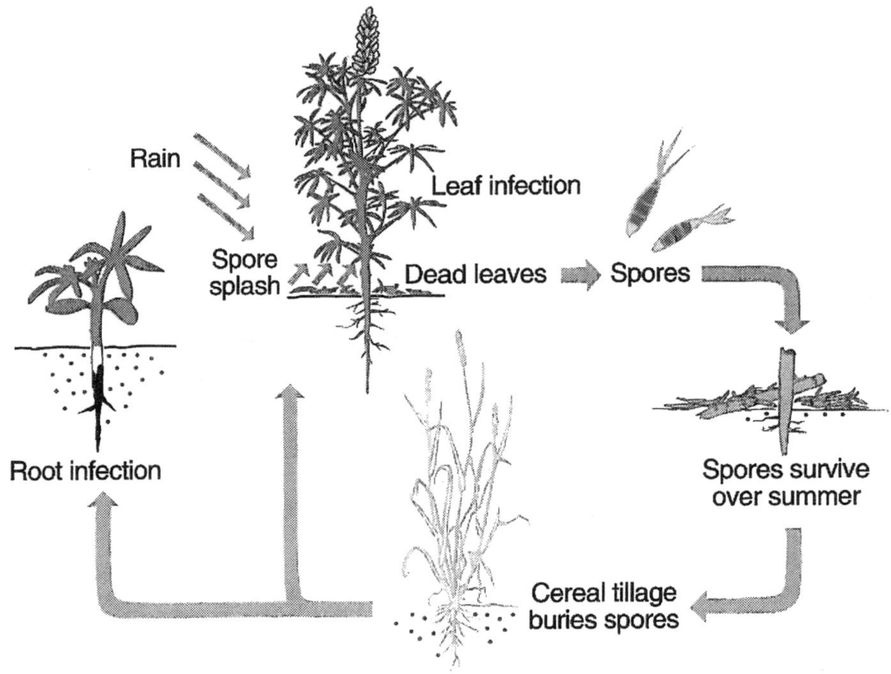

Fig. 9.2. The infection cycle of *Pleiochaeta setosa* in a lupin–cereal rotation in a Mediterranean environment.

is optimal for pleiochaeta root rot control (Sweetingham, 1991). Sowing deeper than 5 cm results in even less root rot, but can reduce or delay plant emergence and predispose seedlings to rhizoctonia hypocotyl rot. Shallow tillage with off-set discs can dramatically increase disease severity. In this situation the spores are redistributed without dilution deeper into the soil, where they can more readily infect lupin roots (Sweetingham, 1990). Soil inversion with a mould-board plough can dilute the spores by mixing with a large volume of soil and can reduce disease severity. However, because of the risk of soil erosion, this practice is not employed in Australia.

In *L. angustifolius*, resistance to pleiochaeta root rot is under polygenic control. However, lines with good resistance to brown spot are not necessarily resistant to root infection (Cowling *et al.*, 1997). The new cv. Kalya has moderate resistance. All lines of *L. luteus* tested thus far are extremely resistant (Sweetingham *et al.*, 1994). *L. albus* cv. Kiev Mutant is very susceptible but resistance is present in *L. albus* landraces from the Azores and Crete. Of the rough-seeded lupins, *L. cosentinni* and *L. pilosis* have resistance but *L. atlanticus* is very susceptible.

Rhizoctonia bare patch (*Rhizoctonia solani* AG 8)

In lupins, *R. solani* (AG 8) causes distinct edged circular patches, 0.3 to 5 metres in diameter, of severely stunted plants. Both tap and lateral roots are pinched-off by distinctive brown 'spear-tipped' lesions. The disease, which is only important in Australia, is most severe in crops established by reduced or zero-tillage methods (MacNish and Neate, 1996). It affects all lupin species as well as cereals, legumes and oilseeds. The fungus over-summers on organic matter in the surface soil; with autumn rains, hyphae grow out and colonize roots.

Because of its wide host range, the fungus cannot be controlled by crop rotation. The disease can be managed by tillage prior to seeding, which is believed to break the hyphal network into less infective pieces. To minimize the risk of soil erosion, seeding machines are being developed to provide soil disturbance in the drill row in a direct-drilling operation. Fungicide seed treatment and in-furrow treatments have proved ineffective.

Eradu patch (*Rhizoctonia* sp.)

Eradu patch is a recently described disease of lupins in Western Australia that resembles rhizoctonia bare patch. It is not known to occur elsewhere and is caused by a previously undescribed species of the *Rhizoctonia* fungus (MacLeod and Sweetingham 1997). For the last ten years the disease has been a consistent problem in *L. angustifolius* crops on sandplain soils in the northern grainbelt of Western Australia. It is becoming more widespread, and is often confused with rhizoctonia bare patch. Fields have been observed more than 50% covered with patches. Above ground it appears identical to rhizoctonia patch. The brown spear-tipped root lesions it causes are also similar, but lesions less commonly pinch-off the tap root. It differs in that wheat is not affected.

Unlike rhizoctonia bare patch, tillage has little impact on eradu patch. The risk is greater in rotation following barley and least following pasture (MacLeod and Sweetingham, 1997). Fungicide seed treatment and in-furrow treatments are ineffective. European cultivars of *L. luteus* are highly resistant. No resistance has been found in *L. angustifolius*, *L. albus* or *L. cosentinii*.

Charcoal rot (*Macrophomina phaseolina* [Maudbl.] Ashby)

Radman (1964) reported considerable damage due to charcoal rot on lupins in Bosnia and Herzegovina. In Western Australia, significant areas of *L. angustifolius* occasionally wilt and senesce rapidly in late spring, associated with the symptoms of charcoal rot. Cutting a section of tap root from an affected lupin reveals abundant microsclerotia (0.25 mm in size) of *M. phaseolina* throughout the conductive and pith tissue, collectively giving a charcoal appearance (Sweetingham, 1986). Grain filling and yields are reduced. It is often difficult to distinguish the disease from drought stress. In France, mild infection can lead to small and malformed seeds in *L. albus* (Gondran *et al.*, 1994).

The causal fungus infects more than 500 hosts but not temperate cereals. Microsclerotia can survive for long periods in dry soil. They germinate on the root surface, and penetrate and block xylem tissues. The fungus has an optimum temperature for growth of 28–35°C and water stress is commonly noted as a predisposing factor for plant infection. Control strategies or resistance have not been studied in lupins but rotation with cereals should be beneficial.

Thielaviopsis black root rot (*Thielaviopsis basicola* [Berk. & Br.] Ferraris)

In Europe, black root rot is only of significance where *L. albus* is grown in close rotation with other susceptible crops including tobacco, peas or tomatoes (Gondran *et al.*, 1994). The pathogen has not been recorded on lupins in Australia. Infection takes place in the seedling phase; roots become progressively lesioned through the season and develop a dark-brown to almost black decay in which the characteristic chlamydospores of *T. basicola* can be observed in abundance under the microscope. Plant tops have no obvious symptom other than lack of vigour.

Virus diseases

Bean yellow mosaic virus (BYMV)

BYMV is the most important virus disease of lupins worldwide. Chamberlain (1935) was the first to detect this virus in lupins when he showed that 'sore shin' of *L. angustifolius* in New Zealand was caused by 'pea mosaic virus'. Other synonyms for BYMV infecting lupins include 'lupin narrow leaf virus', 'lupin mottle virus', 'lupin mosaic virus' and 'lupin virus 1' (Jones and McLean, 1989). BYMV is particularly important in *L. luteus*, *L. albus*, *L. atlanticus* and *L. pilosus*, in which it is seed-borne. It also causes significant losses in *L. angustifolius* when infection spreads into crops from infected alternative hosts (Jones, 1993). In south-eastern USA, seed-borne BYMV infection was responsible for the demise of the *L. luteus* industry (Leuck *et al.*, 1962).

Symptoms of current-season infection with BYMV by vector aphids vary between lupin species. In *L. luteus* narrowing of leaflets, pallor, vein clearing and faint mosaic develop on young leaves. Leaflets tend to fold along the midrib and are held at an acute angle, giving an upright appearance. Mature leaves develop vein clearing or remain unaffected (Norris, 1943). In *L. albus*, vein clearing is first noticed in the youngest leaves. Then, as infection spreads through the plant, severe mottle, leaf deformation, defoliation and stunting follow. Early infected plants may die (Corbett, 1955). In *L. atlanticus, L. pilosus* and *L. cosentinii*, symptoms resemble those in *L. albus* but tend to be less severe (Norris, 1943; Jones and McLean, 1989). In *L. mutabilis*, symptom severity varies but may be little more than vein clearing and mottle (Eppler *et al.*, 1986). In *L. angustifolius*, necrotic streaking develops on the youngest portion of the shoot, which then bends over, producing a characteristic 'shepherd's crook'. The growing tip dies and necrosis spreads, blackening the stem and killing the plant. In older plants the virus may remain localized in some branches, producing similar necrosis and blackening but without the 'shepherd's crook'. When infection occurs after pod set, tips of affected shoots die and pods blacken and fail to fill.

BYMV is transmitted non-persistently by aphids. Early work in New Zealand implicated *Myzus persicae*, *Aphis craccivora* and *Macrosiphum euphorbiae* in the spread to *L. angustifolius* (Chamberlain, 1935). Three colonizing species (*M. persicae*, *A. craccivora* and *Acyrthosiphum kondoi)* and several non-colonizing species (e.g. *Lipaphis erysime*, *Rhopalosiphum maidis* and *R. padi*) have been implicated as vectors to sweet *L. angustifolius* in Australia. Colonizing aphids important as vectors in *L. luteus* or *L. albus* in different parts of the world include *Acyrthosiphum kondoi*, *A. pisum*, *Aphis craccivora*, *A. fabae*, *Macrosiphum albifrons*, *M. euphorbiae*, *Megoura viciae* and *Myzus persicae*. Other aphids that can colonize sweet lupins and are potential vectors include *Aphis lupini, Brachycaudus helichrysi*, *Macrosiphum zionensis* and *Myzus ornatus* (Jones and McLean, 1989).

Seed-infected plants are the most important source of BYMV infection for *L. luteus* and *L. albus* and rates of seed transmission can be as high as 35 and 14%, respectively (Corbett, 1958; Beczner *et al.*, 1983). Strains of BYMV vary in the extent to which they are seed-borne in *L. luteus*. In *L. luteus*, seed-borne infection causes stunted plants with faintly mottled, pale and narrowed leaflets, deformation and necrosis of the inflorescence and proliferation of side shoots. In *L. albus* and rough-seeded lupin species, severe vein clearing on young leaflets, mottle, leaf deformation and stunting develop (Jones and McLean, 1989).

Infected legumes in adjacent pasture or legume weeds also constitute sources for acquisition and spread of BYMV to lupin crops by aphids. These are the primary source for infection of *L. angustifolius* and *L. cosentinii*. Red clover pastures are an important source in many parts of the world (Jones and McLean, 1989) while in mediterranean climatic parts of Australia, subterranean clover pastures are the main source (Jones, 1993). In south-eastern USA, the weeds *Melilotus alba* and *M. indica* are important reservoirs (Corbett, 1958).

The amount and proximity of the source, together with the time of aphid arrival and their abundance, are the principal factors influencing BYMV spread. Yield losses are greater when lupins are grown in aphid-prone regions. In southern Australia, these are warmer and higher rainfall regions. Early spread is associated with greater yield loss and, with *L. albus* and *L. luteus*, greater infection of harvested seed. In *L. luteus*, widespread infection of very young plants can cause almost complete loss of yield, while infection at flowering time decreases seed yield by about 20% (Blaszsack and Gryzb, 1971). In *L. angustifolius,* where BYMV kills plants, little or no seed is set and so the decrease in seed yield tends to be proportional to the numbers of plants infected (Jones, 1993, 1994).

Sowing seed with minimal virus content is essential for *L. luteus* and *L. albus*. Crop agronomy can be modified as part of an integrated approach to BYMV control. Early sowing, high seeding rates and close row spacing, which result in dense stands and early canopy development, shade out and deny aphids access to the stunted seed-infected plants (*L. luteus* and *L. albus)* or early, current-season infected plants (all species). By increasing ground cover, both early canopy development and stubble mulching reduce aphid landings. A cereal buffer can be sown around the perimeter of the lupin crop to decrease BYMV spread from outside (Blaszczak and Grzyb, 1971; Blaszczak, 1981; Jones, 1993).

Early-maturing cultivars tend to escape from BYMV spread occurring late in the growing season. Quantitative variation in resistance to BYMV seed transmission and to infection by aphid vectors exists in *L. luteus* (Pospiezny and Wojtowicz, 1989; Schmidt *et al.* 1990). Similar variation in resistance to BYMV infection by aphid vectors is found in *L. angustifolius.* No immunity is known to BYMV in any lupin species (Jones and McLean, 1989; McKirdy and Jones, 1995). However, genetic engineering for BYMV resistance, using coat protein and defective replicase genes, is being applied in Western Australia and transgenic *L. luteus* and *L. angustifolius* plants with putative BYMV resistance have been produced.

Cucumber mosaic virus (CMV)

Kohler (1935) first showed that the causal agent of 'lupin browning disease' in *L. angustifolius* and *L. luteus* was CMV. This virus is now known to infect both of these species worldwide. It has raised most concern in *L. angustifolius* in Australia, while in eastern Europe its symptoms in *L. luteus* have sometimes been confused with those of BYMV (Jones and McLean, 1989). CMV does not infect *L. albus, L. cosentinii, L. pilosus* or *L. atlanticus* but can infect *L. mutabilis* (Jones and Latham, 1996).

Current-season infection of *L. luteus* by aphid vectors results in chlorosis, faint mottle, bunching and downcurling of young leaves, reduction in leaf size and brown streaking of the stem. Infected plants are stunted but often have normal-looking lower leaves. Late infection may result in little more than chlorosis of the youngest leaves. With *L. angustifolius,* symptoms are similar but somewhat less severe and without stem streaking (Frencel and Pospieszny, 1983; Jones, 1988; Jones and McLean, 1989).

CMV is transmitted non-persistently by aphids. In Australia, *M. persicae, Acyrthosiphum kondoi* and *Aphis craccivora,* which commonly colonize sweet *L. angustifolius*, transmit the virus. The non-colonizing aphids *R. padi, R. maidis* and *Lipaphis erysime* are also vectors. *M. persicae* and *A. craccivora* are the most efficient vectors (Jones and Proudlove, 1991; Bwye *et al.,* 1994, 1995).

CMV is seed-borne in *L. luteus* (Zchau, 1960) and in *L. angustifolius* (Jones, 1988), which accounts for the importance of the virus in these two species. Rates of seed transmission to seedlings vary with genotype; they are normally in the 5–20% range but may reach 75% or more in some genotypes (Jones and Cowling,

1995; Jones and Latham, 1996). Seed-borne infection of *L. luteus* results in three types of reactions: seedlings become necrotic and die shortly after emergence; weak chlorotic seedlings with deformed leaflets die after 6–8 weeks; or seedlings grow into stunted plants with pale downcurled leaflets which flower and set some seed. In *L. angustifolius*, seed-borne infection causes chlorotic, stunted seedlings, with faint mottle, downcurling and bunching of leaflets.

Seed-infected lupin plants act as the main source of CMV for acquisition and spread by aphid vectors. CMV-infected herbaceous weeds may play a role as secondary virus sources in parts of the world where they can survive all year round, but are not significant under dryland farming conditions in Australia (McKirdy and Jones, 1994).

Yield losses and infection of harvested seed are both greater in regions that are more aphid-prone. In Australia, the amount of CMV infection in the seed sown along with the time of arrival and abundance of aphid vectors are the main factors determining the extent of spread in *L. angustifolius*. Early current-season spread results in greater yield loss and infection of harvested seed. Other factors influencing spread include the extent of establishment of seed-infected plants, canopy closure and ground cover. Early canopy closure shades out seed-infected and early current season-infected plants, thereby removing the source of CMV. As with BYMV, late or no canopy closure maximizes the opportunities for aphids to acquire and spread the virus. Ground cover, which may be due to canopy closure, stubble on the soil surface or presence of weeds, decreases the landing rates of incoming aphids, thereby reducing spread (Jones and Proudlove, 1991; Bwye *et al.* 1994, 1995).

Sowing seed with minimal virus content is fundamental for CMV control (Wylie *et al.*, 1993; Bwye *et al.*, 1994). In *L. angustifolius*, early sowing at high seeding rates using narrow row spacing is recommended, to obtain early canopy development, thereby shading out CMV-infected source plants and providing ground cover to reduce aphid landings. This strategy is less effective in years when aphids arrive early, as significant CMV spread can occur before the canopy forms. As with BYMV, stubble mulching is beneficial in reducing early aphid landings. This is most important with wide row spacing (Bwye *et al.*, 1994, 1995). Application of carbamate and organophosphorus insecticides is ineffective in decreasing CMV spread (Bwye *et al.*, 1997) but more promising results have been obtained with newer generation pyrethroids.

Useful resistance to seed transmission occurs in some lines of *L. angustifolius* and *L. luteus*. This resistance is not related to alkaloid content or flowering date and appears to be polygenically controlled. A dominant hypersensitivity gene conferring resistance to CMV is found in some *L. luteus* cultivars (Jones and Cowling, 1995; Jones and Latham, 1996).

Other

In addition to BYMV, four other aphid-transmitted potyviruses are recorded as infecting lupins: clover yellow vein (CYVV), bean common mosaic, peanut mottle and bidens mottle. In *L. luteus* and *L. albus*, CYVV symptoms tend to be more severe than those of BYMV but in *L. angustifolius* both cause similar symptoms. CYVV is present in white clover pastures worldwide and is a threat to nearby lupins. A cucumovirus, peanut stunt (PSV), has been reported infecting *L. luteus, L. albus* and *L. angustifolius* in Poland, usually as a mixed infection with BYMV. Apart from CYVV, these viruses are of minimal economic importance (Jones and McLean, 1989).

Subterranean clover red leaf luteovirus (SCRLV), synonymous with soybean dwarf virus, has been found infecting *L. albus, L. angustifolius, L. luteus* and *L. cosentinii* in New Zealand and Tasmania. Like CYVV, this virus is spread to lupins from white clover pastures by aphid vectors. Heavy losses in yield of *L. albus* plots have been reported in New Zealand (Jones and McLean, 1989). SCRLV is not seed-borne in lupin and requires a green host to over-summer. Therefore it is not a problem in regions with a mediterranean climate. Other viruses that occasionally infect lupins are listed by Jones and McLean (1989).

Phytoplasma

A witches' broom disorder occurs in *L. mutabilis* in Peru, sometimes infecting large numbers of plants (Frey and Yabar, 1983). It also occurs in ornamental lupins in Czechoslovakia (Ulrychova *et al.,* 1980). Symptoms include proliferation of axillary shoots and phyllody with reduced pod set. The presence of a phytoplasma has been confirmed by electron microscopy. A similar disorder is found in Western Australia in *L. angustifolius* which remains green at the end of the growing season when other plants have matured and dried out. It has been confused with manganese deficiency, but does not develop 'split-seed' which is characteristic of that condition (Chapter 5). Few plants are affected and the disorder has little economic importance. The causal role of a phytoplasma has not been confirmed (Jones and McLean, 1989).

Invertebrate Pests

Seedling root and shoot feeders

In the establishment phase, lupins are vulnerable to pest attack from a range of polyphagous insects, mites and molluscs. Sometimes chemical intervention is required to avoid substantial losses (Michael *et al.*, 1982; Ferguson, 1994).

Moth and beetle larvae

Cutworms (*Agrotis* spp.) and the brown pasture looper (*Ciampa aritaria* Guenee) are chewing pests that occur sporadically on lupins in Australia. Both over-summer in pupal diapause, emerging in autumn as moths that lay eggs on soil or plant material. Some cutworm species can over-summer as moths. Several generations of cutworms are possible in one season, whereas pasture looper larvae grow to full size in two months then pupate. Seedlings can be partially or fully defoliated, growing points eaten out or hypocotyls cut off at ground level, resulting in thin stands or large patches of bare ground. Brown pasture loopers are easy to see. Lupins are at greater risk from this pest if sown in rotation after pasture rather than a crop. Cutworms feed mostly at night but during the day can be found just below the soil surface near affected plants. Both can be controlled by application of pyrethroid or organophosphate insecticide shortly after seedling emergence.

In Europe, wireworm larvae (*Agriotes* spp.) attack lupin hypocotyls below the ground and kill seedlings. They are most serious after a pasture phase as it takes four years for the larvae to develop fully. Crop damage can be severe at larval populations of 50 m^{-2} (Gondran *et al.*, 1994). Larvae of the genus *Sitona* can 'eat out' root nodules and have been recorded on lupins in parts of Australia and the Iberian Peninsula. *Lupinus* appears not to be a preferred host.

Bean seedling maggot (*Delia platura* Meigen)

The bean seedling or onion maggot is a common pest of *L. albus* in Europe (Bournoville, 1990) and occasionally causes losses in *L. angustifolius* in the higher rainfall regions of Western Australia. The small black adult fly emerges from pupae in the soil and females lay their eggs on newly cultivated soil. They are attracted to soils rich in organic matter. Larvae emerge, tunnel into the hypocotyl and down into the tap root, feeding on and destroying the inner tissues. They often cause seedlings to wilt and die. In lighter infestations the hypocotyls become swollen and cracked and plants are less vigorous. In France, autumn sowings are more susceptible than spring sowings, due to slower emergence and establishment of the crop. Deep sowing predisposes seedlings to damage and excessive stubble retention may encourage egg lay.

Seed treatments or pre-plant soil applications with organophosphates give effective control. There are no reported differences in resistance to bean seedling maggot between lupin species or cultivars.

Red-legged earth mite (*Halotydeus destructor* Tucker)

The red-legged earth mite is a major pest of legume pastures in temperate Australia. It can also attack young lupin crops. The mite feeds particularly on cotyledons and young leaves, sucking and collapsing the tissue to leave a silvered appearance. High populations sometimes cause significant stunting and even death of seedlings. Diapause eggs hatch after the first autumn rains when soil temperatures are below 21°C (Wallace, 1970). Late-sown crops are more at risk because seedlings are smaller at the onset of infestation. Lupins sown after pasture are at greatest risk. In spring, heavy grazing or spraying insecticide in pastures can reduce diapause egg production, decreasing emergence and consequent damage in the following season. Merton *et al.* (1995) reported reduced numbers of mites in cultivated compared with direct-drilled plots.

Sowing *L. luteus* after pasture is extremely risky as it is a sensitive species, possibly because the cotyledons and growing point remain closer to the ground than with other lupin species. Even cultivars with intermediate alkaloid levels are very susceptible. Variation in susceptibility between cultivars has not been examined within lupin species, but *L. albus* is less susceptible than *L. angustifolius* (Thackray *et al.*, 1997).

Other

The Lucerne flea (*Sminthurus viridis* L.) occasionally chews windows in lupin leaves and cotyledons in Australia. Damage is most severe in wet years on heavier soils where partial waterlogging slows plant growth (Michael *et al.*, 1982). Damage from leather jackets (*Tipula* spp.), larvae of the crane fly, occurs in lupins sown after pasture in parts of Europe (Gondran *et al.*, 1994).

In Europe, slugs (*Deroceras reticulatum* Muller and *Arion hortensis* Fer.) can be notable pests of lupin seedlings, their presence often being indicated by slime trails. Humid and mild conditions favour the slugs. Metaldehyde and methiocarb baits control them (Gondran *et al.,* 1994; Ferguson, 1994).

Pests of the vegetative and reproductive stage

The indeterminate growth habit of lupins allows them to tolerate significant insect predation without necessarily leading to loss of grain yield. However, during flowering, and particularly during pod fill, some insects can reach damaging populations. Again, most of the insect species involved are polyphagous.

Aphids

Aphids cause feeding damage to lupins during flowering and early pod set as well as vectoring virus diseases. In Australia, the green peach (*Myzus persicae*), blue green (*Acyrthosiphum kondoi*) and cowpea (*Aphis craccivora*) aphids frequently colonize lupins. Populations sometimes become large enough to cause wilting and abortion of flowers and young pods. In Europe and North America, lupins are colonized by several other species (see section on Virus diseases). The lupin aphid *Macrosiphum albifrons*, which arrived in Europe from North America in the early 1980s, can establish damaging populations (Stroyan, 1981). It is a large aphid covered with a greyish-white wax and its host range is restricted to *Lupinus. M. albifrons* can colonize bitter lupins. It is tolerant of alkaloids, accumulating them during feeding, which renders it less palatable to predators (Wink and Roemer, 1986).

The breeding of lupin cultivars with low alkaloid contents has increased susceptibility to aphids (Berlandier, 1996). The possibility of developing cultivars that combine low alkaloid levels in seeds with increased levels in the plant, as occurs in naturalized *L. cosentinii* populations in Australia, has been postulated (Gladstones, 1994).

Control of aphids on lupins, other than *M. persicae*, can usually be achieved by a range of common aphidicides but more research is required to define economic thresholds. Insecticide resistance is common in *M. persicae* on lupins (Bwye *et al.,* 1997). High seeding rates to achieve rapid canopy development and direct drilling into a cereal stubble reduce aphid landing rates early in the season. Although this reduces virus spread, it is less likely to reduce direct feeding damage at flowering time because aphids are capable of rapid population expansion.

Large variation in susceptibility to colonization by the common aphid species occurs between sweet *L. angustifolius* and *L. luteus* genotypes. *L. angustifolius* cv. Yorrel and *L. luteus* cv. Weiko III are particularly susceptible. *L. albus* is more susceptible to *M. albifrons* than are *L. angustifolius* or *L. luteus* (Gondran *et al.,* 1994).

Thrips

Various thrip species colonize lupins, especially on the youngest growth at flowering time. Large populations can cause feeding damage on flower buds and leaves. Deformation followed by bronzing of leaflets is typical. In Western Australia, several species including the recently introduced western flower thrip (*Frankliniella occidentalis*) have colonized lupin in experimental plots. Economic damage to crops has been rare. In the UK, *Thrips angusticeps* has the greatest pest potential on lupins (Ferguson, 1994).

Pod and seed feeding insects

Australian native budworm (*Helicoverpa punctigera* Wallengren)

During pod fill, *Helicoverpa punctigera* is by far the most important pest of lupins in Australia. In some seasons losses can be large if insecticide protection is not in place. Only a single generation develops on lupins. This results from eggs laid by immigrant moths which fly out of the interior of the continent after building up during winter on native annual species (Walden, 1992). Regular crop inspections are required to monitor development of larval infestations. Only late instars are capable of penetrating the tough pod wall and spraying may safely be delayed until this stage is reached. One 20 mm long larva m^{-2} feeding on pods can cause approximately 10 kg ha^{-1} grain loss in *L. angustifolius* (Michael *et al.*, 1982). Normally, only a single spray of synthetic pyrethroid is required for control.

H. armigera occasionally causes damage in lupins in eastern Australia (Fitt, 1989) and can be more difficult to control than *H. punctigera* because of insecticide resistance. *H. punctigera* has shown no signs of developing resistance in Australia, probably because the populations originate in unsprayed rangeland areas and negligible diapause and recycling occurs on crops. Budworm are rarely important lupin pests elsewhere in the world.

L. angustifolius is the most resistant of the cultivated species. *L. luteus* is more susceptible and *L. albus* is particularly vulnerable, owing to the slow maturation of pods and seed.

Mirid bugs (*Lygus* spp.)

Sucking mirids cause abortion of young pods on *L. albus* in Europe (Bournoville, 1990). They have also been recorded as a significant pest of *L. albus* in experimental plots in Washington State, USA. Insecticide application to suppress nymph and adult populations during the critical pod set period resulted in large yield increases (Tanigoshi and Babcock, 1989). In Australia, mirids colonize *L. angustifolius* but no economic damage has been reported.

References

Agaev, R.N. (1993) A dangerous disease of lupin. *Zachchita-Rastenii Moskva* 8, 40–41.

Armstrong, G.M. and Armstrong, J.K. (1964) *Lupinus* species – common hosts for wilt *Fusaria* from alfalfa, bean, cassia, cowpea, lupine and US cotton. *Phytopathology* 54, 1232–1234.

Beczner, L., Horvath, J. and Borbely, F. (1983) Susceptibility of *Pisum sativum* and *Lupinus* species and cultivars to some economically important viruses. *Tagungsbericht Akademie der Landwirtschaftwissenschaften* 216, 313–322.

Berlandier, F.A. (1996) Alkaloid level in narrow-leafed lupin, *Lupinus angustifolius*, influences green peach aphid reproductive performance. *Entomologia Experimentalis et Applicata* 79, 19–24.

Blaszczak, W. (1962) The susceptibility of lupins to Fusariosis, Rhizoctoniosis and other diseases in light of experiments. *Review of Applied Mycology* 41, 718 (abstract).

Blaszczak, W. (1981) Virus diseases of lupin, red clover and broad bean in Poland. *Zeszyty Problemowe Postepow Nauk Rolniczynch* 244, 199–208.

Blaszczak, W. and Gryzb, B. (1971) Spreading of narrow-leaf disease virus (*Phaseolus* virus 2 [Pierce] Smith) in yellow lupin plants as related to the age and place of plant inoculation. *Phytopathologische Zeitschrift* 70, 163–177.

Bournoville, R. (1990) Les ravageurs du lupin blanc. *Phytoma* 417, 33–35.

Brown, A.G.P. (1984) Within crop spread of *Phomopsis leptostromiformis* on lupins. *Australasian Plant Pathology* 13, 11–12.

Bwye, A.M., Jones, R.A.C and Proudlove, W. (1994) Effects of sowing seed with different levels of infection, plant density and the growth stage at which plants first develop symptoms on cucumber mosaic virus infection of narrow-leafed lupins (*Lupinus angustifolius*). *Australian Journal of Agricultural Research* 45, 1395–1412.

Bwye, A.M., Jones R.A.C., Proudlove, W. (1995) Cucumber mosaic virus in lupins. *Western Australian Journal of Agriculture, Fourth Series* 36, 124–130.

Bwye, A.M., Proudlove, W., Berlandier, F.A. and Jones R.A.C. (1997) Effects of applying insecticides to control aphid vectors and spread of cucumber mosaic virus in narrow-leafed lupins. *Australian Journal of Experimental Agriculture* 37, 93–102.

Carling, D.E., Rothrock, C.S., MacNish, G.C., Sweetingham, M.W., Brainard, K.A. and Winters, S.W. (1994) Characterisation of anastomosis group 11 (AG-11) of *Rhizoctonia solani*. *Phytopathology* 84, 1387–1393.

Chamberlain, E.E. (1935) Sore-shin of blue lupins. Its identity with pea mosaic. *New Zealand Journal of Agriculture* 51, 86–92.

Collins, D.J., Stevens, C. and Khan, V.A. (1993) Effects of soil solarization on Southern Blight of lupins. In: Frobish, L.T. (ed.) *Grain Research Update 1993*. Alabama Agricultural Experimental Station, Auburn University, Alabama, p. 3.

Collins, D.J., Burch, K. and Williamson, D. (1996) Pearl millet stubble mulch and iprodione seed treatment control brown spot of *Lupinus albus*. Abstract in: *Eighth International Lupin Conference*, Pacific Grove, California.

Corbett, M.K. (1955) Virus diseases of lupines in Florida. *Proceedings of the Soil and Crop Science Society of Florida* 15, 35–39.

Corbett, M.K. (1958) A virus disease of lupines caused by bean yellow mosaic virus. *Phytopathology* 48, 86–91.

Cowling, W.A. (1988) Breeding for resistance to leaf and root resistance caused by *Pleiochaeta setosa* in narrow-leafed lupin. In: *Proceedings of the Fifth International Lupin Conference* PWRiL, Poznań, Poland, pp. 217–229.

Cowling, W.A., Hamblin, J., Wood, P.McR. and Gladstones, J.S. (1987) Resistance to *Phomopsis* stem blight in *Lupinus angustifolius* L. *Crop Science* 27, 648–652.

Cowling, W.A., Sweetingham, M.W., Dieperveen, D. and Cullis, B. (1997) Heritability of resistance to brown spot and root rot of narrow-leafed lupins caused by *Pleiochaeta setosa*. (Kirchn.) Hughes in field experiments. *Plant Breeding* 116(4), 341–345.

Decker, P. (1947) Anthracnose of blue lupin is seed-borne. *Plant Disease Reporter* 31, 270–271.

Dick, M.A. (1994) Blight of *Lupinus arboreus* in New Zealand. *New Zealand Journal of Forestry Science* 24, 51–68.

Dingley, J.M. and Gilmour, J.W. (1972) *Colletotrichum acutatum* Simmds. f. sp. *pineae* associated with terminal crook disease of *Pinus* spp. *New Zealand Journal of Forestry Science* 2, 192–201.

Eppler, A., Hinz, U. and Romer, P. (1986) Virus diseases of *Lupinus mutabilis* Sweet in Germany. *Mededelingen Faculteit Landbouwetenshappen Rijksuniversiteit Gent* 51, 817–826.

Faluyi, M.A., Mather, D.E., Atlin, G.N., Merrick, L.C. and Paulitz, T.C. (1993) Field evaluation of seed, pod and stem rot in white lupine germplasm. *Plant Disease* 77, 926–929.

Ferguson, A.W. (1994) Pests and plant injury on lupins in the south of England. *Crop Protection* 13, 201–210.

Fitt, G.P. (1989) The ecology of *Heliothis* species in relation to agroecosystems. *Annual Review of Entomology* 34, 17–52.

Forbes, I., Wells, H.D. and Edwardson, J.R. (1965) Inheritance of resistance to *Stemphylium solani* in blue lupine endemic to Portugal. *Phytopathology* 55, 627–628.

Frencel, I. and Pospieszny, K. (1985) Viruses of natural infections of yellow lupin (*Lupinus luteus* L.) in Poland. V. Cucumber mosaic virus. *Acta Phytopathologica Academiae Scientarum Hungaricae* 20, 87–90.

Frey, F. and Yabar, E. (eds) (1983) *Enfermedades y plagas de lupinos en el Peru*. Sociedad Alemana de Cooperacion Tecnica, Eschborn, West Germany, 86 pp.

Furgal-Wegrzycka, H. (1984) Occurrence and variation of the fungus *Fusarium oxysporum* f. sp. *lupini* in lupin in the Olsztyn region in the years 1972–1976. *Roczniki Nauk Rolniczych e Ochrona Roslin* 14, 163–174.

Germar, B. (1939) Untersuchungen uber *Ceratophorum setosum* Kirchn. auf *Lupinus albus*. *Zeitschrift Pflanzenkrankheiten, Pflanzenpathologie und Pflanzenschutz* 49, 482–509.

Gladstones, J.S. (1982) Breeding lupins in Western Australia. *Western Australian Journal of Agriculture, Fourth Series* 23, 73–76.

Gladstones, J.S. (1989) Breeding *Phomopsis* resistant lupins. *Western Australian Journal of Agriculture, Fourth Series* 30, 3–7.

Gladstones, J.S. (1994) An historical view of lupins in Australia. In: Dracup, M. and Palta, J. (eds) *Proceedings of the First Australian Lupin Technical Symposium*. Department of Agriculture, South Perth, Western Australia. pp. 1–38.

Gondran, J. (1984) Les maladies du lupin blanc doux en France. *Perspective Agrico* 77, 31–41.

Gondran, J., Bournoville, R. and Duthion, C. (1994) *Diagnostic des maladies, ravageurs et accidents chez le lupin blanc*. INRA Editions, Versailles, France, 47 pp.

Griffiths, D.A. and Isaac, I. (1963) Wilt of lupin and sunflower caused by species of *Verticillium*. *Horticulture Research* 2, 104–114.

Hogetop, C. (1937) Uma doenca fungica do Tremoco. *Revue Agronomie* 1, 346–350.

Hughes, S.J. (1951) Studies on microfungi. III. *Mastigosporium, Camposporium* and *Ceratophorum*. *Mycological Papers* 36, 1–43.

Jones, R.A.C. (1988) Seed-borne cucumber mosaic virus infection of narrow-leafed lupin (*Lupinus angustifolius*) in Western Australia. *Annals of Applied Biology* 113, 507–518.

Jones, R.A.C. (1993) Effects of cereal borders, admixture with cereals and plant density on the spread of bean yellow mosaic potyvirus into narrow-leafed lupins (*Lupinus angustifolius*). *Annals of Applied Biology* 122, 501–518.

Jones, R.A.C. (1994) Effects of mulching with cereal straw and row spacing on spread of bean yellow mosaic potyvirus into narrow-leafed lupins (*Lupinus angustifolius*). *Annals of Applied Biology* 124, 45–58.

Jones, R.A.C. and Cowling, W.A. (1995) Resistance to seed transmission of cucumber mosaic virus in narrow-leafed lupins (*Lupinus angustifolius*). *Australian Journal of Agricultural Research* 46, 1339–1352.

Jones, R.A.C. and Latham, L.J. (1996) Natural resistance to cucumber mosaic virus in lupin species. *Annals of Applied Biology* 129, 523–542.

Jones, R.A.C. and McLean, G.D. (1989) Virus diseases of lupins (a review). *Annals of Applied Biology* 144, 609–637.

Jones, R.A.C. and Proudlove, W. (1991) Further studies on cucumber mosaic virus infection of narrow-leafed lupin (*Lupinus angustifolius*): seed-borne infection, aphid transmission, spread and effects on grain yield. *Annals of Applied Biology* 118, 319–329.

Kalis-Kuznia, R.A., Stewart, E.L. and Meronuck, R.A. (1991) First record and notes on *Pleiochaeta setosa* in Minnesota. *Mycologia* 83, 826–828.

Kiselev, I.I., Kurlovich, B.S. and Strepanova, P. (1988) Sources of resistance to *Fusarium* in lupin. *Sbornik Nauchnykh Trudov po Prikladnoi Botanike, Genetike i Selektsii* 117, 56–60.

Kochman, J. and Kubicka, H. (1974) Aggressiveness and pathogenicity of *Phomopsis leptostromiformis* and the development of the pathogenic process caused by this fungus. *Acta Agrobotanica* 27, 5–17.

Kohler, E. (1935) Ubertragungsversuche mit dem Virus der Lupinenbraune. *Angewardte Botanik* 17, 227–286.

Korniichuk, N.S. (1985) *Methodology for Infection and Measuring Resistance to* Fusarium *in Lupins*. Kiev Institute for Crop Production, Moscow, 12 pp.

Korniichuk, N.S. (1991) Testing lupin for resistance to fusarial wilt. *Zakhist Roslin* 38, 27–33.

Kuznia, R.A., Meronuck, R.A. and Stewart, E.L. (1993) Effect of seed treatments on stand establishment, root necrosis, and yield of *Lupinus albus*. *Plant Disease* 77, 892–895.

Lamberts, H. (1955) Broadening the bases for the breeding of Yellow Sweet Lupine. *Euphytica* 4, 97–106.

Lara, Z.O. and Andrade, V.O. (1983) Una nueve enfermedad 'Mancha cafe' (*Pleiochaeta setosa* Kirchn. Hughes) en cultivos de lupino (*Lupinus albus* L.) en la IX region. *Agricultura Tecnica* 43, 391–395.

Leach, S.S. and Clapham, W.M. (1992) *Rhizoctonia solani* on white lupine. *Plant Disease* 76, 417–419.

Leuck, D.B., Wells, H.D. and Beck, E.W. (1962) Systemic insecticides for indirect bean yellow mosaic virus control in seed fields of yellow lupines. *Plant Disease Reporter* 46, 240–242.

Luttrell, E.S. and Samples, J.W. (1954) Mildew of lupins caused by *Microsphaera diffusa*. *Plant Disease Reporter* 38, 719–720.

MacLeod, W.J. and Sweetingham, M.W. (1997) A root disease of *Lupinus angustifolius* caused by a new species of binucleate *Rhizoctonia*. *Australian Journal of Agricultural Research* 48, 21–30.

MacNish, G.C. and Neate, S.M. (1996) Rhizoctonia bare patch of cereals – an Australian perspective. *Plant Disease* 80, 965–971.

Malan, D.E. and Knox-Davies, P.S. (1990) *Fusarium oxysporum* associated with *Lupinus albus* in the Highveld of South Africa. *Phytophylactica* 22, 109–112.

McKirdy, S.J. and Jones, R.A.C. (1994) Infection of alternative hosts associated with narrow-leafed lupins (*Lupinus angustifolius*) and subterranean clover (*Trifolium subterraneum*) by cucumber mosaic virus and its persistence between growing seasons. *Australian Journal of Agricultural Research* 45, 1035–1049.

McKirdy, S.J. and Jones, R.A.C. (1995) Bean yellow mosaic potyvirus infection of alternative hosts associated with subterranean clover (*Trifolium subterraneum*) and narrow-leafed lupins (*Lupinus angustifolius*): field screening procedure, relative susceptibility/resistance rankings, seed transmission and persistence between growing seasons. *Australian Journal of Agricultural Research* 46, 135–152.

Merton, E., McDonald, G. and Hoffmann, A. (1995) Cultural control of redlegged earth mite, blue oat mite and lucerne flea in canola. *Plant Protection Quarterly* 10, 65–66.

Michael, P.J., Woods, W.M., Richards, K.T. and Sandow, J.D. (1982) Lupins and insects. *Western Australian Journal of Agriculture, Fourth Series* 23, 83–85.

Norris, D.O. (1943) Pea mosaic on *Lupinus varius* L. and other species in Western Australia. *Australian Council for Scientific and Industrial Research Bulletin No. 170*, 27 pp.

Ostazeski, S.A. and Wells, H.D. (1960) A phomopsis stem blight of yellow lupine (*Lupinus luteus* L.). *Plant Disease Reporter* 44, 66–67.

Osman, A.R., Fahim, M.M., Sahab, A.F. and Abd-El-Kader, M.M. (1986) Soil solarization for the control of lupin wilt. *Egyptian Journal of Phytopathology* 18, 75–88.

Pape, H. (1927) Krankheiten und Schadllinge der Lupine. *Review of Applied Mycology* 6, 731 (abstract).

Papineau, J. and Huyghe, C. (1989) Collecting white lupin in the Azores. *FAO/BPGR Plant Genetic Resources Newsletter 88/89*, 77–78.

Paulitz, T.C. and Atlin, G. (1992) First report of brown spot of lupines caused by *Pleiochaeta setosa* in Canada. *Plant Disease* 76, 1185.

Paulitz, T.C. and Cote, E. (1991) First report of *Phoma* spp. on white lupines in North America. *Plant Disease* 75, 862.

Paulitz, T.C., Atlin, G. and Gray, A.B. (1995) First report of *Colletotrichum gloeosporiodes* on lupine in Canada. *Plant Disease* 79, 319.

Pospieszny, H. and Wojtowicz, A. (1989) The influence of yellow lupin (*Lupinus luteus*) cultivars on seed-transmission of the bean yellow mosaic virus. *Lupin Newsletter* 13, 32–35.

Pulselli, A. (1928) Un parasita di alcune specie di *Lupinus* e di *Cytisus* (*Ceratophorum setosum* Kirch.1892). *Review of Applied Mycology*, 7, 582 (abstract).

Radman, L. (1964) Charcoal rot of lupin roots. *Review of Applied Mycology* 44, 28–33 (abstract).

Reed, P.J., Dickens, J.S.W. and O'Neil, T.M. (1996) Occurrence of anthracnose (*Colletotrichum acutatum*) on ornamental lupin in the United Kingdom. *Plant Pathology* 45, 245–248.

Reis, E.M., Dos Santos, H.P. and Ferretto, M.F. (1983) Occurrencia da antracnose do tremoco no rio grande do sul, causada por *Glomerella cingulata*. *Fitopatologia Brasileira* 8, 353–356.

Richter, H. (1941) Lupinefusariosen. *Review of Applied Mycology* 23, 110 (abstract).

Roemer, P. (1988) Disease and pests of *Lupinus mutabilis* in Germany. *Lupin Newsletter* 11, 28–32.

Schmidt, H.E., Meyer, U., Misko, I., Jagoda, G., Hellmund, K.J., Jakuseva, A. and Haack, I. (1990) Quantitative resistance of yellow lupin (*Lupinus luteus* L.) to bean yellow mosaic virus and correlation of its traits. *Archiv für Zuchtungsforschung, Berlin* 20, 81–91.

Shankar, M., Cowling, W.A. and Sweetingham, M.W. (1995) Variation and stability of virulence in isolates of *Diaporthe toxica*. *10th Biennial Australasian Plant Pathology Society Conference*, Abstract No. 213. Lincoln University, Christchurch, New Zealand.

Shankar, M., Cowling, W.A. and Sweetingham, M.W. (1996) The expression of resistance to latent infection by *Diaporthe toxica* in narrow-leafed lupin. *Phytopathology* 86, 692–697.

Shivas, R.G., McClements, J.L. and Sweetingham, M.W. (1998) Vegetative compatibilty among populations of *Colletotrichum* causing lupin anthracnose. *Australian Plant Pathology* (in press).

Smith, G.C. and Wheeler, B.E.J. (1969) The overwintering of cleiostocarps of *Erysiphe polygoni* on different hosts. *Transactions of the British Mycological Society* 52, 437–445.

Snynder, W.C. and Hansen, H.N. (1940) The species concept in *Fusarium*. *American Journal of Botany* 27, 64–67.

Sreenivasapprasad, S., Mills, P.R. and Brown, A.E. (1994) Nucleotide sequence of the rDNA spacer 1 enables identification of isolates of *Colletotrichum* as *C. acutatum*. *Mycological Research* 98, 186–188.

Stroyan, H.L.G. (1981) A North American lupin aphid found in Britain. *Plant Pathology* 30, 253.

Sweetingham, M.W. (1984) *Pleiochaeta setosa* – a root pathogen of lupins. *Australasian Plant Pathology* 13, 21–22.

Sweetingham, M.W. (1986) Research into lupin root diseases. *Western Australian Journal of Agriculture, Fourth Series* 27, 49–52.

Sweetingham, M.W. (1989) Fungi associated with root and hypocotyl diseases of seedling lupins. *Australian Journal of Agricultural Research* 40, 781–789.

Sweetingham, M.W. (1990) Coping with brown spot and root rots of lupins. *Western Australian Journal of Agriculture, Fourth Series* 31, 5–13.

Sweetingham, M.W. (1991) The effect of inoculum distribution and sowing depth on Pleiochaeta root rot of lupins. *Australian Journal of Agricultural Research* 42, 121–128.

Sweetingham, M.W., Loughman, R.L. and Porritt, S.E. (1993) Cereal stubble mulching protects lupins from brown spot (*Pleiochaeta setosa* Kirchn. Hughes). *Australian Journal of Experimental Agriculture* 33, 469–473.

Sweetingham, M.W., Cowling, W.A., Buirchell, B.J. and Henderson, J. (1994) Resistance of *Lupinus* spp. to leaf and root diseases caused by *Pleiochaeta setosa*. In: Neves Martins, J.M. and Beirao da Costa, M.L. (eds) *Advances in Lupin Research*. ISA Press, Lisbon, pp. 368–370.

Sweetingham, M.W., Cowling, W.A., Buirchell, B., Brown, A.G.P. and Shivas, R.G. (1995) Anthracnose of lupins in Western Australia. *Australasian Plant Pathology* 24, 271.

Tanigoshi, L.K. and Babcock, J.M. (1989) Insecticide efficacy for control of lygus bugs (Heteroptera: Miridae) on white lupin, *Lupinus albus* L. *Journal of Economic Entomology* 82, 281–284.

Thackray, D.J., Ridsdill-Smith, T.J. and Gillespie, D.J. (1997) Susceptibility of grain legume species to red legged earth mite (*Halotydeus destructor* Tucker) damage at the seedling stage. *Plant Protection Quarterly* 12, 141–144.

Truszkowska, W. (1977) Plant rotation as a means for controlling wilt disease of yellow lupin (*Lupinus luteus* L.) on the basis of mycological investigations. *Acta Mycologica* 13, 195–218.

Tu, J.C. (1983) Epidemiology of anthracnose caused by *Colletotrichum lindemuthianum* on white bean in southern Ontario: survival of the pathogen. *Plant Disease* 67, 402–404.

Ulrychova, M., Jokes, M. and Kynclova, J. (1980) Mycoplasma-like organisms in *Lupinus* hybrids showing witches' broom symptoms. *Biologia Plantarum* 22, 362–365.

Van Jaarsveld, A.B. (1980) Chemical control of *Pleiochaeta setosa* (Kirchn.) Hughes on lupin in the winter rainfall region. *Elsenberg Joernaal* 4, 14–16.

Van Jaarsveld, A.B. (1983) Resistance of lupin accessions to powdery mildew (*Erisyphe polygoni*). *Phytophylactica* 15, 55.

Van Jaarsveld, A.B. and Knox-Davies, P.S. (1974) Resistance of lupins to *Phomopsis leptostromiformis*. *Phytophylactica* 6, 55–60.

Walden, K. (1992) On the trail of the native budworm. *Western Australian Journal of Agriculture, Fourth Series* 33, 109–113.

Wallace, M.M.H. (1970) The influence of temperature on post-diapause development and survival in the aestivating eggs of *Halotydius destructor* (Acari: Eupodidae). *Australian Journal of Zoology* 18, 315–329.

Weimer, J.L. (1943) A *Botrytis* disease of lupines. *Phytopathology* 31, 319–323.

Weimer, J.L. (1944) Some root rots and a foot rot of lupines in the southeastern part of the United States. *Journal of Agricultural Research* 68, 441–457.

Weimer, J.L. (1952) Lupine anthracnose. *United States Department of Agriculture Bulletin No. 904*, 17pp.

Wells, H.D., Forbes, I., Webb, T.E. and Edwardson, J.R. (1956) Two *Stemphyllium* diseases of blue lupines. *Plant Disease Reporter* 40, 803–806.

Wells, H.D. and Forbes, I. (1967) Effects of temperature on growth of *Glomerella cingulata in vitro* and on its pathogenicity to *Lupinus angustifolius* genotypes *an an* and *An An*. *Phytopathology* 57, 1309–1311.

Welty, R.E. (1984) Blue lupine as a host for *Colletotrichum trifolii* from alfalfa and for *C. fragariae* from strawberry. *Plant Disease* 68, 142–144.

Williamson, P.M., Sivasithamparam, K. and Cowling, W.A. (1991) Formation of subcuticular coralloid hyphae by *Phomopsis leptostromiformis* upon latent infection of narrow-leafed lupins. *Plant Disease* 75, 1023–1026.

Williamson, P.M., Highet, A.S., Gams, W., Sivasithamparam, K. and Cowling, W.A. (1994) *Diaporthe toxica* sp. nov., the cause of lupinosis in sheep. *Mycological Research* 98, 1364–1368.

Wink, M. and Romer, P. (1986) Acquired toxicity – the advantages of specializing on alkaloid-rich lupins to *Macrosiphon albifrons* (Aphididae). *Naturwissenschaften* 73, 210–212.

Wood, P.McR. and Brown, A.G.P. (1975) Phomopsis: the causal fungus of lupinosis. *Western Australian Journal of Agriculture, Fourth Series* 16, 31–32.

Wood, P.McR. and McLean, G.D. (1982) Lupins – the disease problem. *Western Australian Journal of Agriculture, Fourth Series* 23, 86–88.

Wylie, S., Wilson, C.R., Jones, R.A.C. and Jones, M.G.K. (1993) A polymerase chain reaction assay for cucumber mosaic virus in lupin seeds. *Australian Journal of Agricultural Research* 44, 41–51.

Yang, H.A., Sweetingham, M.W. and Cowling, W.A. (1996) The leaf infection process and resistance to *Pleiochaeta setosa* in three lupin species. *Australian Journal of Agricultural Research* 47, 787–799.

Zchau, K. (1960) Zur Ubertragung des Gurkenmosaikvirus durch den Samen von *Lupinus luteus*. *Tagungsbericht* 29, 41–51.

10 Agronomy and Farming Systems

M.W. Perry[1,2], M. Dracup[1,2], P. Nelson[1,3], R. Jarvis[1], I. Rowland[1] and R.J. French[1,2]

[1]*Agriculture Western Australia, 3 Baron-Hay Court, South Perth, Western Australia 6151;* [2]*Centre for Legumes in Mediterranean Agriculture (CLIMA), University of Western Australia, Nedlands, Western Australia 6907;* [3]*Grain Pool of WA, 172 St Georges Terrace, Perth, Western Australia 6000.*

Introduction

Agronomy is crop production in practice. In 1967 a new crop was born when the first domesticated cultivars of narrow-leafed lupin were introduced to agriculture. Over three decades a significant new industry has been created as farmers, scientists and marketers have worked together. Matching the growth of the industry, there has been an equally significant growth in scientific knowledge of the species – its biochemistry, physiology, breeding, production and utilization. Agronomy is the art of applying that knowledge to enhance human interests – profitable and sustainable production from the species.

All crops are a part of more complex farming systems, where the benefits and costs of crop sequences are balanced for the greater profitability of the system as a whole. Lupins are legumes forming nitrogen-fixing symbioses with *Bradyrhizobium lupinii.* They have developed a special role as a rotation crop with cereals, but their contribution to farming systems is much greater than that alone. It is the interaction of the unique physiology and structure of the crop with the special characteristics of the environments of southern Australia that is of greatest interest and the major topic of this chapter. Because of the dearth of information on other species and the wealth of literature on the agronomy of the narrow-leafed lupin, the chapter concentrates on this species and makes only passing mention of the other lupins, cultivated as crop plants.

Agronomy straddles the interface between research and the application of science and so it should not be surprising that insight and advances come from both the science and the practical experience of producers. The role of producer in discovery and adaptation, and the importance of the close relationship between producer and scientist in the development of the industry has been acknowledged by Nelson (1994). This can make it difficult for the writers of a review because much information is undocumented, on-farm application changes rapidly, and the subtle effects of environment are ever present. In this review we try to explain the evolution of agronomic thought and on-farm

 Lupins as Crop Plants: Biology, Production and Utilization (eds J.S. Gladstones, C.A. Atkins and J. Hamblin)

practice against the background of an industry that emerged *de novo* less than 30 years ago.

Industry Background

Southern Australia's narrow-leafed lupin industry is by far the largest and most technically sophisticated agricultural industry based on a species of the genus *Lupinus.* From no commercial production at the time of the release of the first cultivar in 1967, the area sown has increased to 1.3 million ha and a production of about 1.4 Mt in 1995 (Table 10.1). The industry began in south-western Australia and this region has dominated the industry (over 90% of production in most years) throughout its history. To understand the agronomy of the crop, and its significance as a component of farming systems, we begin with some essential background to the industry.

Table 10.1. Growth of the narrow-leafed lupin industry in southern Australia: area sown (ha). Production in Queensland and Tasmania has been less than 1000 ha in any year.

	State				
Year	NSW	VIC	SA	WA	Total
1969–70				5.0	5.0
1970–71				12.0	12.0
1971–72				27.0	27.0
1972–73				44.0	44.0
1973–74				64.0	64.0
1974–75				119.0	119.0
1975–76				122.0	122.0
1976–77	2.4	2.9	13.5	95.9	115.0
1977–78	10.2	6.4	13.4	57.4	87.8
1978–79	8.3	9.1	11.0	39.0	67.6
1979–80	30.7	15.0	13.7	46.2	105.7
1980–81	23.2	21.7	17.8	55.1	117.9
1981–82	17.2	25.0	21.9	96.8	161.0
1982–83	19.1	21.0	19.1	197.4	256.7
1983–84	22.2	16.2	21.3	318.9	378.9
1984–85	35.3	20.1	20.7	517.3	593.7
1985–86	46.5	24.2	23.6	492.9	587.4
1986–87	37.1	28.0	28.9	653.1	748.1
1987–88	57.2	37.0	43.0	876.5	1014.9
1988–89	47.7	35.0	46.1	719.9	849.4
1989–90	46.3	36.0	42.3	676.3	801.6
1990–91	55.3	55.0	38.4	664.8	758.9
1991–92	66.5		47.7	786.6	900.8
1992–93	99.0	37.0	58.0	823.0	1017.0
1993–94	80.0	53.0	57.0	900.0	1090.0
1994–95				912.5	
1995–96	80	30	65	1105	1280

The climate of south-western Australia is typically mediterranean, with hot, dry summers and a cool, wet, winter growing season. The length of growing season varies from four to eight months, delineated by a sharp 'break-of-season' – the onset of autumn rains between April and June, and an equally sharp end to the growing season caused by cessation of rainfall and rapidly rising temperatures in spring. The region's soils reflect the great age of the landscape and are in general coarse-textured, acidic and low in nutrients. Of particular significance to the lupin industry are large areas of deep, coarse-textured siliceous sands of exceptionally low natural fertility (cation exchange capacity (CEC) < 5 cmol kg^{-1}). It is on these soils that narrow-leafed lupin has proved to be exceptionally well adapted and where a distinct and economically significant farming system based on narrow-leafed lupin has developed. Elsewhere in south-western Australia, and in southern Australia generally, narrow-leafed lupin is also grown on acidic to neutral duplex soils with a sandy or loamy surface horizon over a clay subsoil, and on acidic, loamy soils. Productive crops are also grown on soils with an alkaline reaction provided the free lime content is less than 3% (Chapter 8, this volume).

The release of domesticated narrow-leafed lupin cultivars in 1967 and 1971 occurred at the end of two decades of rapid agricultural expansion in south-western Australia. Approximately 10 million ha of land, much of it 'sandplain', had been cleared and sown to cereal crops grown in rotation with the leguminous pasture plant subterranean clover (*Trifolium subterraneum*) or annual medic (*Medicago* spp.). By the early 1970s the subterranean clover–wheat farming system that had developed on the coarser textured, lower fertility sandplain was in economic and physical decline. Grassy weeds had invaded the pastures, competing with the clover and acting as a host for cereal root disease. Clover seed-banks were reduced by too frequent cropping, and as the clovers declined so did fertility, causing lower cereal yields and an economic imperative to crop even more frequently, so reinforcing the cycle of decline.

It was into this environment that narrow-leafed lupin was released. When Uniharvest, the first fully non-shattering cultivar, was released in 1969 it was for 'medium rainfall areas south of Perth' – the prevailing scientific view of the species' adaptation. However, as seed became available, the industry grew initially on the northern sandplains where the need for a rotational crop that could restore soil fertility had become critical. It quickly became clear that the species had found an ecological niche to which it was exceptionally well adapted and an economic need which drove rapid adoption. Thus was born the lupin–wheat rotation which has become the defining farming system for those regions where the deep coarse-textured siliceous sandy soils predominate.

Setbacks were experienced but, after a decade of research and farmer experimentation during the 1970s, adoption accelerated and between 1980 and 1990 the area increased 12-fold from 55,000 to 875,000 ha. The area sown has continued to increase, but Perry (1995) has speculated that the industry is now mature and, based on potential soil types and rotations, is unlikely to exceed 1 million ha in south-western Australia. In southern and south-eastern Australia growth of the industry has been slower as the large areas of deep sandy soils are absent and the choice of farm enterprise is wider. Nevertheless there is probably the potential for production to increase substantially as cereal producers turn to legume crops as their source of nitrogen for cereals and for other rotational benefits.

Development and Morphology

Structure

The general structure of modern narrow-leafed lupin crop plants has been described by Dracup and Kirby (1993, 1996a). The plants develop a strong main stem which terminates in a racemose inflorescence. First-order apical branches appear in the axil of the main-stem leaves and only the uppermost two to four branches usually flower and produce seed. Similarly, second-order branches develop in the axil of leaves of first-order branches, and likewise only the upper ones usually flower and produce seed. This sequence is repeated with third and higher order branching, depending on the length of the growing season. The successive orders of branching decrease in vigour, size and productivity (Dracup and Kirby, 1993). Such an indeterminate pattern of development leads to a large overlap in vegetative and reproductive growth (Fig. 10.1) and strong intra-plant competition for resources exacerbated by environmental stresses such as drought, frost and heat (Chapter 8, this volume). As a consequence there is considerable abortion of flowers, which commonly exceeds 90% (Farrington and Gladstones, 1974; Pigeaire *et al.*, 1992; Dracup, 1994), pods (with or without abscission) (Dracup, 1994) and seeds (Dracup and Kirby, 1996b).

Development

Lupin development has recently been comprehensively described (Dracup and Kirby, 1996a) and here we provide an overview of the subject. Seed germination is highly sensitive to both temperature and soil water, with optimum germination occurring at field capacity and 20°C (Dracup *et al.*, 1993). Under optimum conditions germination occurs in 1–1.5 days and no germination is possible below 0°C or a soil matric potential of about 1.5 MPa. At germination, the seed water content exceeds 60% of the seed weight (Dracup *et al.*, 1993) and the radicle protrudes through the testa near the hilum. Under optimum conditions (20°C, field capacity) the radicle then lengthens at about 24 mm day^{-1} compared with 18 mm day^{-1} for the hypocotyl. At low soil water concentrations, hypocotyl growth decreases more than the radicle; for example, between field capacity (–0.003 MPa) and a matric potential of –0.03 MPa, radicle extension slowed by 40% compared with 75% with the hypocotyl. Emergence is thus delayed, favouring root exploration of the soil for water. In the above example seedlings emerging from soil at field capacity (in six days) had radicles about 8 cm long compared with 23 cm when emerging from soil at –0.03 MPa (in 14 days) (Dracup *et al.*, 1993). This also means that emergence is more sensitive to low soil water contents than is germination (Dracup *et al.*, 1993).

Emergence is epigeal and the hypocotyl forms a hook as it elongates and the cotyledons are folded tightly together and point downwards. The hook protects the young leaves and shoot apex from abrasion as hypocotyl extension draws

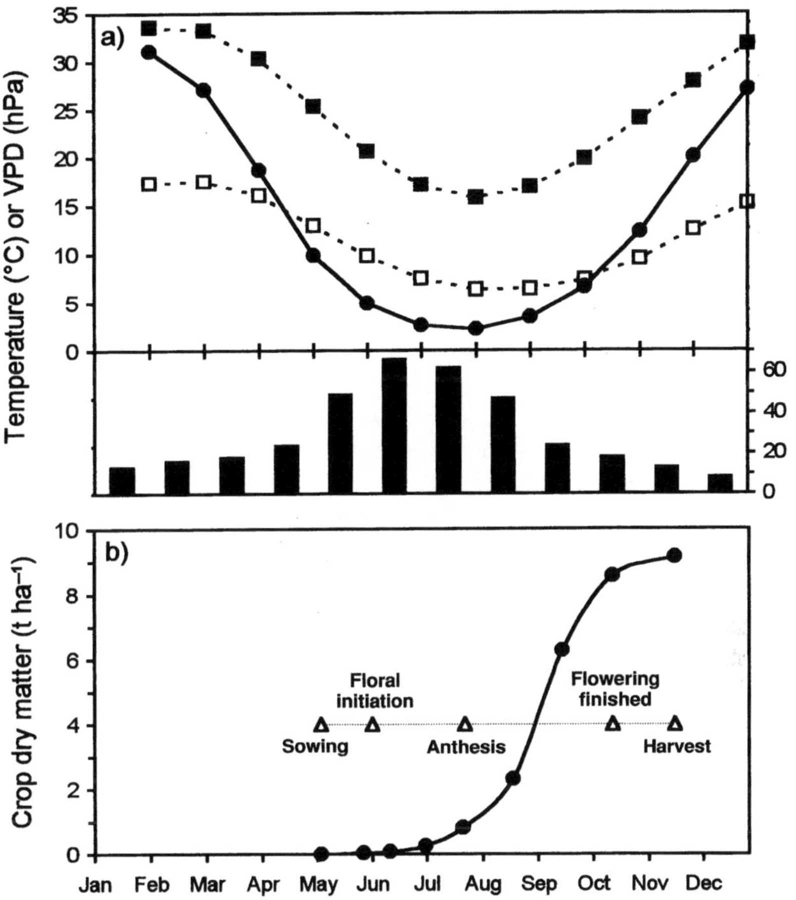

Fig. 10.1. Climate, growth and development data for Wongan Hills (lat. 30°51, long. 116°43) in the grain-growing belt of Western Australia. (a) The climate data are long-term monthly averages for maximum (■) and minimum (□) temperature, average vapour pressure deficit (VPD) (●) and monthly rainfall (columns). (b) Lupin growth and development data are for a crop grown in a typical season in 1988; open triangles represent the timing of the indicated stages.

the cotyledons and the enveloped shoot apex to the soil surface. It takes about 100°Cd for seedlings to emerge from 4 cm in moist soil (i.e. about one week when the seed bed temperature is 15°C) (Dracup and Kirby, 1996a). Since the projected area of cotyledons and the hypocotyl hook is large, there is considerable resistance to their upward movement through the soil; lupins are therefore sensitive to soil crusting (see below). Once above the soil surface and exposed to light, hypocotyl elongation stops, the hypocotyl and cotyledons turn green, the cotyledons expand and open out and the young leaves and shoot grow from between them. Since the shoot apex is taken above the soil surface the seedling is particularly sensitive to predation or mechanical damage, such as by 'sand blasting'.

Five or six leaves have already been initiated within the seed before it reaches maturity (Dracup and Kirby, 1993) and these continue to develop as new leaves are initiated after imbibition. A new leaf is initiated on the shoot apex each 26°Cd (base temperature 0°C) (Dracup and Kirby, 1993), continuing until floral initiation at about 400°Cd after sowing (Dracup and Kirby, 1993), when about 20 leaves have been initiated for typical sowing times in Western Australia (Dracup and Kirby, 1996b). The first pair of leaves begin to unfold from the shoot apex shortly after the cotyledons open, and successive leaves then emerge each 33–42°Cd (Dracup and Kirby, 1996a), corresponding to about two leaves per week at typical temperatures in Western Australia. The leaves are palmate with 5–9 leaflets and arranged spirally up the stem to minimize leaf shading (phyllotaxis is described in Chapter 7, this volume). The number of leaves produced depends on time of sowing and location. In Western Australia lupins sown late experience longer day-lengths and produce fewer leaves.

Flowering in Western Australia starts approximately 1000°Cd after sowing but may vary from 850 to 1150°Cd with extreme sowing dates and different locations (Reader *et al.*, 1995). Using long-term temperature records, for mid-May sowings, flowering can be expected between 75 and 100 days after sowing – earliest with the most northerly sowings, where temperatures are highest and days longest (Reader *et al.*, 1995). Time to first flower is only marginally sooner in late sowings (Reader *et al.*, 1995); thus the time for reproductive growth before the terminal drought and high temperatures is shortened, reducing branch development and yield. This highlights the importance of early sowing in lupins.

Anthesis occurs one or two days before the 'open flower stage' and narrow-leafed lupins are therefore nearly always self-pollinated. Only a small proportion of the flowers set pods, most commonly those at the base of the inflorescence, and this is not related to poor ovule fertilization (Pigeaire *et al.*, 1992). Once pods have 'set', i.e. the pod is 8–10 mm long, the frequency of abscission declines substantially (Pigeaire *et al.*, 1992). However, even pods at comparatively advanced stages of filling can abort, not always with accompanying abscission, so that empty pods remain on the plant at harvest; in one experiment between 10 and 98% of the pods set on the various racemes aborted before maturity (Dracup, 1994). Further adjustment in yield components to accommodate stresses and competition for assimilate during reproductive growth is through seed abortion, and in the above experiment up to 25% of the seeds in productive pods had aborted. Seed weights are comparatively stable, but those on the main stem and lower branch orders are favoured over those on the higher order branches (Dracup and Kirby, 1996b).

During pod and seed growth the nature of the pod wall and seed changes, and together with their colours these characters provide useful indicators of stages of development (Dracup and Kirby, 1996a,b). In the early stages of pod growth there is marked thickening and lengthening of the walls and then septa (cross-walls) develop between the seeds. The thick walls serve as a reservoir of nutrients for the seeds, reaching maximum weight before the seeds are even half filled. Once the walls have reached maximum weight their colour changes from green to khaki through to pale brown, while their weight decreases and that of the seeds continues to increase as assimilate passes from the walls to the seeds.

Similarly, the seeds remain green until they reach maximum weight (physiological maturity), whereupon the cotyledons turn yellow. At physiological maturity seeds contain about 62% water and thereafter lose water rapidly and are generally harvested at below 12% water.

Interactions with the Environment

The lupin growing season is defined mainly by the beginning and end of reliable rains, generally lasting between April–June and September–December in southern Australian environments. Early sowing is probably the most critical management factor in lupin cultivation (Perry and Poole, 1975; Perry *et al.*, 1986) and to achieve this, dry sowing is commonly practised (see below). Very early sowing introduces an element of risk of water deficits early in the crop's life cycle, when many plant organs are being initiated, and so early water deficit could have important consequences to plant structure later on (see below). Frosts can occur during early spring and reproductive tissues are most susceptible to freezing damage, although the wide window of flowering in lupin provides much opportunity for the plant to compensate for frost damage (Chapter 8, this volume). During spring, temperatures rise, rainfall decreases, vapour pressure deficits rise (Fig. 10.1) and soil moisture becomes exhausted; this is often referred to as terminal drought. High temperatures (Chapter 8, this volume) are not likely to occur early enough in spring to have a major impact on flowers on the main yield-bearing stems, but there is a strong chance of seed filling being affected. Temperatures inside pods exposed to direct solar radiation are about 5°C higher than in the surrounding air (Reader and Dracup, 1994); and when combined with water deficit and the resultant senescence of leaves such temperatures are likely to reduce seed filling.

The timing and the rate of development of the terminal drought have major consequences to lupin productivity. The developing water deficit speeds up reproductive growth and crop maturation (Dracup and Kirby, 1996b) and reduces vegetative growth. Crops on shallow soils with little available soil moisture are particularly sensitive. M. Dracup and M. Reader (Perth, 1996, personal communication) found that delaying the terminal drought by just two weeks improved harvest index from 14 to 20%.

Phenology and Plant Improvement

Domestication and breeding improvement of narrow-leafed lupins in Western Australia has required changes in several developmental and structural features, improving the phenological adaptation of the crop and its agronomic management. Flowering time is foremost among these. Flowering needs to occur when the risk of both frost and high temperature is low, and when there is sufficient capacity to meet the demands of growing pods but also still adequate time for pod filling before the end of the season. Early flowering was achieved by removing the vernalization requirement using the dominant gene *Ku* (Gladstones and Hill, 1969); this hastened flowering by about four weeks, reduced the number of main-stem nodes by about 11 (Perry and Poole, 1975) and extended the adaptation of lupin from the south-west to the whole of the grain-growing belt of Western Australia.

Improvements in phenology and assimilate partitioning have been (e.g. Hamblin *et al.*, 1986; Gladstones, 1994) and are being sought through restricting the plant's branching, and the argument for this follows the generalized ideotype model for crop plants proposed by Donald and Hamblin (1983). Genes that restrict the branching of lupins are available and the character leads to shorter branches and fewer orders of branching (Dracup and Kirby, 1995), enhanced main-stem pod set (Delane *et al.*, 1986; Delane and Gladstones, 1988) and generally higher harvest indices (Delane *et al.*, 1988). The greater control this character offers over vegetative growth and the flowering period will allow better matching of genotypes to the environment by varying the degree of branching reduction (Hamblin *et al.*, 1986). Since branching is restricted, more assimilate flows to pods at an earlier stage of development rather than toward further branching, as in normal branching lupins. This seems a more appropriate plant type; partitioning to reproductive growth would be more efficient (and prevent highly vegetative yet low yielding crops), more reproductive growth would occur before the terminal drought and high temperatures (which should also improve yield stability), competition between reproductive and vegetative growth would be reduced (improving pod set), and there is little shading of organs, particularly the pods (if illuminated, pods can make a substantial contribution to their own filling; Pate *et al.*, 1977). It will be important to match appropriate agronomy to this plant type, particularly planting density (Hamblin *et al.*, 1986) and row spacing. Restricted branching plants will also probably confer other benefits to the cropping system; with all pods positioned at or close to the top of the canopy harvestability (see below) will be improved and the earlier maturity will give more flexibility to farming operations in terms of timing of sowing or harvesting, and weed control (see below) prior to sowing or via crop desiccation close to maturity.

Breeding since the release of cv. Unicrop in 1973 has been toward a more restricted growth habit, without consciously seeking 'restricted branching'. Both harvest and canopy heights have decreased by about 8 cm, and the proportion of yield on the main stem and first-order branches has risen (H. Tapscott, M. Dracup and W. Cowling, Perth, 1996, personal communication). A parallel improvement in harvest index from 24 to 29% has also occurred with little change in biomass (Tapscott *et al.*, 1994). In some environments, however, harvestability has now been compromised by the shortness, and so taller plants are being sought. The modern cultivars require higher population densities to realize their yield potential than the earlier cultivars (Cowling and Speijers, 1994). Breeding has also led to synchronous seed filling and maturation, despite large differences in the timing of pod set between the various branch orders (Dracup and Kirby, 1996b).

Agronomy

In this section we consider seed quality, time of sowing, seeding rate and depth of sowing, crop establishment technology, weed control and harvesting. The nutrition of lupins, disease and pest control and the interaction with environmental stresses and animal enterprises are the topics of other chapters of this volume and are considered here only in relation to other aspects of agronomy.

Seed quality

Parameters of seed quality for lupins include permeability, freedom from physical damage, nutrient content, seed size, germinability and freedom from seed-borne fungal or viral infection. Some of these issues are dealt with in other chapters and are discussed only briefly here.

Physical damage

The relatively large size of lupin seed compared with cereals, the flattened angular shape (*L. albus* and *L. luteus*) and the low moisture content at which the crop is harvested make lupin seed susceptible to damage during harvesting and in handling and transport. In samples subjected to physical damage, germination tests find high levels of 'abnormal' seedlings. These may have one or both cotyledons detached, or the shoot or radicle may be broken off above or below the point of attachment of the cotyledons to the axis. Damage of this type does not affect the seed coat and is not detectable by visual examination. Abnormal seedlings generally do not contribute viable plants to the stand and can be an important cause of low plant density in commercial crops. Good harvest technique for seed crops includes machine adjustment to ensure 'soft' threshing (low drum speeds and wide concave gaps) and harvesting at night when humidity is usually greater. Transfer of seed using mechanical augers may also cause physical damage and should be minimized.

Germination

Seed of commercial lupin cultivars has high permeability to water due to the presence of the gene *mollis* for soft seed (Gladstones, 1977) and in hand-harvested samples 99–100% germination can be expected. Seed dormancy appears very shortlived and seed harvested at physiological maturity will germinate immediately. Despite the high inherent germinability, germination testing is still essential to ensure that poor quality or damaged seed does not result in low stand density and thus limit yield. Producer samples submitted for the ISTA germination test have varied in germination from about 20 to 99% in any one year (D. Nicholas, Perth, 1996, personal communication). The germination of commercial seed lots submitted for testing has improved as producers have become aware of the importance of correct harvesting and handling procedures and is now consistently between 80 and 85% normal seedlings (Table 10.2). That only about 25% of commercial seed lots are germination tested in any one year remains a source of serious concern, and probably represents considerable loss to the industry.

Storage of narrow-leafed lupin seed with admixture of 10% green pod material from the cruciferous weed wild radish (*Raphanus raphanistrum*) has been shown to reduce germination from 100% to 25% in as little as one week (A.H. Cheam, Perth, 1996, personal communication), probably due to

Table 10.2. Average germination percentage, seed size and percentage infection with cucumber mosaic virus in seed samples submitted to the Western Australia Department of Agriculture seed testing laboratory and tested using standard ISTA procedures.

Season	Germination (%)	Seed size (g per 1000 seeds)	Range	CMV (% nil)	CMV (% > 0.5%)	Number of tests
1971–75	77.6	—				
1976–80	75.9	—				
1981–85	79.6	—				
1986–90	81.4	—				
1990	84.8	—		46	30	
1991	88.0	142		72	5	1063
1992	80.0	130	77–172	76	10	1263
1993	83.0	143			6	
1994	84.0	148		85	2	
1995	84.0	147		76	2	956

isothiocyanates released from the pod. Solutions involve selection of radish-free seed crops, swathing or dessication, and prompt seed cleaning before storage.

Seed vigour testing

Vigour tests have not been used for narrow-leafed lupin despite their use in soybean and other legume species. This may reflect the good harvesting and storage conditions for seed with a consequent general lack of concern for seed quality by Australian producers. Their primary concerns are with levels of damaged and cucumber mosaic virus-infected seed. Nevertheless, the large and still unexplained differences in field performance of seed lots matched for size and germination percentage, but from different geographic locations, begs the question as to whether further study of seed vigour is required. If a test were able to differentiate between seed lots, and the differences were translated into improved field establishment, yield would almost certainly also increase and adoption of a vigour test would be rapid.

Nutrient content

Low levels of manganese and phosphorus in the grain of narrow-leafed lupin have been shown to reduce emergence and growth, respectively. Low manganese seed (7–8 mg kg^{-1}) has been shown to reduce emergence and early growth of narrow-leafed lupin even in the presence of manganese fertilizer (Crosbie *et al.*, 1993, 1994; Chapter 5, this volume), although in field experiments Tapscott and Cowling (1995) found a small but significant negative correlation between seed manganese and shoot dry weight. Seed phosphorus concentrations of less than about 2.1 g kg^{-1} (Thomson *et al.*, 1992) and low seed cobalt (Robson and Mead, 1980) have also been shown to reduce growth of lupins in glasshouse culture, but have not appeared significant in commercial production. Nutritional effects are covered in more detail in Chapter 5.

Seed size

Seed size is an important facet of the agronomy of narrow-leafed lupin because of its effect on stand density, which is a significant determinant of seed yield. Commercial seeding machinery sows at a constant weight or constant volume of seed per unit area, and thus at a constant seeding rate (kg ha^{-1}) plant density will be greater if individual seed weight is low, and will be lower if individual seed weight is high. Thus, should seeding rates be reduced if small seed is sown (but still achieves the recommended density) and would size-grading and sowing of large seed be beneficial? These questions are addressed in some detail in the section on seeding rate.

Individual seed weight in cultivars of *L. angustifolius* varies from 40 to 240 mg (Gladstones and Crosbie, 1978) but current cultivars are all medium to large-seeded (120–160 mg). Season-to-season variation in average seed size is quite small (Table 10.2), but masks considerable variation in the weight of seed from individual crops. For example, seed of cv. Gungurru collected from 22 variety trials of diverse geographic origin, but sown with the same maternal seed source, varied in weight from 101 to 186 g per 1000 seeds; and from 1263 samples submitted for germination testing from the 1992 season, seed weight varied from 77 to 172 g per 1000 seeds (Table 10.2).

The effect of seed size on crop performance has been examined in some detail in narrow-leafed lupin. At constant plant density, smaller seed produced smaller plants, and was consistently lower yielding than larger seed (R.J. French, unpublished data). In other experiments using 22 seed lots of diverse origin, also sown to produce a constant plant density, there was no statistically significant effect of seed size (101 to 186 g per 1000 seeds) on any measured plant attribute or on grain yield (Tapscott and Abrecht, 1994a). The same conclusion was reached in a second experiment using four seed sources of 112, 124, 132 and 182 mg weight sown at four densities (Tapscott and Abrecht, 1994b), although in this experiment the weight of individual seed produced from the large seed source was 4% larger than that from the smallest lot. Tapscott and Cowling (1995) examined the performance of 97 farm-derived seed lots sown at 100 kg ha^{-1}, but without adjustment for germination or seed size (i.e. commercial practice). Seed size (mean 144 mg, range 120–168 mg) significantly influenced stand density and individual shoot dry weight measured at six weeks – as would be expected – but had no significant effect on grain yield.

Although there may be particular circumstances, such as deep sowing, where large seed may be advantageous, in general the increased stand obtained when small seed is sown appears to compensate for any effect of the smaller individual plants arising from that seed. Only when seed size was exceptionally small would replacement be warranted. Further, there is no convincing evidence that size grading and sowing large seed will consistently increase yield at normal stand density.

Seed source

Seed source is known to affect the growth of grain legumes and may arise from differences in the environment under which the seed (embryo) matures, differing nutritional content, the presence of viral or fungal diseases, or differences in germinability due to physical damage during handling of the seed.

Delane *et al.* (1989) found that plant establishment and grain yield of narrow-leafed lupin cv. Gungurru varied significantly amongst seed lots that had been carefully selected for uniform germination and graded to uniform seed size. Because of the potential commercial significance of this observation, Tapscott and Cowling (1995) compared 97 seed sources selected to represent wide geographic origin and related their field performance to measured physical and nutritional characteristics of the seed. Over four sites, plant density, seedling shoot dry weight and grain yield varied significantly amongst lots with average grain yield ranging from 951 to 1478 kg ha^{-1}. This range was more than double the least significant difference value ($P = 0.05$, 240 kg ha^{-1}), leading to the conclusion that it is highly unlikely that the range in yield occurred by chance.

Of factors measured on the seed before sowing (and thus of potential predictive value), germination percentage and percentage infection with CMV were the major predictors of grain yield, but still accounted for only 40% of the variance. Seed nutrient content (N, P, K, Ca, Mg, S, Co, Cu, Fe, Mn, Zn) had little effect except for a negative association of grain yield with seed Fe ($r = -0.26$) and a negative correlation of shoot dry weight with seed Mn ($r = -0.38$). No consistent regional effects on seedlot yield potential were apparent despite the large range of annual rainfall (250–750 mm) and variation in latitude (28–35°S) of the sources.

Seed obtained from experiments which used varying phosphorus rates and from time of sowing experiments was used by Tapscott and Abrecht (1994a) to compare seed source effects. Although differing in size, there were no effects of source on emergence, plant growth or yield.

As with seed size (which is itself a 'source' effect) there are undoubted effects of seed source which may be expressed in yield in particular circumstances. However, until some predictive measure can be found with a high correlation with performance in the field, it is of no practical value. Given the 60% of variance in yield unaccounted for by pre-sowing factors, 'vigour' testing warrants further attention. Until that time, selection of a healthy seed crop, careful harvesting and routine testing for germination and CMV are the agronomic requisites for selecting seed for the next season's sowing.

Seed-borne diseases

Brown spot (*Pleiochaeta setosa*), which is expressed as both a net-like leaf spot and a root rot (Sweetingham, 1996), may be seed borne in both narrow-leafed lupin and *L. albus*. Seed infection is rare in narrow-leafed lupin and infected seed is usually non-viable, but in *L. albus* seed-borne infection may give rise to infected seedlings (Sweetingham, 1993). Anthracnose (*Colleototrichium gloeosporoides*), potentially a serious disease of both species, is also seed borne. Given the recent indentification of highly virulent strains, careful quarantine surveillance and a highly responsible approach by industry will be required if the Australian industry is to remain free of the disease.

The viral diseases cucumber mosaic cucumovirus (CMV) and bean yellow mosaic potyvirus (BYMV) may be seed transmitted in narrow-leafed lupin and *L. albus* lupin, respectively (Jones, 1994). Seed transmission is particularly important in the mediterranean environments generally used for lupin cultivation because the seed infection acts as a 'bridge' for the virus over the dry summer period, when alternative annual hosts are absent. The importance of seed transmission for the persistence of the virus also means that detection and use of virus-free seeds stocks can usually eliminate the diseases in commercial production (Jones, 1994). Commercial testing of seed samples submitted for germination testing began in 1990. In that year 30% of samples had greater than 0.5% CMV infection, the level considered unsuitable for commercial use; however, wider awareness and the use of seed testing appears to have been highly effective in reducing the levels of infection (Table 10.2).

Seed inoculation

Lupins form a symbiosis with the root-nodule bacterium *Bradyrhizobium lupini*. This bacterium also nodulates serradella (*Ornithopus* spp.) and is closely related to the nodule bacteria of native legumes common to south-western Australia (Chatel and Rowland, 1982; Chapter 6, this volume). Inoculation was originally recommended where lupins had not been grown before, or if there had been an extended (4–5 year) gap between crops. 'Gum slurry' inoculation (Chatel, 1987) has been very successful for lupins. Commercial peat inoculant is mixed with a methyl cellulose adhesive and applied to the seed, which is then dried and sown. Batch methods suitable for small batches have been modified by producers to treat the large quantities of seed needed for broadacre sowing. These usually involve inoculation and mixing of seed in an auger as seed is transferred from storage to the seeding unit, and this has the added advantage that inoculated seed is not stored before sowing. Although the gum used to stick the bacteria to the seed coat offers some protection, inoculated seed should be sown as soon as possible and seed treatment with fungicides and insecticides avoided. Sowing with high rates of acidic superphosphate or with trace element superphosphate containing copper sulphate may also lead to nodulation failure.

It is now known that the lupin *Bradyrhizobium* is highly persistent in the soil (Chapter 6, this volume) and re-inoculation of lupins is rarely necessary. Thus, in practice there is little conflict between inoculation and disease control.

Time of sowing

Time of sowing was a topic of extensive research in the early phase of the narrow-leafed lupin industry in Australia. The indeterminate structure of the species and the relationship of this to yield potential and growing season length was explored by Perry and Poole (1975) and Perry (1975). They showed that yield potential was determined by the production of successive orders of inflorescences produced on axillary branches growing from the two to four nodes below the previous inflorescence. This contrasts with the near synchronous flowering on the tillers of a cereal plant. In southern Australia's mediterranean environments the length of growing season is determined almost entirely by the timing of the break of season and although yields of all crops are greater when the growing season is prolonged, lupin yields decline more rapidly than cereals as the sowing date is delayed and the effective growing season shortened. The sensitivity of narrow-leafed lupin to high temperature in spring (Chapter 8, this volume) is a further reason for early sowing.

'Early' sowing – sowing the narrow-leafed lupin crop at the first opportunity following season opening rains – has therefore been a critical management practice. Its importance is greatest where the season length is least, and in such environments dry sowing before the season break is widely practised.

Despite the emphasis on early sowing, narrow-leafed lupin can be sown too early in some parts of southern Australia. Farmers have enthuiastically adopted the early sowing message and will often sow on suprisingly small amounts of rain (10–15 mm). If this occurs in April there is a significant probability that it will be followed by several weeks without further rain, and the young crop will experience moderate to severe water deficit. This can result in patchy emergence (French and Dracup, 1992). In crops that do emerge fully, crop death can occur under severe conditions but more commonly plant development is delayed, the potential for the crop to branch is impaired, and yields are reduced (French and Abrecht, 1993). As a consequence, very early sown lupin crops (mid-April to early May) in the dry fringe on Western Australia's agricultural belt sometimes yield less than those sown slightly later (Anderson *et al.*, 1994). Improved sowing technology that places the seed more precisely and concentrates rainfall into the seeding furrows, as well as reducing evaporation from the seed bed after sowing, should reduce the risk of this happening, and some producers now claim improved establishment using no-till methods under marginal moisture conditions.

Cultivars of *L. cosentinni, L. albus* and *L. luteus* have greater main-stem pod set and are less dependent on the growth of lateral branches, and yield in these species may be less dependent on early sowing.

Seeding rate

> No matter how many experiments we conduct, farmers will always disagree with us about 'optimum' seeding rates
>
> (The voice of an experienced agronomist)

Seeding rate, or plant density, has indeed been one of the most controversial issues in the agronomy of narrow-leafed lupin. In summarizing the first five years of narrow-leafed lupin agronomy, Walton (1976) found that the plant density at maximum yield varied widely (23–64 plants m^{-2}) in five experiments, but in another six experiments seed yield was still increasing at the highest seeding rate established (generally 25–45 plants m^{-2}). There was no detectable interaction of cultivar with seeding rate, but in three experiments greater plant density compensated for the yield depression associated with later sowing.

Based on further experiments Walton (1979) summarized recommendations as the establishment of 35 plants m^{-2} in southern production areas and 55 plants m^{-2} in the shorter season northern production zone. For seed of average size (0.145 mg) and 80% germination, this translated into seeding rate recommendations of 63 and 100 kg ha^{-1} (Table 10.3).

These results led to a general recommendation to use 90 kg ha^{-1} of 80% germinable seed to ensure the establishment of a minimum 35–40 plants m^{-2}.

Table 10.3. Effect of seed size and seeding rate on established plant density, assuming 80% germination and no effect of density on plant survival.

	Seeding rate (kg ha^{-1})			
Seed size (g per 1000 seeds)	30	60	90	120
100	24	48	72	96
120	20	40	60	80
140	17	34	51	68
160	15	30	45	60
180	13	26	39	52

Few narrow-leafed lupin producers agreed. Cereals were commonly sown at 35–45 kg ha^{-1}; lower seeding rates save on storage, cleaning, fungicide and handling; and the less seed held for sowing, the more can be sold. Further, crops sown at 50 kg ha^{-1} and establishing 20–25 plants m^{-2} appeared to yield equally well. Plants at the lower density had more pods per plant – a significant observation for producers – and where double densities were sown on corners and headlands lupins often lodged with weak stems and disease. This observation was probably due to systemic phomopsis (*Phomopsis leptostromiformis*) infection following mid-season water deficit.

Re-interpretation of the accumulated information on seeding rate now allows many of the observations and experimental results to be reconciled. For a single experiment, the yield–density response can be fitted by an inverse polynomial function and an 'optimum' density can be defined where the return to increasing density is some small multiple of the cost. French *et al.* (1994) fitted functions to approximately 120 yield–density experiments conducted over many years, sites and cultivars and demonstrated that there was a positive linear relationship between yield potential (defined as site yield in the absence of any limitation by plant density) and the optimum plant population. In other words, the greater the yield potential of the site, the greater the economic optimum plant density (Fig. 10.2).

This dependence of optimum plant density on yield potential explains the effects of many other factors thought to influence optimum seeding rate, such as soil type, cultivar and sowing time (French *et al.*, 1994). The consequences are profound: seeding rate can, and should, be adjusted to the yield potential of the site – which may be below or well above the recommendation of 40–45 plants m^{-2}. In practice, location and soil type provide the only useful guide to yield potential (as most crops are sown before the season breaks), but even this probably encompasses a four-fold range in yield, and a theoretical variation in optimum seeding rate of 20–60 plants m^{-2}.

Differences between the experience of producer and researcher can also be explained by the interaction of seeding rate and site yield potential. Experimental work is almost invariably conducted on sites more uniform and of greater yield potential than commercial paddocks, leading inevitably to recommendation of densities greater than optimum for commercial production.

Even so, in addition to increased yield, there are many sound reasons for establishing a high plant density:

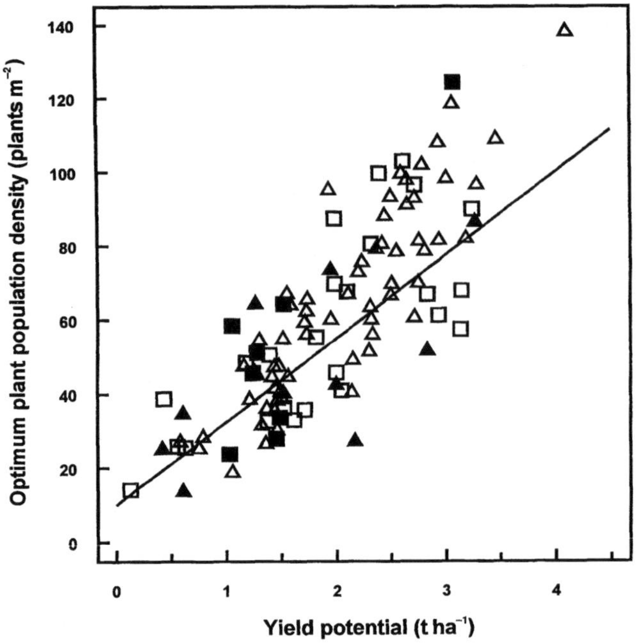

Fig. 10.2. Relationship between site yield potential (defined as the yield achieved in the absence of any limitation by plant density) and economic optimum plant density.

- Greater plant densities are competitive with weeds, and this may become of greater significance as less reliance is placed on herbicides for weed control in the future.
- On susceptible soils greater density will reduce sand-blast damage and wind erosion.
- Greater density reduces rain-splash of the spore of brown leaf spot (*Pleiochaeta setosa*) from the soil surface to the leaves.
- There is a greater safety margin for simazine damage caused by favourable seasonal conditions.
- Dense crops shade out stunted cucumber mosaic virus-infected seedlings (seed-borne infection), reducing sources of infection. Dense crops also appear less attractive to aphids.
- Harvesting is easier because dense crops are taller and more uniform and ripen evenly.
- Greater density at sowing will compensate for loss of plants due to brown spot and rhizoctonia, and insect damage to seedlings.
- Dry seeding and sowing on water-repellent soils are not conducive to good stand establishment, and targeting a high plant density is good insurance at minimal cost.

Seed size and seeding rate

The variation in seed size discussed in an earlier section and the weight, or more correctly volume, calibrated sowing machinery raise the question of whether seeding rate should be adjusted to establish a set optimum density regardless of seed size. In practical terms, 40 plants m^{-2} can be established from 60 kg ha^{-1} of seed weighing 120 mg, or from 80 kg ha^{-1} of seed weighing 160 mg. Over a 400 ha commercial crop this represents 8 t of seed that does not have to be stored or handled, and which could have been sold instead. This question was addressed by Tapscott and Abrecht (1994b) in an experiment using four seed sources of 112, 124, 132 and 182 mg weight sown at 20, 45, 70 and 100 plants m^{-2}. Plant density was the dominant factor in determining crop performance and stands established at 100 plants m^{-2} flowered earlier and were taller and higher yielding. In contrast, at similar stand density, larger seed did not yield more than small.

What emerges from this is that maintaining stand density is by far the most critical agronomic factor. If seed size is below average, stand density will automatically adjust upward provided the usual seeding rate is maintained. Potentially more serious, however, is the lower density and lower yield that may arise if producers do not adjust seeding rate upward for larger than normal seed size.

An interesting finding from the same experiment was that individual seed weight was 6% greater for seed harvested from the 100 plants m^{-2} stands compared with 20 m^{-2}. This contrasts with cereals where greater plant density invariably reduces individual seed size, and is probably due to differing proportions of seed from the main-stem inflorescence and the branches.

Cultivar and seeding rate

Cultivar also influences the plant response to density, and there is evidence that cultivars with a greater yield potential – new cultivars – respond to plant density in a way that increases the economically optimum seeding rate. Cowling and Speijers (1994) measured the yield of eight early-flowering cultivars of narrow-leafed lupin that formed a historical series. The cultivars were grown at seven plant densities and the response of yield to density was fitted using an asymptotic model. Comparing the earliest with the most modern cultivar, the estimated y_{max} had increased by 85% and the curvature of the relationship implied that at low density (20 plants m^{-2}) there was no difference in yield between cultivars, but at 70 plants m^{-2} the yield of the earliest cultivar was 83% of its y_{max}, compared with 64% for the most modern. These results have been repeated with even newer cultivars and a more generalized form of the density response (R.J. French and W.A. Cowling, Perth, 1995, unpublished data). The important implication of these findings is that at low density the benefits of breeding are not expressed, and that to benefit fully from the potential of new cultivars the plant density (seeding rate) must increase.

Row spacing

Because narrow-leafed lupin production was based on technology developed for cereals, crops were originally sown on conventional 180 mm row spacings. Research conducted in the early 1980s confirmed the yield advantage of narrow (180 mm or narrower) row spacing for cereals; results from experiments with narrow-leafed lupin were inconclusive, but on average showed a small decrease in yield. In these first experiments (Table 10.4) cereal stubble was removed by burning and superphosphate fertilizer was broadcast in order ensure that 'valid' comparisons could be made of 180 and 360 mm row spacings. It was later realized that comparisons of this nature which are divorced from the practical management of the narrow-leafed lupin crop have little value and may actually disadvantage wide (285–360 mm) row spacings when compared with narrow ones (Jarvis, 1992). This is because surface-spread superphosphate becomes progressively less effective in wide rows, and because retaining stubbles (considerably easier in wide than narrow rows) is essential in reducing brown leaf spot (*Pleiochaeta setosa*) induced by rain-splash of spores from the exposed soil surface.

Subsequent experiments in a systems context (Jarvis, 1992) suggest that narrow-leafed lupin yield is the same, or marginally greater, when sown in wide (270–380 mm) compared with narrow (180–190 m) row spacings (Table 10.4). Even in environments where neither phosphorus nor brown leaf spot limit production, there has been no disadvantage in wide rows (R. Jarvis, Perth, 1996, personal communication).

The significance of these results is that narrow-leafed lupin producers can take advantage of the substantial systems benefits of using wider than conventional row spacings without the risk of lower yield. The most important of these benefits are the improved ability to retain stubbles, which minimize the risks of wind erosion and brown leaf spot infection, and the use of banding to maximize the effectiveness of phosphate fertilizer without the risk of inducing phosphorus toxicity (see later discussion of phosphorus fertilizer placement).

Table 10.4. Narrow-leafed lupin yields at narrow and wide row spacings, achieved under differing agronomic practices.

			Mean yield (t ha^{-1})	
Year	Trials	Method	Narrow	Wide
1984–1989	10	20–200 kg ha^{-1} seed 10–20 kg ha^{-1} P topdressed Stubbles burned Pre-cultivated 180 mm vs. 360 mm rows	1.20	1.12 (–7%)
1991	10	100 kg ha^{-1} seed 20 kg ha^{-1} P banded Stubbles burned 180 mm vs. 360 mm rows All tines cultivating	1.28	1.33 (+4%)
1992	13	100 kg ha^{-1} seed 12.5–20 kg ha^{-1} P banded Stubbles retained 180–190 mm vs. 360–380 mm rows Seeding tines only cultivating	1.38	1.42 (+3%)

Response to seeding rate appears unaffected by varying row spacing and at constant seeding rate this implies a doubling of plant density within the row (at 90 kg ha^{-1} from approximately eight to 16 plants per linear metre). The greater inter-plant competition induced by this greater density of plants within the row appears to increase pod height at harvest (Jarvis, 1992), and may also contribute to the shading out of weakened cucumber mosaic virus-infected seedlings which may otherwise act as sources of infection within the crop. Disadvantages of the greater within-row density may be weaker individual stems which may break or lodge more readily, and greater harvest losses as plants are concentrated on certain sections of the harvester cutter bar.

As with many aspects of the agronomy of the crop, the effect and significance of row spacing is highly dependent on the environment. Research interest in wider row spacings arose from observations on the increased effectiveness of banded phosphate fertilizer on high fixing, low phosphorus status soils. These advantages, like those of retaining stubbles, are not universal, but for producers, systems that reduce the risk of crop failure and minimize risk of environmental hazards are increasing in importance.

Depth of sowing

Lupins are epigeal germinating and the large cotyledons must be forced through the soil to the surface, where they unfold. Emergence on sandy, non-crusting soils to which the species is well adapted is seldom a problem, but on fine-textured soils this may reduce emergence. Initial advice to growers was to 'sow shallow' at 1–2 cm (compared with 4–5 cm for cereals) to reduce the time to emergence and get the crop away quickly, but the implications of this advice for both herbicide damage and for disease control were not understood.

The root rot caused by the fungus *Pleiochaeta setosa* is caused by spores that collect at the soil surface in leaf litter infected with the aerial phase of the disease known as brown leaf spot. Cultivation mixes these spores through the top of the soil profile and usually results in a gradient of spores which decreases with depth. Sowing of the lupin seed close to the surface places the seed and the newly emerged radicle in a zone of high spore density and increases the incidence of disease. Sowing at a depth of 5 cm places the germinating seed under less disease pressure and reduces disease incidence. This is one part of a strategy to control brown leaf spot, which is discussed in Chapter 9 of this volume. Slightly deeper sowing also increases crop safety with the triazine herbicides, especially where the more soluble triazines are substituted for the relatively insoluble simazine.

Sowing technology

In seasons when break-of-season rainfall events are sufficient to wet the soil surface thoroughly, there are few problems with lupin establishment and differences between establishment systems are small. However, in those seasons where the break-of-season rainfall is ill defined and wetting of the soil surface is intermittent, the system used is often critical for successful establishment.

The evolution of establishment technology for the lupin crop reflects its particular ecological characteristics, the environments in which it became significant, and the farm machinery inventory when the crop was introduced. In the early phases of the narrow-leafed lupin industry, cereal-sowing machinery was also used for the lupin crop. Conventional wisdom, amongst both producers and professional advisers, was to aim for that 'holy grail' of agronomy: the clean, smooth, weed-free seedbed. This was best achieved with tined machines – combines – but their ability to handle stubble was very poor and so stubbles were burned; and to control brome grasses (*Bromus* spp.) the only option was to wait for germination and then to plough. On 'new' land, disc culti-trash machines were used where tree stumps prevented the use of tined implements, but the aim was the same: to prepare that perfect seedbed.

The result of these practices was near disaster. On smooth seedbeds and in the absence of cereal stubble, wind erosion devastated lupin crops year after year on the sandy-surfaced soils. Unlike cereals with their protected, below-ground apex, the narrow-leafed lupin was cut off and could not recover from sand blast, and if diuron or simazine had been applied, the wind-blasted areas could not be re-sown with cereals to avoid further damage. By the late 1970s wind erosion had brought the fledgling narrow-leafed lupin industry to the brink of disaster.

As the problem became increasingly clear, perceptive individuals – both producers and professional agronomists – were beginning to see solutions. On the extensive loose sandy-surfaced soils of the northern wheatbelt, the imperative had to be to retain cereal stubble to reduce wind erosion. Disc implements that could sow through stubble were available on many farms and at its most basic this system involved topdressing of superphosphate, broadcasting of lupin seed on to uncultivated ground, and then a single pass cultivation with a light disc plough. Use of the culti-trash to sow into cereal stubble was strongly advocated and slowly the paradigm of the aesthetically pleasing perfect seedbed was replaced by that of the rough, ridged, straw-covered paddock that provided the needed protection from wind erosion. The cost of controlling wind erosion by use of the culti-trash was poor control of seed depth, low plant establishment and herbicide damage as the discs moved soil and concentrated herbicide in bands. The solutions to these problems had to await further machinery development a decade later.

In the cooler central, eastern and southern areas, tined sowing systems predominated from the start. Disc seeders were uncommon, soils less fragile, the burning of stubbles common, and narrow-leafed lupin crops were often sown on previously pasture land. The tined implements gave more precise and uniform seed placement and applied phosphorus fertilizer with the seed, which was important where soil fixation of phosphorus made surface application inefficient. However, even this system contained the seeds of its own destruction – systems that worked well for the first one or two years developed severe brown spot infections where stubble was not present to reduce rain-splash of spores from soil to seedling; and even the advice to sow as shallow as possible exacerbated the problem by placing the seed in soil with a high concentration of spores (Chapter 9). Only subsequently was it realized that retaining stubble and precise placement of seed below the spore band were key factors in the control of brown spot.

With the benefit of two decades of experience, the machinery requirement for lupin sowing can be summarized (Riethmuller, 1994) as follows.

- Able to handle high levels (2–6 t ha^{-1}) of cereal stubble, both flattened and standing.
- Able to sow into uncultivated soil.
- Placement of seed at a precise depth, below the band of incorporated herbicide (and below the spore band where brown spot is a threat).
- Even incorporation of pre-sowing herbicides into the soil surface layer containing the majority of weed seeds.
- Able to place fertilizer below the seed in some circumstances.
- Work at speeds of 10–12 km h^{-1} to allow the sowing of large areas at critical times.

To achieve these objectives the subsequent history of machinery development has been that of incremental modification and refinement of both the disc and tine systems toward more consistent performance. The most important modifications are:

- Refinements to culti-trash implements to improve seed placement. Careful positioning of the sowing tube behind the rear disc journals can direct seed and fertilizer precisely into a closing furrow of soil.
- Preparation of cereal stubble by reduction in straw length to allow easy passage through seeding machinery.
- Improvement in stubble handling capacity of tined implements through greater frame clearance, wider row spacing, improved tine pattern, and the use of coulters in front of the tines.
- Cultivation below the seed depth, with or without the banding of phosphorus below the depth of seed placement.
- Development of rolling harrows attached behind the sowing machine to effect uniform mixing of pre-sowing simazine and in returning buried stubble to the surface.
- Use of press wheels in certain circumstances.

Although no single optimal lupin sowing machine has evolved, or is ever likely to do so, on current indications the most reliable and consistent system that meets the basic criteria (Riethmuller, 1994) is stubble management at harvest, plus sowing with a tined seeder modified for stubble handling, sowing on wide row spacings and using rotary harrows to mix the soil and maximize stubble cover.

Dry sowing

Sowing into dry soil before the season break has been widely adopted for narrow-leafed lupin. The primary reason has been to ensure that the crop exploits the greatest possible growing season, and adoption of the practice has been greatest in areas where the growing season is short. So critical has this become that as the break of season is delayed so a greater and greater proportion of the narrow-leafed lupin crop – perhaps as much as 80% – will be dry sown, as in years such as 1994. It has the added benefit of ensuring that a part of the enterprise's cropping programme can be undertaken before the time-critical cereal sowing period which follows the break of season, allowing improved utilization of both manpower and machinery.

The practice of dry sowing is an interesting example of the interaction between the characteristics of the species, the environment and technology. Dry sowing as normally practised involves the spraying of simazine on to uncultivated stubbles from a previous crop or pasture, and sowing without prior cultivation. Use of tined or disc implements for sowing incorporates and mixes the simazine in the topsoil, where it will be activated in the root zone of weed seeds when the soil profile is wetted. Lupins are epigeal germinating species and this system could only work on sandy surface soils that do not seal or crust on drying, as the plants would be unlikely to emerge on soils that crust as they dry.

The critical importance of retaining stubbles during dry sowing cannot be overestimated. Initially disc machinery was widely used; however, seed placement was poor and plant establishment often low due to death of seeds left close to or on the surface, or buried too deep to emerge. Improved disc design, the use of tines and the advent of reduced or even zero disturbance seed drills is providing solution to the problems of seed placement. Where simazine incorporation cannot be effectively done using the seed drill, this critical step is now undertaken by rolling chains and other devices which provide soil mixing, reduce surface ridging and bring buried stubble to the surface.

Stubble management in the lupin : cereal rotation

Although not all lupin crops will be sown into cereal stubbles, the significance of the rotation benefits have made the continuous lupin : wheat or lupin : wheat : wheat rotations very important. In these systems cereal stubbles must be retained for wind erosion control, for control of the disease brown spot, and increasingly to maximize the return of organic material to the soil. Treatment of stubbles may be at harvest of the cereal crop, or following harvest, or may be left until the lupin sowing. Each method has its advantages.

Stubble treatment at harvest

Tined sowing machinery has the advantages of more precise seed placement, the ability to separate seed and fertilizer placement, good mixing ability for herbicide incorporation and compatibility with cereal sowing. Its disadvantages are the problems caused by long straws blocking tines in the critical lupin sowing operation, but management of stubble at harvest to reduce straw length can be used to prepare stubbles for lupin sowing. The cereal crop may be cut low to leave only a short anchored stubble and then the harvested straw is chopped and spread from the harvester; or it may be cut high – the heads stripped – and the tall standing stubble is then re-cut using a second cutter bar attached to the harvester, or mown at some later time.

Straw length for tined machinery should be one-third to one-half the tine spacing for easy flow of stubble through the machine. Modifications to machinery to improve stubble flow include changing tine shape, increasing under-frame clearance, widening tine spacing (both laterally and front-to-back) and using coulters in front of tines to cut or move stubble from the tine.

Postharvest stubble treatment

Grazing, cultivation, mowing/slashing of stubbles and burning are methods of postharvest treatment and may be used in particular circumstances. Burning of stubbles is regarded as a highly undesirable practice for sustainability and environmental reasons, but the concentration of stubbles in windrows and the burning of the windrows alone, leaving anchored stubble over the majority of the area, is a compromise.

Stubble management at sowing

Disc implements have the ability to sow into high cereal stubble loads without blockage. However, precise seed placement, fertilizer separation and banding of previously applied herbicides and drying of the soil surface are often problems. An even spread of cereal stubble is important if discs are to be used, as harvester trails often concentrate weed seeds and reduce herbicide efficacy and may harbour insects (Riethmuller, 1994).

New no-till disc seeders are currently being tested for lupin establishment. These are sophisticated machines that generally have good stubble-handling ability and precise seed placement, but like all current lupin systems their use is heavily dependent on herbicides for weed control.

Furrow sowing

The coarse-textured sandy soils favoured for narrow-leafed lupin production are also susceptible to water repellence caused by organic materials of plant origin that form hydrophobic films on the surfaces of sand grains. Coarse-grained sands with the lowest clay content and lowest surface area are most at risk and there is some evidence that legume material contributes more to water repellence than material from cereals or grasses (McGhie and Posner, 1981). Repellency causes uneven infiltration of rainfall, and delayed and uneven germination; it may also leave large areas of the surface dry and susceptible to wind erosion. Dry, water-repellent surfaces may often overlie subsoils that are more uniformly wet and able to support plant growth, but which cannot be tapped by the plant because of the dry surface. It is most severe when break-of-season rainfall is low and fails to saturate the soil.

Spraying surfactants and the incorporation of clays with a high surface area can overcome repellency, but surface management that uses the non-wetting characteristic to concentrate rainfall into furrows is an innovative use of an environmental problem to improve lupin establishment. Furrows may be formed with wide-spaced (360 mm) tines, or with discs (in stubbles) set at about 30 degrees to the direction of travel. Numerous machinery specifications for forming furrows are possible, but the principle is the same: to sow seed at the base of a wide furrow that is used as a catchment for rainfall that does not infiltrate due to the water repellency of the soil forming the ridge top and sides. Water ponding in the base of the furrow above the seed infiltrates due to the increased surface pressure.

Results using the furrow system have been spectacular (Blackwell, 1993), as might be expected where large differences in time of germination and establishment are possible. In contrast to most sowing systems, where the aim is to produce a spatially uniform soil surface (good herbicide incorporation, uniform stubble cover), furrow systems deliberately create separate ridge and furrow environments. Potential problems include erosion risk of water moving downslope in furrows, concentration of herbicide in the furrows as a consequence of water and particulate movement from ridges, and movement of water and nutrients beyond the root zone in the preferred infiltration zone formed by the furrows.

Phosphorus in the lupin : wheat rotation

Narrow-leafed lupin's origin on largely sandy infertile soils suggests an adaptation to, or at least an ability to cope with, low fertility. Lupins are not known to form associations with vesicular-arbuscular mycorrhizae (Robson, 1986) although *L. albus* forms proteoid roots and narrow-leafed lupin has long and abundant root hairs, both of which may be adaptations to assist nutrient uptake. Both species also have a high capacity to release protons into the rhizosphere. The deep rooting habit (Hamblin and Hamblin, 1985) and the paucity of root length in the surface also suggest a species adapted to soils where nutrients are distributed throughout the soil profile.

In the cultivated crop, a quite different situation applies. Nutrients are applied to, or close to, the soil surface, and the crop may be grown on soils with properties markedly different from those present in the natural distribution. For phosphorus, the most important fertilizer applied with the narrow-leafed lupin crop, both of these have considerable significance and interact strongly with establishment agronomy of the crop.

Early investigation of the phosphorus requirement for narrow-leafed lupin showed a higher requirement than for wheat in solution culture (Asher and Loneragan, 1967), in pot experiments (Ozanne *et al.*, 1969), and in field trials to establish the relative response to phosphorus on phosphorus-deficient 'new land' (Bowden, 1979). Recommendations for phosphorus application were developed from these response trials based on the price ratio and the phosphorus status of the site (Bowden, 1979) and these have remained the basis of phosphorus fertilizer recommendations to the present (Nelson and Delane, 1991).

Farmers traditionally faced the choice of topdressing superphosphate in autumn, which allowed flexibility and saved time at seeding, or drilling the fertilizer with the seed. Early research on low-phosphorus 'new land' clearly indicated that drilling was about twice as effective as topdressing (Bowden, 1979). However, this average relative efficiency neglected the differing ability of soils to fix phosphorus by adsorption on to iron and aluminium oxides in the soil and subsequent research has shown that the placement of phosphorus relative to the distribution of roots (and thus the method of application) interacts strongly with soil type in determining the response to phosphorus.

The true significance of placement for phosphorus nutrition was only realized as narrow-leafed lupin production spread on to yellow earth soils with low phosphorus status and high phosphorus fixing capacity in the eastern wheatbelt of Western Australia. In these low rainfall environments, drying of the topsoil may further reduce the availability of phosphorus, particularly during the later part of the growing season. Banding phosphorus up to 9 cm below the seed increased the relative efficiency of phosphorus for wheat by about 20% (Jarvis and Bolland, 1990), but when the experiment was re-cropped to narrow-leafed lupin, the year-old superphosphate banded between 7 and 13 cm doubled the yield achieved where superphosphate had been drilled with the wheat seed at 3 cm depth. Banding at 5 cm was intermediate. Observations on these experiments showed that the greater the rate of phosphorus applied and the deeper the placement of the band, the longer the lupins remained green as they approached maturity.

These results stimulated enormous interest in 'deep banding' of phosphorus for lupin production and the results obtained now allow the construction of a model of lupin response to phosphorus that accounts for the effect of soil properties, seasonal effects, root distribution and the pattern of plant demand.

Because of its indeterminate growth pattern, vegetative growth of narrow-leafed lupin continues until growth is truncated by moisture deficit at the end of the season. Compared with wheat, the demand for phosphorus is not only greater (average phosphorus content of lupin dry matter 0.3% versus 0.2% for wheat), but also much of this demand occurs late in the growing season (J.W. Bowden, Perth, 1996, personal communication). This large demand for phosphorus occurs at a time when phosphorus in the surface soil may or may not be available, depending on the vagaries of end-of-season rainfall. If the surface soil is dry, phosphorus supplied by topdressing or drilling at shallow depth will be unavailable. However, given that water extraction occurs from the top of the profile downward, there is a greater probability of deep-banded phosphorus remaining available to the plant. On soils with a low fixing capacity, phosphorus will leach down the profile; and on such soils the method of applying phosphorus is unimportant. This explains the lack of response to deep placement on the extensive earthy sand of the northern wheatbelt (Bolland and Jarvis, 1996) and on coarse-textured duplex soils (R.J. Jarvis, Perth, 1996, personal communication).

Row spacing, phosphorus toxicity and deep placement

Germinating lupin seedlings are more sensitive to salt effects than those of cereals and application of as little as 10 kg P ha^{-1} with the seed can reduce stand density and yield due to phosphorus and/or salt toxicity (Bowden and Smith, 1984). Toxicity is greater the lower the fixing capacity of the soil and the drier the soil during germination. Deep placement of phosphorus has been shown to be an effective strategy to avoid toxicity (Jarvis and Bolland, 1990, 1991) even in the absence of other beneficial effects of placement below the seed.

Deep placement of phosphorus is also relevant to the practical application of wider row spacing. Increasing row spacing from the traditional 180 mm is one of the most effective ways of improving the stubble handling capability of tined machinery; however, if fertilizer is applied at the same rate per hectare, a doubling of row spacing to 360 mm implies a doubling of the concentration of phosphorus in the seeding row. Since under adverse circumstances as little as 200 kg superphosphate ha^{-1} (18 kg P ha^{-1}) drilled with the seed can cause phosphorus toxicity, normal maintenance rates of about 100 kg ha^{-1} applied in wide rows would risk frequent damage. Fertilizer placement in a band below the seed is effective in avoiding toxicity when seeders are modified to sow on wide row spacings.

Weed control

Development of weed control technology

The need to sow as early as possible, and the use of the lupin crop to effect grass weed control (and as an additional benefit the control of the root disease take-all) in a following cereal crop, has made the crop highly dependent on herbicide use and vulnerable should herbicide use be restricted for reasons of resistance or environmental concern.

The major weeds of lupins are those common to the farming systems of southern Australia: annual ryegrass (*Lolium rigidum*), brome grasses (*Bromus* spp.), wild oats (*Avena* spp.) and the broadleaf species, capeweed (*Arctotheca calendula*), doublegee (*Emex australis*), wild radish (*Raphanus raphanistrum*), wild turnip (*Brassica tournefortii*) and wild mustards (*Sisymbrium* spp.).

Narrow-leafed lupin was quickly shown to be susceptible to the hormone-based post-emergence broadleaf herbicides widely used in cereals. Promising weed control, with acceptable levels of crop damage, was reported with a range of pre-emergence soil-applied chemicals including alachlor, carbetamide, diallate, diuron, linuron, nitralin, pronamide, trifluralin, propazine and simazine in the fertile, high rainfall areas of southern Australia (Reeves and Lumb, 1974; Cuthbertson, 1976; Allen, 1977b). With the exception of the triazine herbicide, simazine, many of these had been used in the field pea industry – the only other legume crop of significance in southern Australia.

In contrast, in Western Australia's northern wheatbelt where the narrow-leafed lupin industry developed most rapidly in the early years from 1967 to 1975, Allen (1977a,b) found no response to herbicide application even on frequently cropped, weed-infested sites. This he attributed to the exceptionally low natural fertility of the soils which favoured the legume over the non-leguminous grasses and broadleaf weeds), the lack of annual ryegrass as a common weed, and the predominance of cruciferous species which presented serious harvesting problems but did not compete strongly with the lupin (Allen, 1977b). The first cultivars (Uniwhite and Uniharvest) were also tall, late-flowering types with an ability to shade out competitors. Herbicide damage was common (Allen, 1977b) and this was later attributed to the combination of post-sowing, pre-emergence herbicide application and shallow sowing of the lupin on these very coarse-textured soils.

Weed control problems began to emerge in lupins as producers adopted the lupin–wheat rotation and the fertility of the sandy soils began to improve, and problems were exacerbated when the first early-flowering cultivar, Unicrop, was released in 1973. Although substantially higher yielding it suffered from weed competition due to its shorter stature. Recommendations (Allen, 1977b) were still to apply simazine or diuron at a rate of 1 kg ha^{-1} post-seeding and, in order to avoid crop damage, to restrict the use of the more effective simazine to medium and heavy-textured soils.

Even in 1980 (Gilbey, 1980) recommendations were still to use diuron post-seeding as a safer product, although simazine, applied immediately after seeding to a level, weed-free soil, was acknowledged to give better weed control. Another strand in the story was attempts to develop trifluralin for control of annual ryegrass in narrow-leafed lupin. Even worse than the application of post-seeding diuron and simazine, trifluralin required incorporation into the well prepared seed bed, with all its consequent hazards of wind erosion.

These recommendations, although correct in themselves, were not compatible with the sustainability of the farming system and the recommendation to apply simazine post-seeding to 'a smooth, trash-free seedbed' led to wind erosion and widespread crop damage, which reached its zenith in 1979. However, research within the industry was discovering solutions and by the early 1980s (Gilbey, 1982, 1984) simazine application pre-sowing and incorporation by the seeding operation were in widespread use, making feasible the retention of stubbles required for control of wind erosion.

Herbicides

Gradually improving knowledge of the mode of action of simazine and the realization of the implications for its effective use have led to the pivotal role of the chemical in narrow-leafed lupin production today. It is particularly useful because of the broad spectrum of annual grasses and broadleaf weeds controlled (Table 10.5). Its use is conditioned by its low solubility in water (5 ppm) and poor absorption through leaves. Effective use of simazine thus depends on soil moisture to solubilize the chemical and allow its uptake by the roots of germinating weeds. Even incorporation of the chemical into the surface soil, where most weed seeds are germinating, is vital and this may be achieved by rainfall, if simazine is applied after sowing, or more commonly by mechanical incorporation as a part of the seeding operation where it is applied to the soil surface before seeding. Release of more simazine-tolerant narrow-leafed lupin cultivars, which has allowed the use of significantly higher rates of simazine, has been a further important development in weed control.

Even with the best technology, if soil moisture is limiting due to low or irregular rainfall, simazine activity and consequently weed control may be poor. Seventy per cent control of annual ryegrass and brome grass and of the staggered

Table 10.5. Weedy species controlled by the triazine herbicide simazine.

Grasses		Broadleaf weeds	
Common name	Scientific name	Common name	Scientific name
Annual ryegrass	*Lolium rigidium*	Capeweed	*Arctotheca calendula*
Brome grass	*Bromus* spp.	Annual clovers	*Trifolium* spp.
Barley grass	*Hordeum murinum*	Annual medics	*Medicago* spp.
Silver grass	*Vulpia* spp.	Docks	*Rumex* spp.
Wheat	*Triticum aestivum*	Doublegee	*Emex spinosa, E. australia*
Barley	*Hordeum vurgare*	Fumitory	*Fumaria* sp.
Oats	*Avena*	Canola	*Brassica napus*
Wild oats	*Avena* spp.	Wild radish	*Raphanus raphistrum*
		Wild turnip	*Brassica tournefortii*
		Wireweed	*Polygonum aviculare*

germinating cruciferous weeds is now accepted by the industry as reasonable and in high weed populations follow-up herbicides are commonly used (Table 10.6).

Conversely, if the soil is continuously moist, simazine activity will be high and weed control good. Excessive rainfall can leach simazine into the deeper lupin root zone and in wet seasons damage is often observed, but plants will only be killed if it is severe. A proportion of simazine may be replaced with the more soluble atrazine (50 ppm solubility) to increase herbicide activity, but at the risk of greater crop damage should environmental conditions favour high herbicide efficacy.

Despite pre-sowing application and incorporation during seeding now being conventional wisdom, it is interesting to speculate that the mixing of simazine into the soil during dry sowing is actually unnecessary and more uniform incorporation would be obtained by spraying immediately after, rather than immediately before, the soil disturbance involved in sowing. Opening rains would then incorporate and activate the simazine in the most spatially uniform manner possible. With this view, the mixing of simazine into the soil is most critical when lupins are seeded into wet soil where weed seeds may be germinating, but where post-seeding applied simazine could remain unactivated on the drying or dry surface until follow-up rainfall at some unknown future date.

Sprayseed® (paraquat/diquat) and glyphosate are widely used for control of summer-germinating weeds on areas to be dry sown to lupins. Where small germinated weeds are present at the time of sowing, tank-mixes of simazine with either Sprayseed® or glyphosate are often used with a delay of one to several days between application and sowing. On coarse-textured soils, the use of glyphosate immediately before sowing has occasionally increased the level of root and hypocotyl rot in lupins caused by *Rhizoctonia solani* (MacLeod, 1993), but not to the level required to modify commercial practice.

Selective in-crop weed control in lupins includes post-emergence application of triazine herbicides (simazine and metribuzin), the grass-selective Group A herbicides (aryloxyphenoxypropionates and cyclohexanediones, known colloquially as 'fops' and 'dims', respectively), and the use of the diflufenican (Group F) and metosulam (Group B) for weedy cruciferous species.

Simazine can also be used as a post-emergence 'top-up' herbicide to replace pre-planting simazine as it is degraded, and this 'top-up' maintains the herbicide's activity in the soil. Because of low leaf absorbtion, the simazine top-up strategy depends on rainfall for activation and root uptake and is less successful in drier areas. Metribuzin, which is more soluble and primarily leaf absorbed, may also be used post-emergence, but only on the tolerant cultivars Gungurru and Merrit. Atrazine is too damaging for post-emergence application. Mixtures of simazine or metribuzin and diflufenican also appear promising for control of broadleaf weeds in lupins, and metosulam can also be used for the control of the important weed wild radish (*Raphanus raphanistrum*), in tolerant narrow-leafed lupin cultivars. The development of herbicide resistance, considered in detail later in this chapter, emphasizes the need to consider the whole rotation in weed control and to develop integrated weed management systems that are flexible enough to cope with the wide range of possible rotations, and the impacts of seasonal

Table 10.6. Herbicides for use in lupin farming systems. Sources: Agriculture Western Australia, 1996 Weed Spraying Chart. Not all chemicals listed are registered for use on narrow-leafed lupin. The table is for information only, and must not be used for recommendations.

Stage	Chemical	Rate (ha^{-1})	Weed spectrum	Comment
Knock-down	Paraquat/diquat (Sprayseed®)	0.8–1.2 l	Non-selective	Tank mix with simazine if required
	Glyphosate (450 g l^{-1})	1.1–1.6 l	Non-selective	Tank mix with simazine if required
Pre-sowing	Simazine (500 g l^{-1})	1–2 l (3–4 l)	Annual grasses, broadleaf weeds (see text)	Severe crop damage on very coarse textures soils Rates vary from 1–4 l depending on soil type
	Simazine +atrazine		Annual grasses, broadleaf weeds (see text)	Part atrazine may be substituted for simazine; improved activity but greater risk of crop damage
	Trifluralin (400 g l^{-1})	1–2 l	Annual ryegrass, fumitory, wireweed	Incorporate before sowing, tank mix with simazine or diuron
	Tri-allate (400 g l^{-1})	2 l	Wild oats	Incorporate before sowing
Post-sowing, pre-emergence	Diuron (500 g l^{-1})	2.0 l	Capeweed, clovers, doublegee, annual ryegrass	Useful for doublegee, poor control of grasses Tank mix with trifluralin for annual ryegrass
Post-emergence – grasses	'Fops' (diclofop-methyl, fluazifop, haloxyfop, quizalofop, propaquizafop)	Various	Annual grasses, self-sown cereals	Potential for development of herbicide resistance
	'Dims' (sethoxydim, clethodim)	Various	Annual grasses, self-sown cereals	Potential for development of herbicide resistance
	Paraquat Gramoxone®	0.5–1.0 l	Annual ryegrass	Dessicant 'crop-top' for control of herbicide resistant ryegrass
Post-emergence – broadleaf	Simazine (500 g l^{-1})	0.5–1 l	Broadleaf weeds and annual grasses	Simazine 'top-up', half to full recommended rate
	Metribuzin (750 g kg^{-1})	75–150 g	Broadleaf weeds and annual grasses	Tolerant cultivars only
	Simazine/diflufenican	0.5–1 l + 0.1 l	Broadleaf weeds and annual grasses	
	Metribuzin/diflufenican	100–130 g + 0.05 l	Broadleaf weeds and annual grasses	
	Diflufenican (500 g l^{-1})	0.1–0.2 l	Capeweed, wild mustard, wild radish, wild turnip	
	Metosulam (714 g kg^{-1})	7–10 g	Wild radish	

conditions. In the future every possible management tool will be brought to bear on weed control as the agronomy of the crop evolves to cope with the emerging problem.

Control of lupins in other crops

Grown in rotation, lupins may volunteer as weeds in the cereal phase. Because they germinate readily, they are usually killed by the combination of knock-down herbicide and tillage used for the establishment of the cereal crop. Surviving plants may be removed from cereals by using broadleaf herbicides (2,4-D, MCPA, bromoxynil) or herbicides from the sulphonyl-urea group (chlorsulfuron, metsulfuron), the choice of product depending on the spectrum of other weeds that may need to be controlled (Pierce, 1995).

Harvesting

Delayed harvesting of narrow-leafed lupin results in losses from lodging, breakage of the peduncle resulting in the loss of the whole pod, and from shattering of the pod itself. Prompt harvesting is therefore critical and is usually carried out when the grain reaches about 12% moisture and the lower stems are still slightly green.

Lupins are generally harvested with machinery designed primarily for cereals. In contrast to the cereals, and despite the presence of two genes for reduced shattering, harvesting losses have remained at 5–10% of crop yield. At harvest, the relatively stiff and rigid structure of the plant and the impact of the harvester cutter bar, shaking the plant, combine to cause both loss of whole pods and pod shattering either at the knife or as the cut material is moved to the central auger of the harvester. Modifications to reduce harvest losses have included double-density knife guards which reduce shaking, extending the harvester platform, air blast assistance, modifications to the reel, and vibrating mats and platform sweeps to assist movement of material across the platform to the auger (Riethmuller and Blanchard, 1995). Improving the removal of material from the platform by modifications to the table auger has also reduced losses.

Swathing of narrow-leafed lupin has been investigated (Snowball *et al.*, 1986) and found to generally reduce harvest losses, though its value has been reduced by the improvements to direct harvesting machinery. In particular circumstances, should wild radish (*Raphanus raphanistrum*) remain green and delay harvesting, or if harvesting is to be delayed for a significant period, then swathing may be justified.

Small seeds and lupin pod material are usually deposited in header trails or spread behind the harvester. This material is of moderate to high digestibility and collection systems have been developed that will gather this material and weed seeds. The value of such systems depends on the structure of the enterprise and whether animals are available to use the material.

Seed lost on the ground at harvest is usually picked up by sheep grazing lupin stubbles, but despite the value of the seed to the sheep, this hardly justifies reducing the harvested grain and the flexibility to sell or feed in a controlled manner.

Farming Systems

Lupins have been cultivated for their seed for at least 3000 years in the Mediterranean Basin and the highlands of South America (Gladstones, 1970; Chapter 1, this volume). The earliest known reference to the use of lupins in agriculture is believed to be that by Theophrastus, who died in 287 BC, and from the earliest times lupins were recognized for their usefulness in soil improvement (Gladstones, 1970). Reeves (1984) reviewed the role of lupins in crop rotations and commented on the scarcity of modern references describing the role of lupins in farming systems and of their effects on soil fertility. This has been the subject of considerable study in the last decade and there is now a wealth of literature describing the impact of lupins in rotation with other crops – principally cereals.

Lupins will be included in rotations when the readily apparent benefits to the farming system are greater than those of alternative rotations (Table 10.7). The relative weighting of systems benefits and costs will vary with the physical and financial structure of the enterprise, but so important are the rotational interactions for narrow-leafed lupin that the industry would be placed in serious doubt should they be jeopardized by other changes in the farming system (Chapter 11, this volume). Many of the rotational issues listed in Table 10.7 are discussed in greater detail elsewhere in this volume: for example, nitrogen fixation (Chapter 6), animal nutrition (Chapters 13 and 14) and stubble value (Chapter 14). In the remainder of this chapter, we consider the evidence for increased cereal yield following lupins, and the major factors contributing to the rotation effects. Herbicide resistance of weeds, and the acidification of soil profiles – the major disadvantages of lupins in the farming systems – are also discussed.

Table 10.7. System benefits and disbenefits determining the inclusion of lupins in farming systems in southern Australia.

Benefits	Costs
Increases the yield and quality of wheat grown in rotation	Low single year gross margin versus alternative crops (wheat, barley, canola)
• provides fixed nitrogen	High year-to-year variability in yield
• break crop for fungal disease	Rapid development of herbicide resistance
• opportunity of grass weed control	More rapid and deeper acidification of the soil
Improved soil structure and nutrient cycling	Increased water repellency of sandy soils
Diversifies sources of income from sale of high protein seed	
Seed can be used in integrated livestock production (pigs, sheep, cattle)	
High value stubbles complement ruminant livestock production	
Improved labour and machinery utilization in crop establishment and harvest	
Encourages the adoption of conservation tillage – reduced cultivation and retention of stubbles	

Effects of lupins on cereal yields

The most important and most widely acknowledged aspect of lupins in rotation is that there is a high probability of increased cereal yield or increased grain protein content in a following cereal crop.

In Australia, this was first demonstrated as a consequence of the use of *L. cosentinii* as a self-regenerating plant used for winter and summer grazing. The use of this naturalized species for soil fertility improvement and development of the sandplain soils was strongly advocated by the prominent farmer Sir Eric Smart (Smart, 1953) on the basis of his experience in the Mingenew district of Western Australia. Evidence of the increased wheat yield and grain protein that could be obtained from rotations including *L. cosentinii* (Table 10.8) had already been documented by Shier and Cullinane (1948).

Widespread cultivation of the narrow-leafed lupin in the 1970s and early 1980s produced a wealth of evidence of the beneficial effects of this species on wheat yield. Rowland *et al.* (1986) summarized the results of 130 comparisons of the yield of wheat or barley following either a previous wheat crop or lupin. From 58 different sites throughout southern Australia from 1973 to 1985, the yield of wheat following wheat averaged 1.29 t ha^{-1} and that of wheat following lupins 1.87 t ha^{-1}, an advantage of 45%.

Many of these results were for comparisons of wheat following wheat, but without added nitrogen. This provides a baseline for the effect of a lupin crop, but does not necessarily reflect farm practice, which is generally to maximize yield at an acceptable protein content. This was addressed by Rowland *et al.* (1988) who reported the results of 26 site–year combinations where wheat : wheat and narrow-leafed lupin : wheat rotations were compared using rates of nitrogen applied to the wheat crop grown in the second year. This allows the interaction of applied nitrogen and rotation history to be determined.

The responses to applied nitrogen differed between sites and years, but the response curves could be summarized as either convergent or parallel, or a combination of both types (Figure 10.3).

If soil nitrogen is the only, or predominant, limiting factor in the response of cereals to rotation with lupins, then it should be possible to replace the nitrogen supplied by the legume with artificial nitrogen. Provided the highest rates used exceed the point at which nitrogen becomes non-limiting in the continuous wheat situation, the response curves will converge, suggesting that residual nitrogen can explain all of the benefit of lupins. Only five of the comparisons resulted in convergence, and in these cases an average of 37 kg N ha^{-1} was

Table 10.8. Average wheat yield (kg ha^{-1}) 1936–1946 for four rotations at two sites in Western Australia, *L. cosentinii.*

Rotation	Chapman	Wongan Hills
Continuous wheat	485	480
Fallow–wheat	620	950
Volunteer pasture–fallow–wheat	730	1200
Lupin–lupin–fallow–wheat	1230	1390

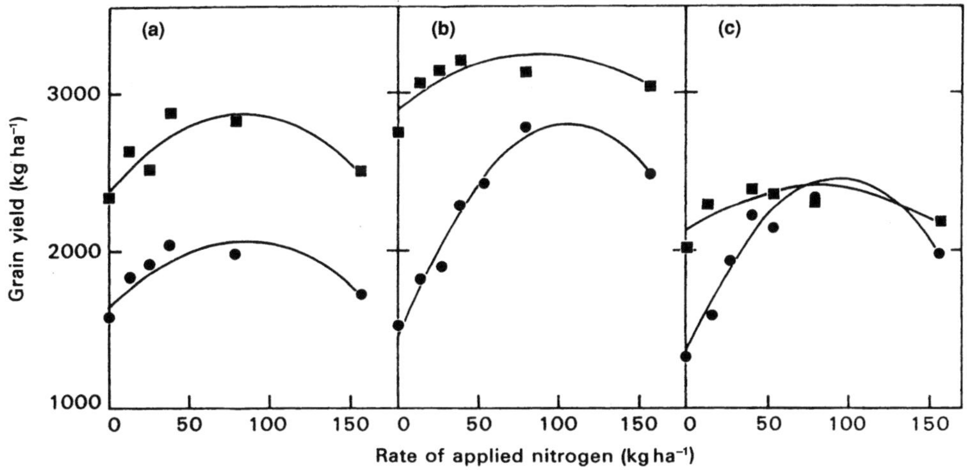

Fig. 10.3. Types of quadratic response curves of wheat yield to applied nitrogen in a wheat : wheat (●) or a wheat : lupin (■) rotation. (a) Parallel response curves: Bunjil, 1984. (b) Converging response curves which do not meet: Pingaring. 1984. (c) Convergent response curves: Pinjaring, 1983.

required to bring the wheat : wheat yield up to that of the narrow-leafed lupin : wheat rotation.

In 12 cases the response curves were parallel, indicating that some factor other than nitrogen was responsible for the yield benefit of narrow-leafed lupin, and in a further five cases the curves converged but did not meet, suggesting that nitrogen contributed to the benefit but other factors were also significant. Overall, narrow-leafed lupin increased wheat yield by 41% (0.85–1.20 t ha^{-1}) with no added nitrogen, and by 10% (1.11–1.22 t ha^{-1}) at the most profitable nitrogen rate. In updating their research (Rowland and Mason, 1994) to a total of 49 lupin : wheat rotation trials, in the absence of nitrogen fertilizer the lupin crop increased the yield of a following wheat crop by 75% (lupin : wheat 1.4 t ha^{-1} versus wheat : wheat 0.80 t ha^{-1}). On the acidic, N-infertile sandy soils of the Pilliga Scrub, Doyle *et al.* (1988) also demonstrated parallel, convergent and mixed responses to applied nitrogen. Wheat following lupins outyielded wheat following wheat by an average 57% in the absence of fertilizer nitrogen, and between 40 and 80 kg of fertilizer nitrogen ha^{-1} was required to raise the wheat : wheat yields to the levels of lupin : wheat.

It might be argued that such responses are to be expected on the acutely nitrogen deficient sandplain soils, but are less likely to occur where soil fertility is high. This appears not to be so. At Rutherglen in north-eastern Victoria, narrow-leafed lupin increased yield of a following wheat crop by 23% (3.16–3.90 t ha^{-1}) averaged over three years; and that of a second wheat crop following lupin by 16% compared with continuous wheat (Reeves *et al.*, 1984). On similar soils in the same region, de Boer *et al.* (1993) recorded a doubling of wheat yield following narrow-leafed lupin in each of three years. Marcellos (1984) and Strong *et al.* (1986a,b) also report substantial increases in wheat yield following lupins on relatively fertile soils.

Factors contributing to cereal yield improvement

Nitrogen

The large increases in cereal yield obtained following a single lupin crop on N-infertile sandy soils and the ability of artificial nitrogen fertilizer to mimic the response in some circumstances suggest that increased nitrogen supply is an important factor in the rotational response to lupins. Nitrogen fixation in lupins has been reported by Nutman (1976) as being in the range of 145–208 kg N ha^{-1} per year, with an average of 176 kg N ha^{-1} per year, and more recent estimates in Australia confirm that nitrogen fixation by narrow-leafed lupin may be substantial (Chapter 6, this volume). The proportion of fixed versus soil-derived N in the plant will depend on the level of soil fertility, and uptake of soil N may account for a major portion of N in the lupin plant (Gibson and Alston, 1986) where fertility is already high. In this circumstance, the removal of soil-derived nitrogen in the seed may even be greater than the return of fixed N, leading to a net reduction of total soil N.

Although lupins appear to fix large quantities of N, the net N benefit will depend on the proportion mobilized to the high protein grain which is usually harvested and removed. For narrow-leafed lupin, the amount of N in 1 tonne of seed is generally 50–60 kg (Schultz and French, 1978; White *et al.*, 1981; Reeves *et al.*, 1984) compared with 20 kg N t^{-1} of cereal grain. N remaining in crop residues will be the source of recycled N and Mason and Rowland (1992) showed that on a N-infertile site, narrow-leafed lupin produced more residue, of lower C/N ratio (33.2), than either oats or medic pasture.

Actual measurements of total soil N (TSN) after lupins and cereals are at best confusing. At three sites at Rutherglen in north-eastern Victoria, there was no significant difference in TSN (average 0.125%) between narrow-leafed lupin : wheat and continuous wheat rotations (Reeves *et al.*, 1984), although sampling procedures that removed large root material before analysis may have reduced the TSN measured in the lupin : wheat treatment. On a red-brown earth in South Australia, King (1984) found no difference in TSN (0.14%) or organic C between pasture : wheat, wheat : wheat or pulse (including narrow-leafed lupin) : wheat rotations; however, Schultz (1985) found that narrow-leafed lupin : wheat rotations could maintain TSN compared with declines in continuous wheat and wheat : fallow rotations.

On the infertile sandy soils common in Western Australia and averaged over eight sites and three years, wheat : narrow-leafed lupin rotations had an additional 0.0053% TSN compared with continuous wheat (I.C. Rowland, Perth, 1996, unpublished data) This may appear small, but the starting TSN at the eight sites was only 0.034% and the additional N in the top 10 cm was approximately 70 kg ha^{-1}. On a sandy surfaced duplex soil at Esperance, Western Australia, over a ten-year period (1974–1984), TSN in a continuous wheat rotation remained unchanged (0.06%) whilst that in a narrow-leafed lupin : wheat rotation increased at 0.0021% TSN per year for the top 10 cm. In contrast, TSN declined at the same rate under lupin : wheat and continuous wheat rotations over an eight-year period (Mason and Rowland, 1990) but wheat yields were substantially greater following lupins.

Thus, although there is little doubt that N cycled from the lupin is a major contributor to increased yield and protein content of a following cereal, the labile nature of the residues and the impact of environment are so great that changes in TSN have little value in predicting the response to rotation. Nevertheless, under low soil fertility, as on the extensive sandy soils of south-western Australia, N fixation is almost certainly supplying most of the narrow-leafed lupin N requirement and transient increases in soil N are sufficient to sustain a following cereal crop.

Rotation aspects of disease control

Take-all (*Gauemannomyces graminis* var. *tricici*) is the most serious root pathogen of cereals in southern Australia and carry-over of disease on the root material of grasses in ley pasture : cereal rotations is a major source of yield loss. Rotation with broadleaf crops, where take-all's grass weed hosts can be removed by herbicides, provides the opportunity to break the disease cycle.

Although there has been a long acceptance of lupins as a 'break' crop, evidence of specific effects in breaking disease cycles is more limited. For narrow-leafed lupin, take-all incidence in wheat was virtually eliminated by a single lupin year in the Rutherglen experiments (Reeves *et al.*, 1984), and pulse crops (including narrow-leafed lupins) reduced take-all infection to 2.5%, compared with 6% following continuous wheat and 18% following pasture (King, 1984). Adding a second break-crop using oats or a *Brassica* was no more effective in reducing incidence of the disease in these experiments. Even more dramatic effects of narrow-leafed lupin on reducing take-all were recorded by de Boer *et al.* (1993), where yields of wheat were more than doubled in each of three years of experimentation.

The break-crop effect of narrow-leafed lupin is primarily that of 'absence of the host', and the opportunity to control a wide range of annual grasses (including self-sown cereals) using both simazine and the 'fop' and 'dim' herbicides has been a widely recognized system benefit of the lupin : wheat rotation (Bayliss and Hamblin, 1986; Heenan *et al.*, 1990). Other mechanisms of disease suppression may also be operative as narrow-leafed lupin was more effective that field pea in reducing take-all inoculum in the field (Cotterill and Sivasithamparam, 1988). Kollmorgen *et al.* (1983) found that soil planted to rotation crops suppressed the fungus to a greater degree than fallow.

For other diseases, the incidence of *Rhizoctonia solani* was lower when wheat followed pulses compared with pasture or continuous wheat (King, 1984), and eyespot (*Pseudocercosporella*) was reduced in a narrow-leafed lupin : wheat rotation (de Boer *et al.*, 1993).

Rotation aspects of weed control

The rotation of narrow-leafed lupin, a dicotyledon, with a cereal opens the possibility of selective grass weed control in the lupin phase and effective broadleaf weed control in the cereal. This has been one of the most important (but poorly documented) aspects of the lupin : cereal rotation. By using the broad-spectrum grass-selective 'fop' and 'dim' herbicides in the lupin, producers have been able to reduce annual grasses to low levels, and then sow the following cereal crop into the lupin stubbles without the need to wait for germination of otherwise uncontrollable annual grasses. Grass control by cultivation or non-selective herbicide in the cereal has thus been transferred to the lupin year, allowing earlier sowing and increased productivity in the cereal.

Significance of rotational crops for grass weed control may become more important as pressures to burn stubbles and to reduce cultivation become greater. For example, Heenan *et al.* (1990) were able to use stubble burning and multiple cultivation to keep brome grass to low levels in continuous wheat, but where wheat was direct drilled and stubbles retained, the inclusion of lupin in the rotation was necessary to control the weed.

Effective broadleaf weed control can be achieved relatively easily in the cereal phase using a wide range of herbicides (Anon., 1996). This is less effective than for grasses because of dormancy, which results in germination being staggered over several years for wild radish, wild turnip and doublegee, the most important broadleaf weeds in lupins.

Soil structure and nutrient cycling

Lupins, grown both for seed production and as a green manure crop, are believed to improve soil structure and aid in soil erosion control (Bayliss and Hamblin, 1986). Changes in total soil N discussed earlier in this chapter suggest that including lupins in rotations can maintain or increase soil organic matter, probably as a consequence of increased productivity (total biomass) from the rotation as a whole. This increased organic matter may have marked effects on properties such as bulk density and water-stable aggregates on a fine-textured soil (Reeves *et al.*, 1984), but be of little significance on the coarse-textured soil where most narrow-leafed lupins are grown. Apart from increased inputs of organic matter, the extent of tillage in a rotation is likely to be most important in affecting soil structure, and in this respect the improved grass weed control in the lupin : wheat rotation has reduced the need for tillage in establishing the cereal crop.

Soil compaction (see also Chapter 8, this volume), the result of implement traffic over moist soil, is particularly severe on the coarse-textured soils widely used for lupin and cereal production (Henderson, 1989a). Growth of the strong tap root of narrow-leafed lupin appears relatively unaffected by the closer packing of soil particles and greater bulk density of compacted layers, compared with the thinner fibrous roots of cereals; and mechanical disruption of the compaction pan seldom improves the yield of lupins. Because of the expense of mechanical disruption of compacted layers, the use of tap-rooted species as a rotational 'biological plough' is attractive, and Henderson (1989b) demonstrated that lupins could reduce the impact of compaction on a following cereal crop.

Potassium cycling

As it is a mobile nutrient, leaching of potassium on coarse-textured soils may lead to potassium deficiency, especially in shallow-rooting species. Narrow-leafed lupin is well known for its ability to thrive on soils where potassium deficiency limits the growth of subterranean clover (Toms and Fitzpatrick, 1961), its deeper rooting habit enabling the lupin to gain access to a greater soil store of potassium.

Recycling of potassium from subsoils to the surface by lupins has been reported in Brazil and the former USSR (Bayliss and Hamblin, 1986); and on deep sandy soils in Western Australia lupins may also increase extractable potassium in the soil surface (Rowland, 1986; Loss *et al.*, 1993) (Table 10.9). Even small increases in potassium in the surface soil are surprising because narrow-leafed lupin have twice the concentration of potassium in the grain compared with wheat (0.85–0.95% vs 0.40–0.44% of grain dry matter, respectively). Thus, for comparable yields, lupins remove twice the amount of potassium for each crop. On soils where total potassium is low, the combined effects of greater removal in lupin grain and the cycling that makes potassium more available to the wheat crop would be expected to accelerate the onset of nutrient deficiency due to depletion of the soil reserves.

Herbicide resistance

For the narrow-leafed lupin industry, with its current high dependence on herbicides, resistance is a critical issue both in managing the now widespread herbicide-resistant annual ryegrass and in preventing the development of resistance in other weeds.

Herbicide resistance in annual grasses

Lupins have been used by producers as the sacrificial lamb on which to build the highly profitable, high yielding wheat crop. Before herbicide resistance became widespread, ryegrass in the lupin crop was challenged by simazine in the germination–establishment phase, and survivors then sprayed with a grass-selective Group A 'fop' or 'dim' herbicide to clean up often low levels of annual grasses. Use of the same (or related) Group A herbicides for control of annual ryegrass in the following wheat phase of the lupin : wheat rotation can now be seen as self-defeating and has led, inevitably, to the development of herbicide resistance in annual ryegrass (*Lolium* spp.) which will reduce, and may eventually eliminate, these chemicals from the production system.

Control of ryegrass through grazing and use of non-selective chemicals in pastures has a role in controlling resistant ryegrass, but at a substantial financial cost. Broadening the weed control options in the lupin phase of the lupin : wheat

Table 10.9. Effects of rotation on soil potassium level in the surface (0–10 cm) on deep sandy soils at five sites in Western Australia (K extracted in 0.1 M HCl)

	Rotation	
Site	Wheat–wheat	Wheat–lupin
1	18	26
2	34	42
3	21	26
4	20	48
5	15	24

rotation is the preferred alternative and in responding to resistance the system has changed to the following.

- Increase the rate of simazine – possible because of selection of cultivars with greater tolerance to triazines.
- Use dinotroaniline herbicides (trifluralin) where incorporation of chemical into wet soil is possible.
- Increase use of cultivation to control weeds when allowed by seasonal conditions.
- Increase the competitive ability of the crop through higher seeding rates.
- Attack the annual ryegrass at two critical points in its life cycle – using non-selective herbicides to desiccate the developing seed, and collection and removal of seed at harvest.

Mechanical seed capture at harvest can reduce ryegrass seed number by 60–80% (Gill, 1994). This can be very significant in reducing the annual ryegrass seed pool, but, in itself, is insufficient for effective weed control and needs to be integrated with other strategies.

Crop-topping (the use of the non-selective desiccant herbicide, paraquat) to prevent seed set in annual ryegrass is showing some promise in lupins. The technique depends on small differences in the timing of the life cycle of the ryegrass and lupin and the differential susceptibility of the grass and broadleaf species to paraquat. Timing is critical and must balance the time of maximum susceptibility of the ryegrass (the late flowering to milky dough stage) and the tolerance of the lupin to the desiccant. The method can be very effective in reducing ryegrass seed set (Gill, 1994) but is at the cost of up to 10% yield reduction in experiments, and perhaps 20–40% in commercial practice where timing is more difficult. This is likely to limit adoption of the technique unless more selective contact herbicides, such as glufosinate, become available.

The response to herbicide resistance in ryegrass is an instructive example of how an industry can respond to an emerging problem. Unlike the 'magic bullet' herbicides of the past, no one solution is likely to emerge, and even those that have emerged may only be interim solutions – for example, seed collection depends on ryegrass seed being held in the seed head at harvest. Collection will exert a selection pressure for ecotypes that may shed seed and reduce the effectiveness of collection systems.

The dependence of narrow-leafed lupin production systems on simazine must also be a cause for concern. Resistance to simazine has developed where high rates have been used in other applications and it is a matter of speculation that the widespread use of Group A herbicides following simazine has prevented the appearance of simazine-tolerant ryegrass by eliminating survivors. As resistance to the Group A chemical spreads, simazine-tolerant survivors will not be challenged by a second herbicide and resistance will emerge. Use of higher simazine rates, post-emergence simazine application and the use of simazine at very high rates in triazine-tolerant canolas will increase selection pressure even further, with the inevitable emergence of resistance to this herbicide.

Resistance in broadleaf weeds

Development of resistance in broadleaf weeds poses even greater threats to the lupin : wheat rotation. Resistance to Group B herbicides (the sulphonylureas, imidazolinones and sulphonamides) has already been reported in some cruciferous weeds, including probably the most serious weed of lupins: wild radish (*Raphanus raphanistrum*). Although these herbicides (with the exception of metosulam) are not used in the lupin phase, they are widely used in cereals, and resistance, once developed, will increase the weed burdens and weed control problems in the lupin crop.

In an innovative approach to weed control, genetically engineered resistance to the previously non-selective herbicides glufosinate and glyphosate are being introduced into lupins (Hamblin and Atkins, 1995; Chapter 3, this volume). Success will add new tools to the weed control armoury and thus herbicide resistance will be a catalyst for ongoing change in the agronomy of the crop.

Soil acidification

Acidification of soils used for agriculture has been well documented in Australia (Williams, 1980) and appears to be an inevitable consequence of farming systems that include legumes, the use of artificial nitrogen on cereals, and the removal of relatively large amounts of grain with a high ash alkalinity. In managing acidification, it is important to know whether narrow-leafed lupin increases or decreases the rate of acidification relative to other rotations, and whether, given the deep rooting habit, acidification occurs at greater depth and is therefore more difficult to ameliorate.

Lupin appears to acidify the soil more rapidly than pasture : wheat or continuous wheat rotations, on acidic red-brown earth soils in south-eastern Australia (Coventry and Slattery, 1991) and on acidic coarse-textured soils in south-western Australia (Loss *et al.*, 1993; Dolling, 1995). The reasons for the more rapid acidification are probably the greater N fixation by lupin compared with legume pastures and consequent greater potential for leaching of nitrate as the fixed organic N is oxidized in the soil. Lupins also absorb less NO_3^- and excrete organic anions that may not increase soil pH when uptake of anions exceeds that of cations (Loss *et al.*, 1994).

All three studies also suggest that subsoil acidification is greater where lupins are included in the rotation. The cycling of potassium from the subsoil to the surface under narrow-leafed lupin noted by Rowland *et al.* (1986) and Loss *et al.* (1993) may provide the mechanism as excess cation uptake at depth is balanced by extrusion of H^+ at a shallower depth in the soil profile.

Non-wetting properties

The coarse-textured soils favoured for lupin production are susceptible to development of non-wetting properties due to the addition of plant organic matter. Annual pasture legumes, in particular subterranean clover, produced highly water-repellent mixtures, whilst addition of cereal residues did not (McGhie and Posner, 1981). Addition of ground material of *L. cosentinii* to sands did not induce water repellence in the study of McGhie and Posner (1981);

however, Loss *et al.* (1993) recorded high water repellency induced by 25 years of regenerating *L. cosentinii* growing on a coarse-textured siliceous sand.

Although water repellence must be regarded as a potential environmental hazard to lupin production on susceptible soils, it is more likely to arise following legume pastures than as a consequence of the lupin : cereal rotation. Both amelioration and management of water-repellent soils are possible (McGhie, 1979; Blackwell, 1993) and it is unlikely to present a significant hazard to the lupin industry.

Challenges for the Future

The emergence of the narrow-leafed lupin industry is undoubtedly one of the success stories of Australian agriculture – in the pioneering domestication of a new species, in the research that assisted its integration into farming systems, and in the rapid adoption and adaptation of technology by the farming community.

Success, however, often sows the seeds of its own destruction and this is the case for the narrow-leafed lupin. The unique ability of the species to thrive on deep sandy soils and to increase their productivity has seen the growth of a new farming system and a new industry; but the increase in fertility has, as consequences, increases in weed burdens, herbicide resistance, more rapid acidification and the development of water repellence.

While the benefits of lupin production have been captured in the short term – notably the improvement in yield and protein of the following cereal – the major system disadvantages have been insidious and of a longer-term nature. The challenge now for the lupin industry is to increase the economic value of the crop in its own right while maintaining its contribution to the lupin–cereal rotation. This imperative reflects the relatively low cash returns for narrow-leafed lupin grain, and the significance attributed by producers to rotational benefits for wheat production. Should either be threatened, the narrow-leafed lupin industry would be extremely vulnerable to substitution by other crops, or the return of pasture-based systems should economics allow.

There can be no doubt that the greatest threat, and the greatest challenge, to the lupin industry is increasing herbicide resistance among the various weeds. Lupins are grown in large measure as the sacrificial lamb for the highly profitable wheat crop that may follow. The dependence on herbicides may be illustrated by a typical spray regime: simazine, pre-sowing, at perhaps double the rates used at the start of the industry, possibly a simazine top-up at four weeks, an application of a 'dim' or 'fop' grass-selective herbicide to clean up ryegrass, and finally diflufenican or metosulam to control the wild radish that escaped the simazine. The result is a healthy, weed-free lupin crop, suppression of take-all where present, and the opportunity to maximize the yield potential of the rotational wheat crop by sowing early, into a low-weed, low-disease, high N-status seedbed. It is also the recipe for herbicide resistance.

Resistance will undoubtedly change the agronomy of the crop. Old herbicides used in new ways, cultivation for weed control, seed collection, crop-topping and genetically engineered resistance to exotic herbicides will all contribute to maintaining the value of the lupin in the farming system. Where these are too costly, or are ineffective, the length of the rotation will inevitably extend to include a pasture phase where weed management by a combination of grazing and non-selective chemicals can control but not eliminate resistant weeds. The development of serradellas (*Ornithopus* spp.) as adapted pasture species for the very soils where narrow-leafed lupin has been the only adapted legume will be a key part of this technology.

Visionaries might say that this is the next step in the evolution of a truly sustainable farming system – one that started with the clearing of native vegetation in the 1950s, a cereal monoculture in the 1960s that brought the farming system to the brink of collapse, the lupin–wheat system that after two decades has sown the seeds of its own demise, and a future of even wider rotations including legume pastures and other crops, but also wheat and its partner: narrow-leafed lupin.

Acknowledgements

The authors gratefully acknowledge the assistance of Dr J.W. Bowden, Dr D. Bowran, Mr D. Gilbey and Dr J. Hamblin for their comments and suggestions on the manuscript.

References

Allen, J.M. (1977a) Weeds in grain lupins. 1. The effect of weeds on grain–lupin crops. *Australian Journal of Experimental Agriculture and Animal Husbandry* 17, 112–117.

Allen, J.M. (1977b) Weeds in grain lupins. 2. The response of narrow-leafed lupins to pre-emergence herbicides. *Australian Journal of Experimental Agriculture and Animal Husbandry* 17, 118–125.

Anderson, W.K., Young, K.J., Seymour, M., French, R.J., Shackley, B., Abbotts, R., Hoyle, F., Kerr, N. and Loss, S.P. (1994) *Flowering Responses of Crops in Western Australia*. Technical Report No. 65, Western Australian Department of Agriculture, Division of Plant Industries.

Anon. (1996) *Cereal, Pulse and Oilseed Weed Spraying Charts*. Bulletin 4323, Agriculture Western Australia, Perth.

Asher, C.J. and Loneragan, J.F. (1967) Responses of plants to phosphorus concentration in solution culture. *Soil Science* 103, 225–233.

Bayliss, J.M. and Hamblin, J. (1986) Lupins in the farming system: the results of a survey on world lupin production in 1985. *Proceedings of the 4th International Lupin Conference*, Department of Agriculture, Western Australia pp. 161–172.

Blackwell, P. (1993) Improving sustainable production from water repellent sands. *Journal of Agriculture, Western Australia* 34, 160–167.

Blackwell, P., Morrow, G. and Nicholson, D. (1994) Wide furrow sowing for improved lupin production on water repellent sands in the West Midlands of Western Australia. In: Dracup, M.N. and Palta, J. (eds) *Proceedings of the First Australian Lupin Technical Symposium*. Western Australian Department of Agriculture, Perth, Western Australia, pp. 221–224.

Bolland, M.D.A. and Jarvis, R.J. (1996) Effectiveness of different methods of applying superphosphate for lupin grown on sandplain soils. *Australian Journal of Experimental Agriculture* 36, 707–715.

Bowden, W.J. (1979) Phosphorus and lupins. In: Poole, M.L. (ed.) *Proceedings of the Lupin Research Review Workshop.* Department of Agriculture, Perth, Western Australia, pp. 633–367.

Bowden, W.J. and Smith, J.A.C. (1984) *Soil Test and Superphosphate Rate.* Farmnote No. 92/84, Department of Agriculture, Perth, Western Australia.

Chatel, D.L. (1987) *Techniques for Inoculation and Lime Pelleting.* Farmnote 6/87, Western Australian Department of Agriculture, Perth, Western Australia.

Chatel, D.L. and Rowland, I.C. (1982) Nitrogen fixation and inoculation by lupins. *Journal of Agriculture, Western Australia* 23, 92–95.

Cheam, A.H. (1986) Seed production and seed dormancy in wild radish and some possibilities for improving control. *Weed Research* 26, 405–413.

Cheam, A.H., Ralph, C., Hamblin, J. and Nelson, P. (1987) The effect of pasture and crop rotations on great brome populations and seed reserves. In: Madin, R.W. and Dodd, J. (eds) *Proceedings of the 7th Australian Weeds Conference.* Australian Weed Society, Sydney, p. 356.

Cotterill, P.J. and Sivasithamparam, K. (1988) Reduction of take-all inoculum by rotation with lupins, oats and field peas. *Journal of Phytopathology* 121, 125–134.

Coventry, D.R. and Slattery, W.J. (1991) Acidification of soil associated with lupins grown in a crop rotation in north-eastern Victoria. *Australian Journal of Agricultural Research* 42, 391–397.

Cowling, W.A. and Speijers, E.J. (1994) Grain yield improvement in narrow-leafed lupin cultivars from 1973 to 1991 as predicted from plant density experiments. In: Dracup, M. and Palta, J. (eds) *Proceedings of the First Australian Lupin Technical Symposium.* Western Australian Department of Agriculture, Perth, Western Australia, p. 262.

Crosbie, J., Longnecker, N.E., Davies, F. and Robson, A.D. (1993) Effects of seed manganese concentration on lupin emergence. *Plant and Soil* 155/156, 449–452.

Crosbie, J., Longnecker, N.E. and Robson, A.D. (1994) Seed manganese affects the early growth of lupins in manganese deficient conditions. *Australian Journal of Agricultural Research* 45, 1469–1482.

Cuthbertson, E.G. (1976) Chemical weed control in lupins. *Australian Journal of Experimental Agriculture and Animal Husbandry* 16, 394–401.

de Boer, R.F., Steed, G.R., Kollmorgen, J.F. and Macauley, B.J. (1993) Effects of rotation, stubble retention, and cultivation on take-all and eyespot of wheat in northeastern Victoria, Australia. *Soil and Tillage Research* 25, 263–280.

Delane, R. and Gladstones, J. (1988) Improving lupin pod setting and yield. *Western Australian Journal of Agriculture* 29, 83–87.

Delane, R.J., Hamblin, J. and Gladstones, J. (1986) Reduced-branching lupins. *Western Australian Journal of Agriculture* 27, 47–48.

Delane, R.J., Matthews, J.C. and Hamblin, J. (1988) Effect of genotype and environment on harvest index of *L. angustifolius.* In: *Proceedings of the 5th International Lupin Conference* p. 527.

Delane, R., French, R.F. and McCarthy, K. (1989) Effect of seed size on lupin establishment and yield. In: Ayling, G.P. (ed.) *Proceedings of the Fifth Australian Agronomy Conference.* The Australian Society of Agronomy, Perth, p. 495.

Delroy, N.D. and Bowden, J.W. (1986) Effect of deep ripping, the previous crop, and applied nitrogen on the growth and yield of a wheat crop. *Australian Journal of Experimental Agriculture* 26, 469–479.

Dolling, P.J. (1995) Effect of lupins and location on soil acidification rates. *Australian Journal of Experimental Agriculture* 35, 753–763.

Donald, C.M. and Hamblin, J. (1983) The convergent evolution of annual seed crops in agriculture. *Advances in Agronomy* 36, 97–143.

Doyle, A.D., Moore, K.J. and Herridge, D.F. (1988) The narrow-leafed lupin (*Lupinus angustifolius* L.) as a nitrogen fixing rotation crop for cereal production. III. Residual effects of lupins on subsequent cereal crops. *Australian Journal of Agricultural Research* 39, 1029–1037.

Dracup, M. (1994) Physiological aspects to lupin improvement for water-limited mediterranean environments. In: Dracup, M. and Palta, J. (eds) *Proceedings of the First Australian Lupin Technical Symposium.* Western Australian Department of Agriculture, Perth, Western Australia, pp. 140–151.

Dracup, M. and Kirby, E.J.M. (1993) Patterns of growth and development of leaves and internodes of narrow-leafed lupin. *Field Crops Research* 34, 209–225.

Dracup, M. and Kirby, E.J.M. (1995) Architectural stability in restricted branching narrow-leafed lupin. In: *Proceedings of the 2nd European Conference on Grain Legumes.* European Association for Grain Legume Research, Copenhagen, July 1995.

Dracup, M. and Kirby, E.J.M. (1996a) Pod and seed growth and development of narrow-leafed lupin in a water limited mediterranean environment. *Field Crops Research* 48, 209–222.

Dracup, M., and Kirby, E.J.M. (1996b) *Lupin Development Guide.* University of Western Australia Press, Perth, Western Australia.

Dracup, M., Davies, M. and Tapscott, H. (1993) Temperature and water requirements for germination and emergence of lupin. *Australian Journal of Experimental Agriculture* 33, 759–766.

Evans, J., O'Connor, G.E., Turner, G.L. and Bergersen, F.J. (1987a) Influence of mineral nitrogen on nitrogen fixation by lupin (*Lupinus angustifolius*) as assessed by ^{15}N isotope dilution methods. *Field Crops Research* 17, 109–120.

Evans, J., Turner, G.L., O'Connor, G.E. and Bergersen, F.J. (1987b) Nitrogen fixation and accretion of soil nitrogen by field grown lupins (*Lupinus angustifolius*). *Field Crops Research* 16, 309–322.

Evans, J., O'Connor, G.E., Turner, G.L., Coventry, D.R., Fettell, N., Mahoney, J., Armstrong, E.L. and Walsgott, D.N. (1989) N_2 fixation and its value to soil nitrogen increase in lupin, field pea and other legumes in south-eastern Australia. *Australian Journal of Agricultural Research* 40, 791–805.

Farrington, P. and Gladstones, J.S. (1974) Effect of genotype on yield of narrow-leafed lupin (*Lupinus angustifolius*) and sandplain lupin (*Lupinus cosentinii*). *Australian Journal of Experimental Agriculture and Animal Husbandry* 14, 742–748.

French, R.J. and Abrecht, D.G. (1993) Effects of early water deficit on narrow-leafed lupin in the field. *Abstract No. 17, 33rd Annual Meeting of the Australian Society of Plant Physiologists.* University of Western Australia, Perth.

French, R.J. and Dracup, M.N. (1992) Seedbed moisture is important in the establishment of lupin crops. *Abstract No. 36, 32rd Meeting of the Australian Society of Plant Physiologists.* La Trobe University, Melbourne.

French, R.J., Smart, W.L. and McCarthy, K. (1994) Optimum plant population densities for lupin (*Lupinus angustifolius* L.) in the wheatbelt of Western Australia. *Australian Journal of Experimental Agriculture* 34, 491–497.

Gardner, W.K. and McDonald, G.K. (1988) Responses by wheat to lupin, soil amelioration and fertilizer treatment in a solodised solonetz soil. *Journal of Experimental Agriculture* 28, 607–615.

Gardner, W.K. and Parbery, D.G. (1982) The aquisition of phosphorus by *Lupinus albus* L. 1. Some characteristics of the soil/root interface. *Plant and Soil* 68, 19–32.

Gibson, P.R. and Alston, A.M. (1986) Nitrogen balances in field-grown crops of *Lupinus angustifolius* L. In: Bacon, P.E., Evans, J., Storrier, R.R. and Taylor, A.C. (eds)

Proceedings of the Symposium on Nitrogen Cycling in Temperate Agricultural Systems, Wagga Wagga, pp. 44–50.

Gilbey, D.J. (1980) *Herbicides for lupins*. Farmnote 44/80, Department of Agriculture, Perth, Western Australia.

Gilbey, D.J. (1982) Weed research pays off. *Journal of Agriculture, Western Australia* 23, 81–82

Gilbey, D.J. (1984) *Lupin herbicide recommendations*. Farmnote 50/84, Department of Agriculture, Perth, Western Australia.

Gill, G. (1994) Integrated weed control in lupins. In: Dracup, M. and Palta, J. (eds) *Proceedings of the First Australian Lupin Technical Symposium*. Western Australian Department of Agriculture, Perth, Western Australia, pp. 324–326.

Gladstones, J.S. (1970) Lupins as crop plants. *Field Crop Abstracts* 23, 123–148.

Gladstones, J.S. (1977) *The Narrow-leafed Lupin in Western Australia*. Bulletin 3990, Western Australian Department of Agriculture, Perth, Western Australia.

Gladstones, J. (1994) An historical review of lupins in Australia. In: Dracup, M. and Palta, J. (eds) *Proceedings of the First Australian Lupin Technical Symposium*. Western Australian Department of Agriculture, Perth, Western Australia, pp. 1–38.

Gladstones, J.S. and Crosbie, G.B. (1978) *Lupin Wild Types Introduced into Western Australia to 1973*. Technical Bulletin No. 43. Western Australian Department of Agriculture, Perth, Western Australia.

Gladstones, J.S. and Hill, G.D. (1969) Selection for economic characters in *Lupinus angustifolius* and *L. digitatus*. 2. Time of flowering. *Australian Journal of Experimental Agriculture and Animal Husbandry* 9, 213–220.

Hamblin J. and Atkins, C.A. (1995) Genetically modified lupins: past, present and future. In: *The Role of Herbicide-Resistant Crops and Pastures in Australian Agriculture*. Australian Government Publishing Service, p. 294.

Hamblin, J. and Hamblin, A.P. (1985) Root characteristics of some temperate legume species and varieties on deep, free-draining entisols. *Australian Journal of Agricultural Research* 36, 63–72.

Hamblin, J., Delane, R., Bishop, A. and Gladstones, J. (1986) Yield potential of reduced-branching lupins (*Lupinus angustifolius*) on sandy soils in a short season environment. *Australian Journal of Agricultural Research* 37, 611–620.

Heenan, D.P. and Chan, K.Y. (1992) The long-term effects of rotation, tillage, and stubble management on soil mineral nitrogen supply to wheat. *Australian Journal of Soil Research* 30, 977–988.

Heenan, D.P., Taylor, A.C. and Leys, A.R. (1990) The influence of tillage, stubble management and crop rotation on the persistence of great brome (*Bromus diandrus* Roth). *Australian Journal of Experimental Agriculture* 30, 227–230.

Heenan, D.P., Taylor, C., Cullis, B.R. and Lill, W.J. (1994) Long term effects of rotation, tillage, and stubble management on wheat production in southern NSW. *Australian Journal of Agricultural Research* 45, 93–117.

Henderson, C.W.L. (1989a) Using a penetrometer to predict the effects of soil compaction on the growth and yield of wheat on uniform sandy soils. *Australian Journal of Agricultural Science* 40, 497–508.

Henderson, C.W.L. (1989b) Lupin as a biological plough: evidence for, and the effects on wheat growth and yield. *Australian Journal of Experimental Agriculture* 29, 99–102.

Herridge, D.F. and Doyle, A.D. (1988) The narrow-leafed lupin (*Lupinus angustifolius* L.) as a nitrogen fixing rotation crop for cereal production. II. Estimates of fixation by field grown crops. *Australian Journal of Agricultural Research* 39, 1017–1028.

Jarvis, R.J. (1992) Wider row spacing for lupins. *Journal of Agriculture, Western Australia*, 33, 8–9.

Jarvis, R.J. and Bolland, M.D.A. (1990) Placing superphosphate at different depths in the soil changes its effectiveness for wheat and lupin production. *Fertilizer Research* 22, 97–107.

Jarvis, R.J. and Bolland, M.D.A. (1991) Lupin grain yields and fertiliser effectiveness are increased by banding superphosphate below the seed. *Australian Journal of Experimental Agriculture* 31, 357–366.

Jones, R.A.C. (1994) *Virus Diseases of Lupins*. Bulletin 4294, Western Australian Department of Agriculture, Perth, Western Australia.

King, P.M. (1984) Crop and pasture rotations at Coonalpyn, South Australia: effect on soil-borne diseases, soil nitrogen and cereal production. *Australian Journal of Experimental Agriculture and Animal Husbandry* 24, 555–564.

Kollmorgen, J.F., Griffiths, J.B. and Walsgott, D.N. (1983) The effects of various crops on the survival and carry-over of the wheat take-all fungus (*Gaeumannomyces graminis* var. *tritici*. *Plant Pathology* 32, 73–77.

Loss, S.P., Ritchie, G.S.P. and Robson, A.D. (1993) Effect of lupins and pasture on soil acidification and fertility in Western Australia. *Australian Journal of Experimental Agriculture* 33, 457–464.

Loss, S.P., Robson, A.D. and Ritchie, G.S.P. (1994) Nutrient uptake and organic acid anion metabilism in lupins and peas supplied with nitrate. *Annals of Botany* 74, 69–74.

Macleod, W.J. (1993) *Herbicide interaction with hypocotyl rot of lupins*. Technote 9/93, Department of Agriculture, Perth, Western Australia.

Marcellos, H. (1984) Influence of prior crops of chickpea, fababeans and lupins on wheat. *Journal of the Australian Institute of Agricultural Science* 50, 111–113.

Mason, M.G. and Rowland, I.C. (1990) Nitrogen fertilizer response of wheat in lupin–wheat, subterranean clover–wheat and continuous wheat rotations. *Australian Journal of Experimental Agriculture* 30, 231–236.

Mason, M.G. and Rowland, I.C. (1992) Effect of amount and quality of previous crop residues on the nitrogen fertilizer response of a wheat crop. *Australian Journal of Experimental Agriculture* 32, 363–370.

McGhie, D.A. (1979) Managing water-repellent soils. Journal of Agriculture, Western Australia 20, 78–81.

McGhie, D.A. and Posner, A.M. (1981) The effect of plant top material on the water repellence of fired sands and water repellent soils. *Australian Journal of Soil Research* 32, 609–620.

Moore, J.K., Herridge, D.F., Starr, D.F. and Doyle, A.D. (1982) Evidence for disease control as a factor in improving wheat yield in a lupin–wheat rotation. In: Narman, M.J.T. (ed.) *Proceedings of the 2nd Australian Agronomy Conference*. Australian Society of Agronomy, Wagga Wagga, p. 327.

Nelson, P. (1983) *Weed Control in Narrow-leafed Lupin in Northern Agricutltural Areas*. Bulletin 4122, Western Australian Department of Agriculture, Perth, Western Australia.

Nelson, P. (1994) The development of the lupin industry in Western Australia and its role in farming systems. In: Neves-Muntins, J.M. and Beirao da Costa, M.L. (eds) *Proceedings of the 7th International Lupin Conference*, Instituto Superior da Agronomia – Technical University of Lisbon, Lisbon, pp. 565–576.

Nelson, P. and Delane, R.J. (1991) *Producing Lupins in Western Australia*. Bulletin 4179, Department of Agriculture, Western Australia.

Nulsen, R.A. and Baxter, I.N. (1982) The potential of agronomic manipulation for controlling salinity in Western Australia. *Journal of the Australian Institute of Agricultural Science* 48, 222–226.

Nutman, P.S. (1976) Alternative sources of nitrogen for crops. *Journal of the Royal Agricultural Society* 137, 86–94.

Ozanne, P.G., Keay, J. and Biddiscombe, E.F. (1969) The comparative applied phosphate requirements of eight annual pasture species. *Australian Journal of Agricultural Research* 20, 809–818.

Pate, J.S., Sharkey, P.J. and Atkins, C.A. (1977) Nutrition of a developing legume fruit – functional economy in terms of carbon, nitrogen and water. *Plant Physiology* 59, 506–510.

Perry, M.W. (1975) Field environment studies on lupins. II. The effect of time of planting on dry matter partition and yield components. *Australian Journal of Agricultural Research* 26, 809–818.

Perry, M.W. and Poole, M.L. (1975) Field environment studies on lupins. I. Developmental patterns in *Lupinus angustifolius* L., the effects of cultivar, site and planting time. *Australian Journal of Agricultural Research* 26, 81–91.

Perry, M.W., Delane, R.J., Tennant, D. and Hamblin, A.P. (1986) The growth and grain yield of lupins in relation to the soil water balance. *Proceedings of the 4th International Lupin Conference.* Western Australian Department of Agriculture, Perth, Western Australia, pp. 112–118.

Pierce, J.R. (1995) *Control of Peas and Lupins as Weeds in Cereals.* Farmnote 14/95, Western Australian Department of Agriculture, Perth, Western Australia.

Pigeaire, A., Delane, R., Seymour, M. and Atkins, C.A. (1992) Predominance of flowers and newly formed pods in reproductive abscission of *Lupinus angustifolius* L. *Australian Journal of Agricultural Research* 43, 1117–1129.

Reader, M. and Dracup, M. (1994) High temperatures in lupin pods could affect seed filling. In: Dracup, M. and Palta, J. (eds) *Proceedings of the First Australian Lupin Technical Symposium.* Western Australian Department of Agriculture, Perth, Western Australia, p. 294.

Reader, M., Dracup, M. and Kirby, E.J.M. (1995) Time to flowering in narrow-leafed lupin. *Australian Journal of Agricultural Research* 46, 1063–1077.

Reeves, T.G. (1984) Lupins in crop rotations. *Proceedings of the 3rd International Lupin Conference, La Rochelle, France*, pp. 208–226.

Reeves, T.G. and Lumb, J.M. (1974). Selective chemical control of annual ryegrass (*Lolium rigidum*) in oilseed rape, field peas and lupins. *Australian Journal of Experimental Agriculture and Animal Husbandry* 14, 771–776.

Reeves, T.G., Ellington, A. and Brooke, H.D. (1984) Effects of lupin–wheat rotations on soil fertility, crop disease and crop yields. *Australian Journal of Experimental Agriculture and Animal Husbandry* 24, 595–600.

Riethmuller, G. (1994) Machinery for successful lupin establishment and harvesting in Western Australia. In: Dracup, M.N. and Palta, J. (eds) *Proceedings of the First Australian Lupin Technical Symposium.* Western Australian Department of Agriculture, Perth, Western Australia, pp. 295–299.

Riethmuller, G. and Blanchard, E. (1995) *Lupin harvesting modifications.* Farmnote 19/95, Department of Agriculture, Perth, Western Australia.

Robson, A.D. (1986) The mineral nutrition of lupins. *Proceedings of the 4th International Lupin Conference.* Western Australian Department of Agriculture, Perth, Western Australia, pp. 84–95.

Robson, A.D. and Mead, G.R. (1980) Seed cobalt in *Lupinus angustifolius. Australian Journal of Agricultural Research* 31, 109–116.

Rowland, I.C. (1986) Potassium in lupin : wheat rotations. In: *Division of Plant Industries, Annual Report, 1985*, Vol. 1. Department of Agriculture, Western Australia, pp. 18–22.

Rowland, I.C. and Mason, M.G. (1994) The nitrogen response of wheat crops following lupins. In: Dracup, M.N. and Palta, J. (eds) *Proceedings of the First Australian Lupin Technical Symposium.* Western Australian Department of Agriculture, Perth, Western Australia, pp. 300–301.

Rowland, I.C., Mason, M.G., and Hamblin, J. (1986) Effects of lupins on soil fertility. *Proceedings of the 4th International Lupin Conference.* Western Australian Department of Agriculture, Perth, Western Australia, pp. 96–111.

Rowland, I.C., Mason, M.G. and Hamblin, J. (1988) Effect of lupins and wheat on the yield of subsequent wheat crops grown at several rates of applied nitrogen. *Australian Journal of Experimental Agriculture* 28, 91–97.

Schultz, J.E. (1985) *The Tarlee Rotation Trial. Effect of Rotations and Stubble Handling Methods on Soil Fertility and Crop Yields.* Sixth Progress Report, Northfield Research Laboratory, South Australian Department of Agriculture, Adelaide, pp. 40–64.

Schultz, J.E. and French, R.L. (1978) The mineral contents of cereals, grain legumes and oilseed crops in South Australia. *Australian Journal of Experimental Agriculture and Animal Husbandry* 18, 579–585.

Shier, F.L. and Cullinane, W.P. (1948) The influence of rotation on the yield and flour strength of wheat. *Journal of Agriculture, Western Australia* 25, 351–361.

Smart, E.F. (1953) Potentialities of the Western Australian sandplain country. *Transformation of Wastelands.* E.F. Smart, Mingenew, Western Australia.

Snowball, R., Nelson, P. and Rowe, D.L. (1986) *Swathing Lupins – A New Harvesting Method.* Farmnote 22/86, Department of Agriculture, Western Australia.

Strong, W.M., Harbison, J., Nielsen, R.G.H., Hall, B.D. and Best, E.K. (1986a) Nitrogen availability in a Darling Downs soil following cereal, oilseed and grain legume crops. 1. Soil nitrogen accumulation. *Australian Journal of Experimental Agriculture and Animal Husbandry* 26, 347–351.

Strong, W.M., Harbison, J., Nielsen, R.G.H., Hall, B.D. and Best, E.K. (1986b) Nitrogen availability in a Darling Downs soil following cereal, oilseed and grain legume crops. 2. Effects of residual soil nitrogen and fertilizer on subsequent wheat crops. *Australian Journal of Experimental Agriculture and Animal Husbandry* 26, 353–359.

Sweetingham, M. (1993) *Brown Spot and Pleiochaeta Root Rot of Lupins.* Farmnote 23/93, Western Australian Department of Agriculture, Perth, Western Australia.

Sweetingham, M. (1996) *Brown Spot and Pleiochaeta Root Rot of Lupins.* Farmnote 5/96, Agriculture Western Australia, Perth, Western Australia.

Tapscott, H.L. and Abrecht, D.G. (1994a) Impact of lupin seed source on lupin performance and seed quality. In: Dracup, M. and Palta, J. (eds) *Proceedings of the First Australian Lupin Technical Symposium.* Western Australian Department of Agriculture, Perth, Western Australia, p. 315.

Tapscott, H.L. and Abrecht, D.G. (1994b) Lupin seed quality – effect of seed size and plant density. In: Dracup, M. and Palta, J. (eds) *Proceedings of the First Australian Lupin Technical Symposium.* Western Australian Department of Agriculture, Perth, Western Australia, p. 316.

Tapscott, H.L. and Cowling, W.A. (1995) Predictors of yield of *Lupinus angustifolius* (cv. Gungurru) seedlots from different sources in Western Australia. *Australian Journal of Experimental Agriculture* 35, 745–751.

Tapscott, H., Cowling, W.A., Dracup, M. and Speijers, E.J. (1994) Effect of genotype, site and plant density on yield components of a historical set of narrow-leafed lupin cultivars. In: Dracup, M. and Palta, J. (eds) *Proceedings of the First Australian Lupin Technical Symposium.* Western Australian Department of Agriculture, Perth, p. 317.

Taylor, R.L. (1962) Weed control in lupins. In: *Proceedings of the Fifteenth New Zealand Weed Control Conference*, Wellington, pp. 160–165.

Thomson, B.D., Bell, R.W. and Bolland, M.D.A. (1992) Low seed phosphorus concentration depresses early growth and nodulation of narrow-leafed lupin (*Lupinus angustifolius* cv. Gungurru). *Journal of Plant Nutrition* 15, 1193–1214.

Toms, W.J. and Fitzpatrick, E.N. (1961) Potassium deficiency in medium rainfall areas. *Journal of Agriculture, Western Australia* 2, 763–773.

Walton, G.H. (1976) Agronomic studies on *Lupinus angustifolius* in Western Australia: effect of cultivar, time of sowing and plant density on seed yield. *Australian Journal of Experimental Agriculture and Animal Husbandry* 16, 893–904.

Walton, G.H. (1979) Lupin agronomy why isn't it working? In: Poole, M.L. (ed.) *Proceedings of the Lupin Research Review Workshop.* Department of Agriculture, Geraldton, Western Australia.

White, P.F., Robson, A.D. and Fisher, H.M. (1981) Variation in nitrogen, sulfur, selenium, cobalt, manganese, copper and zinc contents of grain from wheat and two lupin species grown in a range of mediterranean environments. *Australian Journal of Agricultural Research* 32, 47–59.

Williams, C.H. (1980) Soil acidification under clover pasture. *Australian Journal of Experimental Agriculture and Animal Husbandry* 20, 561–567.

11 Economic Assessment of the Role and Value of Lupins in the Farming System

David J. Pannell

Centre for Legumes in Mediterranean Agriculture (CLIMA), University of Western Australia, Nedlands, Western Australia 6907

Introduction

In the right conditions, lupins can make a major contribution to an agricultural system. The conditions which matter are economic, biological, physical, sociological and technical. This chapter reviews the importance of these various conditions in determining the economic value and role of lupins, and examines the outstanding example of Western Australia, where, for many farmers, the right conditions have coincided.

Following the release of suitable varieties of *L. angustifolius* in Western Australia in the late 1970s (Gladstones, 1994), areas sown to lupins increased rapidly (Table 11.1). Further genetic improvement coupled with improved agronomic practices (identified through research and practical experience) allowed average yields to increase during the early 1980s, and to be maintained despite the spread of lupins to less than ideal soil types (Table 11.1). Marsh *et al.* (1994) studied the pattern of adoption and diffusion of lupins in different shires of Western Australia. They identified clear differences between shires in the timing and level of adoption (e.g. Fig. 11.1). They were able to explain these statistically as being due to factors such as seed yields and the intensity of promotion and extension activities. It has been estimated that the development and uptake of lupins generated net economic benefits for Western Australia of approximately A$331 million (GRDC, 1992).

This apparently simple success story belies some of the complexities and difficulties involved in the introduction of a new legume crop into an established farming system. This chapter has three aims:

- To describe the complexities and difficulties of introducing a new crop in general terms.
- To demonstrate the identification of economically optimal management practices for lupins, using one Western Australian farming region as a case study.

Table 11.1. Total areas and state-wide average yields of lupins in Western Australia (source: Australian Bureau of Statistics).

Year	Area (ha)	Yield (t ha^{-1})
1970	12,500	0.58
1971	23,200	0.01
1972	44,100	0.34
1973	63,700	0.77
1974	117,000	0.65
1975	121,000	0.73
1976	95,700	0.24
1977	57,100	0.41
1978	38,800	0.61
1979	45,900	0.54
1980	54,600	0.88
1981	95,200	0.86
1982	191,000	0.90
1983	318,900	0.99
1984	517,300	0.97
1985	492,800	0.83
1986	652,900	1.05
1987	876,200	0.83
1988	719,600	1.10
1989	676,200	0.92
1990	664,800	0.94
1991	786,600	1.11
1992	822,800	1.00
1993	928,700	1.27
1994	1,152,100	0.85
1995	1,120,000	1.07

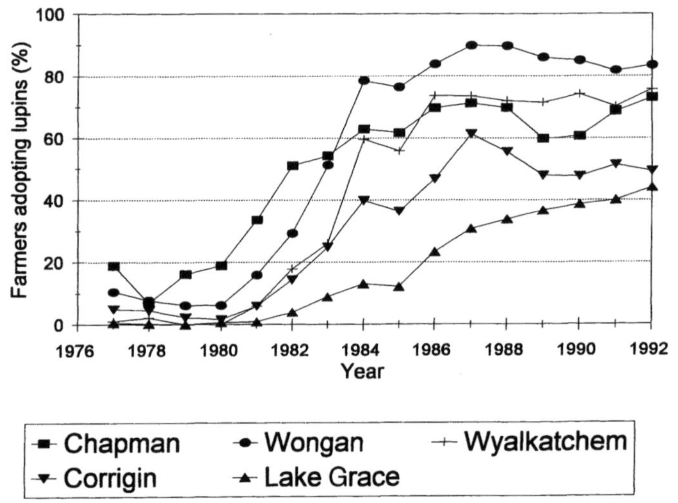

Fig. 11.1. Adoption of lupins in a selection of shires of Western Australia.

- For this example, to estimate the total contribution of lupins to whole-farm profit and the individual contributions of the different products of lupins, these being seed for sale or livestock feed, crop residues for livestock feed, fixed nitrogen and other yield-increasing impacts on cereal crops.

Management Decisions for Lupins in the Farming System

The management decisions which lupin producers must make include the following:

1. Whether lupins should be produced and, if so, the total area of land devoted to lupin production.
2. The soil types on which lupins should be grown.
3. The optimal rotational sequence of lupins and other crops and pastures on different soil types.
4. The quantities of lupin seed which should be sold or fed to livestock.
5. The timing and quantity of feeding of lupin seed to livestock.
6. The timing of grazing of livestock on lupin crop residues.
7. The duration and stocking rate of grazing on lupin crop residues.
8. The timing of sowing for lupins.
9. The method of seedbed preparation for lupins.
10. The types and levels of crop inputs, such as fertilizers, herbicides, pesticides, fungicides, seeding rates and inoculant.
11. Lupin variety or species.

Outside Western Australia, there has been very limited economic analysis of these decisions. The rare exceptions include studies of decision 1 (Hoffman and Baumann, 1990; Schneeberger, 1990) and decision 5 (Purroy *et al.*, 1990). In Western Australia most of these decisions have been analysed in some depth, including 1, 2 and 3 (Ewing *et al.*, 1986, 1987; Abadi, 1994), 4, 5, 6 and 7 (Pannell and Bathgate, 1991), 10 (French *et al.*, 1994) and 11 (Warren *et al.*, 1989; Shea *et al.*, 1998).

Pannell (1995) outlined the many factors that affect optimal management decisions for legumes in a mixed farming system (Table 11.2). The most striking aspect of this list is the large number of considerations that influence a farmer's management decisions for a legume crop like lupins. Pannell (1995) grouped the factors into those relating to short-term profit, dynamic (carry-over) effects, sustainability, risk and the whole farm and there are several factors in each of these groups, as described in the following sections.

Short-term profit factors

These are the factors that affect the direct profitability of lupins (relative to the alternatives) in the year in which they are grown. Such factors are important, of course, but the fact that they constitute only one of five categories of factors in Table 11.2 highlights the dangers of simplistic economic analyses which only consider returns in a single year.

Dynamic factors (short to medium term)

These include most of the positive side effects of lupins and contribute greatly to their popularity in Western Australia. Foremost of these are the improvements in cereal yield and quality following lupins due to nitrogen fixation (Rowland *et al.*, 1986, 1988; Pannell and Falconer, 1988; Unkovich *et al.*, 1994) (although N carry-over from lupins may have been significantly underestimated: Chapter 6,

Table 11.2. Factors affecting decisions on legume production.

1. *Short-term profit factors*
   - 1.1 Legume grain yield (depends on weather, soil type, weeds and so on)
   - 1.2 Legume crop stubble production and quality
   - 1.3 Pasture production level, quality and timing
   - 1.4 Yields of non-legume crops and pastures
   - 1.5 Input costs
   - 1.6 Output prices (for legume crops, livestock, livestock products and non-legume crops)
2. *Dynamic factors (short to medium term)*
   - 2.1 Nitrogen fixation and yield boost from other factors (e.g. disease break, soil structure)
   - 2.2 Pasture density
   - 2.3 Legume crop disease
   - 2.4 Stubble management
   - 2.5 Weed control
   - 2.6 Tillage method
   - 2.7 Carry over of fertilizer
3. *Sustainability factors*
   - 3.1 Herbicide resistance
   - 3.2 Soil degradation (acidification, organic matter decline, erosion, nutrient decline, non-wettability)
   - 3.3 Pasture legume persistence
   - 3.4 Pasture establishment costs
4. *Risk factors*
   - 4.1 Yield variability
   - 4.2 Price variability
   - 4.3 Yield/price covariance
   - 4.4 Flexibility of the enterprise in response to changed conditions
   - 4.5 The farmer's attitude to risk
5. *Whole-farm factors*
   - 5.1 Total crop area
   - 5.2 Machinery
     - 5.2.1 Total capacity
     - 5.2.2 Timing of requirements of different enterprises
   - 5.3 Total feed supply (timing and quality)
   - 5.4 Feed requirements of livestock on hand (timing and quality)
   - 5.5 Finance availability and cost
   - 5.6 Labour availability, quality and cost
   - 5.7 The farmer's objectives (profit, risk reduction, sustainability, leisure)
   - 5.8 The farmer's knowledge and experience

this volume) and other factors (e.g. Henderson, 1989). Choice of rotation also has dynamic impacts (positive and negative) on lupins. One example is disease, which is generally greater in rotations with shorter breaks between lupin phases. Shea *et al.* (1998) cite the greater disease resistance of yellow lupins as a factor which tends to favour them over *L. angustifolius*.

Without the benefits that lupins have for production of cereal crops, their production in Western Australia would be substantially lower than it is. This is illustrated in Fig. 11.2, which shows results from the MIDAS model giving the impact of lupin sale price on the optimal area of lupins (i.e. the farm-level supply curve) for a typical farm in the eastern wheatbelt of Western Australia. (The assumptions underlying all model runs in this chapter are presented by Pannell and Bathgate, 1991). At all reasonable lupin prices, when short- to medium-term interactions are included in the model, the optimal level of lupin production is dramatically higher. Note that the triangle-like area bounded by the supply curve, the price axis and the current price line (at A$145) is an approximate measure of the financial benefits of including lupins on the farm (termed 'producer surplus' by economists). The areas of the two 'triangles' are dramatically different. The smaller one reflects the fact that the benefits of lupin production would be negligible if not for the interactions with wheat, especially the impact of lupins on wheat yield.

Sustainability factors

Sustainability concerns related to legume rotations are not new. Problems of acidification have long occupied scientists (e.g. Loss *et al.*, 1993), but now we have the emergence of herbicide resistance and new threats of long-term soil quality decline in lupin rotations. To the extent that these problems are caused by or are associated with lupin production, there is a tendency to move away from lupins.

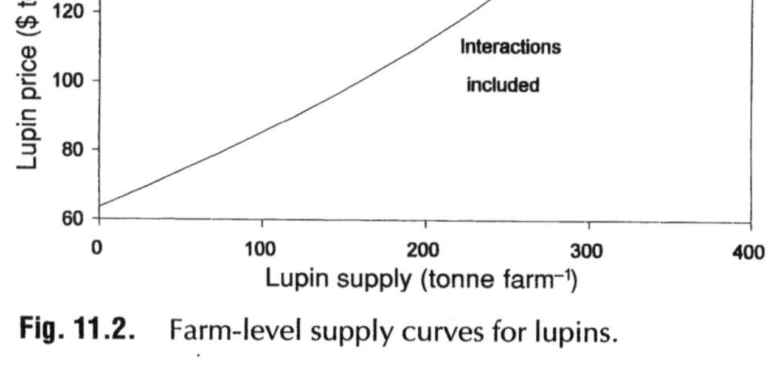

Fig. 11.2. Farm-level supply curves for lupins.

Risk factors

There is strong empirical evidence that most Australian farmers are 'risk averse' in the sense that they would be willing to accept a lower average income in order to reduce the variability or riskiness of their income (Bond and Wonder, 1980; Bardsley and Harris, 1987). Lupins provide an additional opportunity for farmers to diversify in order to reduce income risk. Because returns from different enterprises are not perfectly correlated, a mix of enterprises has a lower variation in total net returns, even if all enterprises have the same mean and variance of income. On the other hand, there is survey evidence that farmers perceive at least some grain legumes to be more risky than other enterprises (Abadi Ghadim *et al.*, 1996) and anecdotal evidence that this applies to lupins. These two separate issues may counterbalance each other to some extent. However, Marsh *et al.* (1994) found that the coefficient of variation of lupin yields is negatively related to farmers' adoption of lupins in different shires of Western Australia. On this evidence, they suggested that the riskiness of lupins may have been more decisive in affecting farmer behaviour than the opportunity to diversify. Different farmers will respond differently to these factors depending on their personal degree of risk aversion, which varies widely between individuals (Bond and Wonder, 1980), and on the relative riskiness of lupins in their particular farming systems.

The other aspect of risk is the extent to which it is possible to benefit from tactical adjustments in response to seasonal conditions. For example, the long-term average returns from lupins may be increased by reducing the area sown to lupins in seasons which have poor yield prospects at seeding time (Kingwell *et al.*, 1992).

Whole-farm factors

Farmers cannot manage individual paddocks independently of the rest of the farm. They have to juggle their available resources of labour, machinery, finance and feed so as to maximize returns over the farm as a whole. It can be extremely difficult to weigh up the required trade-offs but any attempt at a realistic analysis must do so. The total area of crop is of overriding importance in determining what the required trade-offs are. For example, in Western Australia lupins can often be sown earlier than cereals, increasing the total potential area of crop. Whether this is a significant advantage depends largely on whether the total crop area is large relative to the capacity of the available seeding machinery.

The attractiveness of lupins clearly depends on the farmer's objectives. There is strong evidence that profit and risk both figure in farmers' decisions to adopt new enterprises (e.g. Jansen, 1992), but a range of other objectives may be important for different farmers. Examples include:

- short run financial survival;
- debt reduction;
- personal preferences for or against a particular enterprise;
- leisure time;
- land acquisition for inheritance purposes.

These are perfectly legitimate objectives but they complicate economic planning since each can be in conflict with maximization of long-run profits.

The objective of sustainability is also of relevance to farmers, but they do not pursue an ethic of sustainability at any cost. Within the objective of maximizing long-run profit, a sustainable system will be favoured (other things being equal). If the economic sacrifices necessary to adopt a truly sustainable system are too great, farmers may be better off exploiting the best of the available unsustainable systems for a period until its productivity is reduced to that of the sustainable system. For example, a one-year-lupin/one-year-wheat rotation which is reliant on use of selective herbicides for weed control may be unsustainable in the long run due to development of herbicide resistance, but this does not mean that farmers should not employ the rotation while they can. The cost of this approach depends on the extent to which the short-term strategy reduces long-term productivity and management choices.

Apart from the farmer's objectives, lupin production decisions depend on the farmer's knowledge of and experience with lupins. Compared with cereals, lupins require farmers to acquire a substantially greater amount of knowledge (e.g. Jarvis, 1994). During the learning phase of lupin adoption, yields are likely to be lower on average and more variable than those achieved by experienced lupin growers.

Integrating Multiple Factors into Management Decisions

Given this wide range of factors, which operate simultaneously and often in conflicting directions, decision making about lupins can be very complex. Determination of the profit for each possible decision requires quantification of each of the factors – a major and difficult task. In Western Australia, the MIDAS model (Kingwell and Pannell, 1987) has been used successfully for this purpose. During the phase of lupin uptake in Western Australia, MIDAS played an important role in some regions in guiding extension and agronomic research on lupins. MIDAS has been used widely to examine economic aspects of lupin production, including studies by Ewing *et al.* (1986, 1987), Warren *et al.* (1989), Abadi (1994) and Shea *et al.* (1998).

A detailed description of MIDAS is provided by Kingwell (1987) and complete details of data and assumptions for the version used here are in Pannell and Bathgate (1991). In summary, the models use linear programming to represent whole-farm systems, giving equal weight to biological and economic factors. They select optimal farm management practices (e.g. rotations, crop species, livestock numbers, feeding strategies) based on yields and biological parameters considered sustainable, rather than those expected in a given year. Subject to this requirement of sustainability, the farm practices selected by the model maximize whole-farm profit.

Optimal Integrated Role for Lupins

In this section we focus on a series of MIDAS results that illustrate how the various factors outlined earlier combine to determine the optimal role for lupins in the farming decisions. We focus on decisions about the total area of lupins, their allocation to soil types and optimal rotations with cereal crops. All results are for version EWM91-4 of MIDAS, calibrated to a representative farm in the Merredin region in the eastern wheatbelt of Western Australia.

Figure 11.3 shows how profit varies as the area of the model farm is varied from zero up to the maximum allowable level. For this farm, the optimal area of lupins is 417 ha (out of a total farm area of 2500 ha, including 1000 ha considered suitable for lupins and another 875 on which lupins can be grown with poorer yields). It is notable that the economic cost of planting an excessive area of lupins is substantially greater than the cost of planting areas below the optimum. Indeed, excessive areas, involving planting of lupins on relatively unsuitable soil types, can result in lower overall profits than a system with no lupins at all. This may tend to encourage a conservative approach to lupin adoption.

Table 11.3 shows details of the rotations selected for the optimal solution illustrated in Fig. 11.3. Each letter (e.g. L, W, P or F) symbolizes a year of an

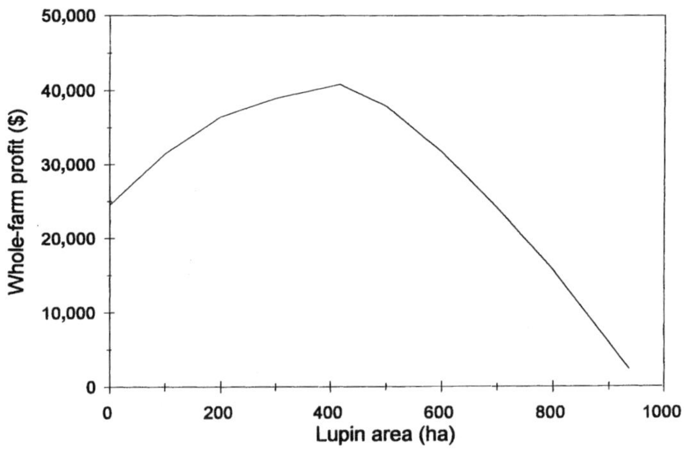

Fig. 11.3. Impact of lupin area on maximum whole-farm profit.

Table 11.3. Gross margins (A$ ha^{-1}) of a selection of rotations[1] (average over length of rotation).

Soil type	1	2	3	4	5	6	7
Optimum	PPPP ($15)	WL ($69)	WWL ($57)	WWL ($66)	WWF ($77)	PPPP ($26)	WWF ($77)
Others	PPPW ($4)	WWL ($67)	WL ($54)	WWWW ($61)	PWWW ($70)	PPPW ($20)	WWWF ($74)
	PPWW (–$5)	WPPWL ($54)	WPPWL ($49)	WWWF ($58)	WWWW ($67)	PPWW ($17)	WF ($64)
	WWL (–$12)	PPWW ($43)	WWWW ($46)	WL ($58)	PPWW ($66)	WWWW ($3)	WWWW ($62)
	WWWW (–$15)	WWWW ($39)	PPWW ($43)	PPWW ($53)	WWL ($58)	WWF ($3)	PPWW ($57)

[1]W = wheat; L = lupins; P = pasture; F = field peas.

Table 11.4. Optimal rotation[1] on soil type 3.

Wheat price (A$ tonne^{-1})	Lupin price (A$ tonne^{-1})	Wool price (A$ kg^{-1})		
		2.50	3.00	3.50
130	162	PPPP	PPPP	PPPP
	187	PPP	PPPP	PPPP
	212	WL	WL	WL
170	162	WWL	WWL	WWL
	187	WWL	WWL	WWL
	212	WWL	WWL	WWL
210	162	WWWW	WWWW	WWWW
	187	WWL	WWL	WWL
	212	WWL	WWL	WWL

[1]W = wheat; L = lupins; P = pasture; F = field peas.

enterprise in the rotational sequence. Lupins are selected on three of the seven soil types represented: S2, S3 and S4. They are included as options on soils S1 and S5 but are not profitable on these soils. The table shows the gross margins for a selection[1] of rotations on each soil type (an annual average over the whole rotation). On soil type S2 the wheat–lupin rotation is substantially more profitable than the best sequence that excludes any lupins (continuous wheat cropping). On soils S3 and S4 the profit advantage of wheat–wheat–lupin over continuous wheat is smaller. On soils S1 and S5, lupins would have to perform substantially better to be included in the optimal plan. These differences in optimal rotation between soil types are due to the different biological responses of lupins and their alternatives to the physical properties of the soils. The main alternative enterprises considered by the model are wheat, field peas and annual pasture for grazing by sheep.

Table 11.4 shows the result of a modelling experiment in which the major relevant prices (those of wheat, lupins and wool) are widely varied. The table shows the optimal rotation for soil type S3. The wide variation in rotations for different prices demonstrates that the optimal selection of rotations does not just depend on biological or physical factors. Economic factors are equally important in determining the optimal balance between the competing influences.

Another aspect of lupin management relates to their contribution to livestock diets. Table 11.5 shows the major constituents of livestock diets over a year in the standard model solution. Lupin crop residues contribute substantially to the summer and autumn diets. With the improvement in resistance to *Phomopsis* in more recent lupin varieties, sheep losses due to 'lupinosis' have diminished and the period over which lupin stubbles can safely be grazed has increased (Warren *et al.*, 1989). Lupin grain also makes an important contribution in the autumn period, when other available feeds have deteriorated to very low qualities. As well as supplying energy and protein directly, supplementation with lupin seed

1 There are approximately 30 rotation options on each soil type.

Table 11.5. Constituents of optimal livestock diet in standard model result.

Feed	Quantity (t)	Energy concentration (MJ kg^{-1})
Green pasture	796.3	9.0–12.0
Dry pasture	326.6	5.8–9.0
Lupin grain	20.1	13.3
Lupin crop residues (pods and spilt grain)	101.7	8.0–8.6
Lupin crop residues (leaf)	131.3	7.0–7.6
Lupin crop residues (stem)	60.7	5.6–6.0
Cereal crop residues (chaff and spilt grain)	43.4	7.0–7.5

allows sheep to make some use of low quality crop residues and dry pastures which otherwise would not be grazed.

The Contribution of Lupins to Farm Profit

Figure 11.3 illustrates the potential for lupins to improve farm profit. Compared with the best system that excludes lupins, the estimated increase in profit for the best system that includes lupins is A$18,000. Of course this benefit is sensitive to the particular biophysical and economic circumstances at hand. Table 11.6 shows how the potential profit improvement from lupins changes with the three main commodity sale prices.

It is interesting to consider which of the biological outputs of lupins contribute most to the overall value of lupins. Using MIDAS it is possible to estimate economic values for lupin seed produced, lupin crop residues grazed by livestock, reduction in nitrogen fertilizer in cereal crops following lupins, and other impacts on subsequent cereal crops due to factors such as reduced disease and improved soil structure. The values are calculated by excluding these factors one at a time and calculating a new optimal solution. Table 11.7 shows the individual values calculated in this way together with the optimal lupin area when each of the factors is excluded individually.

Table 11.6. Potential improvement in whole-farm profit from inclusion of lupins.

Wheat price (A$ t^{-1})	Lupin price (A$ t^{-1})	Wool price (A$ kg^{-1}) 2.50	3.00	3.50
130	162	6,000	6,000	5,000
	187	13,000	12,000	11,000
	212	23,000	21,000	20,000
170	162	11,000	10,000	8,000
	187	20,000	18,000	16,000
	212	28,000	27,000	25,000
210	162	10,000	10,000	9,000
	187	14,000	14,000	13,000
	212	20,000	20,000	19,000

Table 11.7. Impacts of the various lupin products or impacts on whole-farm profit.

Product or impact	Reduction in whole-farm profit if excluded (A$)	Optimal lupin area if excluded (ha)
Lupin seed	18,300	0
Lupin crop residues for grazing	0	417
Lower nitrogen requirement in subsequent cereal crops	3,600	417
High maximum yield of subsequent cereal crops	15,328	250

It is not surprising that seed is the lupin product of highest individual value. In this farming system, without the seed, it is not worth growing lupins for the other three products. However, the impact of lupins on subsequent cereal yields is of only slightly lower total value than is the seed. Even without this impact, it is still worth growing lupins on soil type S2, but the profit is substantially reduced. The values of reduced nitrogen fertilizer and crop residues are substantially lower. The reason for the nitrogen result is that optimal nitrogen fertilizer rates are normally not substantially reduced in cereals following lupins. The fixed nitrogen present in organic matter apparently does not substitute directly for artificial fertilizer. In the case of crop residues, although substantial quantities are grazed, there are sufficient quantities of unused cereal crop residues and dry pasture to substitute for the lupin residues if necessary.

Conclusion

The economic contribution of the lupin to Western Australian agriculture has been very substantial but this success is not necessarily directly transferable to other farming systems. The optimal allocation of land to lupins has been shown to be sensitive to factors such as the physical characteristics of soils, relative commodity prices, and the magnitudes of impacts of lupins on subsequent crops. Other issues not examined in detail here, such as relative yields, weed problems and disease, can also play important roles in determining the potential economic role and value of lupins. Nevertheless, the considerable and continuing success in Western Australia should provide encouragement for those wishing to see a broader role for lupins in world agriculture.

References

Abadi, A. (1994) Key economic factors affecting lupin production in the wheatbelt of Western Australia. In: Dracup, M. and Palta, J. (eds) *Proceedings of the First Australian Lupin Technical Symposium, Perth, Western Australia, 17–21 October 1994*. Department of Agriculture, Western Australia, Perth, pp. 187–190.

Abadi Ghadim, A.K., Pannell, D.J., Bennett, A. and Stewart, V. (1996) Farmers' risk perspective on adoption of legume crops. In: Asghar, M. (ed.) *Proceedings of the 8th Australian Agronomy Conference. The University of Southern Queensland, Toowoomba, Jan 30 to Feb 2, 1996*. Australian Society of Agronomy, Carlton, pp. 56–59.

Bardsley, P. and Harris, M. (1987) An approach to the econometric estimation of attitudes to risk in agriculture. *Australian Journal of Agricultural Economics* 31, 112–126.

Bond, G. and Wonder, B. (1980) Risk attitudes amongst Australian farmers. *Australian Journal of Agricultural Economics* 24, 16–34.

Ewing, M.A., Pannell, D.J. and Morrison, D.A. (1986) The place of lupins in the farm rotation: a whole-farm modelling approach. In: *Proceedings of 4th International Lupin Conference, Geraldton, Western Australia, August 15–22, 1986.* Western Australian Department of Agriculture, Perth, pp. 152–160.

Ewing, M.A., Pannell, D.J. and James, P.K. (1987) The profitability of lupin : cereal rotations. In: Kingwell, R.S. and Pannell, D.J. (eds) *MIDAS, a Bioeconomic Model of a Dryland Farm System.* Pudoc, Wageningen, pp. 82–90.

French, R.J., McCarthy, K. and Smart, W.L. (1994) Optimum plant population densities for lupin (*Lupinus angustifolius* L.) in the wheatbelt of Western Australia. *Australian Journal of Experimental Agriculture* 34, 491–497.

Gladstones, J.S. (1994) An historical review of lupins in Australia. In: Dracup, M. and Palta, J. (eds) *Proceedings of the First Australian Lupin Technical Symposium, Perth, Western Australia, 17–21 October 1994.* Department of Agriculture, Western Australia, Perth, pp. 1–38.

GRDC (1992) *Gains for Grains, Volume 1 – Overview.* Grains Research and Development Corporation, Canberra, 48 pp.

Henderson, C.W.L. (1989) Lupin as a biological plough: evidence for, and effects on wheat growth and yield. *Australian Journal of Experimental Agriculture* 29, 99–102.

Hoffman, H. and Baumann, H. (1990) Economic aspects of growing grain legumes in France. *Bayerisches Landwirtschaftliches Jährbuch* 67, 839–867 (in German)

Jansen, H.C.P. (1992) Inter-regional variation in the speed of adoption of modern cereal cultivars in India. *Journal of Agricultural Economics* 43, 88–95.

Jarvis, R.J. (1994) Agronomic practices for enhancing lupin yields. In: Dracup, M. and Palta, J. (eds) *Proceedings of the First Australian Lupin Technical Symposium, Perth, Western Australia, 17–21 October 1994.* Department of Agriculture, Western Australia, Perth, pp. 213–220.

Kingwell, R.S. (1987) A detailed description of MIDAS. In: Kingwell, R.S. and Pannell, D.J. (eds) *MIDAS, a Bioeconomic Model of a Dryland Farm System.* Pudoc, Wageningen, pp. 15–54.

Kingwell, R.S. and Pannell, D.J. (eds) (1987) *MIDAS, a Bioeconomic Model of a Dryland Farm System.* Pudoc, Wageningen, 207 pp.

Kingwell, R.S., Pannell, D.J. and Robinson, S. (1992) Tactical responses to seasonal conditions in whole-farm planning in Western Australia. *Agricultural Economics* 8, 211–226.

Loss, S.P., Ritchie, G.S.P. and Robson, A.D. (1993) Effect of lupins and pasture on soil acidification and fertility in Western Australia. *Australian Journal of Experimental Agriculture* 33, 457–464.

Marsh, S.P., Pannell, D.J. and Lindner, R.K. (1994) Extension and adoption of lupins in Western Australia. In: Dracup, M. and Palta, J. (eds) *Proceedings of the First Australian Lupin Technical Symposium, Perth, Western Australia, 17–21 October 1994.* Department of Agriculture, Western Australia, Perth, pp. 191–211.

Pannell, D.J. (1995) Economic aspects of legume management and legume research in dryland farming systems of southern Australia. *Agricultural Systems* 49, 217–236.

Pannell, D.J. and Bathgate, A. (1991) *MIDAS, Model of an Integrated Dryland Agricultural System, Manual and Documentation for the Eastern Wheatbelt Model Version EWM91-4.* Miscellaneous Publication 28/91, Western Australian Department of Agriculture, Perth, 162 pp.

Pannell, D.J. and Falconer, D.A. (1988) The relative contributions to profit of fixed and applied nitrogen in a crop–livestock farm system. *Agricultural Systems* 26, 1–17.

Purroy, A., Surra, J., Sebastian, I. and Munoz, F. (1990) Use of leguminous seed crops in lamb fattening diets. 2. Sweet lupin. *Informacian Tecnica Economica Agraria* 86A, 188–193 (in Spanish).

Rowland, I.C., Mason, M.G. and Hamblin, J. (1986) Effects of lupins on soil fertility. In: *Proceedings of 4th International Lupin Conference, Geraldton, Western Australia, August 15–22, 1986.* Western Australian Department of Agriculture, Perth, pp. 96–111.

Rowland, I.C., Mason, M.G. and Hamblin, J. (1988) Effects of lupins and wheat on the yield of subsequent wheat crops grown at several rates of applied nitrogen. *Australian Journal of Experimental Agriculture* 28, 91–97.

Schneeberger, W. (1990) Economic considerations in growing lupins. *Forderungsdienst* 38, 78–80 (in German).

Shea, G.G., Free, N.A., French, R.J. and Sweetingham, M.W. (1998) The yellow lupin – a role for a new crop in the low rainfall eastern wheatbelt farming systems in Western Australia. In: *Proceedings of 8th International Lupin Conference, Looking Toward the 21st Century, Pacific Grove, California, 11–16 May 1996.*

Unkovich, M.J., Pate, J.S. and Hamblin, J. (1994) The nitrogen economy of broadacre lupin in southwest Australia. *Australian Journal of Agricultural Research* 45, 149–164.

Warren, J., Allen, J. and Cowling, W. (1989) Economic impact of growing *Phomopsis*-resistant lupins. *Journal of Agriculture, Western Australia* 30, 8–10.R

12 Composition and Food Uses of Lupins

D.S. Petterson

Agriculture Western Australia, 3 Baron-Hay Court, South Perth, Western Australia 6151

Historical Use of Lupins in the Mediterranean and Andean Regions

Seeds of various species of lupins have been used as a food for over 3000 years around the Mediterranean (Gladstones, 1970) and for as much as 6000 years in the Andean highlands (Uauy *et al.,* 1995), but never have they been accorded the same status as soybeans or dry peas and other pulse crops. The pearl lupin of the Andean highlands of South America, *Lupinus mutabilis*, known locally as *tarw(h)i* or *choco*, was extensively cultivated, but there seems to have been no conscious genetic improvement other than to select for larger and water-permeable seeds. Users soaked the seed in running water to remove most of the bitter (and toxic) alkaloids and then cooked or toasted the seeds to make them edible (Hill, 1977; Aguilera and Trier, 1978), or else boiled and dried them to make *kirku* (Uauy *et al.,* 1995). Andean people had long recognized the benefits of consuming lupins, and associated their use with religious festivals (Lopez-Bellido and Fuentes, 1986; Uauy *et al.,* 1995). However, Spanish domination led to a change in the eating habits of the indigenous peoples, and only recently has interest in using lupins as a food been renewed (Hill, 1977).

Cultivation of *L. albus* may have occurred in ancient Egypt as early as 2000 BC (Zhuvosky, 1929, cited by Gladstones, 1970). Comments about lupins and their uses came from writers such as Columella, Theophrastus, Varro and Virgil, all of whom reported their value in soil improvement and ability to grow on poor soils and under harsh conditions (Gladstones, 1970). *Lupinus albus* has persisted in peasant cultivation in Europe and North Africa until the present day. After debittering in running water the seeds were usually cooked and served as a vegetable, or pickled in brine and served as a snack. Nevertheless, lupins seem to have always been at the fringe of Mediterranean agriculture, and never to have appealed to any society enough for development of food uses beyond their acknowledged value in times of drought and to the poorer classes (Bauhin *et al.,* 1651, and Savi, 1798, cited in Gladstones, 1970).

The beginnings of modern lupin breeding (Chapters 1 and 4, this volume) owe much to German interest in using the then-available bitter lupins for human food, following the blockade of World War I. Technology was developed for extracting alkaloids from lupin seed on an industrial scale. This was announced in 1918, when members of the Association of Applied Botany in Germany were invited to a lupin dinner (Hondelmann, 1984):

> There on a tablecloth made of lupin fibres, a lupin soup was served followed by a lupin steak fried in lupin oil and spiced with lupin extract. In addition, lupin margarine containing 20% of lupin constituents, cheese from lupin proteins, lupin spirits and lupin coffee were offered. Moreover, lupin soap for washing hands, paper and envelopes supplied with lupin adhesives were available.

The discovery of low-alkaloid mutants of lupin species by von Sengbusch in 1928, and subsequent development of lupins as a crop plant in Australia in the 1960s (Gladstones, 1970), and later in South America and Europe, saw renewed interest in their possible uses for human consumption. There have been several reviews of lupins as a potential food (Hill, 1977, 1986; Kyle *et al.*, 1991; Beiarao da Costa, 1990; Gross, 1990; Petterson and Crosbie, 1990; Uauy *et al.*, 1995; Yáñez, 1998). The composition and value of lupins as a food (ingredient) will be discussed, together with a possible future role in filling the needs of an ever increasing world population.

Composition of Lupin Species

Proximate analysis

Proximate analyses for the major lupin species are shown in Table 12.1. The high content of crude fibre is a consequence of the thick coats of lupin seeds, which comprise about 30% of the seed weight for *L. luteus*, 25% for *L. angustifolius,* 15% for *L. albus* and 12% for *L. mutabilis*. Energy storage within the cotyledons comes from variable proportions of oil, oligosaccharides and non-starch polysaccharides. There is virtually no starch in any of the lupin species, in marked contrast to crops such as field peas and chickpeas, which can have 50–70% of the cotyledon weight as starch and have low protein and oil content, and the soybean with 15–20% oil, some starch and a high protein content.

Proximate analyses for cotyledons (splits, meats) and seed coats (hulls) of these lupin species are given in Tables 12.1(a) and 12.1(b), respectively. These values are for fractions derived by hand separation or careful laboratory fractionation. Such separations are not possible with commercial dehulling, where protein values of 35% for the splits and 7–10% for the hulls are more common.

Proteins and amino acids

The major proteins of lupin seed are a group of globulins, called conglutins by Blagrove and Gillespie (1975). The globulin fraction which is soluble in buffers at about pH 7.5 contains three major proteins: conglutin α, conglutin β and a lupin-specific protein called conglutin δ (Blagrove and Gillespie, 1975). They comprise about 85% of the total protein and have similar size and physical properties to the storage proteins of the field pea, soybean and other legumes.

The largest protein, conglutin α or globulin 8/9a (Restani *et al.*, 1981), has a molecular weight (MW) of about 315,000 daltons (315 kDa) and is a complex of several small sub-units. It sediments at about 11–12S and is analogous to legumin of peas and glycinin of soybeans. Duranti *et al.* (1990) showed that the legumin-like protein are glycosylated in lupins, unlike most other species. The 7S protein is made up of conglutin β or globulin 4/5/6/7 (Restani *et al.*, 1981), a complex of MW about 185,000 kDa and conglutin. It is analogous to the vicilin of peas and conglycinins α and β in soybeans. The 2S protein, conglutin δ (about 17 kDA), is analogous to the conglycinin of soybeans (Lilley, 1986). It is a sulphur-rich protein.

Table 12.1(a). Chemical composition of the major lupin species (g kg^{-1} as received).

Component	*L. albus* (Albus lupin, White lupin)	*L. angustifolius* (Australian sweet lupin, Narrow-leafed lupin)	*L. luteus* (Yellow lupin)	*L. mutabilis* (Pearl lupin, Andean lupin)
Moisture	85.8	84.4	94.4	62.0
Protein (N × 6.25)	361.0	321.6	413.6	447.4
Ash	32.9	27.8	37.0	30.0
Crude fat	90.8	58.2	57.4	140.7
Crude fibre	102.7	148.9	127.2	70.4
ADF	142.8	197.3		
NDF	171.5	226.8	192	
Lignin	6.5	7.0	5.0	
Starch	ND	ND	ND	
Calcium	2.0	2.2	2.1	1.8
Phosphorus	3.6	3.0	6.1	8.8
Sulphur	2.4	2.3	2.5	
Gross energy (MJ)	19.0	18.6	19.6	

Sources: Hill (1977); Schoeneberger *et al.* (1982); Petterson and Mackintosh (1994); Zdunczyk *et al.* (1994).

Table 12.1(b). Chemical composition of the kernels of lupin species (g kg^{-1} as received) (data obtained from GRAILE database in Australia and Zdunczyk *et al.*, 1994).

Component	*L. albus*	*L. angustifolius*	*L. luteus*	*L. mutabilis*
Moisture		100.1	83.0	
Protein (N × 6.25)	400.9	400.3	525.7	454
Ash	33.0	26.9	43.4	
Crude fat	114.0	65.5	71.6	148
Crude fibre	18.0	87.1	17.1	105
ADF		70.2	31.2	
NDF		71.2	47.9	
Lignin		6.7		
Calcium		1.0	1.2	
Phosphorus		5.1	9.7	
Sulphur		2.5	5.0	
Gross energy (MJ)	20.4	18.9	19.7	

The remaining 15% of proteins are albumins, which are soluble at pH 5 and vary in size from about 6 to 117 kDa. Salmanowicz (1995) found that the 2S proteins, including conglutin δ, formerly considered as globulins were actually albumins. Depending on the species, the 2S albumin class consists of four (*L. luteus*), five (*L. albus*) or six (*L. angustifolius*) proteins when separated by reverse-phase HPLC (Salmanowicz, 1995).

Typical amino acid profiles for the domesticated lupin species show the deficiencies in lysine and methionine, compared to FAO standards, common to most legume seeds (Table 12.2). As with all grain species recorded (Mosse *et al.*, 1987), the proportion of lysine and methionine in each species falls as the total nitrogen content increases, whereas that of some non-essential amino acids rises. The relationships for *L. albus*, *L. angustifolius* and *L. luteus* in Australia are shown in Table 12.2.

Fatty acids

Fatty acid profiles for various lupin species are shown in Table 12.3. There are reports of that in *L. albus* varying with growing conditions (Williams and McGibbon, 1980). Jimenez *et al.* (1991) studied *L. albus* grown at two locations in Spain and found both genotype and location effects on content of oil, which had a negative correlation with protein content, and oleic acid. Covariance analysis showed a strong positive correlation between oleic acid and oil content at the genotype level, and a strong negative correlation with both linoleic and linolenic acids. Oliveira and Ferreira (1988) also found environmental influences on the composition of oil from *L. albus*. They reported that concentrations of oleic and linoleic acids were similar to other edible oils, and of those of erucic acid 1.5–2.7% were below the 5% EEC limit. Bhardwaj *et al.* (1998) also noted environmental influences on oil quantity and profile in *L. albus* grown at different sites in the United States.

No such effect was observed for *L. angustifolius* in a survey across the lupin-growing regions of Western Australia in 1984 (F.M. Legge and D.S. Petterson, Perth, 1985, unpublished results). These findings are not necessarily contradictory, because although the north–south extremes were about 500 km apart, and the east–west 300 km, the actual maturing conditions throughout are very similar (J.S. Gladstones, Perth, 1985, personal communication). Extracts of *L. angustifolius* oil in the latter study were stable for three months at 51°C, indicating a high level of antioxidant activity in this material. Since the tocopherol content of lupins is low (Abel-Nabey and Shehata, 1991) there may be other unidentified antioxidant components in lupin seed.

Carbohydrates

The seed hulls and cotyledons contain different types of carbohydrate. Lupin hulls are predominantly composed of structural polysaccharides: cellulose, hemicelluloses and pectins (Brillouet and Riochet, 1983; Evans 1994). The main carbohydrate reserves of the cotyledons are the non-structural polysaccharides of the cell walls, with the main components being galactose, arabinose and uronic acid (Brillouet and Riochet, 1983; Evans 1994). These complex compounds,

Table 12.2. Essential amino acid profile for major lupin species.

	g 16 gN^{-1}				% in seed				
Amino acid	*L. albus*[1]	*L. angustifolius*[1]	*L. luteus*[1]	*L. mutabilis*[2]	*L. albus*	*L.angustifolius*	L. luteus	FAO: Infant	WHO: Other
Arginine	12.4	12.0	9.1	10.5	4.68	3.65	4.37		
Cystine	1.30	1.48	2.47	1.6	0.50	0.46	0.88		
Histidine	1.72	2.41	3.3	3.0	0.65	0.76	1.05	2.6	1.9
Isoleucine	3.72	3.97	3.50	4.8	1.41	1.23	1.42	4.6	2.8
Leucine	6.06	6.61	7.61	7.0	2.30	2.08	3.06	9.3	6.6
Lysine	4.20	4.66	4.10	5.9	1.57	1.43	2.07	6.6	5.8
Methionine	0.65	0.72	0.59	0.8	0.24	0.22	0.27	(4.2)	(2.5)
Phenylalanine	3.27	3.65	3.51	p	1.23	1.12	1.56	(7.2)	(6.3)
Threonine	3.13	3.36	2.71	3.8	1.19	1.04	1.36	4.3	3.4
Tryptophan	0.97	1.06	0.84	0.7	0.37	0.32	0.84	1.7	1.1
Tyrosine	5.09	3.46	3.1	3.4	1.71	1.07	1.12		
Valine	3.64	3.91	3.65	4.2	1.36	1.18	1.33	5.5	3.5

[1]Petterson and Mackintosh (1994).
[2]Schoeneberger *et al.* (1982).
Values in brackets are the sums of methionine + cystine, and phenylalanine + tyrosine. Values are from FAO/WHO scoring patterns (1990).

Table 12.2a. The content of lysine, threonine and methionine as a function of crude protein in major lupin species (raw data from GRAILe database, Australia).

	L. albus		*L. angustifolius*		*L. luteus*	
	a	b	a	b	a	b
lysine	0.0191	0.9012	0.0406	0.2100	0.0041	1.8414
Threonine	0.0043	1.0754	0.0305	0.1456	0.0085	0.9466
Methionine	0.0038	0.1064	0.0037	0.0072	0.0005	0.2521

Table 12.3. Fatty acid profile of major lupin species (% fatty acid in hexane extract).

Fatty acid	*L. albus*[1]	*L. angustifolius*[1]	*L. luteus*[1]	*L. mutabilis*[2]
14:0	p	p	p	
16:0	7.7	10.3	5.3	0.6
16:1	0.3	p	p	13.9
16:2				0.2
18:0	1.1	4.8	1.5	3.0
18:1	51.0	34.9	31.5	41.7
18:2	17.3	37.0	46.9	38.8
18:3	11.1	6.2	4.2	2.6
20:3	1.0	0.7	2.4	p
20:1		0.3		p
20:4	0.4	p	2.2	
22:0	3.3	1.3	7.2	p
22:1	2.5	p	p	
24:0	p	p		
Sterols		2.2		

[1]Petterson and Mackintosh (1994).
[2]Gross (1990).
p – present, not accurately quantified.

where structure is not yet fully defined, are referred to as non-starch polysaccharides (NSP). The water-soluble portion, about 5% of the seed, is considered to have an antinutritional effect due to it's viscous nature and effect on intestinal transit time and changes in hormonal regulation due to differential nutrient absorption rates. The insoluble NSP, about 30% of the seed, has a minimal effect on nutrient utilization by monogastrics. An important attribute of insoluble NSP is their ability to hold large quantities of water, about eightfold by weight for lupins, and maintain normal gut motility.

Brillouet and Riochet (1983) found similar ratios of monosaccharide, in *L. albus* and *L. angustifolius*, but about one-half as much galactose and twice as much glucose in *L. luteus*, compared with arabinose. The absence of absolute values makes it difficult to interpret their data. In contrast, one analysis of the cell wall material of seed from eight lupin species, including the above three, showed little difference between the carbohydrate profiles (A.J. Evans, Sydney, 1994, personal communication). The profiles for various species are shown in Table 12.4.

Some authors have given starch values as high as 11% in lupin seed, but the majority have reported traces, at most. An examination of methods of analysis suggests that the high values are an artefact of the analytical procedures (D.J. Harris and A.J. Evans, Perth, 1994, personal communications).

Oligosaccharides are discussed later in the chapter as anti-nutritional factors.

Minerals

The calcium content of lupin seed (1.5–2.2 g kg^{-1}) is higher than for peas but lower than soybean meal, and the phosphorus content (3.0–5.1 g kg^{-1}) is similar to that of peas but much lower than for soybean meal. The percentage of phytate phosphorus in lupin species is comparable with the others. Magnesium content (1.4–2.1 g kg^{-1}) is similar and potassium (8.1–9.8 g kg^{-1}) is similar to peas but much lower than soybean meal.

The trace mineral content of lupin species is influenced by genotype but also tends to reflect the soil types on which they are grown. The accumulation of manganese by *L. albus* is well known, with Hung *et al.* (1987) in Victoria,

Table 12.4(a). The carbohydrates in lupin hulls (% dry weight).

Species	Cultivar	Cell wall content	Cellulose	Lignin	Starch	Sucrose	Oligo-saccharides
L. angustifolius	Gungurru	88	49	1.5	0.4	1.0	0.4
L. angustifolius	Yorrel	90	52	1.2	0.3	0.9	0.3
L. angustifolius	Danja	91	51	1.3	0.4	1.4	0.4
L. angustifolius	Lang 61		50	1.7			
L. angustifolius	Uniwhite	92	51	0.3		0.7	

Table 12.4(b). The carbohydrates in lupin cotyledons (% dry weight).

Species	Cultivar	Dietary fibre	Cellulose	Lignin	Starch	Sucrose	Oligo-saccharides
L. angustifolius	Gungurru	29	1.7	0.8	0.6	3.8	7.7
L. angustifolius	Yorrel	27	0.8	0.9	0.6	3.1	8.0
L. angustifolius	Danja	31	1.3	0.9	0.7	3.5	7.4
L. angustifolius	Lang 61	21	0.8				
L. albus		18–22	1.7–1.0				
L. luteus	LL29	12	1.3				
L. mutabilis		11	1.4				

Adapted from Evans (1994) with permission from the publishers.

Table 12.4(c). Individual sugars in total, soluble and insoluble NSP from hull and cotyledon of *L. angustifolius.*

	Monosaccharide, as % total polysaccharides					
	Hull			Cotyledon		
Sugar	Total	Soluble	Insoluble	Total	Soluble	Insoluble
Arabinose	8.7	11.8	8.5	13.0	10.0	13.8
Galactose	2.2	10.0	1.4	67.0	67.7	64.6
Glucose	59.4	4.0	62.9	4.6	2.9	4.5
Mannose	2.5	13.9	1.7	0.7	3.5	0.3
Rhamnose	nd	1.3	nd	2.6	3.7	2.9
Uronic acids	12.3	17.9	12.8	9.9	10.2	10.8
Xylose	14.9	41.2	12.5	2.3	1.9	3.2

Adapted from Evans *et al.*, 1993.
nd – Not detected.

Australia, reporting contents of up to 4600 mg kg^{-1} in seed from crops grown on heavy red clays. The reported values in the Australian database for grain legumes range from 19 to 3800 mg kg^{-1}, with most of the low values coming from crops grown on the sandy soils of Western Australia (mean 399, range 58 to 1184 mg kg^{-1}). In contrast the manganese content of *L. angustifolius* is low. One study comparing the composition of *L. albus* and *L. angustifolius* grown at three sites in Western Australia found values of 2313 cf. 20, 867 cf. 13, and 745 cf. 23 mg kg^{-1}. The same comparisons for selenium were 3382 cf. 44, 88 cf. 33, and 328 cf. 133 mg kg^{-1} (R.J. Hughes, Roseworthy, 1997, personal communication). In this same study there were no differences between the two species for content of magnesium, aluminium, boron, iron and zinc, where all were close to the typical values shown in Table 12.5. Manganese contents of up to 326 mg kg^{-1} have been found in *L. cosentinii* and 60 mg kg^{-1} in *L. luteus* (D.S. Petterson, Perth, 1997, unpublished results).

The selenium content of lupin seed grown in Western Australia varies according to species and rainfall zone. Where the species were grown together, the order of selenium content was *L. albus* > *L. luteus* > *L. angustifolius*. For *L. angustifolius*, the higher the rainfall, the lower was the selenium content. This is regarded as a consequence of the leaching from the ancient soils in Western Australia with time (Underwood, 1977). Similar effects apply to the content of copper, cobalt and zinc in lupins, to a lesser degree, and in the cereal crops grown in the State (White *et al.*, 1981).

An extensive survey of the cadmium and lead content of *L. angustifolius* (178 samples) and *L. albus* (12 samples) grown in Australia found no samples exceeded the Codex Alimentarius limits of 0.1 and 2.0 mg kg^{-1}, respectively (Petterson and Harris, 1995). In the same study, similar levels of the two elements were found in cereals grown at the same or adjacent sites. Gross (1990) reported 0.1 ppm Cd and 1.0 ppm Pb in seed of *L. mutabilis* and 0.15 and 1.2 ppm,

Table 12.5. Minerals content of major lupin species (means of data sets).

Mineral	*L. albus*[1]	*L. angustifolius*[1]	*L. luteus*[1]	*L. mutabilis*[2]
Calcium (g kg^{-1})	20	22	15	18
Magnesium (g kg^{-1})	14	16	21	21
Phosphorus (g kg^{-1})	36	30	51	88
Potassium (g kg^{-1})	98	81	97	
Sulphur (g kg^{-1})	24	23		
Sodium (g kg^{-1})	5	5	3	
Copper (mg kg^{-1})	5	5	9	8
Iron (mg kg^{-1})	26	75	93	54
Manganese (mg kg^{-1})	835	17	86	26
Molybdenum (mg kg^{-1})	2	2	7	
Zinc (mg kg^{-1})	30	35	56	
Cobalt (mg kg^{-1})	320	112		
Selenium (μg kg^{-1})	57	82		

[1]Adapted from Petterson and Mackintosh (1994).
[2]Gross (1990).

respectively, in seed of *L. albus* grown in Chile. There is some evidence in the literature for the accumulation of cadmium by *L. luteus* (Korol and Matyka, 1988). This may be a reflection of the different root architecture in this species (Chapter 5, this volume).

Anti-nutritional factors

Legumes are well known as containing a range of compounds with apparent untoward effects on any species ingesting them. Traditionally these have been known as anti-nutritional factors (ANFs) or anti-nutrients. Their known clinical effects are discussed in Chapter 4. Recent epidemiological work suggests that some of these compounds may have beneficial effects for humans. For example, phytate, some tannins and trypsin inhibitors also have an antioxidant effect and may help to prevent the onset of cancer (G.H. McIntosh, Adelaide, 1995, personal communication). Table 12.6 shows typical content of major ANFs in different lupin species.

Alkaloids

The lupin alkaloids are usually bicyclic (e.g. lupinine), tricyclic (e.g. angustifoline) or tetracyclic (e.g. sparteine) derivatives of quinolizidine (Fig. 12.1). An important exception is gramine, which is found in some cultivars of *L. luteus*.

The major alkaloids of commercial cultivars are:

- *L. albus*: lupanine, 13-hydroxylupanine, sparteine (some European cultivars).
- *L. angustifolius*: lupanine, 13-hydroxylupanine, angustifoline.
- *L. mutabilis*: lupanine, sparteine, 13-hydroxylupanine.
- *L. luteus*: gramine (some cultivars only), lupanine, cytosine.

The rangeland lupins of North America have quite different profiles and some of them contain anagyrine, the alkaloid considered responsible for crooked calf disease (Keeler, 1973).

Table 12.6. Anti-nutritional factors in the major domesticated lupin species.

Factor	Units	*L. albus*[1] (Albus lupin, White lupin)	*L. angustifolius*[1] (Australian sweet lupin, Narrow-leafed lupin)	*L. luteus*[2] (Yellow lupin)
Total alkaloids	mg kg^{-1}	< 100	200	500[3]
Oligosaccharides	%	6.7	5.2	11.87
Saponins	mg kg^{-1}	nd[4]	573	
Condensed tannins	%	0.01	< 0.01	0.02
Lectins		nd	nd	
Trypsin inhibitors	mg g^{-1}	0.13	0.14	0.29
Phytate	%	0.79	0.58	0.96

[1]Petterson and Mackintosh (1994).
[2]Zdunczyk *et al.* (1994).
[3]Newer cultivars from Poland and Australia have even lower content of alkaloids.
[4]nd, Not detected.

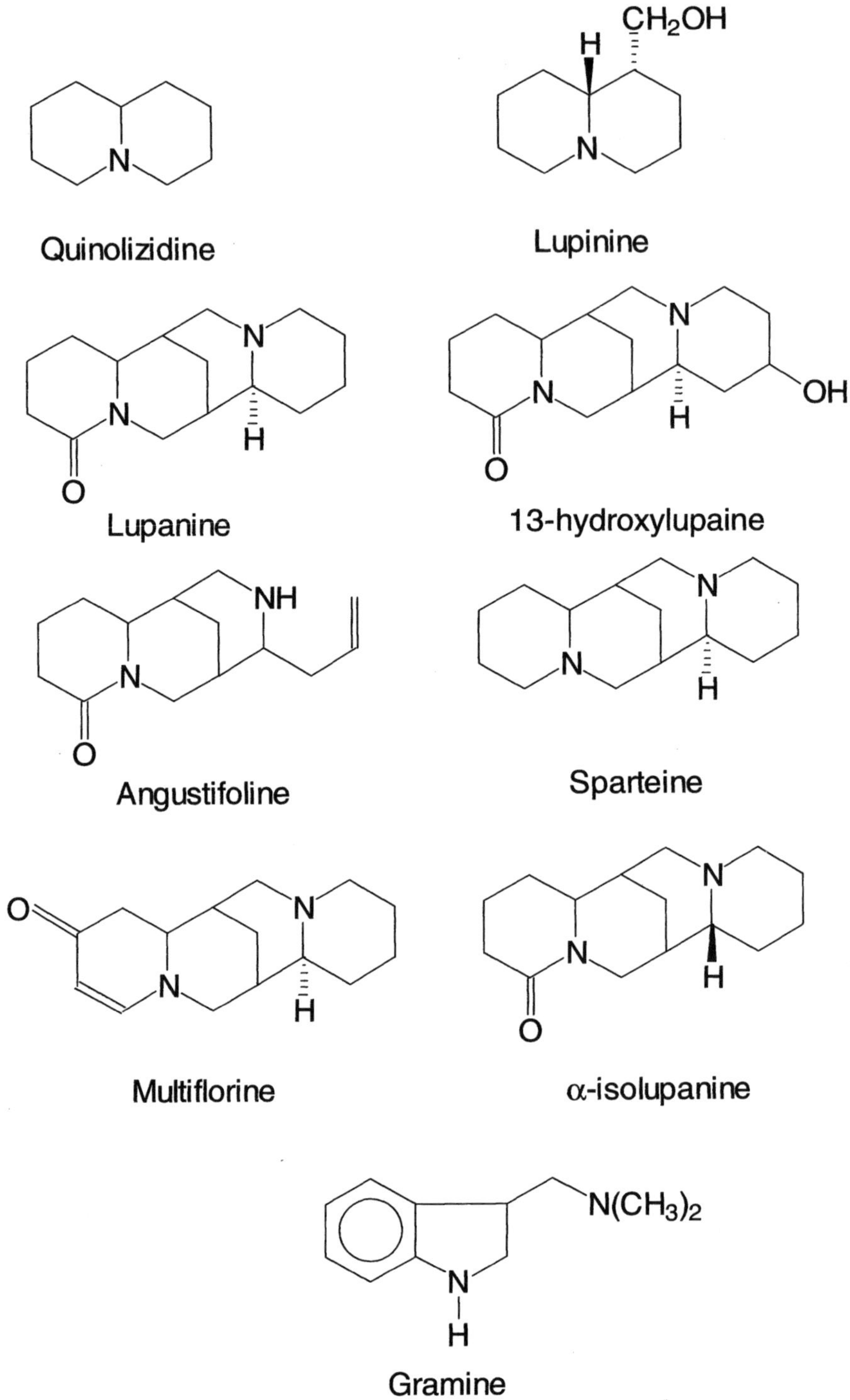

Fig. 12.1. Structure of lupin alkaloids.

Modern cultivars of *L. angustifolius* typically contain < 200 mg alkaloids kg^{-1} (Harris and Jago, 1984) and are commonly referred to as 'sweet' lupins to differentiate them from the 'bitter' wild and green-manuring types, which may contain from 5 to 40 g alkaloid kg^{-1} depending on the species and growing conditions. The typical alkaloid profile of sweet *L. angustifolius* is lupanine (42–59%), 13-hydroxylupanine (24–45%), angustifoline (7–15%), α-isolupanine (1–1.5%) and traces of other alkaloids (< 1% total). Table 12.7 shows concentrations of alkaloids in commercial *L. angustifolius* from years and areas in Western Australia, the world's largest producer of grain lupins. It illustrates the effects of growing environment on seed alkaloid content, but also the influence of the new cultivars Gungurru and Yorrel (released in 1989), which are lower in alkaloid than the cultivars they replaced. Some typical profiles for other species are shown in Table 12.8.

Table 12.7. Alkaloid content of commercial crops for Australian sweet lupins in Western Australia (mg kg^{-1}): zone-weighted averages 1985–1992.

Growing zone[1]	Year 1985	1986	1987	1988	1989	1990	1991	1992
1	190	300	300	100	470	140	70	60
2	120	240	200	120	360	140	80	60
3	230	260	170	90	290	150	70	30
4	130	210	70	90	230	110	70	50
5	120	120	100	80	220	150	150	60
State	170	270	240	110	400	140	70	60

[1]Bulk handling authority collection zones.

Table 12.8. Quinolizidine alkaloids in the major lupin species[1].

Alkaloid	% in seed *L. albus*	*L. angustifolius*	*L. luteus*	*L. mutabilis*
Albine	15[2]	–	–	–
Lupanine	70	70	–	46
Multiflorine	3	–	–	–
13-hydroxylupanine	8	12	–	12
Angustifoline	–	10	–	1
Lupinine	–	–	60	–
Sparteine	–	–	30	16
Ammodendrine	–	–	–	2
Tetrahydrorhombifoline	–	–	–	2
3-hydroxylupanine	–	–	–	12
13-angeloyloxylupanine	–	–	–	2

[1]Adapted from Wink *et al.* (1995).
[2]Percentage of total alkaloids.
– Not detected in chromatogram.

Wink *et al.* (1995) give a fuller review of alkaloids in lupin species but they do not acknowledge all the changes that have followed breeding of low-alkaloid cultivars. For example, albine is not present in the commercial cultivars of *L. albus* grown in Australia, and gramine is present in new cultivars of *L. luteus* being developed in Poland and Australia. For a discussion on the toxicology of these compounds see Chapter 14, this volume.

Phytate

Phytic acid is inositol hexaphosphate. Together with the lesser substituted homologues and their collective salts these compounds are referred to as phytate. Phytate can form insoluble complexes with divalent cations, particularly calcium and zinc, thus rendering them less available for absorption and utilization. The net effect of phytates depends on the overall composition of the food or diet, particularly the protein content and characteristics and the total mineral content. The amount of phytate in lupins does not seem to be of concern. For example, Petterson *et al.* (1994) found the absorption of Zn from a range of lupin-based (*L. angustifolius*) foods to be higher overall than from comparable soy products. They concluded that lupin milk could be an attractive alternative to soy milk for infant formulae (Petterson *et al.,* 1994). The phytate content of lupin seed can be lowered by germination (Dagnia *et al.,* 1992) and fermentation (Fudiyansyah *et al.,* 1995).

Tannins

Tannins are compounds of plant origin with molecular weights ranging from about 500 to 2000 Da, and with one to two phenolic hydroxyl groups per 100 Da. This enables them to form cross-linkages between proteins and other macromolecules (Griffiths, 1991). The two sub-groups are hydrolysable tannins, which typically have a central glucose core with the hydroxyl groups being wholly or partly esterified with gallic acid or hexahydrodiphenic acid, and condensed or ion-hydrolysable tannins, which are higher oligomers of flavan-3-ols with varying degrees of substitution. The astringent taste and the ability to precipitate proteins, resulting especially in the inactivation of gut enzymes, give the tannins their anti-nutritional role. In lupins, as with other grain legumes, the tannins are concentrated in the seed coats and the simple act of dehulling minimizes any effects. The concentration of condensed tannins – those most responsible for protein-binding – is so low in lupins that it is unlikely to affect human nutrition. Concentrations in *L. albus* and *L. angustifolius* are typically < 0.01% (Petterson and Mackintosh, 1994).

Saponins

Saponins have the ability to produce a soapy foam even at very low concentrations, and can lyse red blood cells. They have a bitter taste. Saponins are glycosides with the non-sugar moiety being known as a sapogenin, which may be a steroid or a triterpenoid compound. The concentration of saponins in *L. albus* appears to be negligible, whilst that in *L. angustifolius* ranged from about 480 to 730 mg kg^{-1} (Ruiz *et al.,* 1995). An earlier study on seed from Spain and Australia found similar levels in both species (Muzquiz *et al.,* 1993). These are lower than earlier reports by authors using less sophisticated and less precise techniques. Saponins are generally harmless to humans, and some might even be beneficial in lowering blood cholesterol levels, conferring protection against coronary heart disease (Fenwick *et al.,* 1991). The concentrations in lupins are lower than in many other species.

Protease inhibitors

Trypsin inhibitor activity is very low in lupins, ranging from < 0.01 to 0.28 mg g^{-1} in *L. angustifolius* and from 0.1 to 0.2 mg g^{-1} in *L. albus* (Petterson and Mackintosh, 1994). Chymotrypsin inhibitor activity was reported at < 0.01 to 0.59 mg g^{-1} in *L. angustifolius.*

Lectins

The conventional assay procedure agglutination of at least some of a wide range of different types of red blood cells has not detected any sign of lectin activity in *L. albus* or *L. angustifolius* (Liener, 1989; B.N. Greirson and D.S. Petterson, Perth, 1990, unpublished results). A slight activity can be induced if the red cells are specially treated (A. Pustzai, Aberdeen, 1995, personal communication), but this is not thought to be significant. Duranti *et al.* (1995) isolated conglutin γ from *L. albus* seed and showed it was capable of binding to various *N*-glycosylated proteins. Immunological homology with two bean lectins suggests there could be some lectin activity by this protein, which is almost certainly the activity observed by Pustzai (cited above).

Oligosaccharides

The oligosaccharides of *Lupinus* species belong to the raffinose family. Raffinose has one galactose moiety linked to a sucrose molecule through an α-1,4 bond, stachyose has an α-1,4 linkage to the galactose unit of raffinose, and so on through verbascose and ajugose. These compounds cannot be metabolized by monogastrics and they pass through to the colon, where bacterial digestion breaks them down to produce carbon dioxide, methane and hydrogen. This causes abdominal discomfort and cramps and leads to flatulence. These factors seem to be a major reason for the low interest in consuming pulses in many societies. The oligosaccharides also have a role in osmotic regulation in the gastrointestinal tract which may be beneficial.

Attributes of Lupins for Human Consumption

Seed of the various species of lupin contain about 30–40% protein (N × 6.25). In common with most pulse or grain legume crops, the protein fraction is low in lysine (4–5 g per 16 g N) and methionine (0.3–0.6 g per 16 g N) and the protein digestibility – corrected amino acid scores (PD–CAAS) are around 0.7 (R.R. Bell and D.S. Petterson, Perth, 1994, unpublished results) compared with 1.0 for casein and 0.7 for field peas.

All studies reported on rats show that whilst the protein efficiency ratio (PER) of whole seed is around 1.0, supplementing with about 0.2% DL-methionine brings the PER value up to about that of casein, 2.5 (Yáñez *et al.,* 1983; Zdunczyk *et al.*, 1994; R.R. Bell and D.S. Petterson, Perth, 1995, unpublished data). Supplementation with methionine also improves the apparent protein digestibility (Ballester *et al.*, 1980). Protein from *L. albus* seeds has similar qualities. In addition feeding whole-seed diets to rats lowered plasma glucose, cholesterol and triglycerides by 13, 28 and 27%, respectively (Samish *et al.*, 1990).

The protein quality of two lupin species has been tested in humans. López de Romaña *et al.* (1983) reported apparent protein digestibility values of 81.8% for *L. mutabilis*, and 84.6% when supplemented with 0.2% DL-methionine, compared with 87.1% for casein when fed to children. Based on these and retention data, they suggested a biological value of 55.6% for lupins to humans. Egaña *et al.* (1992) reported digestibilities of 70.2–78.8% for *L. albus* protein across a range of protein intakes. The *L. albus* biological value, i.e. regression for N retained vs. N absorbed, was 91% that of egg protein, and the net protein utilization was 77%.

Oil contents range from about 4% in *L. luteus* to about 15% in *L. mutabilis*, of which unsaturated fats range from about 72% in *L. mutabilis* to 84% in *L. albus*. The major fatty acid is oleic in *L. albus* and *L. mutabilis* (around 50%) and linoleic in *L. angustifolius* and *L. luteus* (35–50%). The balance of omega 6 : omega 3 acids gives a profile similar to soy oil, and it is likely to have similar health benefits (Uauy *et al.*, 1995). López de Romaña *et al.* (1983) found *L. mutabilis* oil more digestible than soy oil. The fatty acid profiles of the major lupin species are shown in Table 12.3.

Fibre in the cell wall material from *L. angustifolius* is likely to have a low glycaemic index, and to be a useful food for diabetics (Evans, 1994). The monosaccharide contents of the NSP fraction of other lupin species are similar to that of *L. angustifolius*. It is likely that they will have the same properties (A.J. Evans, Sydney, 1994, personal communication).

The contents of phytate, oligosaccharides, saponins, hydrolysable tannins and condensed tannins are all low and comparable with those of other grain legume species. The big advantage of lupins over other species is their low level of protease inhibitors and the negligible content of lectins. Unlike other grain legume species, lupins do not have to be cooked before being eaten (see Chapter 14 for more detail on the effects of these compounds).

Safety of Lupins for Human Consumption

The safety of foods can be assessed: historically, from long-term feeding and multi-generation studies in test animals; from a range of short- and long-term tests on specific constituents likely to be of concern; and from their use in animal nutrition. Regulatory requirements for novel foods or processes generally cover all of these considerations (e.g. *Novel Foods Guidelines*, UK Ministry of Health). Additionally, those foods having a history of food use with no significant evidence of any untoward effects on humans may be ‘generally recognized as safe, GRAS’.

The use of de-bittered lupins in Europe and South America over thousands of years provides indicative evidence of safety, provided the alkaloids are mostly moved. It is generally considered that de-bittering will reduce the alkaloid content to around 500 mg kg^{-1} (R. Gross, Perth, 1984, personal communication).

Grant *et al.* (1995) conducted a 700+ day feeding study in rats using seed of *L. angustifolius* (cv. Unicrop) and concluded that lupins were a good source of protein and that there were no apparent untoward effects from this long-term exposure to the feed source. A further 100 days feeding did not elicit any adverse effect (Grant *et al.*, 1993). At the end of this period pancreatic enlargement was evident in rats fed raw soybeans or cowpeas but not in those fed lupins or kidney beans.

Ballester *et al.* (1980, 1984a) conducted a multi-generation study and a 90-day feeding study in rats fed lupin (*L. albus* cv. Multolupa) as the main protein source and found no significant differences in biochemical, histological or haematological measures compared with controls fed a casein-based diet.

More information for the safety of lupins comes from the pig industry in Western Australia. Petterson (1985) reviewed the production records for the Medina Research Station of the then Department of Agriculture, covering more than eight generations and a total of over 1100 litters. During this time, starter, grower and finisher diets contained from 10 to 40% *L. angustifolius* seed, and the sow diets 10–30%. Production figures were comparable to commercial operations that did not use lupins. There were no indications of any teratological effects, and no reports of significant lesions in the sows when slaughtered. Alkaloid levels in the diets were estimated to range from 10 to 80 mg kg^{-1}.

Alkaloids

In vitro studies on an alkaloid preparation from *L. angustifolius* were negative for mutagenic effects in normal and stimulated cultures of *Salmonella typhimurium* (BIBRA, 1986).

In a 90-day feeding study, Butler *et al.* (1996) reported altered foci of hepatocytes in liver of eight out of 40 rats fed from 250 to 5050 mg alkaloids kg^{-1} derived from *L. angustifolius* in the diet. The authors suggested this was compound related, but as the incidence was not dose related there remains some doubt as to these observations.

Robbins *et al.* (1996) conducted a 90-day feeding study in rats using the alkaloids derived from *L. angustifolius* (cv. Fest, which has a virtually identical alkaloid profile to the modern low-alkaloid cultivars), at dose rates up to 5 g alkaloids kg^{-1} of feed. They found an appetite suppression at 1 and 5 m kg^{-1}, with a concomitant growth reduction of about 5% and 10%, respectively. At 330 mg kg^{-1} there was a slight enhancement of appetite and growth rate. This might be attributed to a mild pharmacological effect of the alkaloids. There were no untoward effects on histological or clinical parameters at any dose rate and the authors speculated that a no-untoward-effect level of 1 g kg^{-1} was reasonable. They used an alternative procedure to Butler *et al.* (1996) for blending the alkaloids preparation into the rat diets and considered that it was possible that the affected rats in the earlier study may have ingested significant amounts of alkaloid gum that had not blended adequately into the rat diets.

The pharmacological effects of lupanine and 13-hydroxylupanine are similar to those of sparteine in that they block ganglionic transmission, decrease cardiac contractility and contract uterine smooth muscle (Mazur *et al.,* 1966), have nicotinic cholinergic receptor blocking activity (Yovo *et al.,* 1984) and are weak antagonists at the muscarinic cholinergic receptor. However, the toxicity of lupanine and 13-hydroxylupanine is much less than that of sparteine (Yovo *et al.,* 1984; Petterson *et al.,* 1987), and these alkaloids are only considered to be 'moderately' toxic (Klassen and Doull, 1980).

Schmidlin-Menszaros (1973) found only seven cases of human poisoning due to quinolizidine alkaloids in the literature. Three of the four fatal cases involved the ingestion of lupin seed, containing about 2% alkaloids (i.e. about 100 times greater than in the modern low-alkaloid cultivars) by children, and the fourth was an infant who died within three hours of ingesting ten tablets containing a total of 413 mg sparteine. Two non-fatal poisonings involved the ingestion of lupin seed by adults, with an estimated intake of 2–3 g alkaloids. The subjects went into a coma, before recovering fully. In the third poisoning, a girl ingested tablets containing a total of 750 mg sparteine and displayed clinical signs of alkaloid poisoning before recovering.

There have been three further reports of human poisoning from lupin alkaloids. A young male displayed chronic signs of poisoning after he drank water that was used to 'debitter' seed with a high alkaloid content (Marquez *et al.,* 1991). Cooper and Johnson (1984) cited a case where the accidental ingestion of unripe lupin seeds resulted in nausea, vomiting, migraine, abdominal pain, bradycardia and respiratory depression. A woman in Australia ate some salted bitter lupini beans imported from Chile and required intensive care before recovering from classical signs of alkaloid poisoning (Lowen *et al.,* 1995).

The absence of clinical signs in human subjects given single oral doses of 10–70 mg of lupanine or 13-hydroxylupanine suggests that these amounts were below any threshold for pharmacological activity (Petterson *et al.,* 1994). Furthermore, both compounds were excreted largely unchanged (> 90%) in the majority of subjects, and their half-lives were the same for fast and slow metabolizers of the closely related alkaloid, sparteine. Consumption of any product derived from modern cultivars of 'sweet' lupins is therefore unlikely to have any adverse effects. For example, a standard 100 g serving of pasta fortified with 20% lupin flour would only contain about 4 mg alkaloids. Tempe production has been found to remove about 85% of the alkaloids (Fudiyansyah *et al.,* 1995), and germinating (sprouting) the seeds about 80% (Dagnia *et al.,* 1992). Lupin cell wall (fibre) preparations contain only traces of alkaloids (D.J. Harris, Perth, 1992, personal communication).

Phomopsins

One of the major impediments to the development of the lupin industry has been the (sometimes) lethal condition that occurred in sheep grazing lupin pastures (for reviews see Gardiner, 1967; Gladstones, 1970; Chapter 14, this volume). It is caused by phomopsins, a group of toxic secondary metabolites produced by the weakly parasitic and saprophytic fungus *Diaporthe toxica* (previously known as

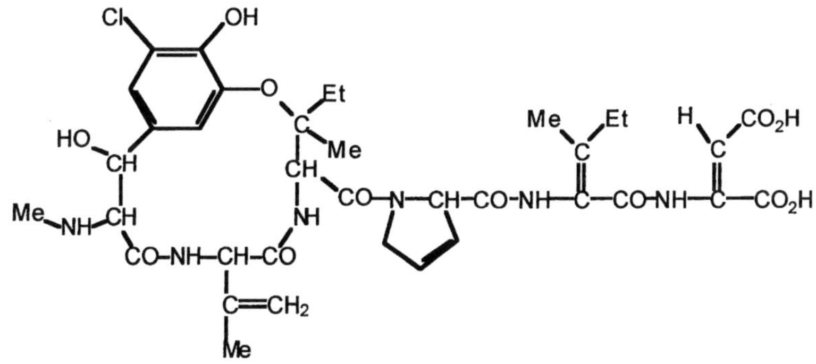

Fig. 12.2. Structure of phomopsin A.

Phomopsis rossiana, *P. leptostromiformis* and *D. woodii*). The phomopsins are linear hexapeptides (Fig. 12.2) with the three terminal amino acids able to form a 13-membered ring by an ether bridge from the 5–OH group of the phenylserine unit to the 3-hydroxyisoleucine unit (Edgar, 1991). They can be hydrolysed to their respective phomopsinamines. It is likely that other compounds make up the group of phomopsins present in nature, because the correlation between measures of biological activity, such as sheep and rat bioassays (Allen *et al.*, 1985), varies between samples and because the correlations between either of these pests and chemical assays is also variable (Chapter 14). The phomopsins act by binding to tubulin and preventing the formation of microtubules (Tonsing *et al.*, 1984; Lacey *et al.*, 1987). Their action is unusually specific in that their primary target *in vivo* is liver parenchymal cells. *In vitro* studies showed them to be amongst the most powerful anti-microtubule agents known (Edgar, 1991). The clinical effects of the phomopsins – functional failure of the liver and fatty infiltrations – are direct consequences of the effects of the phomopsins on microtubular assembly. For a discussion on the toxicology of the phomopsins, see Chapter 14.

Only discoloured seeds contain significant amounts of these toxins (Wood and Petterson, 1986; Wood *et al.*, 1987). The *Diaporthe*-discoloured seeds can be easily removed by grading, using screens, gravity tables and aspirators, because such seeds are smaller and less dense than normal-looking, non-toxin-containing seed (Wood and Petterson, 1986). The same can be achieved by colour-sorting (K.H. Than, Melbourne, 1996, personal communication). A recent survey (K.H. Than, Melbourne, 1996, personal communication) showed that the amount of toxins in commercial harvests in Western Australia from 1991 to 1993 was quite small, with only samples representing less than 10% of the crop containing more than 5 μg phomopsins kg^{-1} seed (the maximum permitted for human consumption by the National Food Authority in Australia and the Department of Health in the UK). Some newer cultivars, and especially pre-release lines of *L. angustifolius*, are more resistant to *D. toxica* invasion than earlier cultivars (W.A. Cowling, Perth, 1996, personal communication) so the situation should be even better in the future. The phomopsins are concentrated initially in the seed coats, and are not found in the cotyledons until there has been a heavy invasion and discoloration of the seed. Since discoloured seeds are readily removed, both by grading and colour-sorting (K.H. Than, Melbourne, 1996, personal communication) the only possible risk of phomopsin ingestion would seem to come from the consumption of very lightly discoloured seed coats, possibly when they have been removed from the cotyledons and used as a high-fibre supplement in breads and the like. An inexpensive and rapid enzyme-linked immunosorbent assay (ELISA) for the phomopsins is readily available to the food industry (K.H. Than, Melbourne, 1996, personal communication).

No other toxicogenic fungi have been detected in lupin seed from Western Australia (P. McR. Wood, Perth, 1995, personal communication). The environment of the Mediterranean climate does not favour fungal activity, added to which the moisture content of seed is almost invariably below 10%.

Allergenicity of lupins

There is a small percentage of people with a food sensitivity to lupins. We have evidence in Australia that fewer than 1% of the population that have eaten lupin-based foods show an immediate allergic reaction (Petterson and Crosbie, 1990), which is less than that commonly found in Western societies towards most food groups. In this case all those who reacted to the lupins were known to react towards other foods such as peanuts and soybeans (D. Barnett, cited in Petterson and Crosbie, 1990). The only other reference to an allergic response to the consumption of lupins in the literature described a typical histamine production in a young girl who consumed a lupin-fortified pasta (Hefle *et al.*, 1994). Gross (1990) reported a skin-prick test on over 200 Chilean children with extracts from lupin and other foods. The sensitivity to lupins (3%) was similar to that for egg (3%) and wheat (2%), and much less than for cow's milk (8%), peanuts (20%) and soybeans (22%). It is feasible that the incidence of acquired allergenicity or food intolerance could rise as more people consume lupin-based foods more often but there is nothing yet to indicate any potential problem. Indeed the Health Department in the UK suggested that, while the public and health officials needed to be aware of this, there was no need for any more stringent action.

Use of Lupins in Foods

The use of lupins in traditional fermented foods

Lupins have been shown to be an excellent substrate for both bacterial and fungal fermentations, used in making foods such as tempe, miso and traditional soy sauces.

Tempe

Tempe is a traditional Indonesian food, made by two successive fermentations using soybeans as substrate (Steinkraus, 1983). Initial bacterial fermentation during the soaking of cooked dehulled soybeans is followed by a solid-state fermentation of the bean mass by the mould *Rhizopus oligosporus* for about 24–48 h at 30–38°C (Steinkraus, 1983). The beans become held together by dense white, cotton-like mycelia. The process has the effect of removing the beany flavour of raw soybeans, increasing B group vitamins, and reducing the content of phytate and oligosaccharides.

Lupins (*L. angustifolius*) were first reported to be suitable for tempe production by Kidby *et al.* (1977). Hung *et al.* (1990) later found them to be a better substrate than soybeans for fermentation, due to greater breakdown of the proteins and complex carbohydrates into simpler and more digestible structures. Agosin *et al.* (1989) found that *L. albus* was also a good substrate for making tempe, and a taste panel found the product to be acceptable. Fudiyansyah *et al.* (1995) found lupin (ASL) tempe to be as acceptable as soy tempe to a group of Australian and Indonesian panellists. Fermentation lowered the contents of oligosaccharides, alkaloids and phytate but there was also a reduction in protein quality, probably due to the greater turnover of the limiting sulphur amino acids during the switch from storage proteins to structural proteins and enzymes. In contrast, Chango *et al.* (1993a) did not find any breakdown of alkaloids or of phytate with fermentation by *R. oligosporus*, possibly a different strain. They did observe a significant breakdown of carbohydrate (30–50%) including the α-galactosides where the greatest breakdown occurred with the highest homologue. In the same laboratory Chango *et al.* (1993b) showed the PER value of the seed (0.83) was lowered by heat treatment (0.58) and that the fermentation partly compensated for this (0.74). They suggested that these variations reflected changes in lysine availability.

While working in Indonesia, Fairbrother and Boughton (1994) learned that lupin tempe was considered quite tasty by many of the local people, but they found the product to be much firmer than the traditional soy tempe. It is uneconomic to cook lupins long enough to soften the seed to the texture of cooked soybeans (Coorey, 1996). In a more recent study in Indonesia, tempe made from *L. albus* splits (dehulled seed) was acceptable to about 200 subjects who regularly consumed tempe (C. Brady-Robeau, T. Sudardi, A. Diah, A.H. Fairbrother, T.J. Boughton and D.S. Petterson, Yogyakarta, 1996, unpublished results). Although the texture of lupin tempe was firmer than that of soy tempe, the taste was sometimes preferred. Overall acceptability was good in that over 40% of the taste panellists indicated that they would buy the product on the market at the same price as soy tempe. Tempe made from a 50 : 50 blend of *L. angustifolius* splits and soybeans was also found to be acceptable by a similar number of people.

The use of splits had the added benefit of cutting labour time (dehulling) and the amount of water needed (dehulling and separating seed coats from cotyledons) for the manufacturing process. A variation on the theme of reducing labour costs was devised by Samish and coworkers in Israel (Y. Birk, Tel Aviv, 1996, personal communication). Whole seeds were hydrated with acidified water then steamed for 10 minutes. This softened the seed coats sufficiently to eliminate the dehulling step with the losses of nutrient-rich waste water. The normal inoculation and incubation resulted in an acceptable (to Israelis) product, which had the same nutritional value, as measured by PER studies in rats, as tempe prepared by traditional methods.

Miso

Miso is a fermented paste made from soybeans, usually with rice. The grains are soaked, cooked and inoculated with the moulds *Aspergillus oryzae* or *A. soyae* and sometimes other organisms. Fermentation occurs over varying periods and the addition of salt with further fermentation results in a thick paste. The paste is then used as a flavouring agent in products such as soups (Shurtleff and Aoyagi, 1976). A traditional process would involve preparing a rice koji by fermenting cooked rice with a culture of *A. oryzae* (tane-koji), then adding cooked soybeans and salt to the koji. This mixture is inoculated with lactic acid bacteria and yeasts, frequently from a previous culture. The colour and flavour of miso can vary with the cereal used (e.g. wheat or barley in lieu of rice), the proportions of salt and cereal in the mash, total cooking time for the soybeans, fermentation time and temperatures, and the inoculum used. Coffey (1989) reported that an expert panel of Japanese found lupin (*L. angustifolius*) miso to be as acceptable as soy miso for flavour and texture and more acceptable in colour and overall appearance. Cunha and Beirao da Costa (1990) reported a satisfactory product from *L. luteus,* with the desired characteristics found in the traditional product.

Other pastes and sauces

Lupins could be used to make sauces similar in flavour and texture to the traditional soy sauces of Japan (Worm and Beira da Costa, 1990) and China (Hung *et al.,* 1990). Lee *et al.* (1982) found lupin a better substrate than soybeans for fermentation of *A. oryzae* to make Korean paste and sauce, but the flavour of the lupin products was slightly less acceptable (Oh and Lee, 1983).

There are no published reports on the use of lupins to make tauco and kecap, though the similarity between their manufacture and that of miso (Steinkraus, 1983) suggests that suitable products could be made. In a recent study in Indonesia, lupins proved a good substrate for fermentation at both the koji and moromi stages of kecap manufacture (T. Sutardi, Yogyakarta, 1997, personal communication).

Bread and pasta products

Breads

Up to about 10% lupin flour can be mixed in with wheat or wholemeal flours to make a more nutritious bread. The blend of cereal and legume helps to balance out the amino acid profile and make it a more complete food. It is generally not possible to add more than about 5% into most wheat flours without a loss in loaf size, because lupin proteins lack the strength and elasticity of wheat gluten (Lucisano and Pompeii, 1981; Petterson and Crosbie, 1990). When stronger flours (e.g. hard red wheat from Canada) are used it is possible to add 15% lupin flour and still retain loaf integrity and a high quality product (L. Brown, Toronto, 1996, personal communication). There is an increase in water-holding capacity and the texture, flavour and yellow colour of the lupin–wheat flour is appealing to many consumers (Petterson and Crosbie, 1989, 1990). In Australia, some bread manufacturers use lupin hull flour in high fibre breads. A lupin loaf which contains about 30% kernel flour, and is free of wheat gluten, is on sale in some states; and a wheaten bread with 5–10% lupin kernel flour is also available (E. Nottage, Perth, 1996, personal communication). Ultrafine kernel flour has an attractive yellow colour, good dispersion in aqueous systems and good emulsifying properties, and in France one company produces lupin kernel and hull flours for the bakery trade (Leterme and Fenart, 1997).

Camacho *et al.* (1989) found that adding up to 12% lupin flour (*L. albus*) to maize flour for production of humitas, a traditional Chilean food, had adverse effects on sensory qualities. It did, however, improve the biological value of the protein. Best overall effects were with an 8% inclusion rate.

Wittig de Penna *et al.* (1989) added 6, 9 or 12% *L. albus* flour to wheat flour to make the traditional Chilean breads 'hallula' and 'marraqueta'. The best products overall had a 6% lupin inclusion, and all were acceptable to children. Yáñez *et al.* (1985) showed that adding lupin (*L. albus*) flour to wheat flour improved protein concentration, amino acid score and PER in rats. Oat flour was not improved by the addition of lupin flour. Corre-Gammere (1993) produced a micronized flour from cotyledons of *L. albus* by pre-grinding with a hammer mill and refining the meal through cooled, smooth cylinders that were differentially geared to achieve the necessary high quality grinding. The end-product was used in breads, biscuits and cakes. For soft breads they could include up to 10% lupin flour without adversely reducing loaf volume. All products were enhanced by protein quality, colour and taste. The added lupin flour retarded stalking, probably by increased water retention and better emulsification properties.

Pasta products

Adding about 10% lupin flour to wheat flours will give a product with similar properties to the full wheat product, but with an improvement in amino acid score from under 40% to over 70% relative to egg albumin (Uauy *et al.,* 1995). Other studies have shown a high acceptability of the lupin-enriched pasta in many countries (Lucisano and Pompei, 1981; Pompei *et al.*, 1985; Yates, 1990) but the product only seems to be used in Australia, where lupini pasta including 15% of *L. angustifolius* kernel flour with durum wheat is sold by three companies, and North America, where pasta flour is fortified with flour from *L. albus* by some companies in the mid-west (D. Gray, Perth, 1995, personal communication).

Rayas-Duarte (1993) found up to 15% substitution of *L. albus* flour for semolina or durum flours enhanced the protein quality and appearance of spaghetti noodles without affecting sensory qualities. Hung and Nithianandan (1993) blended flour from *L. albus* with noodle wheat flour to make unsalted white noodles. Protein and dietary fibre contents were increased, and the colour and texture were improved according to their consumer study. Wittig de Penna *et al.* (1982) recommended the incorporation of 10% sweet lupin (*L. albus*) flour in cookies.

Milk products

Milk

Camacho and Sierra (1988) reported that for maximum solubilization of the lupin proteins it was necessary to extract at pH 8.5 (cf. around neutral pH for soybeans).

These variations are due to the different protein profiles of the two raw materials (Cerletti, 1983; Oomah and Bushuk, 1983).

A lupin-based milk was used in the Chilean programme for children's nutrition for several years (E. von Baer, Monterey, 1996, personal communication). It is still used by households in some parts of the country.

Tofu (tahu)

Hung *et al.* (1990) showed it was possible to incorporate up to 30% lupin milk (*L. angustifolius*) with soy milk before the coagulation stage and produce an acceptable product, with the advantage of a lower unit cost of production. Higher levels of incorporation were not possible because the proteins lack the necessary tertiary structures to produce a good matrix. Ho (1996) worked with newer cultivars of this species and different solubilization techniques but could not improve on yield or quality. Sensory evaluations showed that tofu from a 30/70 lupin/soy blend was equally as acceptable as a tofu made from soy milk (Ho, 1996).

Fermented milks

Yoghurt-type products made from soy milk are being called sogurts (Cheng *et al.*, 1993). Soy milk, calcium lactate, gelatine and lactose are fermented with the appropriate organisms and yield a product that is similar to dairy yogurts, but yellower and firmer. Preliminary results indicate it will be possible to make a lupin-based yoghurt (H. Zhang, Perth, 1996, personal communication).

Sprouts

Lupins can be germinated to make a big sprout suitable for vegetable or salad use. Dagnia *et al.* (1992) germinated seed of *L. angustifolius* for up to six days and found reductions in alkaloids, phytate and oligosaccharides on a dry-weight basis, but also a reduction in the overall content of the protein fraction and its PER value for rats. They reported a fivefold weight increase, whereas Yu (1988) found a 7.5-fold increase compared with fivefold for mung and soybean sprouts. The lupin sprout compared favourably with soy and mung sprouts for taste and texture, though a slight beany and bitter aftertaste made it less acceptable to some panellists at taste sessions (P. Yates, Perth, 1992, personal communication; Kaur, 1994). Unlike most species, there was little evidence for vitamin production on germination. Trugo *et al.* (1993) reported a 90% reduction in the α-galactoside content of both *L. albus* and *L. angustifolius* with germination.

Other food uses

In the Middle East, the seed of *L. albus* is used to make snack foods, most commonly at the time of Ramadan. The seeds are usually boiled whole and spiced, with the hull being removed before eating. Another use is to make a paste similar to the traditional falafel from faba beans.

In Europe, seed of bitter types of *L. albus* (known as lupini beans) is substantially debittered before being pickled and bottled or canned for use as a snack food. Seed of *L. angustifolius* is used by various ethnic groups in the UK as a low-cost alternative to the chickpea. There have been recent newspaper reports of the use of *L. albus* in Germany to make some foodstuffs, including smoked 'cheeses', but there were no indications of quantities used. In South America, the traditional use of tarwi as staple food continues in many Andean communities.

Some Possible Future Food Uses for Lupins and Lupin Products

Whole seed and splits

The best short-term prospects for an increased use for lupins seems to be as a base for fermented foods. There is a lot of interest in using both *L. albus* and *L. angustifolius* in Indonesia for making tempe (T.J. Boughton, Perth, 1996, personal communication). Splits of *L. angustifolius* are being trialled as an alternative to pigeon pea (*Cajanus cajan*) dhal, and could be used as such by many communities.

Immature seeds do not yet contain any alkaloids and have a similar taste and nutritional value to immature (green) peas. There may be a niche market for them, either canned or cooked as a vegetable.

Whole seeds may be deep fried or made into spreads similar in nature to humus (traditionally made from chickpeas) and falafel (traditionally made from faba beans) and used with salads or breads.

Malted lupins could be used as a health drink base. The presence of the beneficial isoflavones genistein and daidzein in the hypocotyls of developing lupin sprouts was shown by Watanabe *et al.* (1993), Angeles Ferrer and Ros Barcelo (1994) and Gagnon *et al.* (1995). These isoflavones also occur in green leaf material.

Processed fractions

The cell wall material from cotyledons of both commercial species is virtually colourless, odourless and tasteless (A.J. Evans, Sydney, 1994, personal communication) and has a very high hydration capacity (seven- to eightfold, compared with three- to fivefold for soy and sugar beet fibres). It is expected to be widely used in breads and baked goods, particularly biscuits (D. Gray, Perth, 1996, personal communication). This fraction was found to have cholesterol-lowering properties when fed to rats (Evans, 1994), and a stool-bulking and transit-time reduction in humans (P. Clifton, Adelaide, 1996, personal communication). Adding 5% lupin cotyledon fibre to white breads enhanced their waterholding capacity and improved acceptability (Evans, 1994).

The fibre can also be used to enrich mousses, jellies and the like to provide tasty desserts. These could form part of the diets of dysphagics and other groups who might otherwise have an inadequate dietary fibre intake.

Consumption of a dietary fibre prepared from *L. albus* (82% purity) was found to lower blood glucose in a panel of eight healthy subjects and give a marked reduction in blood glucose in a panel of four non-insulin diabetics. While more work is needed these results could mean a role for lupin fibre in the treatment of diabetes (Y. Birk, Tel Aviv, 1996, personal communication).

Inclusion of 2% lupin protein isolate in frankfurters improved their processing quality and overall acceptability (Alamanou *et al.*, 1996). A component of the isolate from *L. angustifolius* was found to have excellent whipping properties and could be useful in the confectionery industry (S.J. Johnson, Melbourne, 1996, personal communication).

Lampart-Szczapa (1991) used the plastein reaction to modify a defatted meal from *L. luteus* seed. The food value, as indicated by amino acid profiles, and functional properties, as measured by water and fat absorption, emulsifying capacity and stability, and foam production and stability, were improved.

Extruded foods

It is well known that adding legume flours to cereal flours will give a product with a greater protein quality than either of the component parts. By extruding a lupin–cereal blend it ought to be possible to prepare a tasty snack food of high nutritional quality. Sousa *et al.* (1993a) used a single-screw laboratory extruder to compare the extrusion properties of lupin (*L. luteus*) and soy flours and their respective protein isolates. They concluded that lupin protein extruded in a similar manner to soy protein, and found the lupin product to be lighter and better expanded. In contrast, Sousa *et al.* (1993b) compared the products from extruding (twin-screw) lupin (*L. luteus*) flow and defatted soy grits. Unlike the soy product, the lupin product was not texturized and there was no decrease in protein solubility. This suggested that the higher oil content of the lupin flour (4.9%) could account for this behaviour, and they concluded that with better configuration it should be possible to make a suitable product by twin-screw extrusion. In work with flour from *L. angustifolius,* Fairbrother (1996, unpublished results) produced a satisfactory product with a twin-screw extruder, and was able to produce a high quality product with a corn–lupin blend. Preliminary tests have shown that a high quality product can be made by extruding a (high lysine) lupin – corn preparation (L. Brown, Toronto, 1996, personal communication).

Chajuss (1990) described a process of subjecting an aqueous slurry of defatted lupin meal to high hydraulic or mechanical shear forces at pH 7.0–9.2 as a means of increasing protein solubility. This might be the basis for preparing a better quality milk than has been achieved to date.

Genetically modified lupins and their products

Genetic engineers have already increased the methionine content of *L. angustifolius* (cv. Warrah) seed by a factor of about 2.5 (Chapter 3, this volume). If that is translated into a stronger protein sub-unit with more disulphide bonds then we could expect the lupin protein fraction to have functional properties closer to that of soybeans. The prospects for use in vegetable milks, meat extenders and the like would then appear to be considerable. Whether consumers would want food products from such a source remains to be tested.

Acknowledgements

I thank Professor Yehudith Birk (the Hebrew University of Jerusalem, Israel) for kindly reviewing the manuscript. Drs J.G. Allen, J.S. Gladstones and D.W. Peter kindly reviewed specialist parts of the text. Dr Peter assisted with preparing the tables.

References

Abel-Nabey, A.A. and Shehata, A.A. (1991) Fatty acids, glycerides and unsaponifiables of lupin oil. *Revista Italiana delle Sostange Grasse* 68, 481–485.

Agosin, E., Diaz, D., Aravena, R. and Yáñez, E. (1989) Chemical and nutritional characteristic of lupin tempe. *Journal of Food Science* 54, 102–104.

Aguilera, J.M. and Trier, A. (1978) The revival of the lupin. *Food Technology* 32, 70–76.

Alamanou, S., Bloukas, J.G., Paneras, E.D. and Doxastakis, G. (1996) Influences of protein isolates from lupin seeds (*Lupinus albus* ssp. *graecus*) on processing and quality characteristics of frankfurters. *Meat Science* 42, 79–93.

Allen, J.G., Petterson, D.S. and Ellis, Z.L. (1985) A comparison of two bioassays used to estimate the toxicity of lupin material infected with *Phomopsis leptostromiformis*. *Journal of Applied Toxicology* 5, 293–294.

Angeles Ferrer, M. and Ros Barcelo, A. (1994) Genistein as an endogenous natural substrate of acidic peroxidases in lupin hypocotyls. *Annals of Applied Biology* 125(1), 173–178.

Bailey, R.W., Mills, S.E. and Hove, E.L. (1974) Composition of sweet and bitter lupin seed hulls with observations on the apparent digestibility of sweet lupin seed hulls by young rats. *Journal of the Science of Food and Agriculture* 25, 955–961.

Ballester, D. and Yáñez, E. (1979) Valor nutritivo de harina de trigo enriquecida con harina de lupino (*Lupino albus* y cv. Multolupa). III Reunión del Lupino en Chile, Temuco, p. 11.

Ballester, D., Yáñez, E., Erazo, S., López, F., Haardt, E., Cornejo, S., López, A., Pokniack, J. and Chichester, C.O. (1980) Chemical composition, nutritive value and toxicological evaluation of two species of sweet lupine (*Lupin albus* y *luteus*). *Journal of Agricultural and Food Chemistry* 28, 402–405.

Ballester, D., Brunser, O., Saitúa, M.T., Egaña, J.I., Yáñez, E. and Owen, D. (1984a) Safety evaluation of sweet lupine. II. Nine-month feeding and multigeneration study in rats. *Food and Chemical Toxicology* 22, 45–48.

Ballester, D., Zacarías, I., García, E. and Yáñez, E. (1984b) Baking studies and nutritional value of bread supplemented with full-fat sweet lupin flour (*L. albus* cv. Multolupa). *Journal of Food Science* 49, 14–16.

Beirao da Costa, M.L.D.M. (1980) Aspects of lupin composition as food. In: Birk, Y., Dovrat, A., Waldman, M. and Uzureau, C. (eds) *Proceedings of the Joint CEC–NCRO Workshop*, Israel, 1989, pp. 94–105.

Bhardwaj, H.L., Hamama, A.A. and Merrick, L.C. (1998) Lupin seed composition under different environments. In: Hill, G.D. (ed.) *Proceedings 8th International Lupin Conference, Asilomar, May 1996*. International Lupin Association, Lincoln, (in press).

BIBRA (1986) *A Summary of the Results of Mutagenicity Tests with Alkaloids Isolated from Western Australian Lupins. Report on BIBRA Project No. 3.0585*. The British Industrial Biological Research Association, Carshalton, UK. 50 pp.

Birk, Y., Dourat, A., Imablman, M. and Uzureau, C. (1990) Lupin production and bioprocessing for feed, food and other by-products. In: *Proceedings of the Joint CEC–NCRO Workshop, Israel, January 1989*, Israel SRD, 215 pp.

Blagrove, R.J. and Gillespie, J.M. (1975) Isolation, purification and characterisation of the seed globulins of *Lupinus angustifolius*. *Australian Journal of Plant Physiology* 2, 13–27.

Brillouet, J. and Riochet, D. (1983) Cell wall polysaccharides and lignin in cotyledons and hulls of seeds from various lupin (*Lupinus* L.) species. *Journal of the Science of Food and Agriculture* 34, 861–868.

Butler, W.H., Ford, G.P. and Creasy, D.M. (1996) A 90-day feeding study of lupin (*Lupinus angustifolius*) flour spiked with lupin alkaloids in the rat. *Food and Cosmetics Toxicology* 34, 531–536.

Camacho, L. and Sierra, C. (1988) Lupin processing for the obtainment of a tofu-like product. In: Twardowski, T. (ed.) *Proceedings of the 5th International Lupin Conference, Poznań, July 1988*. International Lupin Association, Poznań, pp. 682–685.

Camacho, L., Banados, E. and Fernandez, E. (1989) Canned 'humitas' prepared from opaque-2 maize with sweet lupin (*Lupinus albus* var. Multolupa), nutritional and quality changes. *Archivos Lantinoamericanos de Nutricion* 39, 185–199.

Cerletti, P. (1983) Lupin seed proteins. *Developments in Food Proteins* 2, 133–171.

Chajuss, D. (1990) *Increasing Solubility of Proteinaceous Material*. UK Patent Application, GB 2 230 932 A.

Chango, A., Bau, H.M., Villaume, C., Mejean, L. and Nicolas, J.P. (1993a) Effects of fermentation by *Rhizopus oligosporus* on chemical composition of sweet white lupin, sweet yellow lupin and bitter yellow lupin. *Sciences des Aliments* 13, 285–295.

Chango, A., Bau, H.M., Villaume, C., Schertz, A., Nicolas, J.P. and Mejean, L. (1993b) Effets de traitements (chauffage et fermentation par *Rhizopus oligosporus* SP-T3) de la graine de lupin blanc doux sur certains facteurs de son utilisation nutritionelle. *Reproduction et Nutrition Development* 33, 89–98.

Cheetham, N.W.H., Cheung, P.C.K. and Evans, A.J. (1993) Structure of the principal non-starch polysaccharide from the cotyledons of *Lupinus angustifolius* (cultivar Gungurru). *Carbohydrate Polymers* 22, 37–47.

Cheng, Y.L., Thompson, L.D. and Britton, H.C. (1993) Sogurt, a yogurt-like soybean product; development and properties. *Journal of Food Science* 55, 1178–1179.

Coffey, R.S. (1989) Lupins as an energy-rich protein source for feed and food. In: Applewhite, T.H. (ed.) *Proceedings of the World Congress on Vegetable Protein Utilisation in Human Foods and Animal Feedstuffs*. American Oil Chemists' Society, Champaign, Illinois, pp. 410–414.

Cooper, M.R. and Johnson, A.W. (1984) In: *Poisonous Plants in Britain and Effects on Animals and Man*. Ministry of Agriculture, Fisheries and Food, London, pp. 149–151.

Coorey, R. (1996) An evaluation of the pilot scale production and shelf-life of lupin tempe. MSc thesis, Curtin University of Technology, Perth, Australia.

Corre-Gammere, V. (1993) The production of micronised sweet white lupin flour for use as a human food. *Advances in Lupin Research* 7, 482–491.

Cunha, A.C. and Beira da Costa, M.L. (1990) *Lupinus luteus* as a potential raw material for miso production. In: von Baer, D. (ed.) *Abstracts of Poceedings 6th International Lupin Conference, Temuco-Pucon, November 1990*. Associacion Chilena del Lupino, Temuco, Chile, p. 32.

Dagnia, S.G., Petterson, D.S., Bell, R.R. and Flanagan, F.V. (1992) Germination alters the nutritional value of lupin seed. *Journal of the Science of Food and Agriculture* 60, 419–423.

Duranti, M., Gorinstein, S. and Cerletti, P. (1990) Synthesis of storage proteins and seed development in *L. albus* under different growth conditions. *Journal of Food Biochemistry* 14, 327–330.

Duranti, M., Gius, C. and Scarafoni, A. (1995) Lectin-like activity of lupin seed conglutin γ, a glycoprotein previously referred to as a storage protein. *Journal of Experimental Botany* 46, 725–728.

Edgar, J.A. (1991) Phompsins, antimicrotubule mycotoxins. In: Keeler, R.F. and Tuy, A.T. (eds) *Handbook of Natural Toxins, Vol. 6, Plant and Fungal Toxins.* Marcel Dekker, New York, pp. 371–395.

Egaña, J.I., Uauy, R., Cassorla, X., Barrera, G. and Yáñez, E. (1992) Sweet lupin protein quality in young adult males. *Journal of Nutrition* 122, 2341–2347.

Evans, A.J. (1994) The carbohydrates of lupins, composition and uses. In: Dracup, M. and Palta, J. (eds) *Proceedings of the First Australian Lupin Technical Symposium.* Western Australian Department of Agriculture, South Perth, pp. 110–114.

Evans, A.J., Cheung, P.C.K. and Cheetham, N.W.H. (1993) The carbohydrate composition of cotyledons and hulls of cultivars of *Lupinus angustifolius* from Western Australia. *Journal of the Science of Food and Agriculture* 61, 189–194.

Fairbrother, A.H. and Boughton, T.J. (1994) The use of lupins in the manufacture of tempe in Indonesia. In: Dracup, M. and Palta, J. (eds) *Proceedings of the First Australian Lupin Technical Symposium.* Western Australian Department of Agriculture, South Perth, pp. 106–109.

Feldheim, W. (1990) Use of lupins as sources of lipids and dietary fibre in human nutrition. In: von Baer, D. (ed.) *Proceedings 6th International Lupin Conference, Temuco-Pucón, November 1990.* Associacion Chilena del Lupino, Temuco, pp. 124–131.

Fenwick, G.R., Price, K.R., Tsukamoto, C. and Okubo, K. (1991) Saponins. In: D'Mello, J.P.F., Duffus, C.M. and Duffus, J.H. (eds) *Toxic Substances in Crop Plants.* The Royal Society of Chemistry, Cambridge, pp. 285–327.

Fudiyansyah, N., Petterson, D.S., Bell, R.R. and Fairbrother, A.H. (1995) A nutritional, chemical and sensory evaluation of lupin tempe. *International Journal of Food Science and Technology* 30, 297–306.

Gagnon, H., Tahara, S.L. and Ibrahim, R.K. (1995) Biosynthesis, accumulation and secretion of isoflavonoids during germination and development of white lupin (*Lupinus albus* L). *Journal of Experimental Botany* 46, 609–616.

Gardiner, M.R. (1967) Lupinosis. *Advances in Veterinary Science* 11, 85–138.

Gattás, V., Barrera, G., Yáñez, E. and Uauy, R. (1990) Evaluación de la tolerancia y aceptabilidad crónica de la harina de lupino (*Lupino albus* var. Multolupa) en la alimentación de adultos jóvenes. *Archivos Latinoamericanos de Nutricion* 40, 490–501.

Gladstones, J.S. (1970) Lupins as a crop plant. *Field Crop Abstracts* 23, 123–148.

Gladstones, J.S. (1982) Breeding the first modern crop lupins. *Journal of Agriculture, Western Australia* 23, 70–72.

Grant, G., Dorward, P.M. and Pustzai, A. (1993) Pancreatic enlargement is evident in rats fed diets containing raw soybeans (*Glycine max*) or cowpeas (*Vigna unguiculata*) for 800 days but not in those fed diets based on kidney beans (*Phaseolus vulgaris*) or lupin seed (*Lupinus angustifolius*). *Journal of Nutrition* 123, 2207–2213.

Grant, G., Dorward, P.M., Buchan, W.C., Armour, J.C. and Pustzai, A. (1995) Consumption of diets containing raw soybeans (*Glycine max*), kidney beans (*Phaseolus vulgaris*), cowpeas (*Vigna unguiculata*), or lupin seeds (*Lupinus angustifolius*) by rats for up to 700 days: effects on body composition and organ weights. *British Journal of Nutrition* 73, 17–29.

Griffiths, D.W. (1991) Condensed tannins. In: D'Mello, J.P.F., Duffus, C.M. and Duffus, J.H. (eds) *Toxic Substances in Crop Plants.* The Royal Society of Chemistry, Cambridge, pp. 180–201.

Gross, R. (1990) Nutrients and anti-nutrient substances in lupins for human consumption. In: Birk, Y., Dourat, A., Waldman, M. and Uzureau, C. (eds). *Proceedings of the Joint CEC–NCRD Workshop, Israel, 1989*, pp. 164–176.

Harris, D.J. and Jago, J. (1984) *Report on Chemical Composition of Sweet Lupin Seed from Western Australia*. Government Chemical Laboratories, Perth, 12 pp.

Hefle, S.L., Lemanske, R.F. and Bush, R.K. (1994) Adverse reaction to lupine-fortified pasta. *Journal of Allergy and Clinical Immunology* 94, 161–172.

Hill, G.D. (1977) The composition and nutritive values of lupin seed. *Nutrition Abstracts and Reviews, B.* 47, 511–519.

Hill, G.D. (1986) Recent developments in the uses of lupins in animal and human nutrition. In: Lawson, E. (ed.) *Proceedings 4th International Lupin Conference, Geraldton, August 1986*. International Lupin Association and Western Australian Department of Agriculture, South Perth, pp. 40–63.

Ho, R.W.L. (1996) Incorporation of lupin into tofu. MSc thesis, Curtin University of Technology, Perth, Australia.

Hondelmann, W. (1984) The lupin – ancient and modern crop plant. *Theoretical and Applied Genetics* 68, 1–9.

Hung, T.V. and Nithianandan, V. (1993) Preparation and evaluation of noodles supplemented with chickpea and lupin flours. *ASEAN Food Journal* 8, 26–31.

Hung, T.V., Handson, P.D., Amenta V.C., Kyle, W.S.A. and Yu, R.S.T. (1987) Content and distribution of manganese in lupin seed grown in Victoria and in lupin flour, spray-dried powder and protein isolate prepared from the seeds. *Journal of the Science of Food and Agriculture* 41, 131–139.

Hung, T.V., Papolais, M., Nithianandan, V., Jiang, H.H. and Versteeg, K. (1990) Utilisation of lupin seeds for human consumption. In: *'Food Pacific' Convention, Gold Coast, May 1990*. Australian Institute of Food Science and Technology, Australia, pp. 13–15.

Jimenez, M.D., Cubero, J.I. and Jaro, A. de (1991) Genetic and environmental variability in protein, oil and fatty acid composition in high-alkaloid white lupin (*Lupinus albus*). *Journal of the Science of Food and Agriculture* 55, 27–35.

Kaur, S.B. (1994) Comparative evaluation of commercially grown lupin and mung sprouts. MSc thesis, Edith Cowan University, Perth, Australia, 227 pp.

Keeler, R.F. (1973) Lupin alkaloids from teratogenic and non-teratogenic lupins. 1. Correlation of crooked calf disease incidence with alkaloid distribution determined by gas chromatography. *Teratology* 7, 23–30.

Kidby, D., McComb, J.R., Snowden, R.L., Garcia-Webb, P. and Gladstones, J.S. (1977) Tempe production from *Lupinus angustifolius* L. In: *Symposium on Indigenous Fermented Foods, Bangkok, Thailand, November 1977*. UNEP/UNESCO/ICRO.

Klassen, C.D. and Doull, J. (1980) Evaluation of safety: toxicological evaluation. In: Doull, J., Klassen, C.D. and Amdier, M.D. (eds) *Cassaret and Doull's Toxicology. The Basic Science of Poisons*, 2nd edn. Macmillan, New York, p. 12.

Korol, W. and Matyka, S. (1988) Cadmium and lead in dried, whole, coarse-seeded legume plants. *Biuletyn-Informacyjng-Przemyslu-Paszwego* 27, 56–59.

Kyle, W.S.A., Petterson, D.S. and Evans, A.J. (1991) The use of lupins in Asian food products. In: *Food Ingredients Asia*. Expoconsultant Publications, The Netherlands, pp. 176–179.

Lacey, E., Edgar, J.A. and Culvenor, C.C.J. (1987) Interaction of phomopsin A and related compounds with purified sheep brain tubulin. *Biochemical Pharmacology* 36, 156–164.

Lampart-Szczapa, E. (1991) Improving the food value of lupin proteins by means of the plastein reaction. *Przemysl Spozywczy* 45, 104–106.

Lee, C.H., Oh, S.H. and Kim, C.S. (1982) Studies on the fermentation of lupin seed. I. Determination of the growth rate of *Aspergillus oryzae* on beans. *Korean Journal of Applied Microbiology and Bioengineering* 10, 227–232.

Leterme, P. and Fenart, E. (1997) New outlets in the food industry. *Grain Legumes*, No. 15, p. 24. European Grain Legume Association.

Liener, I.E. (1989) Antinutritional factors in legume seeds: state of art. In: Huisman, J., van der Poel, A.F.B. and Liener, I.E. (eds) *Recent Advances of Research in Antinutritional Factors in Legume Seeds*. Pudoc, Wageningen, The Netherlands, pp. 6–13.

Lilley, G.G. (1986) The sub-unit structure and stability of conglutin delta, a sulfur-rich seed protein of *Lupinus angustifolius* L. *Journal of the Science of Food and Agriculture* 37, 895–907.

López de Romaña, G., Graham, C.G., Morales, E., Massa, E. and McLean, W.C. Jr (1983) Protein quality and oil digestibility of *Lupinus mutabilis*: metabolic studies in children. *Journal of Nutrition* 113, 773–778.

Lopez-Bellido, L. and Fuentes, M. (1986) Lupin crop as an alternative source of protein. *Advances in Agronomy* 40, 239–295.

Lucisano, M. and Pompei, C. (1981) Baking properties of lupin flour. *Food Science* 14, 323–326.

Lowen, R.J., Alam, F.K.A. and Edgar, J.A. (1995) Lupin bean toxicity. *The Medical Journal of Australia* 162, 256–257.

Marquez, L.R., Guitierrez-Rave, M. and Miranda, F.I. (1991) Acute poisoning by lupine seed debittering water. *Veterinary and Human Toxicology* 33, 265–267.

Mazur, M., Polakowski, P. and Szadowski, A. (1966) Pharmacological investigations of 17-oxolupanine, lupanine aminooxide and 17-hydroxylupanine. *Acta Physiologica Polonica* 17, 244–253, 254–262.

Mosse, J., Huet, J.-C. and Baudet, J. (1987) Relationships between nitrogen, amino acids and storage proteins in *Lupinus albus* seeds. *Phytochemistry* 26, 2453–2458.

Muzquiz, M., Ridout, C.L., Price, K.R. and Fenwick, G.R. (1993) The saponin content and composition of sweet and bitter lupin seed. *Journal of the Science of Food and Agriculture* 63, 47–52.

Oh, S.H. and Lee, C.H. (1983) Studies on the fermentation of lupin seed. II. Preparation of traditional Korean fermented bean sauce and paste. *Korean Journal of Applied Microbiology and Bioengineering* 11, 241–248.

Oliveira, M.B.P.P. and Ferreira, M.A. (1988) Study of lupin (*Lupinus albus* L.) seed oil. *Revista Portuguesa de Farmacia* 38, 25–40.

Oomah, B.D. and Bushuk, W. (1983) Characterization of lupine proteins *Lupinus albus, Lupinus angustifolius*. *Journal of Food Science* 48, 38–41.

Petterson, D.S. (1985) *Lupins as a Food for Man and Livestock.* Western Australia Department of Agriculture, Perth.

Petterson, D.S. and Crosbie, G.B. (1989) Some food uses for lupins. *Proceedings 39th Australian Cereal Chemistry Conference.* Royal Australian Chemical Institute, 12 Vale Street, North Melbourne, Victoria 3051, pp. 158–162.

Petterson, D.S. and Crosbie, G.B. (1990) Potential of lupins as food for humans. *Food Australia* 42, 266–268.

Petterson, D.S. and Mackintosh, J.B. (1994) *The Chemical Composition and Nutritive Value of Australian Grain Legumes.* Grains Research and Development Corporation, Canberra.

Petterson, D.S., Ellis, Z.L., Harris, D.J. and Spadek, Z.E. (1987) Acute toxicity of major alkaloids of *Lupinus angustifolius* to rats. *Journal of Applied Toxicology* 7, 51–53.

Petterson, D.S., Greirson, B.N., Allen, D.G., Harris, D.J., Power, B.M., Dusci, L.J. and Ilett, K.F. (1994) Disposition of lupanine and 13-hydroxylupanine in man. *Xenobiotica* 24, 933–941.

Petterson, D.S., Sandström, B. and Cederblad, Å. (1994) Absorption of zinc from lupin-based foods. *British Journal of Nutrition* 72, 865–71.

Pompei, C., Lucisano, M. and Ballini, N. (1985) Use of lupin flour in pasta manufacture. *Sciences des Aliments* 5, 665.

Rayas-Duarte, P. (1993) Cooling and nutritional quality of lupin-containing spaghetti noodles. In: Neves-Martins, J.M. and Beirao da Costa, M.L. (eds) *Advances in Lupin Research. Proceedings of the 7th International Lupin Conference, Evora, April 1993.* Instituto Superior de Agronomia and International Lupin Association, Lisbon, pp. 537–541.

Restani, P., Durant, M., Cerletti, P. and Simonetti, P. (1981) Subunit composition of the seed globulins of *Lupinus albus. Phytochemistry* 20, 2077–2083.

Robaina, L., Izquierdo, M.S., Moyano, F.J., Socorro, J., Vergara, J.M., Montero, D. and Fernandez Palacios, H. (1995) Soybean and lupin seed meals as protein sources in diets for gilthead seabream (*Sparus aurata*): nutritional and histological implications. *Aquaculture* 130, 219–233.

Robbins, M.C., Brantom, P.G. and Petterson, D.S. (1996) A 90-day feeding study on the alkaloids of *Lupinus angustifolius* in the rat. *Food and Chemicals Toxicology* 34, 679–686.

Ruiz, R.G., Price, K.R., Rose, M.E., Arthur, E.A., Petterson, D.S. and Fenwick, G.R. (1995) The effect of cultivar and environment on saponin content of Australian sweet lupin seed. *Journal of the Science of Food and Agriculture* 69, 347–351.

Salmanowicz, B.P. (1995) Comparative studies of seed albumins in the Old World *Lupinus* species (*Fabaceae*) by reversed-phase HPLC. *Plant Systematics and Evolution* 195, 77–86.

Samish, S., Borges, G., Madar, Z. and Birk, Y. (1990) Nutritional and metabolic aspects of lupin seeds and their fractions in feeding experiments with rats. In: Birk, Y., Dourat, A., Imablman, M. and Uzureau, C. (eds) *Lupin Production and Bioprocessing for Feed, Food and Other By-products. Proceedings Joint CEC–NCRO Workshop,* Israel, 1989, Israel SRD, pp. 134–141.

Schmidlin-Meszaros, J. (1973) Eine Natrungsmittelvergiftung. *Mitteilungen aus dem Gebeite der Lebensmitteluntersuchung und Hygiene* 64, 194–205.

Shurtleff, W. and Aoyagi, A. (1976) *The Book of Miso.* Autumn Press, Kangawa-ken, Japan.

Sousa, I.M.N., Mitchell, J.R., Hill, S.E. and Beirao da Costa, M.C. (1993a) Texturizatin by extrusion cooking of lupin flour and lupin protein isolates. In: Neves-Martins, J.M. and Beirao de Costa, M.L. (eds) *Advances in Lupin Research. Proceedings 7th International Lupin Conference, Evora, April 1993.* Instituto Superior de Agronomia and International Lupin Association, Lisbon, pp. 529–533.

Sousa, I.M.N., Beirao da Costa, M.L., Cabodevila, O., Hill, S.E., Mitchell, J.R. and Mohammed, Z. (1993b) A comparison between the extrusion behaviour of lupin and soya. *Advances in Lupin Research. Proceedings 7th international Lupin Conference, Evora, April 1993.* Instituto Superior de Agronomia and International Lupin Association, Lisbon, pp. 473–481.

Steinkraus, K.H. (1983) Tempe manufacture. In: Steinkraus, K.H. (ed.) *Handbook of Indigenous Fermented Foods.* Marcel Dekker, New York, pp. 11–26.

Tonsing, E.M., Steyn, P.S., Osborn, M. and Weber, K. (1984) Phomopsin A, the causative agent of lupinosis, interacts with microtubules *in vivo* and *in vitro. European Journal of Cell Biology* 35, 156–164.

Trugo, L.C., Farrah, A. and Trugo, N.M.F. (1993) Germination and debittering lupin seeds reduce alpha-galactoxidase and intestinal carbohydrate fermentation in humans. *Journal of Food Science* 58, 627–630.

Uauy, R. and Yáñez, E. (1983) Plant foods for human protein nutrition: studies on soy, lupin and mixed vegetable sources. *Qualitas Plantarum/Plant Foods for Human Nutrition* 33, 17–28.

Uauy, R., Gattas, V. and Yáñez, E. (1995) Sweet lupins in human nutrition. In: Simopoulos, A.M. (ed.) *World Review of Nutrition and Dietetics* 77, 75–88.

Underwood, E.J. (1977) *Trace Elements in Human and Animal Nutrition*, 4th edn. Academic Press, New York.

Watanabe, K., Kinjo, J. and Nohara, T. (1993) Three new isoflavonoid glycosides from *Lupinus luteus* and *L. polyphyllus* × *arboreus. Chemical and Pharmaceutical Bulletin* 41, 394–396.

White, C.L., Robson, A.D. and Fisher, H.M. (1981) Variations in nitrogen, sulphur, selenium, cobalt, managanese, copper and zinc contents of grain from wheat and two lupin species grown in a range of Mediterranean environments. *Australian Journal of Agricultural Research* 32, 47–59.

Williams, W. and McGibbon, R. (1980) Environmental effects on seed oil percentage and fatty acid composition in white lupin (*Lupinus albus*). *Journal of Agricultural Science* 95, 597–602.

Wink, M., Meissner, C. and Witte, L. (1995) Patterns of quinolizidine alkaloids in 56 species of the genus *Lupinus. Phytochemistry* 98, 139–153.

Wittig de Penna, E., Carreno, P., Urrutia, X., Lopez, L. and Ballester, D. (1982) Sensory evaluation and acceptability of cookies enriched with sweet lupine flour (*Lupinus albus* cv. Multolupa). *Journal of Food Science* 52, 1434–1435.

Wittig de Penna, E., Castro, X., Cerda P. and Ballester, D. (1989) Evaluation of the sensory quality and acceptability of 'hallulla' and 'marraqueta' breads enriched with sweet lupin flour (*Lupinus albus* cv. Multolupa). *Resvistade Agroquimica y Tecnologia de Alimentos* 28, 127–134.

Wood, P.McR. and Petterson, D.S. (1986) *Phomopsis leptostromiformis* infection and phomopsin A content of lupin seed in Western Australia. *Australian Journal of Experimental Agriculture* 26, 583–6.

Wood, P.McR., Petterson, D.S., Hancock, G.R. and Brown, G.A. (1987) Distribution of seed infected with *Phomopsis leptostromiformis* and of phomopsin A within a lupin crop. *Australian Journal of Experimental Agriculture* 27, 77–79.

Worm, M.A. and Beirao de Costa, M.L. (1990) Lupin as a raw material for shoyu production. In: von Baer, D. (ed.) *Proceedings 6th International Lupin Conference, Temuco-Pucón, November 1990.* Associacion Chilena del Lupino, Temuco, pp. 60–67.

Yáñez, E. (1990) Lupin as a source of protein in human nutrition. In: von Baer, D. (ed.) *Proceedings 6th International Lupin Conference, Temuco-Pucón, November 1990.* Association Chilena del Lupino, Temuco, pp. 115–123.

Yáñez, E. (1998) Lupins in human nutrition. In: Hill, G.D. (ed.) *Looking Towards the 21st Century. Proceedings of the Eighth International Lupin Conference.* International Lupin Association, Lincoln, (in press).

Yáñez, E., Gattás, V. and Ballester, D. (1979) Valor nutritivo de lupino y su potencial como alimento humano. *Archivos Latinoamericanos de Nutricion* 29, 510–520.

Yáñez, E., Ivanovic, D., Owen, D.F. and Ballester, D. (1983) Chemical and nutritional evaluation of sweet lupines. *Annals of Nutrition and Metabolism* 27, 513–520.

Yáñez, E., Ballester, D. and Ivanovic, D. (1985) Wheat and oat fortification with sweet lupin flour (*L. albus* cv. Maleolupa). *Nutrition Reports International* 31, 493–499.

Yates, P. (1990) *Product Development and Sensory Evaluation of Lupin Pasta.* Edith Cowan University Report, Perth, Australia, 88 pp.

Yovo, K., Huguet, F., Pothier, J., Durand, M., Breteau, M. and Narcisse, G. (1984) Comparative pharmacological study of sparteine and its ketonic derivative lupanine from seeds of *Lupinus albus. Planta medica* 50, 420–424.

Yu, R. (1988) Incorporation of lupin into human foods. In: Saipin Maneepur, Pivan Varangoon and Budan Phithakpol (eds) *Proceedings of the Food Conference '88.* Institute of Food Research and Product Development, Kaselsart University, Thailand, pp. 24–26.

Zdunczyk, Z., Juskiewicz, J., Frejnagel, S., Flis, M. and Godycka, I. (1994) Chemical composition of the cotyledons and seed coat and nutritional value of whole and hulled seeds of yellow lupin. *Journal of Animal and Feed Sciences* 3, 141–148.

13 Lupins for Livestock and Fish

A.C. Edwards[1] and R.J. van Barneveld[2]

[1]ACE Livestock Consulting Pty Ltd, PO Box 108, Cockatoo Valley, South Australia 5351; [2]South Australian Research and Development Institute – Pig and Poultry Production Institute, Nutrition Research Laboratory, The University of Adelaide, Roseworthy, South Australia 5371

Introduction

Lupins are widely used as a source of protein and energy in livestock feeds. Their high protein content makes them a valuable resource for monogastric and ruminant production systems as they are cost-competitive with a wide range of other protein sources. Their low levels of starch and high levels of fermentable carbohydrate make them a highly desirable ruminant feed due to the low incidence of acidosis when they are fed and the fact that they are an excellent microbial fermentation substrate.

All modern livestock systems have a strong emphasis on the efficiency of production. The comparatively high levels of soluble and insoluble non-starch polysaccharides can influence the utilization of other nutrients in lupins and hence they must be used strategically if livestock production responses are to be optimized. In addition, comparatively low levels of the sulphur amino acids, methionine and cystine, influence the way lupins are used in livestock diets.

The aim of this chapter is to present detailed information on the nutritional value of lupins for different classes of livestock. Emphasis on the differing nutritional roles of lupins, depending on the livestock production system, will present them as a highly versatile feed ingredient.

Quantity and Species of Lupins Used in Livestock Production

The potential for lupin use in livestock feeds in Australia is enormous. Edwards (1994) estimated total manufactured stockfeed production (pigs, poultry, dairy, feedlot beef) to be 7.85 Mt, which could include up to 1.36 Mt of grain legumes. Of this, lupins have the potential to be the predominant component with possible use in the vicinity of 1.13 Mt. Beyond that used in intensive livestock production, it has been estimated that grazing sheep consume up to 350,000 t annually (Murray, 1994).

In addition to traditional farming systems, there is potential for lupin use in aquaculture. Aquaculture systems are developing rapidly, with an estimated

 Lupins as Crop Plants: Biology, Production and Utilization (eds J.S. Gladstones, C.A. Atkins and J. Hamblin)

global production of 24 Mt by the year 2000 (Chamberlain and Rosenthal, 1995) and a need to reach 65 Mt by 2020 if the current modest per capita consumption is to be maintained. Like other forms of intensive animal production, pelleted feed costs in many aquaculture systems (freshwater trout, prawns, crocodiles, Atlantic salmon, ocean trout and barramundi) represent over 40% of the total operating costs (Treadwell *et al.*, 1991). For this reason, it is important that cost-competitive feed ingredients can be identified. We must also consider that the use of fishmeal in aquaculture systems is not sustainable due to the fact that global wild fisheries are estimated to peak at 84 Mt (FAO, 1994) with only 29 Mt being used as sources of fishmeal and oils. As a consequence, alternative protein sources must be identified.

A major limitation to the achievement of potential lupin use in stockfeeds in Australia is a distinct geographic dislocation of supply and demand. For example, livestock consumption of lupins in Queensland could be up to 254,000 t, but the growing regions are well beyond the economic transport boundaries of these end-users. Aquaculture enterprises in Asia, however, may represent a large future market for lupins, especially considering the dominance of herbivorous fish species in these systems.

In Australia, *L. angustifolius* and, to a smaller extent, *L. albus* are the species currently available for use in livestock feeds (chapters 11 and 12). *L. luteus* is a further species with potential for Australia and its nutritional value for different classes of livestock is being assessed.

With the exception of pigs, *L. angustifolius* and *L. albus* can be used with equal success in all classes of livestock. *L. albus* is currently not recommended for use in pig diets. Commercial nutritionists formulating diets for large Australian piggeries state categorically that it is uneconomical to include *L. albus* in pig diets due to the resulting depressions in growth rates. The main reason for reduced growth rates when *L. albus* is included in pig diets at levels above 15% (Standing Committee on Agriculture, 1987) is reduced feed intake. Our knowledge of the nutritional value of lupins for aquaculture species, compared with terrestrial animals is limited – largely due to the infancy of the aquaculture industry.

Nutritional Value of Lupins for Monogastrics

Composition relative to monogastric animal production

Protein

PIGS AND POULTRY. Pigs and poultry do not have a requirement for crude protein *per se* but, rather, specific levels of individual amino acids. Despite this, crude protein level is often used as a guide to the amino acid content of lupins, rather than direct amino acid analysis. Hence, variation in crude protein content can affect the efficiency of use of lupins in pig and poultry diets. Petterson and Mackintosh (1994) reported a variation in the protein content of *L. angustifolius* of 272 to 376 g kg^{-1} (air-dry basis, $n = 3351$). Similar variation was reported for *L. albus* (291–403 g kg^{-1}). Not only will variation of this magnitude affect price, but it emphasizes the need for intensive livestock producers to measure the crude protein content of all lupins used in stockfeeds.

AQUACULTURE SPECIES. The digestibility of nitrogen (i.e. crude protein) in *L. angustifolius* and *L. albus* has been assessed in silver perch (G. Allan, Pt Stephens Research Centre, Salamander Bay, New South Wales, 1994, unpublished data). Nitrogen digestibility was extremely high for both lupin species with hulls depressing N digestibility by 3–4% in the whole seed (Table 13.1). These data are supported by Robiana *et al.* (1995) who reported apparent protein digestibility coefficients of 95.5, 94.6 and 93.0% for diets containing 10, 20 or 30% lupin-seed meal, respectively. Methodology used to determine these digestibility coefficients may be artificially elevating the values, resulting in lower than expected performance.

Amino acids

The amino acid composition of lupin protein resembles that of many other legume proteins in being low in methionine (0.59–0.87 g 16 g N^{-1}), but it is also relatively low in lysine (4.21–5.21 g 16 g N^{-1}) and very high in arginine (10.60–13.50 g 16 g N^{-1}). Liebholz (1984) showed a growth response in pigs (21–35 kg) to additional methionine when lupin-seed meal was used as the sole protein supplement to a cereal-based diet. Even in commercial mixed diets containing multiple protein sources, where the deficits and excesses of one ingredient can be matched by the opposite in others, supplementation with synthetic methionine is often required. Genetic engineering of *L. angustifolius* has generated a new line in cv. Warrah that expresses a sunflower seed albumen and, as a consequence, has double the methionine content of the protein component (Chapters 3 and 5, this volume).

The chemical composition of a feed ingredient is only the first step in describing the nutritional value. When dealing with monogastrics, the nutritional value of a feed ingredient is commonly based on the available nutrient content, or the best estimate thereof. Available amino acids represent the proportion of amino acids in a feedstuff that are in a form suitable for use by the animal for growth and metabolism. Ileal digestibility of amino acids is often equated to availability on the assumption that if an amino acid is absorbed from the digestive tract it is utilized. In addition, it is comparatively quick and inexpensive to measure and all amino acids can be assessed simultaneously. In contrast, direct

Table 13.1. Digestibility of dry matter, energy and nitrogen in the whole seed and kernels of *L. angustifolius* and *L. albus* fed to silver perch (from G. Allan, Pt Stephens Research Centre, Salamander Bay, NSW, 1994, unpublished data).

	Digestibility coefficient $\left(\frac{\text{intake} - \text{output}}{\text{intake}}\right)$		
Source	Dry matter	Gross energy	Nitrogen
L. angustifolius			
Whole seed	0.50	0.59	0.97
Kernel	0.68	0.74	1.00
L. albus			
Whole seed	0.65	0.73	0.96
Kernel	0.78	0.85	1.01

amino acid availability measurements determined using slope-ratio analysis are time-consuming and expensive; only one amino acid can be assessed at a time and the standard errors of measurement are high.

PIGS. To date, evaluations of the nutritional value of lupins for pigs have suggested that estimates of apparent ileal amino acid digestibility significantly overestimate amino acid availability, and hence apparent ileal amino acid digestibility values are unsuitable for use in diet formulations. The apparent ileal digestibility of amino acids in the whole seed and kernels of *L. angustifolius* and *L. albus* is high (Table 13.2). In contrast, Batterham *et al.* (1984) reported the availability of lysine in three samples of *L. angustifolius* cv. Uniharvest for pigs to be low and highly variable (Table 13.3). To further confound the issue, the current recommended value for lysine availability in lupins of 0.55 (Standing Committee on Agriculture, 1987) is not supported by results achieved commercially. Industry experience has shown that values of 0.70 and 0.80 can be applied as the availability of lysine in the whole seed and kernels of *L. angustifolius*, respectively. Godfrey and Payne (1987) also suggested that availability of lysine in *L. angustifolius* exceeds 0.70. Despite this, nutritionists could not be certain that lupins in pig diets were being used with maximum efficiency without direct evidence.

Table 13.2. Apparent ileal digestibility[1] of amino acids (proportion of total) of the whole seed and kernels of *L. angustifolius* and *L. albus* determined in growing pigs (35–40 kg) fitted with simple T-piece cannulas (from R.J. van Barneveld, Northfield Pig Research Unit, Northfield, South Australia, 1995, unpublished data).

	L. angustifolius		*L.albus*	
Amino acid	Whole	Kernel	Whole	Kernel
Aspartic acid	0.84	0.93	0.86	0.89
Threonine	0.72	0.86	0.73	0.78
Serine	0.77	0.89	0.81	0.84
Glutamic acid	0.87	0.97	0.90	0.90
Proline	0.60	0.60	0.66	0.63
Glycine	0.60	0.66	0.64	0.68
Alanine	0.67	0.85	0.69	0.74
Valine	0.78	0.91	0.80	0.84
Isoleucine	0.79	0.90	0.83	0.85
Leucine	0.81	0.91	0.83	0.86
Tyrosine	0.79	0.91	0.83	0.83
Phenylalanine	0.82	0.92	0.80	0.86
Lysine	0.82	0.92	0.81	0.84
Histidine	0.84	0.86	0.82	0.87
Arginine	0.88	0.96	0.93	0.95
Methionine	0.68	0.86	0.62	0.84
Cystine	0.84	0.92	0.78	0.90

[1]Apparent ileal digestibility = proportion of total amino acids assumed to be digested by the time digesta reaches the terminal ileum, uncorrected for any endogenous secretions.

Recent experiments (R.J. van Barneveld, SARDI – Pig and Poultry Production Institute, Roseworthy, South Australia, 1996, unpublished data) using a modified slope-ratio analysis suggest that lysine availability is significantly higher in lupins than the estimates provided by Batterham *et al.* (1984) and is closely represented by apparent ileal digestibility estimates (Table 13.4). The differences in estimates between the two experiments may be due to differences in the cultivars tested, or differences in methodology.

POULTRY. Comparisons by Batterham (1992) have shown that there are species differences in the ability to utilize available lysine from lupins and that lysine in lupins has a high availability for poultry (Table 13.3).

Energy

PIGS. Like amino acids, estimates of the available energy content of a feedstuff are important if diet formulations are to match the pig's requirements accurately. The digestible energy content (DE, which equals gross energy, GE, minus energy in faeces) is commonly used as an estimate of available energy in feedstuffs for pigs. Net energy (NE, which equals DE minus energy in urine, gas and the heat increment) is the most desirable method for assessing the 'true' energy content of a feedstuff for both pigs and poultry, but it is very difficult to measure routinely (Edwards and Campbell, 1991).

Wigan *et al.* (1994) summarized recent estimates of the DE content of lupins for pigs. Estimates ranged from 12.3 to 15.3 MJ kg^{-1} for lupin-seed meal and 15.4 to 16.6 MJ kg^{-1} for lupin kernels. This range in DE estimates may be due to the method of preparation of the lupins prior to inclusion in experimental diets. It appears that digestibility is inversely proportional to particle size; that is, the finer

Table 13.3. Comparison of the availability[1] of lysine (proportion of total) in three samples of *L. angustifolius* cv. Uniharvest for pigs, rats and chicks (from Batterham, 1992).

Sample	Pig	Rat	Chick
Lupin seed meal No. 1	0.37	0.81	0.98
Lupin seed meal No. 2	0.54	0.70	0.93
Lupin seed meal No. 3	0.54	0.80	0.84

[1]Availability = proportion in a form suitable for use by the animal.

Table 13.4. Preliminary estimates of lysine availability in the whole seed and kernels of *L. angustifolius* and *L. albus* for growing pigs (25–55 kg) determined using a modified slope-ratio analysis (from R.J. van Barneveld, SARDI – Pig and Poultry Production Institute, Roseworthy, South Australia, 1996, unpublished data).

Lupin	Mean
L. angustifolius – whole seed	0.73
L. angustifolius – kernel	0.79
L. albus – whole seed	0.65
L.albus – kernel	0.77

the grind, the higher the DE. As well as decreasing the accuracy of diet formulations, this range in DE estimates can have a considerable bearing on the price of lupins. For example, differences in DE of 2 MJ kg^{-1} could be worth up to A$30.00 $tonne^{-1}$.

Due to high levels of non-starch polysaccharides in lupins, DE may not be the most appropriate measure of the available energy content of lupins for pigs. This is due to a large proportion of the carbohydrates from lupins being digested in the hindgut (Taverner *et al.*, 1983). Van Barneveld *et al.* (1995b) reported that the inclusion of lupin kernels at graded levels in pig diets resulted in a significant linear decrease in the ileal digestibility of diet dry matter, energy and diet DE (Table 13.5). There was, however, no significant difference in faecal digestibility. This would result in a significant decrease in the efficiency of use of lupin energy by the pig due to its recovery as volatile fatty acids from hindgut fermentation, rather than absorption as monosaccharide units in the small intestine. Hence, when lupin kernels are included in pig diets, compensation should be made in the diet formulation for the lower contribution of available dietary energy to the pig.

POULTRY. Johnson and Eason (1991) reported the apparent metabolizable energy of Victorian and Western Australian lupins to be 9.6 and 7.2 MJ kg^{-1} DM, respectively, using a rapid broiler assay technique in which lupins were included in diets at 475 g kg^{-1}. Bryden *et al.* (1994) reported the apparent metabolizable energy of lupin meal to be 8.66 MJ kg^{-1} DM. These estimates are considerably lower than for field peas (10.8 MJ kg^{-1} DM), faba beans (11.0 MJ kg^{-1} DM) and vetch (10.8 MJ kg^{-1} DM). The data suggest that apparent metabolizable energy of lupins for poultry is inferior to that of other legumes due to the absence of any appreciable hindgut recovery of energy.

Table 13.5. Effect of graded levels of lupin kernels on ileal and faecal apparent digestibility coefficients (proportion of total) for dietary dry matter (DDM), gross energy and diet DE (MJ kg^{-1}, air-dry basis) in growing pigs (from van Barneveld *et al.*, 1995a).

Proportion of kernel included in diet (%)	0	12	24	36	Diet	Linear	SEM
Ileal digestibility							
DDM	0.85	0.82	0.77	0.67	**	***	0.027
Energy	0.90	0.87	0.83	0.76	**	***	0.021
Diet DE	16.85	16.47	15.84	14.47	*	**	0.399
Faecal digestibility							
DDM	0.90	0.91	0.90	0.90	NS	NS	0.008
Energy	0.94	0.94	0.93	0.93	NS	NS	0.006
Diet DE	17.72	17.67	17.64	17.67	NS	NS	0.119

*$P < 0.05$.
**$P < 0.01$.
***$P < 0.001$.
DE = digestible energy.
NS = not significant.

The apparent metabolizable energy of lupin kernels has recently been evaluated (R.J. Hughes, R.J. van Barneveld and M. Choct, SARDI – Pig and Poultry Production Institute, Nutrition Research Laboratory, Roseworthy, South Australia. 1996, unpublished data). The apparent metabolizable energy of *L. angustifolius* cv. Gungurru kernels ranges from 8.0 to 11.0 MJ kg^{-1} DM whereas the apparent metabolizable energy of *L. angustifolius* cv. Danja ranges from 6.5 to 10.5 MJ kg^{-1} DM. The apparent metabolizable energy of *L. albus* cv. Kiev Mutant kernels was considerably higher at 12.5 MJ kg^{-1} DM.

AQUACULTURE SPECIES. The digestibility of dry matter and energy in the whole seeds and kernels of *L. albus* substantially exceeds that of *L. angustifolius* when fed to silver perch (G. Allan, Pt Stephens Research Centre, Salamander Bay, NSW, 1994, unpublished data; Table 13.1).

Complex carbohydrates

Crude fibre is still used as a nutritional constraint when formulating diets for monogastrics. Lupins have a crude fibre content of 10–15%, but a dietary fibre content of up to 40% (Bach Knudsen, 1997). Hence, when lupins are used in monogastric diets, there is a need to account for the specific physiological effects of other dietary fibre components such as oligosaccharides and soluble non-starch polysaccharides. Oligosaccharides are very difficult to define. When two to ten monosaccharide residues are joined by glycosidic linkages the resulting polymer is referred to as an oligosaccharide. The levels of these oligosaccharides in lupins are not dissimilar to those contained in soybean meal (Table 13.6).

PIGS. Little work has been done to quantify the anti-nutritional effects of non-starch polysaccharides from lupins in growing pigs. It has been suggested that variable production responses to lupins may be due to the high levels of lupin non-starch polysaccharides interfering with the action of digestive enzymes and influencing microbial activity (van Barneveld and Hughes, 1994).

Van Barneveld *et al.* (1995b) showed that the addition of graded levels of isolated lupin non-starch polysaccharides to sorghum-based diets resulted in a significant increase in digesta viscosity and reduced the ileal digestibility of energy, lysine and dry matter (Table 13.7). Total non-starch polysaccharide digestion was minimal and there were no significant differences among diets. Non-starch polysaccharide inclusion levels had no significant effect on the ATP content of the digesta in any part of the digestive tract. From these results it can be concluded that increased digesta viscosity is the cause of reduced ileal

Table 13.6. Raffinose, stachyose and verbascose content (% DM) of lupins and soybean meal (from Bach Knudsen, 1997).

	Legume	
Oligosaccharide	Soybean meal	Lupins
Raffinose	1.0	1.0
Stachyose	4.7	5.3
Verbascose	0.3	1.4

Table 13.7. Effect of graded levels of isolated lupin kernel non-starch polysaccharides (NSP) on digesta viscosity, the apparent ileal digestibility of energy, lysine, non-starch polysaccharides and dry matter and adenosine triphosphate (ATP) concentration in the small intestine (S.I.), caecum and large intestine (L.I.) of growing pigs (from van Barneveld *et al.*, 1995b).

		Digestibility (proportion of total)				ATP (mg g^{-1} digesta)		
Diet	Viscosity (mPa.s)	Energy	Lysine	NSP	DM	S.I.	Caecum	L.I.
1	1.43[a]	0.85[a]	0.91[a]	0.05	0.81[a]	0.72	4.78	1.97
2	2.28[a]	0.71[b]	0.90[a]	0.12	0.66[b]	0.61	5.73	1.66
3	2.33[a]	0.50[c]	0.82[b]	–0.31	0.37[c]	0.88	7.14	1.98
4	4.89[b]	0.58[c]	0.85[b]	–0.07	0.52[d]	0.47	5.45	3.38
Diet	***	***	**	NS	***	NS	NS	NS
SEM	0.304	0.031	0.014	0.104	0.030	1.185	1.185	1.185

SEM – mean standard error.
DM – dry matter.
NS – not significant.
*$P < 0.05$.
**$P < 0.01$.
***$P < 0.001$.
Values in a column with different superscripts differ significantly ($P < 0.05$).

digestibility of lysine and energy when high levels of lupins are fed. This may be due to interference with the action of digestive enzymes. Lupin non-starch polysaccharides do not affect microbial activity in the digestive tract.

Wigan *et al.* (1994) suggested that wide variation in oligosaccharide concentration both within and between species of lupins may influence nutritive value and could be responsible for the highly variable performance of pigs when lupins are fed. Recent research (van Barneveld *et al.*, 1996) suggests that oligosaccharides may play a major role in the variable performance of pigs fed either *L. angustifolius* or *L. albus*. Van Barneveld *et al.* (1996) used an ethanol extraction procedure to remove oligosaccharides from lupins and examined subsequent pig responses. The ethanol extraction removed 73 and 67% of the oligosaccharides from *L. angustifolius* and *L. albus*, respectively, but did not change the gross energy content (Table 13.8). Ethanol extraction did not influence the ileal energy digestibility of diets containing *L. angustifolius*, but improved the ileal and faecal energy digestibility of diets containing *L. albus* and the faecal energy digestibility of those containing *L. angustifolius*. Ethanol extraction improved the digestible energy of diets containing *L. angustifolius* and *L. albus* by 0.5 and 0.7 MJ kg^{-1}, respectively.

The extraction of oligosaccharides from *L. albus* had a greater impact on energy digestibility in the small intestine than in the large intestine. As a consequence, extraction of oligosaccharides will have an even greater impact on lupin net energy contributions. The higher oligosaccharide content of *L. albus* compared with *L. angustifolius* and the subsequent effect on net energy may help to explain the comparatively poorer performance of pigs when they are fed *L. albus*.

Table 13.8. The effect of ethanol extractions of *L. angustifolius* and *L. albus* on the oligosaccharide content, gross energy content, energy digestibility in the ileum and whole tract and diet DE when fed to growing pigs (from van Barneveld *et al.*, 1996).

	Dehulled *L. angustifolius*		Dehulled *L.albus*		Statistics	
Treatment	Nil	Extracted[1]	Nil	Extracted[1]	Diet	SEM
Oligosaccharides[2] ($g\ kg^{-1}$ DM)	40.5	11.1	63.2	20.4	NS	NS
Gross energy ($MJ\ kg^{-1}$)	18.73	18.75	17.33	17.34	NS	NS
Energy digestibility (ileum)	0.63[a]	0.67[a]	0.68[b]	0.76[c]	*	0.017
Energy digestibility (faeces)	0.85[a]	0.87[b]	0.87[b]	0.90[c]	**	0.005
Diet DE ($MJ\ kg^{-1}$ DM)	13.1[a]	13.6[b]	14.1[c]	14.8[d]	***	0.077

Values within a row with different superscripts differ.
*$P < 0.05$.
**$P < 0.01$.
***$P < 0.001$.
DM – dry matter.
DE – digestible energy.
SEM – standard error of the mean $n = 4$).
NS – not significant.
[1]Modified ethanol extraction.
[2]Sum of raffinose, stachyose and verbascose.

POULTRY. There is little evidence from the literature to suggest that oligosaccharides or non-starch polysaccharides in lupins have the same anti-nutritive effects in poultry. This is despite the fact that non-starch polysaccharides from cereals are known to have detrimental effects in broiler chicken diets (Annison and Choct, 1991). Annison *et al.* (1994) determined the apparent metabolizable energy of lupins at three levels of inclusion (100, 200 and 300 $g\ kg^{-1}$) in a basal sorghum/casein diet. The apparent metabolizable energy was calculated to be 10.26 $MJ\ kg^{-1}$ DM regardless of dietary lupin inclusion level.

In contrast to the responses observed in pigs, the extraction of oligosaccharides from lupins decreases dietary apparent metabolizable energy to poultry (R.J. Hughes, R.J. van Barneveld and M. Choct, Adelaide, 1997, unpublished data). Ethanol extraction of oligosaccharides significantly reduced dietary apparent metabolizable energy, by 0.8 $MJ\ kg^{-1}$ DM, and subsequent chicken performance when diets contained 30% *L. angustifolius* cv. Danja or Gungurru. Viscosity of ileal digesta was increased by extraction of oligosaccharides; however, moisture content of excreta was unaffected.

Despite evidence to suggest that non-starch polysaccharides from lupins have minimal anti-nutritive effects in poultry, the addition of commercial enzymes to poultry diets containing lupins has been shown to influence the nutritive value (Annison *et al.*, 1996). A commercial enzyme containing primarily xylanase, pentosanase and hemicellulase increased the apparent metabolizable energy of lupins from 10.01 to 11.65 $MJ\ kg^{-1}$ DM when added at 0.5 $g\ kg^{-1}$. In contrast, a second enzyme containing primarily β-glucanase, hemicellulase and pectinase activities did not affect the apparent metabolizable energy of lupins, nor the ileal digestibility of other nutrients, but caused an increase in the concentrations of soluble non-starch polysaccharides in the digesta and an increase in the ileal digesta viscosity. These results demonstrate the sensitivity of poultry to changes in dietary non-starch polysaccharide content and structure, rather than any anti-nutritive effects of lupin non-starch polysaccharides, and emphasize the need to specifically target exogenous enzyme supplementation of poultry diets.

AQUACULTURE SPECIES. Morales *et al.* (1994) studied the utilization of energy from a range of feed ingredients. They found that the carbohydrates from lupin-seed meal had a very low digestibility compared with those of cottonseed meal and corn gluten meal.

Anti-nutritional factors

Lupins contain many potential anti-nutritional factors, including alkaloids, saponins, tannins and α-galactosides (Chapters 12 and 14), with alkaloids having the greatest potential impact on the nutritional value of lupins for monogastrics. Modern lupin cultivars, however, contain only marginal levels of alkaloids and hence they are rarely a concern in monogastric diets when lupins are fed.

PIGS. Alkaloids are often cited as the reason for reduced acceptance of *L. albus* by pigs (Hill and Pastuszewska, 1993). As the amount of alkaloid in a pig diet increases above 0.03% there is a decrease in intake, which in turn reduces liveweight gain. This level, however, is seldom exceeded in pig diets containing lupins. In addition, *L. angustifolius* has similar levels of alkaloids to *L. albus* but is readily accepted by pigs.

To date, there has been limited information available in relation to the saponin content of lupins, particularly for *L. albus*. Recent research suggests that the saponin content of *L. angustifolius* ranges from 379 to 743 mg kg^{-1} and that there are no detectable saponins in *L. albus* (R. Ruiz, Norwich, UK, 1996, unpublished data). Cuardrado *et al.* (1995) also reported *L. albus* to be free of saponins and *L. luteus* with saponins levels of 55 mg kg^{-1}. Cuadrado *et al.* (1995) suggested that saponin level is correlated with the alkaloid content of lupins. On the basis of these data, it can be concluded that saponins are unlikely to be the cause of reduced acceptance of *L. albus* by pigs.

POULTRY. As with pigs, levels of potential anti-nutritional factors in lupins present few problems in poultry diets. Huisman and Tolman (1992) have also reported that poultry appear less sensitive to alkaloids than pigs.

Monogastric production responses to lupins

Pigs

Based on existing data (Standing Committee on Agriculture, 1987; King, 1990) and commercial experience, the maximum inclusion levels of *L. angustifolius* recommended for pig diets are listed in Table 13.9.

Barnett and Batterham (1981) replaced soybean meal in wheat-based diets (equal energy and lysine content) with *L. angustifolius* and found that weaner

Table 13.9. Recommended maximum inclusion levels of *L. angustifolius* grain in pig diets (from Standing Committee on Agriculture, 1987; King, 1990).

Diet type	Inclusion levels (%)
Starter/weaner diet (5–25 kg liveweight)	10–15
Grower diet (25–60 kg liveweight)	20–25
Finisher diet (60 kg to slaughter)	30–35
Dry sow diet (gestation)	20
Lactating sow diet	20

pigs (6–20 kg) could tolerate up to 43% lupin-seed meal without adversely affecting growth. Pearson and Carr (1976) included *L. angustifolius* at levels of up to 37% at the expense of more conventional protein concentrates without reducing the growth of grower/finisher pigs. Taverner (1975) and Batterham (1979) reported similar responses.

In commercial mixed diets where lupins are included in combination with a range of other protein and energy sources, pig performance is very consistent. Responses to lupins under experimental conditions, however, have been highly variable and have caused some confusion in relation to their use. In particular, highly variable responses have been shown to occur when various lupin cultivars are fed to pigs in either a sugar or a wheat base (Fernandez and Batterham, 1992; Wigan, 1995). Due to the high level of non-starch polysaccharides in lupins, and the large difference between net energy content and digestible energy content compared with other legumes, it appears that pig performance is improved when diets containing high levels of lupins also contain a highly digestible energy substrate. As sugar-based diets are rarely used in commercial pig production, responses to lupins determined with these diets should be interpreted with caution.

Poultry

Both *L. angustifolius* and *L. albus* support acceptable production levels in broiler and layer birds. Hughes and Orange (1976) offered laying hens *L. angustifolius* cv. Uniwhite supplemented with methionine at levels of 10 and 20% in the diet with no subsequent loss in laying performance. Prinsloo *et al.* (1992) offered up to 30% *L. albus* cv. Buttercup to Hisex laying hens without a significant effect on egg production, egg mass, food conversion efficiency, egg shell thickness, Haugh units or yolk colour.

Broiler chickens can tolerate up to 25% of low-alkaloid lupin-seed meal without adversely affecting growth, provided there are adequate supplements of lysine and methionine (Brenes *et al.*, 1993). This supports research by Karunajeewa and Bartlett (1985) who concluded that broiler starter diets could contain 22% *L. albus* cv. Hamburg with no adverse effect on growth performance. Johnson and Eason (1991) did observe a decrease in broiler performance when *L. angustifolius* was included in diets at a rate of 150 g kg^{-1} at the expense of soybean meal. Although this difference was not significant, they claim that this liveweight reduction of 45 g and increase in feed conversion by 6 points would have resulted in an economic loss. When interpreting these results, however, it should be noted that the diets were formulated on a total amino acid basis, and hence no allowance has been made for differences in the digestibility of nutrients in lupins and soybean meal.

Constraints on the maximum inclusion level of lupins in poultry diets are not necessarily due to drops in production above this level. In broiler chicken diets, inclusion of either *L. angustifolius* or *L. albus* should not exceed 10%. This is due to the incidence of wet-sticky droppings that may be promoted by high levels of lupin non-starch polysaccharides. Wet-sticky droppings pose a health risk to broilers through adverse effects on litter moisture levels, respiratory stress from high ammonia and potential for an increase in the prevalence of coccidiosis. In addition, carcass quality can be lowered due to breast blisters resulting from high ammonia levels and environmental issues can arise when it comes to disposal of the litter.

The concern over wet-sticky droppings is obviously not as great for caged laying birds and hence, based on nutritional research, a maximum inclusion level of *L. angustifolius* or *L. albus* of 25–35% will not affect laying performance.

Aquaculture species

It is difficult to interpret the nutritional value of lupins for aquaculture species. Due to a comparatively limited knowledge of the nutritional requirements of aquaculture species, it is highly likely that the diets are formulated to contain more nutrients than the fish requires. Hence, comparisons between lupins and other vegetable protein sources may be misleading. Despite this, there is evidence that lupins have a potential role in the nutrition of aquaculture species.

Robiana *et al.* (1995) examined the partial substitution of fishmeal with soybean meal and lupin-seed meal in diets for gilthead seabream (*Sparus aurata*). Mean feed intake and growth rate were not influenced by the type or level of plant protein in the diet. Feed conversion ratio (FCR) and protein efficiency ratio were also unaffected. It is important to note that fish fed lupin-seed meal also had reduced intestinal trypsin activities and a higher peak ammonia excretion rate, which appeared two hours later than the fish fed diets containing fishmeal. These results and the fact that gilthead seabream are essentially carnivores are encouraging for the potential of lupins in aquaculture diets.

Lupin-seed meal has also been assessed against pea meal and faba bean meal as a replacement for brown fishmeal in diets for rainbow trout (*Oncorhynchus mykiss*) (Gouveia *et al.*, 1993). All vegetable protein sources were included to provide 20% of the dietary protein. The fish fed the vegetable protein sources performed better than those fed the control diet with the best performance achieved with lupins. Gomes and Kaushik (1990) reported no effect on growth rates, FCR or apparent digestibility coefficients when lupin-seed meal provided 10 or 20% of the dietary protein. As lupins have a higher protein content than peas and faba beans, they would have had the lowest inclusion level in the diets. Moyano *et al.* (1992) showed that rainbow trout diets containing 50 or 70% lupins supported growth rates, fish acceptance and nutritive utilization similar to fish fed a whole-fish diet or a commercial trout feed. Further support for the use of lupins in rainbow trout diets has been provided by Hughes (1988, 1991). In contrast, Higuera *et al.* (1988) and Gomes and Kaushik (1990) reported that growth rates, FCR and apparent digestibility coefficients were depressed when lupin-seed meal provided more than 30% of the protein in a rainbow trout diet.

Lupins have also shown potential for use in diets for carp (*Cyprinus carpio*; Viola *et al.*, 1988), pink snapper or red sea bream (*Pagrus auratus*; Jenkins *et al.*, 1994) and marron (*Cherax tenuimanus*; Morrissy, 1992).

Value of processing lupins for monogastrics

Lupins can be a cost-effective feed ingredient for use in monogastric diets because they can be fed following little or no processing (apart from adequate crushing). The value of added processing has been investigated for both pigs and poultry (van Barneveld and Hughes, 1994) and, to a lesser extent, for aquaculture species. As yet the cost of processing may not be matched by the subsequent improvement in animal performance. For example, the lupin hull, though not toxic, is of limited nutritional value to monogastrics. In pigs and poultry the hull yields a modest energy contribution following hindgut fermentation, but it is utilized far more efficiently in ruminant diets. Despite this, commercial experience suggests that there is little benefit in dehulling lupins for monogastrics unless an opportunity exists to utilize the hull fraction in ruminant diets. That is, the net value of lupins is elevated by dehulling when the kernel is directed to monogastric diets and the hull fraction to ruminant diets. In pigs, where an alternative use for the hulls does not exist there is the option to direct the hull fraction to sow diets and/or finisher diets as a fibre source or energy diluent, while directing the kernels to the young growing pig. However, the economics of this are less favourable.

Pigs

Godfrey and Payne (1987) reported lupin kernel to be a superior protein and energy source to lupin seed meal, while Wigan *et al.* (1994) also reported that dehulling led to a significant improvement in gross energy digestibility and an observed improvement in apparent ileal digestibility of lysine.

The higher energy, protein and digestibility aspects of lupin kernels are fully appreciated by commercial pig producers and feedmillers and the majority of lupins used by the large Australian piggeries are dehulled. Typical nutritive values ascribed to the lupin fractions for pigs are displayed in Table 13.10.

Heating lupins in an attempt to improve their nutritive value for pigs is not deemed beneficial. Apart from the potential negative effects of heat on amino acid availability (van Barneveld, 1993), Batterham *et al.* (1986) showed that autoclaving lupin seed at 121°C for 5 min to reduce the action of potential anti-nutritional factors had no effect on pig growth.

Poultry

Additional processing of lupins may be highly beneficial for poultry. Brenes *et al.* (1993) demonstrated that the nutritive value of alkaloid-free lupins can be significantly increased by dehulling or autoclaving. Chicks fed diets containing

Table 13.10. Typical nutrient content (g kg^{-1}, air-dry basis) of lupin kernels and hulls (*L. angustifolius* cv. Gungurru) for pigs.

Factor	Kernels	Hulls
Approximate yield*	750	250
Crude protein	350	120
Crude fibre	32	490
Acid-detergent fibre	108	540
Neutral-detergent fibre	121	620
Gross energy (MJ kg^{-1})	18.7	16.5
Digestible energy (MJ kg^{-1})	15.7	10.5
Net energy (MJ kg^{-1})	9.5	4.0

*The actual kernel to hull ratio is approximately 80 : 20 but in the process of dehulling, some kernel fragments and flour are removed with the hulls during aspiration.

high concentrations of autoclaved or dehulled lupins (47.5%) are able to support weight gains and feed-to-gain ratios similar to those obtained with corn–soybean meal diets. These results are supported by Marquardt (1993). Dehulling will have removed the energy diluent effect the hulls have on the lupin seed, while autoclaving may have impaired the action of some anti-nutritional factors in the kernel. Brenes *et al.* (1993) suggest that additional research is required to identify the nature of the anti-nutritional factors in alkaloid-free seeds.

Aquaculture species

Higuera *et al.* (1988) and Gouveia *et al.* (1993) have both reported no change in the performance of rainbow trout fed with lupins subjected to varying degrees of heat treatment.

Although there is little evidence in the literature, it is likely that dehulling lupin will improve the performance of most aquaculture species when lupins are included in their diets except in those fish species adapted to a herbivourous diet and hence capable of utilizing fibrous substrates.

Nutritional Value of Lupins for Ruminants

Composition relative to ruminant nutrition

Lupins are a particularly useful feedstuff for ruminant livestock, being high in both protein and metabolizable energy and relatively low risk compared with alternative feed supplements.

Protein

The protein (amino acid) needs of ruminants are essentially derived from either microbial protein synthesized in the rumen, or ingested protein which escapes rumen fermentation (bypass protein). Since the microbial protein yield can becoming limiting to higher rates of performance, especially in high-producing dairy cows, a significant dietary contribution is often required in the form of bypass protein. Lupins have acquired an anecdotal reputation as a good source of bypass protein for ruminants but this feature is in fact quite variable, depending on how the seed is prepared and the method of feeding. In most instances the bypass protein content of lupins is quite low (Dixon and Hosking, 1992; Margan, 1994).

Three basic properties are required for bypass proteins to be of value:

- A significant proportion of the protein must escape rumen degradation.
- This fraction must subsequently be digestible.
- The profile of absorbed amino acids contributed must either match the requirements of the animal or complement other protein sources to achieve this goal.

The degree of protein degradation in the rumen varies from 42 to 95%, based on work by Orskov and McDonald (1979) (Table 13.11), using a synthetic-fibre bag technique. The lower levels of rumen protein degradation were associated with whole seed at high fractional outflow rates (high feeding levels), while the high degradation rates (low bypass values) were associated with fine grinding and/or low fractional outflow rates (low-level feeding). Dixon and Hosking (1992) point out, however, that the nylon bag technique may not reflect the true situation with whole or coarsely ground lupins since these fragments are generally too large to exit the rumen readily and would normally be retained until extensively fermented. They suggest that most studies of ruminant degradability of lupins under practical conditions report a high rate of degradation of 80% or more.

Consequently, lupins cannot be viewed as a major source of bypass protein since the non-rumen-degradable proportion is often small, and although highly digestible the amino acid profile of lupin protein is quite modest in terms of essential amino acid supply relative to other protein sources, e.g. fishmeal, soybean meal and rumen bacterial protein. The amino acid profile of lupins relative to that of rumen bacterial protein and the 'ideal' amino acid profile for post-duodenally absorbed protein for high-producing dairy cows is presented in Table 13.12. The shortfall in essential amino acid density is quite obvious and the shortfall in the relative proportion of methionine in the overall balance is also quite marked. There is also the question of whether the amino acid profile of that proportion of the protein which escapes rumen degradation differs from that in the original material. Mathers *et al.* (1979) demonstrated that the amino acid composition of the bypass fraction of sunflower meal was markedly inferior to that in the original meal.

Apart from the demands of high-producing dairy cows and young lot-fed cattle, microbial protein synthesis can supply the protein needs of ruminants in

Table 13.11. Typical nutrient content of lupins for ruminant livestock (as-fed basis) (from Orskov and Macdonald, 1992; Petterson and Mackintosh, 1994).

	L. angustifolius	*L.albus*	*L.luteus*
ME cattle (MJ kg^{-1})	12.0	–	–
ME sheep (MJ kg^{-1})	12.2	12.5	–
Rumen degradable protein (%)	42–95	–	–

Table 13.12. Essential amino acid profile of lupin protein, rumen bacterial protein, and ideal amino acid balance in post-duodenally absorbed protein for high producing dairy cows. (Values expressed as % of protein (g 16 g^{-1} N) and balance expressed relative to lysine).

	L. angustifolius[1]		*L. albus*		Rumen bacterial protein[2]		Ideal protein[3]
Amino acid	g 16 g^{-1} N	Rel	g 16 g^{-1} N	Rel	g 16 g^{-1} N	Rel	Rel
Lysine	4.66	100	4.20	100	8.0	100	100
Methionine	0.72	16	0.65	16	2.5	31	31
Cystine	1.48	32	1.30	31	N/A	N/A	N/A
Threonine	3.36	72	3.13	75	5.8	73	61
Isoleucine	3.97	85	3.72	89	5.9	74	77
Tryptophan	1.06	23	0.97	23	N/A	N/A	N/A
Arginine	12.03	263	12.44	296	4.9	61	58
Leucine	3.97	85	6.06	144	7.7	96	103
Valine	3.91	84	3.64	87	6.2	78	81
Histidine	2.41	52	1.72	41	1.8	23	32
Phenylalanine	3.65	78	3.27	78	5.3	66	58

[1]From Petterson and Mackintosh (1994); [2]from Rulquin and Verite (1993); [3]from Chalupa and Sniffen (1993).
N/A – not analysed.

most situations. Lupin protein is utilized well as a substrate for microbial protein synthesis by virtue of its orderly fermentation rate.

Energy

Although lupins contain little or no starch, the carbohydrate profile is nonetheless readily fermentable and contributes to the high metabolizable energy value for ruminants (Margan, 1994) (Table 13.11). The dominance of the carbohydrate profile by β-galactan and oligosaccharides in the endosperm and cellulose in the hull results in a fermentation pattern which is less rapid and less prone to lead to lactic acidosis.

For this reason lupins can be fed freely as a supplement to grazing ruminants without the need for a period of introduction. Similarly, in prepared mixes for dairy cows or feedlot cattle the replacement of cereal grain with lupins lowers the risk of acidosis, as well as providing additional protein. The use of high lupin levels in lieu of cereal grain in feedlot starter diets can achieve the adaptation to high grain intakes without incurring the slower growth usually witnessed in the introductory period (Callow, 1993).

Although sheep are often fed whole unmilled lupins and utilize them most effectively, due to their thorough mastication of the grains, it is generally recommended that lupins be milled or at least cracked to ensure good utilization in cattle (May and Barker, 1984).

In feedmilling operations servicing multiple animal species, advantages have been recognized in dehulling lupins to derive a greater benefit in monogastric diets to the kernels, while the hulls are directed into ruminant feeds where they represent good value, due to their highly fermentable nature. If this latter aspect were not the case, the dehulling process would be hard to justify.

Ruminant production responses to lupins

Sheep

By far the greatest form of lupin utilization in Australia is as a whole-grain feed for grazing sheep, to supplement low-grade roughage diets. Responses vary, depending on the quality of the roughage on offer. Dixon and Hosking (1992) report that the efficiency of utilization of lupin grain varies from liveweight change of 0.8 g g^{-1} of lupin dry matter for low quality roughages where sheep are losing weight rapidly, to liveweight gains of 0.2–0.3 g g^{-1} lupin dry matter where roughage quality is adequate to support maintenance or slow growth. Lupin grain supplements generally result in higher intake, liveweight change and wool growth than comparable supplements of cereal grain, primarily due to their protein contribution as a nitrogen source for microbial protein synthesis, but also possibly due to bypass protein effects, higher metabolizable energy content and less disturbance to fibre digestion which often accompanies the fermentation of cereal starch (Dixon and Hosking, 1992).

The intake of roughage is often stimulated by the use of lupin supplements via the nitrogen stimulation of rumen activity and the reduced rumen dwell time involved (Smith and Kenney, 1987).

Lupins have been shown to be most useful in a range of sheep production situations such as high grain diets for early weaned lambs, growing sheep, lot-fed sheep, pregnant and lactating ewes fed for production or fed restrictedly for survival in droughts, and for superfine wool production (Hodge *et al.*, 1981; Kenney and Smith, 1985; Kenney, 1986). They are also commonly fed to grazing sheep to improve fertility and fecundity. Lupins fed at 200 g day^{-1} to rams for 8 weeks prior to joining increases testicular size and fertility, while ewes fed lupins at 400–500 g day^{-1} for 2 weeks prior to and after joining demonstrated higher lambing rates (Murray, 1994).

Lupins are established as a valuable asset in crop rotation programmes, with the stubble and fallen grain representing a substantial feed resource for grazing sheep. In fact lupins are often broadcast into cereal stubbles to enhance their utilization by grazing. The effects of lupin supplementation to breeding ewes is demonstrated in the results of Kenney and Roberts (1987) from a trial at Rutherglen Research Institute, Victoria, in which ewes were grazed on lupin stubble or dry pasture with and without a daily supplement of 400 g of lupin grain (Table 13.13).

The voluntary feed intake of lupins in grazing and penned sheep varies widely (Murray and Rowe, 1986; Curtis *et al.*, 1994) but this problem is not exclusive to lupins.

The responses which can be expected from lupin grain supplements depend largely on the composition and level of feeding of the base diet, and the ability of the lupin contribution to address the prime constraints to performance. Where energy is the prime constraint, lupins may not perform any better than cereal grains (Rowe *et al.*, 1989), and where fermentable N in the rumen is the prime

Table 13.13. Production of ewes grazing lupin stubble or dry pasture ± lupin grain supplementation (from Kenney and Roberts, 1987).

	Treatment*		
Production parameter	Pasture	Pasture + lupins	Lupin stubble
Liveweight change (g d^{-1})	−194[a]	41[b]	83[b]
Ewe lambing/ewe mated	0.80[a]	0.80[a]	0.85[a]
Lambs born/ewe mated	0.97[a]	1.07[ab]	1.25[b]
Fecundity (lambs born/ewe)	1.21	1.34	1.47
Relative wool production	100[a]	162[b]	158[b]

*Values in a row with different superscripts differ significantly ($P < 0.05$).

constraint lupins may not perform any better than urea or some other form of nitrogenous supplement. (Davis *et al.*, 1987).

However, in many instances a combination of lupins and cereal grain may deliver superior performance by achieving a closer to optimal balance of nutrients in the total diet (Hodge *et al.*, 1981). Kenney (1980), in a study with lambing ewes under drought feeding conditions, determined the optimum proportion of lupin grain in a lupin–cereal grain supplement to be 30%.

Dairy cows

Lupin grain is used commonly in southern Australia as a supplement for dairy cows. It is seldom fed as a sole supplement but more generally in mixtures with cereal grains and other feedstuffs. The level of lupins included in the diet is determined by the protein content of the pasture or conserved fodder on offer, the stage of lactation and level of production, and the cost-competitiveness of lupins relative to other protein meals and cereal grains.

The role of lupins as a source of bypass protein in dairy diets remains debatable, but if coarsely ground and incorporated as part of a high feeding level for high-producing cows, there may be a greater proportion of the protein which escapes rumen degradation than at a maintenance level of feeding. Yet when lupins have been compared with other protein sources in situations likely to respond to bypass protein, they have been found to be inferior to other protein sources (Lemerle *et al.*, 1985).

With the recent voluntary ban on the use of animal protein supplements in dairy diets in response to consumer concerns with ‘mad cow’ disease, the emphasis on vegetable protein to provide the required bypass protein for high-producing cows may see lupins lose some ground to canola meal and cottonseed meal, depending on price relativities. This is due to the fact that lupins, unlike the oilseed meals, are not subjected to any form of heating prior to feeding, which improves bypass protein content. The competitiveness of lupins in this area would be greatly enhanced by any process or plant breeding that could enhance the rumen bypass characteristics of its protein content, and the biological value of the protein.

The value of lupin grain in dairy diets is not normally assessed on any singular aspect but more on the aggregate of its inherent properties, e.g. high energy, high protein, orderly fermentation rate, low acidosis risk, etc. One positive attribute is its apparent ability to maintain milk fat at high levels of supplementation, in contrast to the problems often encountered with similar levels of cereal grain supplementation (Bartsch *et al.*, 1986; Sinclair and Gooden, 1989; Valentine and Bartsch, 1990; Hough and Jacobs, 1994). The explanation for this may involve the orderly rate of fermentation of lupins and the relatively high lipid content of lupins rather than a manipulation of the acetate : propionate ratio in the rumen (Dixon and Hosking, 1992).

In situations where lactating dairy cows have a basal diet of low to medium quality pasture or roughage, lupin grain generally produces a better response than barley grain and non-protein nitrogen mixtures (Bartsch *et al.*, 1986; Valentine and Bartsch, 1990). Yet where cows are fed high quality pasture, lactational responses to lupin grain have been similar to those for oat or barley grain (Moate *et al.*, 1984; Valentine and Bartsch, 1989). Further to this, where pastures are high in fermentable nitrogen the addition of lupins to the diet may result in excessive rumen ammonia levels, leading to high blood urea nitrogen levels which are known to have a negative effect on fertility.

Consequently, the economic use of lupin grain in dairy diets will require a judgement as to how well the nutrient profile of lupins balances the nutritional contribution of the basal feedstuffs, relative to the cost of meeting these from alternative sources.

Beef cattle

With the exception of young, fast-growing cattle in a feedlot situation, most cattle can meet their protein requirements for full growth potential from rumen microbial protein synthesis. Consequently in most production situations the protein and energy attributes of lupins need to compete with the energy of cereal grains and non-protein nitrogen sources such as urea, or other ruminally degraded proteins. Unless lupin grain and cereal grains are similarly priced or the convenience and safety of feeding is highly valued, lupin grain may not be competitive. Even in young feedlot cattle where some bypass protein is indicated, the modest bypass characteristics of lupin grain may render alternatives such as canola, cottonseed meal, bloodmeal, fishmeal, etc. more attractive.

Lupins may have an economic role in the introductory phase of feedlotting, where a gradual transition from lupin grain to cereal grain allows higher initial levels of feeding and subsequent growth than the traditional build-up of cereal grain in lieu of roughage (Callow, 1993).

Lupins probably demonstrate their greatest value when supplementing low quality pasture or roughage, where their energy and protein can be employed concurrently to good effect. Conversely, their role as a supplement to high quality pasture or silage may be limited due to the redundancy of their protein contribution. The extent to which lupins will be utilized in beef cattle feeding will depend on the composition and abundance of the basal feedstuff (pasture or conserved fodder) and the price and availability of lupins relative to alternative feedstuffs.

Limitations to lupin use in ruminant diets

Lupinosis

Lupinosis is a disease involving the ingestion of toxins produced by the fungus *Diaporthe toxica* (formerly known as *Phomopsis leptostromiformis*), which inhabits lupin plants (Chapter 14) and was a major concern for growers and livestock owners prior to the introduction of the *Diaporthe*-resistant variety *L. angustifolius* cv. Gungurru. This new variety and the adoption of appropriate livestock husbandry practices has restored confidence in feeding lupin grain and lupin stubbles to ruminant livestock (Murray, 1994).

Nutrient imbalances

In total mixed diets or blended milled supplements the shortfalls of sulphur, calcium, methionine, rumen non-degradability of protein, etc. that characterize lupins can be balanced by inputs from other sources. The extent to which these might limit responses when lupins are fed as a sole supplement will depend on the composition of the base feed involved and the extent to which this and lupins complement each other. In situations where energy and/or fermentable nitrogen are prime limits, lupins will prove a most effective and convenient supplement. Where supplemental bypass protein is required, lupins will prove to be of limited value.

Relative cost

The ultimate determinant in the decision to feed lupins is their cost relative to the anticipated responses and relative to alternative feedstuffs. In Australia, one of the factors limiting lupin use is the geographic dislocation between supply and demand and the prohibitive freight cost involved in reconciling this. Consequently a large proportion of the lupins grown is exported. A significant component of this imbalance is the density of sheep and cattle numbers in northern New South Wales and Queensland. (potential consumers) yet the unsuitability of agronomic conditions for the cultivation of current lupin varieties in these areas.

Conclusions

In addition to being a very useful feedstuff for most classes of livestock, lupins also store well, are easy to handle and are readily accepted by most stock. They are largely free of any significant anti-nutritional factors and the prime restriction to their use is the ability of each livestock species to accommodate the somewhat complex carbohydrate profile within the seed. This profile is of no concern to ruminants and in fact imparts the advantage of orderly rumen fermentation and low acidosis risk. In monogastrics, however, the complex carbohydrate profile is the main constraint to use as it influences the net energy yield, and has been shown to affect the utilization of other nutrients in the diet.

Although lupins are relatively high in protein, the biological value of the protein is limited by the fact that the amino acid profile is low in methionine and it is also proportionally lower in lysine than many other grain legumes. However, low levels of methionine and lysine are of little or no consequence to ruminants where the protein is mostly ruminally fermented, while in pig and poultry diets these shortfalls can be made up from other proteins or synthetic amino acids.

Lupins are becoming increasingly available across southern Australia and, where competitively priced, they will be used extensively across a broad range of feeding applications, with the greater tonnages being used in ruminant and pig diets than in poultry and aquaculture diets.

References

Annison, G. and Choct, M. (1991) Anti-nutritive activities of cereal non-starch polysaccharides in broiler diets and strategies minimizing their effects. *World's Poultry Science Journal* 47, 232–242.

Annison, G., Choct, M. and Hughes, R.J. (1994) Apparent metabolizable energy determination and its application to lupins. In: Australian Poultry Science Symposium Editorial Committee (eds) *Proceedings of the Australian Poultry Science Symposium*, University of Sydney, pp. 92–96.

Annison, G., Hughes, R.J and Choct, M. (1996) Effects of enzyme supplementation on the nutritive value of dehulled lupins. *British Poultry Science* 37, 157–172.

Bach Knudsen, K.E. (1997) Carbohydrates and lignin contents of plant material used in animal feeding. *Feed Science Technology* 67, 319–338.

Barnett, C.W. and Batterham, E.S. (1981) *Lupinus angustifolius* (cv. Unicrop) as a protein and energy source for weaner pigs. *Animal Feed Science and Technology* 6, 27–34.

Bartsch, B.D., Twigger, C.F. and Valentine, S.C. (1986) Lupins, beans, peas and barley as alternative grains for dairy cows fed hay. In: Dove, H. (ed.) *Proceedings of the Australian Society of Animal Production* 16, 404.

Batterham, E.S. (1979) *Lupinus albus* cv. Ultra and *Lupinus angustifolius* cv. Unicrop as protein concentrates for growing pigs. *Australian Journal of Agricultural Research* 30, 369–375.

Batterham, E.S. (1992) Availability and utilisation of amino acids for growing pigs. *Nutrition Research Reviews* 5, 1–18.

Batterham, E.S., Murison, R.D. and Andersen, L.M. (1984) Availability of lysine in vegetable protein concentrates as determined by the slope-ratio assay with growing pigs and rats and by chemical techniques. *British Journal of Nutrition* 51, 85–99.

Batterham, E.S., Andersen, L.M., Lowe, R.F. and Darnell, R.E. (1986) Nutritional value of lupin (*Lupinus albus*)-seed meal for growing pigs: availability of lysine, effect of autoclaving and net energy content. *British Journal of Nutrition* 56, 645–659.

Brenes, A., Marquardt, R.R., Guenter, W. and Rotter, B.A. (1993) Effect of enzyme supplementation on the nutritional value of raw, autoclaved and dehulled lupins (*Lupins albus*) in chicken diets. *Poultry Science* 72, 2281–2293.

Bryden, W.L., Gill, R.J. and Balnave, D. (1994) Feed enzyme supplement improves the apparent metabolisable energy of lupins for broiler chickens. In: Australian Poultry Science Symposium Editorial Committee (eds) *Proceedings of the Australian Poultry Science Symposium,* University of Sydney, p. 115.

Callow, C.R. (1993) Lupins in introductory rations for lot-fed steers. In: Farrell, D.J. (ed.) *Recent Advances in Animal Nutrition in Australia.* University of New England, Armidale, NSW, p. 9B.

Chalupa, W. and Sniffen, C.J. (1993) Protein and amino acid nutrition of lactating dairy cattle – today and tomorrow. *Proceedings of the Fourteenth Western Nutrition Conference, Calgary, Alberta*, pp. 87–98.

Chamberlain, G. and Rosenthal, H. (1995) Aquaculture in the next century: opportunities for growth, challenges of sustainability. *World Aquaculture* 26, 21–25.

Cuadrado, C., Ayet, G., Burbano, C., Muzquiz, M., Camacho, L., Cavieres, E., Lovon, M., Osagie, A. and Price, K.R. (1995) Occurrence of saponins and sapogenols in Andean crops. *Journal of the Science of Food and Agriculture* 67, 169–172.

Curtis, K.M.S., Holst, P.J. and Murray, P.J. (1994) Measuring supplement intake in the field using ytterbium as a marker. *Australian Journal of Experimental Agriculture* 34, 339–343.

Davis, J.J., Bird, S.H. and Leng, R.A. (1987) Lupins as a supplement for sheep. In: Farrell, D.J. (ed.) *Recent Advances in Animal Nutrition in Australia*. University of New England, Armidale, NSW, p. 5A.

Dixon, R.M. and Hosking, B.J. (1992) Nutritional value of grain legumes for ruminants. *Nutrition Research Reviews* 5, 19–43.

Edwards, A.C. (1994) Domestic markets for lupins. In: Dracup, M. and Palta, J. (eds.) *Proceedings of the First Australian Lupin Technical Symposium*. Department of Agriculture, Western Australia, pp. 238–243.

Edwards, A.C. and Campbell, R.G. (1991) Energy–protein interactions in pigs. In: Haresign, W. and Cole, D.J.A. (eds) *Recent Advances in Animal Nutrition*, pp.1–19.

FAO (1994) *Aquaculture Production 1986–1992*. FAO Fisheries Circular No. 815, Revision 6. Food and Agricultural Organisation of the United Nations, Rome, p. 216.

Fernandez, J.A. and Batterham, E.S. (1992) Lupin-seed meal for pigs – how good are they? In: *Proceedings of the Fourth Biennial Pig Industry Seminar*. Wollongbar Agricultural Institute, pp. 59–64.

Godfrey, N.W. and Payne, H.G. (1987) The value of dietary lupin kernel meal for pigs. In: Australian Pig Science Association Committee (eds) *Manipulating Pig Production*. Australian Pig Science Association, Werribee, p.153.

Gomes, F.E. and Kaushik, S. (1990) Incorporation of lupin seed meal, colzapro or triticale as protein/energy substitutes in rainbow trout diets. In: Takeda, M. and Watanabe, T. (eds) *The Current Status of Fish Nutrition in Aquaculture. The Proceedings of the Third Internationl Symposium on Feeding and Nutrition in Fish, Toba, Japan*, pp. 315–324.

Gouveia, A., Olivia-Teles, A., Gomes, E. and Rema, P. (1993) Effect of cooking/expansion of three legume seeds on growth and food utilization by rainbow trout. In: Kanshick, S.J. and Luquet, P. (eds) *Fish Nutrition in Practice*. Institut National de la Recherche Agronomique, Paris, 933–938.

Higuera, M. de la, Garcia-Gallego, M., Sanz, A., Cardenete, G., Suarez, M.D. and Moyano, F.J. (1988) Evaluation of lupin seed meal as an alternative protein source in feeding of rainbow trout (*Salmo gairdneri*). *Aquaculture* 71(1–2), 37–50.

Hill, G.D. and Pastuszewska, B. (1993) Lupin alkaloids and their role in animal nutrition. In: van der Poel, A.F.B., Huisman, J. and Saini, H.S. (eds) *Recent Advances in Research in Anti-nutritional Factors in Legume Seeds*. Wageningen Press, Wageningen, pp. 343–362.

Hodge, R.W., Bogdanovic, B. and Sweatman, D. (1981) Wool production of merino sheep fed daily or twice weekly on oats or lupins. *Australian Journal of Experimental Agriculture and Animal Husbandry* 23, 19–23.

Hough, G.M. and Jacobs, J.L. (1994) The use of lupins as a feed for dairy and beef cattle. In: Dracup, M. and Palta, J. (eds) *Proceedings of the First Australian Lupin Technical Symposium*. Department of Agriculture, Western Australia, pp. 58–66.

Hughes, R.J. and Orange, K.S. (1976) The laying performance of hens fed Uniwhite lupin seed and supplementary DL-methionine. *Australian Journal of Experimental Agriculture and Animal Husbandry* 16, 367–371.

Hughes, S.G. (1988) Assessment of lupin flour as a diet ingredient for rainbow trout (*Salmo gairdneri*). *Aquaculture* 71(4), 379–385.

Hughes, S.G. (1991) Use of lupin flour as a replacement for full-fat soy in diets for rainbow trout (*Oncorhynchus mykiss*). *Aquaculture* 93(1), 57–62.

Huisman, J. and Tolman, G.H. (1992) Antinutritional factors in the plant proteins of diets for non-ruminants. In: Garnsworthy, P.C., Haresign, W. and Cole, D.J.A. (eds). *Recent Advances in Animal Nutrition*. School of Agriculture, University of Nottingham, pp. 3–31.

Jenkins, G.I., Waters, S.P., Hoxey, M.J. and Petterson, D.S. (1994) Lupin seed (*Lupinus angustifolius*) as an alternative to soybean meal in the diet of juvenile snapper (*Pagrus auratus*). In: Dracup, M. and Palta, J. (eds) *Proceedings of the First Australian Lupin Technical Symposium*. Department of Agriculture, Western Australia. pp. 74–78.

Johnson, R. and Eason, P.J. (1991) The effect of dietary combinations of grain legumes on the growth performance of broiler chickens. In: Australian Poultry Science Symposium Editorial Committee (eds) *Proceedings of the Australian Poultry Science Symposium*, University of Sydney, pp. 64–67.

Karunajeewa, H. and Bartlett, B.E. (1985) The effects of replacing soybean meal in broiler starter diets with white lupin seed meal of high manganese content. *Nutrition Reports International* 31, 53–58.

Kenney, P.A. (1980) Intake and production of lambs fed rations of oats with varying amounts of lupins. *Proceedings of the Australian Society of Animal Production* 13, pp. 253–256.

Kenney, P.A. (1986) Productivity of early weaned lambs fed high grain diets of wheat, oats or barley with or without lupin grain. *Australian Journal of Experimental Agriculture* 26, 279–284.

Kenney, P.A. and Roberts, G.B. (1987) Productivity of ewes grazing lupin stubbles at mating in north-eastern Victoria. *Australian Journal of Experimental Agriculture* 27, 619–624.

Kenney, P.A. and Smith, R.S. (1985) Effects of including lupins with cereal grain rations on the production of lambing ewes during drought. *Australian Journal of Experimental Agriculture* 25, 529–535.

King, R.H. (1990) Lupins. In: Thacker, P.A. and Kirkwood, R.N. (eds.) *Nontraditional Feed Sources for Use in Swine Production*. Butterworths Boston, pp. 237–246.

Lemerle, C., Etheridge, M.D. and Trigg, T.E. (1985) Protein supplementation of dairy cattle on maize silage diets. In: Sinclair, A.J. (ed.) *Proceedings of the Nutrition Society of Australia, Volume 10*, p.158.

Liebholz, J. (1984) A note on methionine supplementation of grower pig diets containing lupin-seed meal. *Animal Production* 38, 515–517.

Margan, D.E. (1994) Energy and protein value of lupin seed as a production ration or as a supplement for sheep fed chaffed wheaten hay. *Australian Journal of Experimental Agriculture* 34, 331–337.

Marquardt, R.R. (1993) Use of legume seeds (peas, faba beans, lupins and grass peas) in poultry diets: influence of antinutritional factors. In: *Ninth Australian Poultry and Feed Convention: Gold Coast, 1993*, pp. 47–51.

Mathers, J.C., Thomas, R.J. and Gray, N.A.M. (1979) The nutritive value of feed proteins which escape degradation in the rumen. *Proceedings of the Nutrition Society* 38, 122a.

May, P.J. and Barker, D.J. (1984) Milling barley and lupin grain in diets for cattle. *Animal Feed Science and Technology* 12, 57–64.

Moate, P.J., Rogers, G.L. and Robinson, I.B. (1984) Lupins or oats as supplements for cows fed pasture in early lactation. *Proceedings of the Australian Society of Animal Production* 15, 721.

Morales, A.E., Cardenete, G., De-la-Higuera, M. and Sanz, A. (1994) Effects of dietary protein source on growth, feed conversion and energy utilization in rainbow trout, *Oncorhynchus mykiss*. *Aquaculture* 124, 117–126.

Morrissy, N.M. (1992) Feed development for marron, *Cherax tenuimanus*, in Western Australia. In: Allan, G.L. and Dall, W. (eds) *Proceedings of Aquaculture Nutrition Workshop, Salamander Bay, 15–17 April, 1991*. NSW Fisheries, Brackish Water Fish Culture Research Station, Salamander Bay, Australia, pp.72–76.

Moyano, F.J., Cardenete, G. and de-la-Higuera, M. (1992) Nutritive value of diets containing a high percentage of vegetable proteins for trout, *Oncorhynchus mykiss*. *Aquaculture and Living Resources*, 5, 23–29.

Murray, P.J. (1994) The use of lupins as a feed for sheep. In: Dracup, M. and Palta, J. (eds) *Proceedings of the First Australian Lupin Technical Symposium*. Department of Agriculture, Western Australia, pp. 67–73.

Murray, P.J. and Rowe, J.B. (1986) A comparison of lupin and barley grains as supplementary feeds. *Proceedings of the Australian Society of Animal Production* 15, 724.

Orskov, R.E. and Mcdonald, I. (1979) The estimation of protein degradability in the rumen from incubation measurements weighted according to rate of passage. *Journal of Agricultural Science* 92, 499–503.

Pearson, G. and Carr, J.R. (1976) Lupin-seed meal (*Lupinus angustifolius* cv. Uniwhite) as a protein supplement to barley based diets for growing pigs. *Animal Feed Science and Technology* 1, 631–642.

Pettersen, D.S. and Mackintosh, J.B. (1994) *The Chemical Composition and Nutritive Value of Australian Grain Legumes*. Grains Research and Development Corporation, Canberra.

Prinsloo, J.J., Smith, G.A. and Rode, W. (1992) Sweet white *Lupins albus* (cv. Buttercup) as a feedstuff for layers. *British Poultry Science* 33, 525–530.

Robiana, L., Izquierdo, M.S., Moyano, F.J., Socorro, J., Vergara, J.M., Montero, D. and Fernandez-Palacios, H. (1995) Soybean and lupin seed meals as protein sources in diets for gilthead seabream (*Sparus aurata*): nutritional and histological implications. *Aquaculture* 130 (2,3) 219–233.

Rowe, J.B., Brown, B., Ralph, I.G., Ferguson, J. and Wallace, J.F. (1989) Supplementary feeding of young merino sheep grazing wheat stubble with different amounts of lupin, oat or barley grain. *Australian Journal of Experimental Agriculture* 29, 29–35.

Rulquin, H. and Verite, R. (1993) Amino acid nutrition of dairy cows: productive effects and animal requirements. In: Garnsworthy, P.C. and Cole, D.J.A. (eds) *Recent Advances in Animal Nutrition. Proceedings of the 27th. University of Nottingham Feed Manufacturers Conference*, Nottingham University Press, Nottingham, pp. 55–77.

Sinclair, S.E. and Gooden, J.M. (1989) Effects of barley and lupin supplementation on milk composition and plasma metabolites in lactating ewes. In: Farrell, D.J. (ed.) *Recent Advances in Animal Nutrition in Australia*. University of New England, Armidale, NSW, p. 17A.

Smith, G.H. and Kenney, P.A. (1987) Lupin grain supplements for sheep and cattle. In: Farrell, D.J. (ed.) *Recent Advances in Animal Nutrition in Australia*. University of New England, Armidale, NSW, pp. 72–79.

Standing Committe on Agriculture (1987) *Feeding Standards for Australian Livestock. Pigs*. CSIRO, East Melbourne.

Taverner, M.R. (1975) Sweet lupin seed meal as a protein source for growing pigs. *Animal Production* 20, 413–419.

Taverner, M.R., Curic, D.M. and Rayner, C.J. (1983) A comparison of the extent and site of energy and protein digestion of wheat, lupin and meat and bone meal by pigs. *Journal of the Science of Food and Agriculture* 34, 122–128.

Treadwell, R., McKelvie, L. and Maguire, G. (1991) Profitability of selected aquaculture species. *Australian Bureau of Agricultural and Resource Economics Discussion Paper No. 91.11*, p. 85.

Valentine, S.C. and Bartsch, B.D. (1989) Milk production by dairy cows fed hammermilled lupin grain, hammermilled oaten grain or whole oaten grain as supplements to pasture. *Australian Journal of Experimental Agriculture* 29, 309–313.

Valentine, S.C. and Bartsch, B.D. (1990) Milk production by dairy cows fed legume grains or barley grain with or without urea as supplements to a cereal hay based diet. *Australian Journal of Experimental Agriculture* 30, 7–10.

van Barneveld, R.J. (1993) Effect of heating proteins on the digestibility, availability and utilisation of lysine by growing pigs. PhD thesis, The University of Queensland, Brisbane, Australia.

van Barneveld, R.J. and Hughes, R.J. (1994) The nutritive value of lupins for pigs and poultry. In: Dracup, M. and Palta, J. (eds) *Proceedings of the First Australian Lupin Technical Symposium*. Department of Agriculture, Western Australia, pp. 49–57.

van Barneveld, R.J., Baker, J., Szarvas, S.R. and Choct, M. (1995a) Effect of lupin kernels on the ileal and faecal digestibility of energy by pigs. In: Hennessy, D.P. and Cranwell, P.D. (eds) *Manipulating Pig Production V. Proceedings of the Fifth Biennial Conference of the Australasian Pig Science Association*. Australasian Pig Science Association, Canberra, p. 30.

van Barneveld, R.J., Baker, J., Szarvas, S.R. and Choct, M. (1995b) Effect of lupin non-starch polysaccharides (NSP) on nutrient digestion and microbial activity in growing pigs. In: Brand Miller, J.C. (ed.). *Proceedings of the Nutrition Society of Australia*, Vol. 19. Australasian Pig Science Association, Melbourne, p. 43.

van Barneveld, R.J., Olsen, L.E. and Choct, M. (1996) Effect of lupin oligosaccharides on energy digestion in growing pigs. In: Brand Miller, J.C. (ed.) *Proceedings of the Nutrition Society of Australia*, Vol. 20. Australasian Pig Science Association, Sydney, p. 114.

Viola, S., Arieli, Y. and Zohar, G. (1988) Unusual feedstuffs (tapioca and lupin) as ingredients for carp and tilapia feeds in intensive culture. *Israel Journal of Aquaculture and Bamidgeh* 40, 29–34.

Wigan, G.C., Batterham, E.S. and Farrell, D.J. (1994) Nutritive value of *Lupinus angustifolius* (cv Gungurru) for growing pigs. In: *Proceedings of the Fifth Biennial Pig Industry Seminar,* Wollongbar Agricultural Institute. NSW Agriculture, pp. 38–46.

Wigan, G.C. (1995) Determintion of, and factors affecting, nutrient digestibility and nutritive value of *Lupinus angustifolius* (cv Gungurru) for growing pigs. Master of Rural Science thesis, The University of New England, Armidale, Australia.

14 Toxins and Lupinosis

Jeremy G. Allen

Agriculture Western Australia, 3 Baron-Hay Court, South Perth, Western Australia 6151

Introduction

Temperate legumes are notorious for the variety of toxic constituents they may contain (Smolenski *et al.*, 1981) and lupins are no exception. Quinolizidine alkaloids, oligosaccharides, phytic acid and polyphenolics are some of the undesirable compounds found in lupins. Furthermore, lupins are also associated with a mycotoxicosis called lupinosis. However, as a result of nature or breeding programmes, toxins and anti-nutritional factors associated with commercially grown lupins have generally been reduced to manageable levels or present a minimal risk.

This chapter looks primarily at the pharmacological effects of such compounds in laboratory and farm animals. For their contents in lupins, significance for human nutrition and greater detail of their chemical nature, see Chapter 13.

Quinolizidine Alkaloids

All plants of the *Lupinus* genus contain quinolizidine alkaloids which occur as tertiary bases, N-oxides and esters of the bases. Different species have different alkaloid profiles, and within a species there are usually four or five major alkaloids and several minor ones (Chapter 13). The total content and relative proportions of these can vary between strains, between growing sites and between seasons. Total concentrations range from 0.01 to 4%, depending on the plant species and the part of the plant. The major sites of synthesis of the alkaloids have been assumed to be the green leaves and shoots, with large amounts transferred to the flowers, pods and seeds as the plant matures. However, Bäumel *et al.* (1995) have modelled the flows of quinolizidine alkaloids in *L. albus* using the same techniques outlined in Chapter 7. These authors conclude that synthesis is mainly in the root, with xylem transportation mediating deposition in shoot organs. These findings have been supported by recent studies using reciprocal grafts of high and low alkaloid lines of *L. angustifolius*. The highest concentrations are usually found in the mature seeds. These alkaloids have a bitter taste and it is suspected that they are produced as a general repellent to insects and grazing mammals (Petterson, 1985; Culvenor and Petterson, 1986; Harborne, 1991).

Intoxication by the quinolizidine alkaloids of lupins is referred to as lupin alkaloidal poisoning (Bennetts, 1957), or lupine (Gardiner, 1967a) or lupin (Seawright, 1989) poisoning, and it is always acute (Bennetts, 1957; Humphreys, 1988). Intoxication necessarily involves eating a large amount of lupin within a short time because the alkaloids are excreted rapidly (Petterson, 1985; Cheeke and Shull, 1985). Wittenberg and Nehring (1965) found that about 50% of lupanine, one of the lupin alkaloids, was excreted in the urine and faeces of rats within 24 hours of administration, and the remainder within a few days.

Under natural conditions lupin alkaloidal intoxication occurs mainly in sheep, but has also occurred in horses, cattle, goats, swine and deer (Gardiner, 1967a). Kingsbury (1964) reported that lupins killed more sheep than any other poisonous plant in Montana, Idaho and Utah. The principal toxic species there are *Lupinus leucophyllus*, *L. leucopsis*, *L. argenteus* and *L. sericeus* (Cheeke and Shull, 1985). A lethal dose for these species is 0.25–0.5% of body weight for seeds, or about 1.5% of body weight for pods and seeds (Cheeke and Shull, 1985). In Western Australia, where this poisoning occurs mostly on *L. cosentinii*, mortality is negligible; 360–680 g of seed from this lupin, which has a relatively low alkaloid content, have been fed daily to sheep for up to eight months without fatality (Bennetts, 1957). There are no remarkable or specific necropsy findings in fatally affected animals (Seawright, 1989).

The pharmacological effects of the quinolizidine alkaloids include respiratory depression and failure (Couch, 1926; Lu, 1964; Schmidlin-Meszaros, 1973), a general hypotensive action and reduction of the amplitude of cardiac muscle contractions (Mazur *et al.*, 1966a,b), neuromuscular blockage (Lu, 1964), and fibrillations and cyanosis (Couch, 1926; Lu, 1964; Petterson *et al.*, 1987). In humans, the symptoms and signs of intoxication include nausea, respiratory and visual disturbances, ataxia, progressive weakness and coma (Cheeke and Shull, 1985). In sheep, signs include dyspnoea, trembling, staggering, wild galloping, frothing at the mouth, convulsions and coma (Bennetts, 1957; Cheeke and Shull, 1985). Death results from respiratory paralysis (Cheeke and Shull, 1985), but many sheep may recover without treatment and suffer no ill affects (Gardiner, 1967a). Intoxicated horses exhibit twitching of subcutaneous muscles, constipation, depression and a tendency to lift the forefeet high when walking (Seawright, 1989), while pigs generally reduce their voluntary feed intake (Pearson and Carr, 1977; Godfrey *et al.*, 1985) before clinical signs appear; but if they do eat enough they will vomit (A.R. Mercy, Perth, 1978, personal communication).

There have been numerous studies on the acute toxicity of the various lupin alkaloids. Many of these have been reviewed by Petterson (1985) and Culvenor and Petterson (1986). In general, the lupin alkaloids are six to eight times more toxic following intraperitoneal, rather than oral, administration, and sparteine and lupanine appear to be the two most toxic alkaloids. Intraperitoneal administration of sparteine to laboratory animals has resulted in a minimum lethal dose or LD_{50} of less than 70 mg kg^{-1} in several studies, and the LD_{50} for lupanine administered intraperitoneally to the rat is about 175 mg kg^{-1}. The oral LD_{50}s for sparteine in the mouse, and for lupanine in the rat, are 350–510 mg kg^{-1} and about 1460 mg kg^{-1}, respectively. For lupanine, the descending order of susceptibility in laboratory animals is mice, rats and guinea pigs. The lethal dose of lupin alkaloids in children is estimated to be 11–25 mg total alkaloids kg^{-1} body weight, while adult humans have suffered non-fatal poisonings following the ingestion of 25–46 mg total alkaloids kg^{-1} body weight. These results are further reviewed in Chapter 13 and indicate that humans are far more susceptible to the lupin alkaloids than laboratory animals.

The effects of long-term consumption of the lupin alkaloids have been reviewed by Petterson (1985) and Culvenor and Petterson (1986). In single and multi-generation studies with both rats and pigs, the only attributable effect has been reduced appetite, or feed rejection, with a resultant depression of growth rate when alkaloid levels get too high. Appetite depression was seen in rats at 1.8% alkaloid in the total diet, and in pigs at 0.033%.

Teratogenicity of quinolizidine alkaloids

The only quinolizidine alkaloid in *Lupinus* that is known to be teratogenic is anagyrine. It is confined to American species such as *L. caudatus*, *L. sericeus* and *L. laxiflorus*. Consumption of lupins containing anagyrine by pregnant cows between the 40th and 70th days of gestation may result in congenital deformities in their calves. The condition is referred to as 'crooked calf disease' and is characterized by arthrogryposis, scoliosis, torticollis and cleft palate (Cheeke and Shull, 1985). Anagyrine is not teratogenic in sheep or goats (Panter *et al.*, 1990).

At least one case of teratogenesis in humans has been tentatively traced to anagyrine. A child with severe bilateral limb deformities was born to a woman who had consumed, during pregnancy, milk from lactating goats that had grazed *L. latifolius*. Kids born to these goats had similar deformities, as did pups from a dog fed the same milk. Anagyrine comprised 86% of the alkaloids present in the lupins eaten, and milk from the goats contained high levels of anagyrine (Kilgore *et al.*, 1981; Ortega and Lazerson, 1987).

Lupinus formosus, another New World lupin, is teratogenic to both cattle and goats, causing a syndrome similar to crooked calf disease. This lupin contains only very small amounts of anagyrine, and it is suspected that the piperidine alkaloid ammodendrine is the teratogenic agent in this plant (Keeler and Panter, 1989, 1992; Panter and Keeler, 1991). Piperidine alkaloids in *Conium maculatum* and a pyridino-piperidine alkaloid in *Nicotiana glauca* cause similar congenital deformities in calves following maternal ingestion (Cheeke and Shull, 1985).

Over the years there have been numerous cases of hemimelia (incomplete development of the distal part of the limb) in lambs in Western Australia. In all cases investigated, the mothers of these lambs have grazed *L. cosentinii* plants or seeds at some time during mating or gestation. No specific alkaloid has been incriminated, although it has been noted that multiflorine, the predominant quinolizidine alkaloid in *L. cosentinii*, has a structural resemblance to anagyrine (Allen *et al.*, 1983a). Hawkins (1994) has also reported congenital joint laxity and dwarfism ('Acorn' calves) in 20 out of 145 calves born to cows that had grazed *L. cosentinii* during the middle trimester of their pregnancy.

Anagyrine does not occur in *L. albus, L. angustifolius, L. luteus* or *L. mutabilis*, the major cultivated species of lupins, while multiflorine is present in *L. mutabilis* in trace amounts (Petterson, 1985; Culvenor and Petterson, 1986).

Low alkaloid ('sweet') lupins

Low alkaloid varieties of *L. albus, L. angustifolius* and *L. luteus* have been available for many years, and major breeding programmes continue to produce more. No low alkaloid varieties of *L. mutabilis* existed until the 1980s when the first stable line, Inti, was bred in Chile (Roemer and Jahn-Deesbach, 1988; von Baer and von Baer, 1988). Other low alkaloid lines have now been produced through crossings with Inti. However, there are still no sweet and early-flowering varieties of *L. mutabilis* available for commercial production (Cowling, 1994). Low alkaloid breeding lines have now also been developed in several of the 'rough-seeded' group of lupin species (Chapter 4, this volume), which may in time yield commercial cultivars. These breeding programmes have largely eliminated intoxication by the quinolizidine alkaloids as a potential problem of cultivated lupins.

Lupinosis and the Phomopsins

Lupinosis is a mycotoxicosis caused by the ingestion of toxins, known as phomopsins, produced by the fungus *Diaporthe toxica* (discussed below) which colonizes lupin plants (van Warmelo *et al.*, 1970; Gardiner and Petterson, 1972; Culvenor *et al.*, 1977, 1978). It is recognized primarily as a disease of sheep, but natural outbreaks have also been reported in cattle, goats, donkeys, horses and pigs, and it has been produced experimentally in rabbits, guinea pigs, mice, rats, dogs, ducklings and chickens (Allen, 1986). As a clinical entity, lupinosis was first recognized in 1872 in Germany when numerous sheep deaths were associated with ingestion of the plant (Zurn, 1879). Since then lupinosis has been reported in the USA, Poland, New Zealand, Australia, the Republic of South Africa and Spain (Allen, 1986; Rodriguez *et al.*, 1991). However, it is only in Australia, and particularly Western Australia, where the dead lupin plant has been extensively used as a fodder, that lupinosis has assumed major importance.

Lupinosis is mainly characterized by severe liver damage, which results in inappetence, loss of condition, lethargy, jaundice and often death. The most distinctive feature of the microscopic pathology is arrested and abnormal mitosis of hepatocytes (Gardiner, 1967b; Peterson, 1978). The course of the disease may be acute, subacute or chronic, depending on the quantity of phomopsins consumed in relation to time. The adrenal glands, pancreas, kidneys, rumen and muscle may also be affected. The disease has been associated with abortion in late pregnant sheep and cattle, death of embryos in early pregnant sheep, up to a 50% reduction in lambing percentage in ewes, reduced wool production and staple strength in sheep and starvation ketosis in cattle (Allen, 1986; Barnes, 1993). Photosensitization is uncommon in sheep (Gardiner, 1967a), but is often seen in cattle (Allen, 1981). Phomopsins are embryotoxic (Peterson, 1983) and carcinogenic (Peterson, 1990) in rats.

The causative fungus

Approximately 25 years ago it was determined that *Phomopsis leptostromiformis* (Kühn) Bubák (syn. *Phomopsis rossiana* (Sacc.) Sacc. & D. Sacc.) was the cause of lupinosis (van Warmelo *et al.*, 1970; Gardiner and Petterson, 1972). The teleomorph of this fungus was later found, described and named *Diaporthe woodii* Punith. (Punithalingam, 1974; Punithalingam and Gibson, 1975; Wood and Sivasithamparam, 1989). Examination of the morphology and toxicogenicity of different *P. leptostromiformis* isolates in Western Australia led to the belief that there existed two biotypes. These were designated Biotype A (highly toxicogenic) and Biotype B (non-toxicogenic) (Allen *et al.*, 1986a; Wood, 1986; Wood and Sivasithamparam, 1989). Subsequent extensive biochemical and cultural studies resulted in Shivas *et al.* (1991) raising these biotypes to varieties. Biotypes A and B were named *P. leptostromiformis* var. *leptostromiformis* and *P. leptostromiformis* var. *occidentalis* Shivas, Allen and Williamson var. nov., respectively. Shivas *et al.* (1991) also showed that var. *occidentalis* (formerly Biotype B) was in fact lowly toxicogenic.

Recently, a new and distinct teleomorph of the toxicogenic variety has been found and described (Williamson *et al.*, 1994). It has been named *Diaporthe toxica* P.M. Williamson, A.S. Highet, W. Gams & K. Sivasithamparam sp. nov. These workers established that the original description of *D. woodii* was correct, but that it was in fact the teleomorph of the lowly toxicogenic *P. leptostromiformis* var. *occidentalis*, which was the fungus that Punithalingam (1974) originally described as *P. leptostromiformis*. Williamson *et al.* (1994) have therefore concluded that *D. woodii* (anamorph *P. leptostromiformis*) is not the cause of lupinosis, but that *D. toxica* (anamorph *Phomopsis* sp.) is.

In situations where there is nearby infected lupin trash, *D. toxica* infects the new lupin plant throughout the growing season. Pycnidiospores are the main source of localized inoculum, although ascospores may be involved in spread of infection over longer distances. Defoliated infected leaves within a lupin crop may provide some supplementary pycnidiospore inoculum. Infected lupin seeds have a minimal role to play in infecting lupin plants where the fungus is already established, but may introduce it to new areas (Wood, 1986).

Infection of the live plant is latent (Williamson *et al.*, 1991) and signs of the infection do not normally become apparent until the plant matures and senesces (Wood and Brown, 1975; Wood, 1989). Very small amounts of phomopsins may be detected at the sites of latent infection (Williamson *et al.*, 1995), but toxin production in quantities sufficient to cause disease is associated with the presence of visible lesions on the lupin stems, pods or seeds. Lupinosis is therefore a disease of summer and autumn, when the dead plants are eaten (Allen *et al.*, 1985).

D. toxica infection of lupin seed is not a major toxicological problem, but it is a concern. Infected seed may be either discoloured or asymptomatic (Wood, 1986). The important point when considering the potential toxicity of harvested seed is that virtually all phomopsins produced are in discoloured seed (Petterson and Hancock, 1986; Petterson and Wood, 1986; Wood *et al.*, 1987; Than *et al.*, 1994a). The risk of phomopsins in any seed sample can therefore be largely avoided by selecting only those samples with very small proportions of discoloured seed, or by removing the discoloured seed by commercial seed cleaning processes (Wood and Petterson, 1986). The frequency of seed infection (Cowling and Wood, 1989) and the concentration of phomopsins in commercial seed lots (Than *et al.*, 1994a) are less in *Phomopsis*-resistant compared with *Phomopsis*-susceptible lupin cultivars (established usage is *Phomopsis*-resistance, rather than *Diaporthe*-resistance). However, the correspondence between resistance to stem infection and that to seed infection is incomplete (J.S. Gladstones, Western Australia, 1996, personal communication).

The fact that lupin seed provides minimal risk of lupinosis needs to be emphasized. The natural outbreak of lupinosis in South Africa that incriminated *P. leptostromiformis* as the cause of the disease resulted when sheep ate lupin seeds and pods visually infected by the fungus. The lupin plants were only 23 cm tall, the pods were discoloured externally, and contained inside a conspicuous, white fungal mycelium; the seeds were also infected and discoloured (van Warmelo *et al.*, 1970). This was an atypical crop. The only other two reports of lupinosis in livestock fed naturally infected lupin seed were also in atypical situations. One involved two long-term pen trials when sheep were fed lupins as a sole diet for 34–45 weeks. Seven and 21% of the seeds fed were infected with *D. toxica* (Allen *et al.*, 1983b). The other involved the feeding of lupins containing 32–45.8% discoloured seeds (an extraordinarily high level) to sheep that had concurrent chronic pyrrolizidine alkaloidosis (another liver disease) as a result of eating heliotrope (*Heliotropium europaeum*) (Links and Walker, 1989). In the third, young sheep in a feedlot were fed lupins containing a very high proportion of discoloured seed and 5400 mg kg^{-1} of phomopsin A (Seward, 1993).

The phomopsins

The phomopsins are a group of low molecular weight natural toxins produced by *D. toxica*. The first two, phomopsins A and B, were isolated in crystalline form about 20 years ago (Culvenor *et al.*, 1977). Phomopsin A made up almost 80% of the toxic material extracted. It was therefore assumed to be the principal toxin responsible for causing lupinosis (Culvenor *et al.*, 1978), and most effort has been made to understand its chemistry and toxicity. Phomopsin A has the chemical formula $C_{36}H_{45}ClN_6O_{12}$ and a molecular weight of 788, and is a linear hexapeptide modified by an ether bridge to form a 13-membered macrocyclic ring (Culvenor *et al.*, 1989). Phomopsin B is the des-chloro analogue of phomopsin A (Edgar, 1991).

From the time when phomopsins A and B were discovered there was the suspicion that other phomopsins existed, because there was residual bioactivity in extracts from which the two known phomopsins had been removed (Culvenor *et al.*, 1977, 1978). Subsequently, Allen and Hancock (1989) designated a peak in a chromatogram of a *D. toxica* culture extract as phomopsin C. They presented evidence of toxicity and reported that the small amount they were able to isolate produced mitotic disturbances in the hepatocytes of nursling rats similar to those produced by phomopsin A. The structure of phomopsin C has not been determined, but it does contain the 3,4-didehydroproline unit that is unique to the phomopsins.

Recently two more phomopsins, D and E, were identified (Cockrum *et al.*, 1994). Phomopsin D has a molecular weight of 790 and the only difference from phomopsin A is that the terminal amino acid residue 2,3-didehydroaspartic acid is changed to aspartic acid. The structure of phomopsin E remains to be determined, but it has many features in common with phomopsin A. Both these phomopsins induce the mitotic abnormalities in hepatocytes of nursling rats characteristic of phomopsin A.

More phomopsins will undoubtedly be found in the future. More work is needed to elucidate the chemistry and relative toxicity of all phomopsins and their metabolites produced in the body. For example, phomopsin C was found in over 100 lupin stubbles examined in the late 1980s (J.G. Allen, W.A. Cowling, G.R. Hancock, P.W. Morcombe and D.S. Petterson, Western Australia, 1989, personal communication; Morcombe *et al.*, 1992a), but the contribution it made in field intoxications is unknown because there is no information on how toxic it is. Also, Allen (1992) has presented evidence that, although the phomopsins are extremely toxic primary toxins, the metabolites produced in the liver are also toxic.

The phomopsins are very potent cytotoxic and anti-mitotic agents (Gardiner, 1967b; Papadimitriou *et al.*, 1973; Papadimitriou and Petterson, 1976; Peterson and Lanigan, 1976; Peterson, 1978; Allen, 1985, 1989). They exert these effects through their ability to bind with microtubule sub-unit proteins (tubulin) (Tonsing *et al.*, 1984; Lacey *et al.*, 1987; Luduena *et al.*, 1989, 1990; Li *et al.*, 1992) and inhibit the formation (polymerization) of the microtubules that are cytoskeletal macromolecules vital to the functioning of cells. Intracellular transport mechanisms and cell division are two processes that require the microtubules. There are two distinct binding sites of phomopsin A on tubulin, a high affinity site and a low affinity site, and phomopsin A both inhibits the polymerization of tubulin, and depolymerizes the polymerized tubulin (Li *et al.*, 1992). Phomopsin A is the most potent inhibitor of microtubule assembly among the known anti-mitotic agents (Li *et al.*, 1992) and is one of only two mycotoxins known to have anti-mitotic activity (Iwasaki, 1992). The other is ustiloxin, produced by the rice smut ball fungus *Ustilaginoidea virens* (Iwasaki, 1992). Interestingly, ustiloxin is a tetrapeptide containing a 13-membered ring including an ether linkage (Koiso *et al.*, 1992), a structure very similar to that of phomopsin A.

Other mechanisms in phomopsin intoxication have been suggested. The earliest pathological changes in rats dosed with phomopsins include swelling of the mitochondria and accumulation of lipids in hepatocytes, accompanied by reduced histochemical activities of mitochondrial enzymes (Papadimitriou and Petterson, 1976). These changes suggest impairment of mitochondrial respiration, and it has been demonstrated that phomopsins inhibit complex I of the respiratory chain, resulting in inhibition of uptake of inorganic phosphate and oxygen consumption by mitochondria (van Aswegan *et al.*, 1980). Changes in the activities of some membrane-associated enzymes (Peterson, 1986) and redistribution of Golgi apparatus membranes (Tonsing *et al.*, 1984) indicate that cell membranes may be directly affected by these mycotoxins. Phomopsins can also induce chromosomal aberrations (Brown and Bick, 1986), and it has been suggested that they react directly with DNA – a possible mechanism for the karyomegaly (Allen, 1989) and carcinogenicity (Peterson, 1990) associated with their intake. Finally, histochemical and chemical studies have also demonstrated considerable disruption of various enzyme systems and metabolic pathways in the liver and kidney of intoxicated mice (Papadimitriou *et al.*, 1974), rats (Papadimitriou and Petterson, 1976) and sheep (J.G. Allen, Western Australia, 1986, personal communication). It is not known whether these latter changes are primary or secondary effects of the phomopsins.

Phomopsins are also produced by *Phomopsis emicis* (Shivas *et al.*, 1994), a natural pathogen of the annual winter weed *Emex australis* (double gee). *P. emicis* has a considerable capacity to produce phomopsins in laboratory cultures, but to date no field cases of a lupinosis-like disease have been reported in sheep known to be eating this weed.

Toxicity of the phomopsins

The toxicity of the phomopsins is affected by the animal species intoxicated, the age and pregnancy state of the affected animal, and the route and pattern of administration of the phomopsins. Sheep are the most susceptible species studied, followed by pigs, rats and mice in decreasing order of susceptibility (Allen, 1985; Peterson, 1986). Cattle are less susceptible than sheep (Allen, 1981); and although lupinosis is not often seen in horses they may be highly susceptible, the disease usually being acute and fatal (Gardiner and Seddon, 1966; Nelson *et al.*, 1986). Goats are believed to have intermediate susceptibility between sheep and cattle (Nelson *et al.*, 1986).

The *per os* LD_{50}s of a single dose of phomopsin A to sheep (6–9 months old) and rats (2 weeks old) are 1000–1350 $\mu g\ kg^{-1}$ and 35 $mg\ kg^{-1}$, respectively. Toxicity is considerably greater when the phomopsins are administered intraperitoneally or subcutaneously, with the LD_{50}s of a single dose of phomopsin A to sheep and rats being 10–27 $\mu g\ kg^{-1}$ and 1.6 $mg\ kg^{-1}$, respectively. The LD_{50} of a single subcutaneous dose of phomopsin A to young adult rats (10 weeks old) is approximately 5.5 $mg\ kg^{-1}$, indicating that older animals, no longer rapidly growing, are less susceptible. Rats in early pregnancy (days 6–10) appear more susceptible than those in later stages of pregnancy (days 11–15) (Peterson, 1983, 1986; Allen 1985; Peterson *et al.*, 1987).

The pattern of dosing (or intake) of phomopsins also influences toxicity. For example, one out of three sheep died after a single intra-ruminal injection of 1000 $\mu g\ kg^{-1}$ of phomopsin A, whereas three out of three died after receiving either five doses of 200 $\mu g\ kg^{-1}$ or 20 of 50 $\mu g\ kg^{-1}$ (Peterson, 1986; Peterson *et al.*, 1987). Multiple smaller doses appear more toxic than a single large dose, but there is an apparent limit to how small the multiple doses can be for an effect to be seen. In the same experiment all three sheep that received 80 doses of 12.5 $\mu g\ kg^{-1}$ survived, and showed minimal clinical signs of disease.

Prevention of lupinosis: *Phomopsis*-resistant lupins and vaccination

Allen (1987) and Allen and Chapman (1988) reviewed management procedures to avoid lupinosis in livestock grazing lupin stubbles or fed *D. toxica* infected seed. These require a sound understanding of the epidemiology of the disease in the field, or involve agronomic practices to reduce the amount of phomopsins ingested. But it has been the release of *Phomopsis*-resistant lupins that has most reduced the annual prevalence of lupinosis in Western Australia, the area of the world with the greatest incidence of the disease.

Gladstones (1989a) described the history of the successful breeding programme that developed *Phomopsis*-resistant *L. angustifolius* varieties for Australia. Successful transfer of resistance from wild to domesticated *L. angustifolius* (Cowling *et al.*, 1986a, 1987) was followed firstly by laboratory studies (Allen *et al.*, 1985; Cowling *et al.*, 1986a,b, 1988) and then field grazing experiments (Allen and Cowling, 1986, 1992; Cowling *et al.*, 1986a; Allen *et al.*, 1989), which established that the transferred resistance greatly reduced phomopsin production in lupin stems. This work culminated in the registration (Gladstones, 1989b,c; Hawthorne and Gladstones, 1989) and commercial release of the first three *Phomopsis*-resistant lupin cultivars, Gungurru, Yorrel and Warrah, in 1989. Since then a comparative study on farms growing both *Phomopsis*-resistant and susceptible cultivars found flock prevalence of lupinosis to be reduced from 56 to 8%, and sheep mortality from 4 to 0.2%, by growing resistant varieties (Morcombe and Allen, 1990). As might be expected, planting of *Phomopsis*-resistant lupins has substantial economic benefits (Warren *et al.*, 1989). A further cultivar with substantially stronger resistance is currently being considered for release (Chapter 4, this volume).

While most work on *Phomopsis*-resistant lupins has concentrated on *L. angustifolius*, it needs also to be pointed out that the Ultra (Wood and Allen, 1980; Hamblin *et al.*, 1981) and Kiev Mutant (McDonald *et al.*, 1991; Cowling, 1994) varieties of *L. albus* are also resistant.

Although the currently released *Phomopsis*-resistant lupins have made an enormous impact on the incidence of lupinosis, they are not completely free of the fungus. They do sometimes develop enough toxicity to cause lupinosis (McDonald *et al.*, 1994). In Western Australia the summer of 1995/96 was the worst for lupinosis since the release of the *Phomopsis*-resistant lupins and cases of lupinosis in sheep grazing their stubbles were reported throughout the agricultural area. More than 200 deaths occurred in each of two cases in the south of the state (Animal Health Laboratories' records, Agriculture Western Australia, South Perth, 1996). Higher levels of resistance in future cultivars will doubtless reduce this risk, but complete elimination of lupinosis by breeding cannot yet be guaranteed. There were also many cases of lupinosis in sheep and cattle grazing *L. cosentinii*, the sandplain lupin, in 1995/96.

Attempts to immunize animals against low molecular weight pasture-associated toxins have been reported, but have generally not been successful (Payne *et al.*, 1992; Edgar, 1994). However, during 1988/89 an experimental vaccine was developed that successfully protected sheep in a pen trial from a total oral challenge of 1000 μg phomopsins kg^{-1} liveweight administered over four weeks (Payne *et al.*, 1992). This was the first example of a protective response against a low molecular weight toxin that is not naturally antigenic, and provided encouragement that a vaccine against lupinosis could be developed.

Subsequent improved formulations of the anti-lupinosis vaccine have been tested in the field (Allen *et al.*, 1994; Than *et al.*, 1994b). The trials showed that the majority of sheep, but unfortunately not all, can be effectively vaccinated against lupinosis. The level of protection depends upon the anti-phomopsin IgG response induced. In some sheep the response is not sufficient to give them protection. Work is continuing, with expectations that a successful commercial vaccine will be developed. Such a vaccine would be beneficial in some areas where the resistance of existing crop cultivars to *D. toxica* remains insufficient, and for use in livestock grazing *L. cosentinii* and perhaps varieties of *L. luteus* that are shortly to be released for commercial growing in Western Australia. Some of the latter may not be resistant to all strains of *D. toxica* (W.A. Cowling, Perth, 1996, personal communication; Cowling and Buirchell, 1996; Sweetingham *et al.*, 1996).

Testing for phomopsins

The Australian Food Standards Code specifies a maximum concentration of 5 μg phomopsins kg^{-1} in 'food' for human consumption (National Food Authority, 1992). The aflatoxins are the only other mycotoxins controlled to this degree in food on sale in Australia (National Food Authority, 1992). An ELISA test is available that detects all phomopsins, with a maximum sensitivity of 10 pg phomopsins per well and an 80–90% recovery (Than *et al.*, 1992, 1994a). Being rapid, inexpensive and sensitive (Than, 1994), the test appears ideal for monitoring phomopsins in lupin products for human consumption.

Using this ELISA, Iyer *et al.* (1994) reported that nearly 18% of lupin seed samples from Victoria and 15% of those from New South Wales contained more than 500 μg phomopsins kg^{-1}. Furthermore, 8% of the Victorian samples had more than 1000 μg kg^{-1} and 19% had 100–500 μg kg^{-1}. These are alarming figures, but Iyer *et al.* (1994) do not report how or from what cultivars the samples were selected. They may have been sent for testing because problems were suspected.

Than *et al.* (1994a) examined 88 random lupin samples from Western Australian receival points. These contained a mixture of seed from *Phomopsis*-resistant and susceptible cultivars. The average phomopsins content was 6.1 μg kg^{-1}, with a range of 0–86 μg kg^{-1}, and 32% of the samples had more than 5 μg kg^{-1}. These results are far more reassuring, because once such seed is incorporated in 'food' the resultant phomopsins concentration would be less than 5 μg kg^{-1} in nearly all situations.

While the present ELISA seems ideal for commercial use, this writer is concerned that it may be too broadly inclusive, and thus over-sensitive. It detects all phomopsins (Than *et al.*, 1994a); but some phomopsins have already been found less toxic than phomopsin A (Culvenor and Petterson, 1986), and indeed it has yet to be established that all phomopsins are toxic. Using this ELISA, Gallagher *et al.* (1994) reported phomopsin concentrations of up to 88 and

217 mg kg^{-1} in lupin stubbles of *Phomopsis*-resistant and susceptible cultivars, respectively. These levels are many times greater than those previously found in lupin stubbles, many of which were associated with clinical lupinosis, using chemical methods (Wood, 1986; Allen *et al.*, 1986b; Hancock and Allen, 1987; Croker and Allen, 1990; Morcombe *et al.*, 1992a,b). A 30 kg sheep would need to consume only 340 and 138 g, respectively, of these stubbles to die (Peterson *et al.*, 1987). It thus seems possible that ELISA values may over-estimate toxicity if applied to established relationships based on chemical assays. However, it is possible that the results provided by the ELISA simply reflect an improved efficiency of extraction. A recent comparative study of the phomopsins concentrations in lupin stubbles, as determined by the ELISA and a nursling rat bioassay (Peterson, 1978; Petterson *et al.*, 1985), showed that a reduced extraction efficiency was responsible for the consistently lower results provided by the bioassay (Allen *et al.*, 1998).

Other Toxic and Anti-nutritional Factors

Anti-nutritional factors are naturally occurring compounds in plants that diminish the nutritive value to humans and animals. Their primary role is thought to be defence of the plant. Chapter 13 of this volume describes the chemical characteristics and typical contents of the main ones in cultivated lupins. Here, mainly their pharmacological effects are presented.

Sweet lupins are generally regarded as very safe with respect to anti-nutritional factors. The seed contains several common anti-nutrients (Table 14.1, Petterson *et al.*, 1986; Chapter 13, this volume), but mostly at such low levels that, unlike whole soybeans and many other grain legumes, they do not need heat treatment to destroy the factors that can reduce protein digestion and availability (Petterson and Mackintosh, 1994a,b). Apart from oligosaccharides, seed from sweet lupin cultivars has a content of anti-nutritional factors more like that in cereal grains than in other legumes (Sipsas, 1994).

Table 14.1. Anti-nutritional factors in some *Lupinus* spp. and other legumes (from Haraszti and Vetter, 1980; Petterson and Mackintosh, 1994a,b; Petterson, 1996).

Anti-nutritional factor	*L. albus*	*L. angustifolius*	*L. luteus*	*Glycine max* (soybean)	*Pisum sativum* (field pea)	*Cicer arietinum* (chickpea Desi)
Oligosaccharides (%)	6.69	5.16		5.75	3.69	
Phytic acid (%)	0.79	0.58		1.59	0.48	0.63
Tannins total (%)	0.37	0.32	0.32	0.57	0.25	0.49
Tannins condensed (%)	0.01	< 0.01	0.02		0.06	0.03
Saponins (mg kg^{-1})	10–23	573	5	3500	1800	
Trypsin inhibitors (mg g^{-1})	0.13	0.14	0.29	17.9	1.01	4.79
Chymotrypsin inhibitors (mg g^{-1})		0.08	ND		1.60	7.72
Lectins (dilutions)	ND	ND	ND	> 4	4	

ND = not detected.

Oligosaccharides

Lupin seed from the four main cultivated lupin species, at least, contain moderately high concentrations of oligosaccharides (Gross, 1988; Chapter 13, this volume). These are α-galactosyl derivatives of sucrose that must be cleaved by the enzyme α-galactosidase before they can be digested. Monogastric animals lack this enzyme and so the oligosaccharides pass unchanged through the small intestine to the lower gut, where they are cleaved by bacterial α-galactosidase. The cleavage product is then rapidly converted to carbon dioxide, hydrogen and methane, and may result in flatulence, diarrhoea, nausea, cramps and general discomfort (Saini, 1989; Petterson and Mackintosh, 1994a). This is a relatively insignificant problem and the concentrations of oligosaccharides in lupins are only slightly higher than the reported averages for other pulse crops widely used for human consumption (Petterson, 1985; Allen *et al.*, 1986c; Chapter 13, this volume). Traditional debittering processes with water, besides removing the alkaloids from lupins, also remove the oligosaccharides (de Almeida *et al.*, 1986). In Western Australia, some horse owners soak lupins for 24 hours prior to feeding to avoid flatulence and colic problems.

Phytic acid

Plant seeds contain phytic acid, an organic compound containing six phosphate groups. Phytic acid chelates various mineral cations to form phytates, rendering the mineral nutritionally unavailable. The decreasing order of stability of the phytate–mineral complexes is zinc, copper, cobalt, magnesium and calcium (Cheeke and Shull, 1985). The phytic acid concentration in cultivated lupins is less than in barley, wheat or soybeans (Cheeke and Shull, 1985; Petterson and Mackintosh, 1994a; Chapter 13, this volume), and the bioavailability of zinc from soybeans is lower than from cereal and animal products (Petterson *et al.*, 1991). The bioavailability of zinc from lupin milk and bread made with lupin flour is similar to that from comparable soybean products. The availability from composite meals made from lupin flour and lupin protein isolate is significantly greater than from comparable soybean products (Petterson *et al.*, 1991).

Tannins

Tannins are polyphenolic compounds that bind with digestive enzymes to inhibit their activity, and with dietary proteins to prevent their digestion. They may also bind with proteins in the intestinal mucosa to affect general nutrient absorption, and with vitamin B_{12} to make it unavailable for absorption. Reaction of the tannins with salivary proteins and glycoproteins in the mouth creates a strong astringent taste. These properties combine to result in reduced palatability and feed intake, irritation of the gastrointestinal tract, lowered energy conversion efficiency from foods and increased nitrogen excretion (Cheeke and Shull, 1985; Petterson and Mackintosh, 1994a). Lupins have a lower concentration of tannins than soybeans (Petterson and Mackintosh, 1994a; Chapter 13, this volume).

Saponins

Saponins are glycosides that are characterized by a bitter taste, foaming properties and the ability to haemolyse red blood cells. They may irritate the mucous membranes of the mouth and digestive tract, resulting in a depressed voluntary feed intake, and, in monogastrics particularly, they can cause growth inhibition and increased excretion of cholesterol (Cheeke and Shull, 1985). The concentration of saponins in lupins is very low compared with other pulses (Petterson and Mackintosh, 1994a,b; Chapter 13, this volume).

Protease inhibitors

A wide range of plants contain protein fractions that inhibit protein digestion by animals through their ability to bind very strongly with the active sites of certain proteolytic enzymes. The two enzymes mainly affected are trypsin and chymotrypsin, and the protease inhibitors are more commonly referred to as trypsin or chymotrypsin inhibitors. If the concentrations of these inhibitors in plants are sufficiently high they can result in poor growth, reduced feed intake, a reduction in protein digestibility, pancreatic hypertrophy and a deficiency of sulphur amino acids (Cheeke and Shull, 1985). Lupins have very low concentrations of protease inhibitors compared with other legumes (Petterson and Mackintosh, 1994a,b; Chapter 13, this volume). Hove *et al.* (1978) found that of eight different types of uncooked legume seed, only sweet lupin was safe as a protein source for rats.

Lectins (haemagglutinins)

Lectins are also referred to as haemagglutinins or phytohaemagglutinins, because they cause the clumping or agglutination of red blood cells *in vitro*. They are proteins or glycoproteins and are found in many plant seeds and vegetative tissues. Their anti-nutrient effect results from their binding to specific sugars or glycoproteins, particularly on the mucosal surface of the intestines. This leads to altered intestinal permeability, impaired transport of nutrients across the intestinal wall, modified gut immune function, reduced growth, diarrhoea and increased bacterial infections (Cheeke and Shull, 1985; Sissons and Tolman, 1991). The trace amounts of lectins in lupins are not considered significant (Petterson and Mackintosh, 1994a,b; Chapter 13, this volume).

Allergens

These are usually proteins that induce allergic responses when eaten by sensitive subjects, resulting in morphological changes in the intestinal mucosa, gastrointestinal disturbances, impaired nutrient absorption and growth retardation. Systemic allergic responses may occur if the antigens are absorbed (Cheeke and Shull, 1985; Petterson and Mackintosh, 1994a). Few people have reported allergic reactions to lupin foods (Petterson and Mackintosh, 1994a). A similarly small number report allergic reactions to lupin dust, displaying symptoms of swollen mucosa of the pharynx, itchy eyes with increased lacrimation, sneezing, tightening of the chest with dyspnoea and raised rashes on the skin. People allergic to lupins tend also to be allergic to other foods (Petterson and Mackintosh, 1994a) or have other allergy problems. There has been a report of ill-thrift caused by an allergic enteritis in goats fed lupin seed and canola meal (Richards, 1996).

Cyanogens

Cyanogens are glycosides with a cyanide-containing aglycone, and occur in over 1000 different species of plant. They can be hydrolysed by enzymatic action to hydrogen cyanide, a potent toxin which inhibits the terminal respiratory enzyme cytochrome oxidase (Cheeke and Shull, 1985). Very small concentrations of cyanogens (0.53–2.89 mg HCN per 100 g dry matter) have been reported in *L. albus* and *L. mutabilis* (Schoeneberger *et al.*, 1982), these levels being well below the legislated maximum of 20 mg per 100 g dry matter for food legumes that applies in some European countries and the USA (Petterson, 1985).

Heavy metals

Many reports confirm the ability of *L. albus* to concentrate manganese in its seed (Hill, 1986; Chapter 5, this volume). The extent of accumulation seems to be related to the soil type on which the crop is grown (Petterson and Mackintosh, 1994a), with concentrations in excess of 4 g kg^{-1} being reported (Oram *et al.*, 1979). Pigs fed rations containing 500 mg Mn kg^{-1} may develop a reduced appetite and growth rate (Grummer *et al.*, 1950), and this author is aware of one situation in which pigs fed a ration with *L. albus* containing just over 4 g Mn kg^{-1} showed these signs.

The cadmium and lead contents of commercial lupin seed (*L. angustifolius*) ranged from < 0.01 to 0.08 mg kg^{-1} and < 0.01 to 0.25 mg kg^{-1}, respectively, in samples from Western Australia and Victoria (Petterson and Harris, 1995). The cadmium concentration in 17 of the 178 samples exceeded the maximum permitted by the National Food Authority (0.05 mg kg^{-1}), but none exceeded the proposed Codex Alimentarius limit (0.1 mg kg^{-1}). Recent work has indicated that new commercial varieties of *L. luteus* accumulate cadmium (mean of 0.36 mg kg^{-1} as received) in Western Australia, and this is being investigated further (W. Cowling, G. Shea and D. Petterson, Perth, 1997, personal communication). All the lead concentrations were well below the National Food Authority limit (2.0 mg kg^{-1}) for legume foods (Petterson and Mackintosh, 1994a). Elsewhere, *L. albus* and *L. mutabilis* have been shown to accumulate lead (to concentrations of 1.0–1.6 mg kg^{-1}) and mercury (*L. albus*, 0.07 mg kg^{-1}; *L. mutabilis*, 0.036–0.088 mg kg^{-1}) to a greater extent than other grain legumes (Hill, 1986).

Isoflavonoids

Isoprenylated flavonoids have been found in all parts of the plant in up to eight lupin species (Harborne *et al.*, 1976; Ingham *et al.*, 1983; Tahara *et al.*, 1984, 1989). At least 38 have been identified and they are all anti-fungal, and probably provide some protection against microbial disease. Some also deter insects from feeding (Sutherland *et al.*, 1980). Many isoflavones have weak oestrogenic activity in mammals (Cheeke and Shull, 1985; Price and Fenwick, 1985) and eating enough can cause varying degrees of infertility in females. However, there have been no reports of oestrogenic effects in livestock or humans from eating lupins and so the concentrations of these compounds in lupins must be considered insignificant. Furthermore, all the isoflavonoids so far characterized in lupins are 5-hydroxy substituted, and so are probably deactivated by rumen metabolism (Harborne, 1991).

Roridin A

The roridins are trichothecene mycotoxins usually produced by *Myrothecium* spp. (Smith, 1987). They are generally radiomimetic in nature, tending to attack rapidly dividing cells, and are recognized for their ability to cause lesions on the skin and mucous membranes (Smith, 1987). Samples *et al.* (1984) reported the isolation of roridin A from cultures of a *Phomopsis* sp. isolated from tobosagrass (*Hilaria mutica*), which they morphologically identified as from *P. leptostromiformis*, and *P. leptostromiformis* (Strain ATCC26115 – American Type Culture Collection). The toxic isolate produced acute inflammation in the abomasum of sheep, but no liver changes. Such findings are consistent with roridin A intoxication (Mortimer *et al.*, 1971). The trichothecenes and the phomopsins are distinctly different chemicals and cause distinctly different pathology in intoxicated animals. In Australia, *D. toxica* has been grown on a range of media, including the seeds of rape, sorghum, wheat, sweet lupin, vetch, linseed and rumer bean (Wood *et al.*, 1973), and there has never been any suggestion that toxins other than phomopsins were produced. There has never been a report of suspicion of trichothecene intoxication associated with lupin consumption, nor has there been any report elsewhere of any trichothecene being isolated from cultures of *P. leptostromiformis* or *Phomopsis* sp. (anamorph of *Diaporthe toxica*). In Western Australia several samples of toxic lupin stubble were specifically examined for roridin A, and none was found (Petterson, 1985).

Rumenitis factor

Lupins contain little or no starch (Evans, 1994). This is the carbohydrate responsible for grain poisoning or rumen acidosis when cereal grain is fed to ruminants, so lupins are regarded as a completely safe feed for sheep and cattle that can be provided with no requirement for a gradual introduction (Rowe, 1995). Nevertheless, during 1995 and 1996 almost 20 cases of lupin seed-associated deaths in sheep and cattle were reported to the Animal Health Laboratories in Western Australia (Allen and Tudor, 1996; Animal Health Laboratories' records). Mortality rates ranged from 1/6 to 25/430. Many of these cases involved the sudden introduction of lupins, often fed via self-feeders, to hungry stock that had been on a falling plane of nutrition. In all cases where the rumen was examined, there was a severe rumenitis (Allen and Tudor, 1996; Animal Health Laboratories' records). The rumenitis has been experimentally reproduced in hungry sheep on a falling plane of nutrition, drenched with 41–64 g of hammer-milled lupins kg^{-1} liveweight. This resulted in a rumen acidosis that persisted for up to 3 days, and substantial increases in the concentrations of rumen lactic acid, volatile fatty acids, ammonia and urea. The rumenitis was not reproduced in sheep on a maintenance ration of sheep cubes and chaff (J.G. Allen, Perth, 1996, personal communication). Evans (1994) reported that dietary fibre in the lupin cotyledon is readily fermented to volatile fatty acids in the hindgut of the rat, causing a reduction in the pH. To test whether dietary fibres in the kernel of lupins may be involved in the pathogenesis of the rumenitis, sheep were dosed with residual cell wall material (the dietary fibre fraction) after wet-milling the kernels to extract a protein concentrate. Acidosis of the rumen and rumenitis were produced (J.G. Allen and D.S. Petterson, 1996, personal communication). Clearly, under specific conditions the rate and extent of fermentation of these dietary fibres is sufficient to cause acidosis of the rumen and rumenitis.

Conclusions

Breeding has had a major impact on the quality of lupin seed and on the value of lupins in farming systems. Alkaloid levels in 'sweet' cultivars of *L. angustifolius* are sufficiently low for the grain to be acceptable for animal and human consumption and problems related to lupinosis in sheep grazing lupin stubbles have been greatly reduced. Although lupin seed contains a range of other antinutritional factors (polyphenolics, phytic acid, saponins, etc.) their levels are low and have little impact on their nutritional value.

References

Allen, J.G. (1981) An evaluation of lupinosis in cattle. *Australian Veterinary Journal* 57, 212–215.

Allen, J.G. (1985) The toxicity of phomopsin in a crude extract of *Phomopsis leptostromiformis* to sheep and pigs. In: Seawright, A.A., Hegarty, M.P., James, L.F. and Keeler, R.F. (eds) *Plant Toxicology.* Queensland Poisonous Plants Committee, Yeerongpilly, pp. 596–603.

Allen, J.G. (1986) Lupinosis, a review. In: *Proceedings of the Fourth International Lupin Conference.* International Lupin Association and Western Australian Department of Agriculture, South Perth, pp. 173–187.

Allen, J.G. (1987) Lupinosis. In: *Proceedings No. 103, Veterinary Clinical Toxicology.* The Post-Graduate Committee in Veterinary Science, Sydney, pp. 113–131.

Allen, J.G. (1989) Studies of the pathogenesis, toxicology and pathology of lupinosis and associated conditions. PhD thesis, Murdoch University, Perth, Australia.

Allen, J.G. (1992) The effect of hepatic microsomal enzyme inducers and inhibitors on the toxicity of phomopsins in the rat. In: James, L.F., Keeler, R.F., Bailey, E.M. Jr, Cheeke, P.R. and Hegarty, M.P. (eds) *Poisonous Plants, Proceedings of the Third International Symposium.* Iowa State University Press, Ames, pp. 245–250.

Allen, J.G. and Chapman, H.M. (1988) Conditions associated with lupins. In: *Proceedings No. 110, Sheep Health and Production.* The Post-Graduate Committee in Veterinary Science, Sydney, pp. 431–456.

Allen, J.G. and Cowling, W.A. (1986) Preventing lupinosis with Phompsis-resistant lupins. *Journal of Agriculture, Western Australia* 27, 89–90.

Allen, J.G. and Cowling, W.A. (1992) A preliminary report on the use of new *Phomopsis*-resistant cultivars of *Lupinus angustifolius* to prevent lupinosis. In: James, L.F., Keeler, R.F., Bailey, E.M. Jr, Cheeke, P.R. and Hegarty, M.P. (eds) *Poisonous Plants, Proceedings of the Third International Symposium.* Iowa State University Press, Ames, pp. 539–541.

Allen, J.G and Hancock, G.R. (1989) Evidence that phomopsins A and B are not the only toxic metabolites produced by *Phomopsis leptostromiformis*. *Journal of Applied Toxicology* 9, 83–89.

Allen, J. and Tudor, G. (1996) Giving stock unlimited access to lupins can be dangerous. In: *Proceedings of the 4th Western Australian Lupin Update Meeting*. Agriculture Western Australia, South Perth, pp. 132–133.

Allen, J.G., Fenny, R.E., Buckman, P.G., Hunt, B.R. and Morcombe, P.W. (1983a) Hemimelia in lambs. *Australian Veterinary Journal* 60, 284.

Allen, J.G., Moir, R.J. and Mackintosh, J.B. (1983b) Ovine lupinosis resulting from the ingestion of lupin seed naturally infected with *Phomopsis leptostromiformis*. *Australian Veterinary Journal* 60, 206–208.

Allen, J.G., Wood, P.McR., Croker, K.P., Cowling, W.A. and Sawkins, D.N. (1985) The prevention of lupinosis in sheep. In: Seawright, A.A., Hegarty, M.P., James, L.F. and Keeler, R.F. (eds) *Plant Toxicology*. Queensland Poisonous Plants Committee, Yeerongpilly, pp. 80- 88.

Allen, J.G., Wood, P.McR., Hancock, G.R. and Ellis, Z. (1986a) The toxicogenicity of different isolates of *Phomopsis leptostromiformis*. In: *Australian Advances in Veterinary Science*. Australian Veterinary Association, Artarmon, pp. 94–95.

Allen, J.G., Croker, K.P., Hancock, G.R. and Southey, I.N. (1986b) Control of ovine lupinosis: the treatment of lupin stubble with alkali. *Australian Veterinary Journal* 63, 351–352.

Allen, D.G, Greirson, B.N. and Wilson, N.L. (1986c) Chemical analysis of antinutritional factors in sweet lupinseed. In: *Proceedings of the Fourth International Lupin Conference*. International Lupin Association and Western Australian Department of Agriculture, South Perth, p. 291.

Allen, J.G., Cowling, W.A. and Warren, J.M. (1989) A preliminary report on the use of new *Phomopsis*-resistant cultivars of *Lupinus angustifolius* to prevent lupinosis. In: *Australian Advances in Veterinary Science*. Australian Veterinary Association, Artarmon, pp. 191–193.

Allen, J.G., Than, K.A., Edgar, J.A., Doncon, G.H., Dragicevic, G. and Kosmac, V.H. (1994) Field evaluations of vaccines against lupinosis. In: Colegate, S.M. and Dorling, P.R. (eds) *Plant-associated Toxins – Agricultural, Phytochemical and Ecological Aspects*. CAB International, Wallingford, Oxon, pp. 427–432.

Allen, J.G., Than, K.A., Speijers, E.J., Ellis, Z., Croker, K.P., McDonald, C.L. and Edgar, J.A. (1998) A comparison of a nursling rat bioassay and an ELISA used to determine the concentration of phomopsins in lupin stubbles. In: *Proceedings of the 5th International Symposium on Poisonous Plants*, (in press).

Barnes, A.L. (1993) The effect of lupinosis on reproduction and wool growth of the Merino ewe. PhD thesis, Murdoch University, Perth, Australia.

Bäumel, P., Jeschke, W.D., Räth, N., Czygan, F.-C. and Proksh, P. (1995) Modelling of quinolizidine alkaloid net flows in *Lupinus albus* and between *L. albus* and the parasite *Cuscuta reflexa*: new insights into the site of quinolizidine alkaloid synthesis. *Journal of Experimental Botany* 46, 1721–1730.

Bennetts, H.W. (1957) Lupin poisoning of sheep. *Australian Veterinary Journal* 33, 277–282.

Brown, J.K. and Bick, Y.A.E. (1986) Comparison of cellular effects of phomopsin and colcemid in Chinese hamster cells *in vitro*. *Cytobios* 46, 79–91.

Cheeke, P.R. and Shull, L.R. (1985) *Natural Toxicants in Feeds and Poisonous Plants*. AVI Publishing Company, Westport, CO.

Cockrum, P.A., Petterson, D.S. and Edgar, J.A. (1994) Identification of novel phomopsins in lupin seed extracts. In: Colegate, S.M. and Dorling, P.R. (eds) *Plant-associated*

Toxins – Agricultural, Phytochemical and Ecological Aspects. CAB International, Wallingford, Oxon, pp. 232–237.

Couch, J.F. (1926) Relative toxicity of lupine alkaloids. *Journal of Agricultural Research* 32, 51–67.

Cowling, W.A. (1994) Breeding for yield and adaptation in *Lupinus* species. In: Dracup, M. and Palta, J. (eds) *Proceedings of the First Australian Lupin Technical Symposium.* Western Australian Department of Agriculture, South Perth, pp. 129–139.

Cowling, W. and Buirchell, B. (1996) Lupin breeding. In: *Proceedings of the 4th Western Australian Lupin Update Meeting.* Agriculture Western Australia, South Perth, pp. 3–8.

Cowling, W.A. and Wood, P.McR. (1989) Resistance to Phomopsis stem and pod blight of narrow-leafed lupin in a range of environments and its association with reduced Phomopsis seed infection. *Australian Journal of Experimental Agriculture* 29, 43–50.

Cowling, W.A., Wood, P.McR., Allen, J.G., Gladstones, J.S. and Hamblin, J. (1986a) Effects of genetic resistance on *Phomopsis* incidence and potential stem and seed toxicity. In: *Proceedings of the Fourth International Lupin Conference.* International Lupin Association and Western Australian Department of Agriculture, South Perth, pp. 230–239.

Cowling, W.A., Allen, J.G., Wood, P.McR. and Hamblin, J. (1986b) Phomopsis-resistant lupins: breakthrough towards the control of lupinosis. *Journal of Agriculture, Western Australia* 27, 43–46.

Cowling, W.A., Hamblin, J., Wood, P.McR. and Gladstones, J.S. (1987) Resistance to Phomopsis stem blight in *Lupinus angustifolius* L. *Crop Science* 27, 648–652.

Cowling, W.A., Allen, J.G. and Wood, P.McR. (1988) Resistance to *Phomopsis* stem blight reduces the lupinosis toxicity of narrow-leafed lupin stems. *Australian Journal of Experimental Agriculture* 28, 195–202.

Croker, K.P. and Allen, J.G. (1990) Treating lupin stubbles with alkali is unlikely to prevent lupinosis. *Australian Veterinary Journal* 67, 230–232.

Culvenor, C.C.J. and Petterson, D.S. (1986) Lupin toxins – alkaloids and phomopsins. In: *Proceedings of the Fourth International Lupin Conference.* International Lupin Association and Western Australian Department of Agriculture, South Perth, pp. 188–198.

Culvenor, C.C.J., Beck, A.B., Clarke, M., Cockrum, P.A., Edgar, J.A., Frahn, J.L., Jago, M.V., Lanigan, G.W., Payne, A.L., Peterson, J.E., Petterson, D.S., Smith, L.W. and White, R.R. (1977) Isolation of toxic metabolites of *Phomopsis leptostromiformis* responsible for lupinosis. *Australian Journal of Biological Sciences* 30, 269–277.

Culvenor, C.C.J., Smith, L.W., Frahn, J.L. and Cockrum, P.A. (1978) Lupinosis: chemical properties of phomopsin A, the main toxic metabolite of *Phomopsis leptostromiformis.* In: Keeler, R.F., van Kampen, K.R. and James, L.F. (eds) *Effects of Poisonous Plants on Livestock.* Academic Press, New York, pp. 565–573.

Culvenor, C.C.J., Edgar, J.A., Mackay, M.F., Gorst-Allman, C.P., Marasas, W.F.O., Steyn, P.S., Vleggaar, R. and Wessels, P.L. (1989) Structure elucidation and absolute configuration of phomopsin A, a hexapeptide mycotoxin produced by *Phomopsis leptostromiformis. Tetrahedron* 45, 2351–2372.

de Almeida, D.C.F., Trugo, L.C. and Gross, R. (1986) The variability of oligosaccharides in the seeds of cultivated lupins. In: *Proceedings of the Fourth International Lupin Conference.* International Lupin Association and Western Australian Department of Agriculture, South Perth, p. 292.

Edgar, J.A. (1991) Phomopsins: antimicrotubule mycotoxins. In: Keeler, R.F. and Tu, A.T. (eds) *Handbook of Natural Toxins.* Marcel Dekker, New York, pp. 371–395.

Edgar, J.A. (1994) Vaccination against poisoning diseases. In: Colegate, S.M. and Dorling, P.R. (eds) *Plant-associated Toxins – Agricultural, Phytochemical and Ecological Aspects.* CAB International, Wallingford, Oxon, pp. 421–426.

Evans, A.J. (1994) The carbohydrates of lupins, composition and uses. In: Dracup, M. and Palta, J. (eds) *Proceedings of the First Australian Lupin Technical Symposium.* Western Australian Department of Agriculture, South Perth, pp. 110–114.

Gallagher, P.F., Walker, B. and Edgar, J.A. (1994) Quantitative analysis of phomopsins in *Phomopsis*-resistant and *Phomopsis*-susceptible lupin stubbles. In: Colegate, S.M. and Dorling, P.R. (eds) *Plant-associated Toxins – Agricultural, Phytochemical and Ecological Aspects.* CAB International, Wallingford, Oxon, pp. 155–160.

Gardiner, M.R. (1967a) Lupinosis. *Advances in Veterinary Science* 11, 85–138.

Gardiner, M.R. (1967b) Liver cell abnormalities induced in mice by the fungal toxin causing lupinosis. *Journal of Pathology* 94, 452–455.

Gardiner, M.R. and Petterson, D.S. (1972) Pathogenesis of mouse lupinosis induced by a fungus (*Cytospora* spp.) growing on dead lupins. *Journal of Comparative Pathology* 82, 5–13.

Gardiner, M.R. and Seddon, H.D. (1966) Equine lupinosis. *Australian Veterinary Journal* 42, 242–244.

Gladstones, J.S. (1989a) Breeding Phomopsis-resistant lupins. *Journal of Agriculture, Western Australia* 30, 3–7.

Gladstones, J.S. (1989b) Register of Australian grain legume cultivars. *Lupinus angustifolius* L. (narrow-leafed lupin) cv. Gungurru. *Australian Journal of Experimental Agriculture* 29, 913–914.

Gladstones, J.S. (1989c) Register of Australian grain legume cultivars. *Lupinus angustifolius* L. (narrow-leafed lupin) cv. Yorrel. *Australian Journal of Experimental Agriculture* 29, 915–916.

Godfrey, N.W., Mercy, A.R., Emms, Y. and Payne, H.G. (1985) Tolerance of growing pigs to lupin alkaloids. *Australian Journal of Experimental Agriculture* 25, 791–795.

Gross, R. (1988) Lupins in human nutrition. In: Twardowski, T. (ed.) *Proceedings of the Fifth International Lupin Conference.* PWRiL, Poznań, pp. 51–63.

Grummer, R.H., Bentley, O.G., Phillips, P.H. and Bohstedt, G. (1950) The role of manganese in growth, reproduction and lactation of swine. *Journal of Animal Science* 9, 170–175.

Hamblin, J., Wood, P.McR. and Allen, J.G. (1981) The use of interspecific mixtures of *Lupinus* species to simulate the effects of different levels of *Phomopsis leptostromiformis* resistance in *L. angustifolius. Euphytica* 30, 203–207.

Hancock, G.R. and Allen, J.G. (1987) A comparison of a sheep bioassay and a chemical assay used to estimate the toxicity of lupin stubbles infected with *Phomopsis leptostromiformis. Journal of Applied Toxicology* 7, 219–220.

Harastzi, E. and Vetter, J. (1980) [Examination of the factors influencing the suitability for feeding as well as the tastefulness of various species and varieties of lupine (*Lupinus*).] *Botanikai Kozlemenyek* 67, 125–133 (in Hungarian).

Harborne, J.B. (1991) Alkaloids and other toxins of lupin. In: D'Mello, J.P.F. and Duffus, C.M. (eds) *Toxic Factors in Crop Plants, Proceedings of the Second Spring Conference.* The Scottish Agricultural College, Edinburgh, pp. 95–103.

Harborne, J.B., Ingham, J.L., King, L. and Payne, M. (1976) The isopentenyl isoflavone luteone as a pre-infectional antifungal agent in the genus *Lupinus. Phytochemistry* 15, 1485–1487.

Hawkins, C.D. (1994) 'Acorn' calves and retained placentae following grazing on sandplain lupins (*Lupinus cosentinii*). In: Colegate, S.M. and Dorling, P.R. (eds)

Plant-associated Toxins – Agricultural, Phytochemical and Ecological Aspects. CAB International, Wallingford, Oxon, pp. 351–356.

Hawthorne, W.A. and Gladstones, J.S. (1989) Register of Australian grain legume cultivars. *Lupinus angustifolius* L. (narrow-leafed lupin) cv. Warrah. *Australian Journal of Experimental Agriculture* 29, 911–912.

Hill, G.D. (1986) Recent developments in the use of lupins in animal and human nutrition. In: *Proceedings of the Fourth International Lupin Conference.* International Lupin Association and Western Australian Department of Agriculture, South Perth, pp. 40–63.

Hove, E.L., King, S. and Hill, G.D. (1978) Composition, protein quality and toxins of seeds of the grain legumes *Glycine max*, *Lupinus* spp., *Phaseolus* spp., *Pisum sativum* and *Vicia faba. New Zealand Journal of Agricultural Research* 21, 457–462.

Humphreys, D.J (1988) *Veterinary Toxicology*, 3rd edn. Baillière Tindall, London.

Ingham, J.L., Tahara, S. and Harborne, J.B. (1983) Fungitoxic isoflavones from *Lupinus albus* and other *Lupinus* species. *Zeitschrift fuer Naturforschung, Section C – Biosciences* 38c, 194–200.

Iwasaki, S. (1992) Chemistry and biological activity of the mycotoxins interfering with tubulin function. In: *Proceedings of the Japanese Association of Mycotoxicologists* 35, 1–6.

Iyer, L., Jackson, K. and Leclerc, J. (1994) Commercial measurement of phomopsin A in lupin seeds and stubble. In: Dracup, M. and Palta, J. (eds) *Proceedings of the First Australian Lupin Technical Symposium.* Western Australian Department of Agriculture, South Perth, pp. 274–277.

Keeler, R.F and Panter, K.E. (1989) The piperidine alkaloid composition and relation to crooked calf disease-inducing potential of *Lupinus formosus. Teratology* 40, 423–432.

Keeler, R.F and Panter, K.E. (1992) Induction of crooked calf disease by the piperidine alkaloid-containing plant *Lupinus formosus.* In: James, L.F., Keeler, R.F., Bailey, E.M. Jr, Cheeke, P.R. and Hegarty, M.P. (eds) *Poisonous Plants, Proceedings of the Third International Symposium.* Iowa State University Press, Ames, pp. 239–244.

Kilgore, W.W., Crosby, D.G., Craigmill, A.L. and Poppen, N.K. (1981) Toxic plants as possible human teratogens: circumstantial evidence points to lupine toxin in goats' milk as cause of human birth defects. *California Agriculture* 35, 6.

Kingsbury, J.M. (1964) *Poisonous Plants of the United States and Canada.* Prentice-Hall, Englewood Cliffs, NJ.

Koiso, Y., Natori, M., Iwasaki, S., Sato, S., Sonoda, R., Fujita, Y., Yaegashi, H. and Sato, Z. (1992) Ustiloxin: a phytotoxin and a mycotoxin from false smut balls on rice panicles. *Tetrahedron Letters* 33, 4157–4160.

Lacey, E., Edgar, J.A. and Culvenor, C.C.J. (1987) Interaction of phomopsin A and related compounds with purified sheep brain tubulin. *Biochemical Pharmacology* 36, 2133–2138.

Li, Y., Kobayashi, H., Tokiwa, Y., Hashimoto, Y. and Iwasaki, S. (1992) Interaction of phomopsin A with porcine brain tubulin. Inhibition of tubulin polymerization and binding at a rhizoxin binding site. *Biochemical Pharmacology* 43, 219–224.

Links, I.J. and Walker, R.J. (1989) Hepatotoxicity in weaner sheep with access to mouldy lupin grain and pyrrolizidine alkaloids from *Heliotropium europaeum.* In: *Proceedings of the Annual Conference of the Australian Society for Veterinary Pathology.* Australian Society for Veterinary Pathology, Walkerville, pp. 37–38.

Lu, G.G. (1964) The cause of death and resuscitation in sparteine poisoning. *Toxicology and Applied Pharmacology* 6, 328–333.

Luduena, R.F., Prasad, V., Roach, M.C. and Lacey, E. (1989) The interaction of phomopsin A with bovine brain tubulin. *Archives of Biochemistry and Biophysics* 272, 32–38.

Luduena, R.F., Roach, M.C., Prasad, V. and Lacey, E. (1990) The effect of phomopsin A on the alkylation of tubulin. *Biochemical Pharmacology* 39, 1603–1608.

Mazur, M., Polakowski, P. and Szadowska, A. (1966a) Pharmacology of lupanine and 13-hydroxylupanine. *Acta Physiologica Polonica* 17, 299–304.

Mazur, M., Polakowski, P. and Szadowska, A. (1966b) Pharmacolgy of 17-oxolupanine, lupanine N-oxide and 17-hydroxylupanine. *Acta Physiologica Polonica* 17, 311–320.

McDonald, C., Croker, K. and Allen, J. (1991) *Phomopsis*-resistant lupin stubbles as feed for weaner sheep. *Journal of Agriculture, Western Australia* 32, 100–104.

McDonald, C.L., Croker, K.P. and Allen, J.G. (1994) Optimising sheep and wool production from new *Phomopsis*-resistant lupin stubbles. *Final Report (Project DAW 32) to the Australian Wool Research and Promotion Organisation.* Western Australian Department of Agriculture, South Perth.

Morcombe, P.W. and Allen, J.G. (1990) The incidence of lupinosis in sheep grazing *Phomopsis*-resistant lupin stubbles. *Journal of Agriculture, Western Australia* 31, 3–4.

Morcombe, P.W., Allen, J.G., Hancock, G.R., Jacob, R. and Hopkins, D.L. (1992a) Monitoring the toxicity of lupin (*Lupinus angustifolius*) stubbles has limited value in the prevention of lupinosis. *Australian Journal of Experimental Agriculture* 32, 707–712.

Morcombe, P.W., Allen, J.G., Hancock, G.R., Jacob, R. and Hopkins, D.L. (1992b) Phomopsin A levels in lupin stubbles. In: James, L.F., Keeler, R.F., Bailey, E.M. Jr, Cheeke, P.R. and Hegarty, M.P. (eds) *Poisonous Plants, Proceedings of the Third International Symposium.* Iowa State University Press, Ames, pp. 599–601.

Mortimer, P.H., Campbell, J., di Menna, M.E. and White, E.P. (1971) Experimental myrotheciotoxicosis and poisoning in ruminants by verrucarin and roridin A. *Research in Veterinary Science* 12, 508–515.

National Food Authority (1992) *Australian Foods Standards Code.* Australian Government Publishing Service, Canberra.

Nelson, P., Hamblin, J. and Williams, A. (1986) *Lupins in the Geraldton Region.* Bulletin No. 4106, Western Australian Department of Agriculture, South Perth.

Oram, R.N., David, D.J., Green, A.G. and Read, B.J. (1979) Selection of *Lupinus albus* L. for lower seed manganese concentration. *Australian Journal of Agricultural Research* 30, 467–476.

Ortega, A. and Lazerson, J. (1987) Anagyrine-induced red cell aplasia, vascular anomaly and skeletal dysplasia. *Journal of Pediatrics* 111, 87–89.

Panter, K.E. and Keeler, R.F. (1991) Toxicologic and teratogenic comparison of *Lupinus formosus, L. arbustus* and *L. caudatus* in goats. *Journal of Animal Science* 69 (Suppl. 1), 393.

Panter, K.E., Keeler, R.F., Bunch, T.D. and Callan, R.J. (1990) Congenital skeletal malformations and cleft palate induced in goats by ingestion of *Lupinus, Conium* and *Nicotiana* species. *Toxicon* 28, 1377–1385.

Papadimitriou, J.M. and Petterson, D.S. (1976) The pathogensis of lupinosis in the rat. *Journal of Pathology* 118, 35–47.

Papadimitriou, J.M., Walters, M.N.-I., Petterson, D.S. and Gardiner, M.R. (1973) Hepatic ultrastructural changes in murine lupinosis. *Journal of Pathology* 111, 221–228.

Papadimitriou, J.M., Bradshaw, R.D., Petterson, D.S. and Gardiner, M.R. (1974) A histological, histochemical and biochemical study of the effect of the toxin of lupinosis on murine liver. *Journal of Pathology* 112, 43–53.

Payne, A.L., Than, K.A., Stewart, P.L. and Edgar, J.A. (1992) Vaccination against lupinosis. In: James, L.F., Keeler, R.F., Bailey, E.M. Jr, Cheeke, P.R. and Hegarty, M.P. (eds) *Poisonous Plants, Proceedings of the Third International Symposium.* Iowa State University Press, Ames, pp. 234–238.

Pearson, G. and Carr, J.R. (1977) A comparison between meals prepared from the seeds of different varieties of lupin as protein supplements to barley based diets for growing pigs. *Animal Feed Science and Technology* 2, 49–58.

Peterson, J.E. (1978) *Phomopsis leptostromiformis* toxicity (lupinosis) in nursling rats. *Journal of Comparative Pathology* 88, 191–203.

Peterson, J.E. (1983) Embryotoxicity of phomopsin in rats. *Australian Journal of Experimental Biological and Medical Science* 61, 105–115.

Peterson, J.E. (1986) The toxicity of phomopsin. In: *Proceedings of the Fourth International Lupin Conference*. International Lupin Association and Western Australian Department of Agriculture, South Perth, pp. 199–208.

Peterson, J.E. (1990) Biliary hyperplasia and carcinogenesis in chronic liver damage induced in rats by phomopsins. *Pathology* 22, 213–222.

Peterson, J.E. and Lanigan, G.W. (1976) Effects of *Phomopsis rossiana* toxin on the cell cycle and on the pathogenesis of lupinosis in mice. *Journal of Comparative Pathology* 86, 293–306.

Peterson, J.E., Jago, M.V., Payne, A.L. and Stewart, P.L. (1987) The toxicity of phomopsin for sheep. *Australian Veterinary Journal* 64, 293–298.

Petterson, D.S. (1985) *Lupin Seed as a Food for Man and Livestock – a Literature Review*. Western Australian Department of Agriculture, South Perth.

Petterson, D.S. (1996) Progress into the quality of and end-uses for lupin seed (grain). In: *Proceedings of the 4th Western Australian Lupin Update Meeting*. Agriculture Western Australia, South Perth, pp. 83–91.

Petterson, D.S. and Hancock, G.R. (1986) Phomopsin A in seed of *Lupinus angustifolius*. In: *Proceedings of the Fourth International Lupin Conference*. International Lupin Association and Western Australian Department of Agriculture, South Perth, p. 328.

Petterson, D.S. and Harris, D.J. (1995) The cadmium and lead content of lupin seed grown in Western Australia. *Australian Journal of Experimental Agriculture* 35, 403–407.

Petterson, D.S. and Mackintosh, J.B. (1994a) The chemical composition of lupin seed grown in Australia. In: Dracup, M. and Palta, J. (eds) *Proceedings of the First Australian Lupin Technical Symposium*. Western Australian Department of Agriculture, South Perth, pp. 39–48.

Petterson, D.S. and Mackintosh, J.B. (1994b) *The Chemical Composition and Nutritive Value of Australian Grain Legumes*. Grain Research and Development Corporation, Canberra.

Petterson, D.S. and Wood, P.McR. (1986) *Phomopsis* infection of lupin seed. *Journal of Agriculture, Western Australia* 27, 53–54.

Petterson, D.S., Peterson, J.E., Smith, L.W., Wood, P.McR. and Culvenor, C.C.J. (1985) Bioassay of the contamination of lupin seed by the mycotoxin phomopsin. *Australian Journal of Experimental Agriculture* 25, 434–439.

Petterson, D.S., Allen, D.G., Greirson, B.N., Hancock, G.R., Harris, D.J. and Legge, F.M. (1986) The suitability of *Lupinus angustifolius* seed for human consumption. In: *Proceedings of the Nutrition Society of Australia* 11, 118.

Petterson, D.S., Ellis, Z.L., Harris, D.J. and Spadek, Z.E. (1987) Acute toxicity of the major alkaloids of cultivated *Lupinus angustifolius* seed to rats. *Journal of Applied Toxicology* 7, 51–53.

Petterson, D.S., Evans, A.J., Cederblad, A. and Sandström B. (1991) Bioavailability of zinc from lupin products. In: Momcilovic, B. (ed.) *Trace Elements in Man and Animals 7*. IMI, Zagreb, pp. 10-2–10-3.

Price, K.R. and Fenwick, G.D. (1985) Naturally occurring oestrogens in foods – a review. *Food Additives and Contaminants* 2, 73–106.

Punithalingam, E. (1974) Studies on *Sphaeropsidales* in culture. II. *Mycological Papers, No. 136*. CAB, Commonwealth Mycological Institute, Kew, pp. 1–63.

Punithalingam, E. and Gibson, I.A.S. (1975) *Diaporthe woodii. CMI Descriptions of Pathogenic Fungi and Bacteria, No. 476*. Commonwealth Mycological Institute, Kew.

Richards, B. (1996) Disease roundup. *Animal Health, Animal Disease Surveillance and Preparedness Newsletter*, June. Agriculture Western Australia, South Perth, p. 3.

Rodriguez, F.S., Santiyan, M.P.M., Zamorano, J.D.P. and Cordero, V.R. (1991) An outbreak of lupinosis in sheep. *Veterinary and Human Toxicology* 33, 492–494.

Roemer, P. and Jahn-Deesbach, W. (1988) Developments in *Lupinus mutablis* breeding. In: Twardowski, T. (ed.) *Proceedings of the Fifth International Lupin Conference*. PWRiL, Poznań, pp. 40–50.

Rowe, J.B. (1995) New horizons for using cereal grain in the nutritional management of sheep. In: *Proceedings of the Australian Sheep Veterinary Society*. Australian Sheep Veterinary Society, Artarmon, pp. 78–83.

Saini, H.S. (1989) Legume seed oligosaccharides. In: Huisman, J., van der Poel, T.F.B. and Liener, I.E. (eds) *Recent Advances of Research in Antinutritional Factors in Legume Seeds*. Pudoc, Wageningen, pp. 329–341.

Samples, D., Hill, D.W., Bridges, C.H. and Camp, B.J. (1984) Isolation of a mycotoxin (roridin A) from *Phomopsis* spp. *Veterinary and Human Toxicology* 26, 21–23.

Schmidlin-Meszaros, J. (1973) Eine Nahrungsmittelvergiftung mit Lupinenbohnen. *Mitteilungen aus dem Gebiete der Lebensmitteluntersuchung und Hygiene* 64, 194–205.

Schoeneberger, H., Ildefonso, C., Cremer, H.-D., Gross, R. and Elmadfa, I. (1982) Some antinutritional substances in lupins compared with other legumes. In: Gross, R. and Bunting, E.S. (eds) *Agricultural and Nutritional Aspects of Lupines*. German Agency for Technical Cooperation (GTZ), Eschborn, pp. 553–568.

Seawright, A.A. (1989) *Animal Health in Australia, Volume 2, Chemical and Plant Poisons*, 2nd edn. Australian Government Publishing Services, Canberra.

Seward, D. (1993) Phomopsin-toxicity following feeding of lupin grain. *Veterinary Pathology Report* No. 38, 6–7.

Shivas, R.G., Allen, J.G. and Williamson, P.M. (1991) Infraspecific variation demonstrated in *Phomopsis leptostromiformis* using cultural and biochemical techniques. *Mycological Research* 95, 320–323.

Shivas, R.G., Allen, J.G., Edgar, J.A., Cockrum, P.A., Gallagher, P.F., Ellis, Z. and Harvey, M. (1994) Production of phomopsin A by a potential mycoherbicide, *Phomopsis emicis*. In: Colegate, S.M. and Dorling, P.R. (eds) *Plant-associated Toxins – Agricultural, Phytochemical and Ecological Aspects*. CAB International, Wallingford, Oxon, pp. 161–166.

Sipsas, S. (1994) Australian sweet lupin (*L. angustifolius*) variations in anti-nutritional values. In: Dracup, M. and Palta, J. (eds) *Proceedings of the First Australian Lupin Technical Symposium*. Western Australian Department of Agriculture, South Perth, pp. 308–309.

Sissons, J.W. and Tolman, H. (1991) Anti-nutritional properties of soybean antigens in calves. In: D'Mello, J.P.F. and Duffus, C.M. (eds) *Toxic Factors in Crop Plants, Proceedings of the Second Spring Conference*. The Scottish Agricultural College, Edinburgh, pp. 62–85.

Smith, B.L. (1987) Mycotoxicoses of cattle and horses in Australia and New Zealand. In: *Proceedings No. 103, Veterinary Clinical Toxicology*. The Post-Graduate Committee in Veterinary Science, Sydney, pp. 270–276.

Smolenski, S.J., Kinghorn, A.D. and Balandrin M.F. (1981) Toxic constituents of legume forage plants. *Economic Botany* 35, 321–355.

Sutherland, O.R.W., Russell, G.B., Biggs, D.R. and Lane, G.A. (1980) Insect feeding deterrent activity of phytoalexin isoflavanoids. *Biochemical Systematics and Ecology* 8, 73–75.

Sweetingham, M., French, B., Carr, S. and Shea, G. (1996) Yellow lupins for low rainfall light-land farms. In: *Proceedings of the 4th Western Australian Lupin Update Meeting*. Agriculture Western Australia, South Perth, pp. 13–14.

Tahara, S., Ingham, J.L., Nakahara, S., Mizutani, J. and Harborne, J.B. (1984) Fungitoxic dihydrofuranoisoflavones and related compounds in white lupin, *Lupinus albus*. *Phytochemistry* 23, 1889–1900.

Tahara, S., Orihara, S., Ingham, J.L. and Mizutani, J. (1989) Seventeen isoflavonoids from *Lupinus albus* roots. *Phytochemistry* 28, 901–911.

Than, K.A. (1994) Applications of phomopsin ELISA in lupin research. In: Dracup, M. and Palta, J. (eds) *Proceedings of the First Australian Lupin Technical Symposium*. Western Australian Department of Agriculture, South Perth, p. 318.

Than, K.A., Payne, A.L. and Edgar, J.A. (1992) Development of an enzyme immunoassay for the phomopsin mycotoxins. In: James, L.F., Keeler, R.F., Bailey, E.M. Jr, Cheeke, P.R. and Hegarty, M.P. (eds) *Poisonous Plants, Proceedings of the Third International Symposium*. Iowa State University Press, Ames, pp. 259–263.

Than, K.A., Tan, R.A., Petterson, D.S. and Edgar, J.A. (1994a) Phomopsin content of commercial lupin seed from Western Australia in 1991/2. In: Colegate, S.M. and Dorling, P.R. (eds) *Plant-associated Toxins – Agricultural, Phytochemical and Ecological Aspects*. CAB International, Wallingford, Oxon, pp. 62–65.

Than, K.A., Anderton, N., Cockrum, P.A., Payne, A.L., Stewart, P.L. and Edgar, J.A. (1994b) Lupinosis vaccine: positive relationship between anti-phomopsin IgG concentration and protection in Victorian field trials. In: Colegate, S.M. and Dorling, P.R. (eds) *Plant-associated Toxins – Agricultural, Phytochemical and Ecological Aspects*. CAB International, Wallingford, Oxon, pp. 433–438.

Tonsing, E.M., Steyn, P.S., Osborn, M. and Weber, K. (1984) Phomopsin A, the causative agent of lupinosis, interacts with microtubules *in vivo* and *in vitro*. *European Journal of Cell Biology* 35, 156–164.

van Aswegan, C.H., Anderson, L.A.P., Potgieter, D.J.J., Venter, A.M., Neitz, A.W.H. and Vermeulen, N.M.J. (1980) Influence of a toxic fraction from *Phomopsis leptostromiformis* on mitochondrial respiration. *South African Journal of Science* 76, 516–519.

van Warmelo, K.T., Marasas, W.F.O., Adelaar, T.F., Kellerman, T.S., van Rensburg, I.B.J. and Minne, J.A. (1970) Experimental evidence that lupinosis of sheep is a mycotoxicosis caused by the fungus *Phomopsis leptostromiformis* (Kühn) Bubák. *Journal of the South African Veterinary Medical Association* 41, 235–247.

von Baer, E. and von Baer, D. (1988) *Lupinus mutablis*: cultivation and breeding. In: Twardowski, T. (ed.) *Proceedings of the Fifth International Lupin Conference*. PWRiL, Poznań, pp. 237–247.

Warren, J., Allen, J. and Cowling, W. (1989) Economic impact of growing *Phomopsis*-resistant lupins. *Journal of Agriculture, Western Australia* 30, 8–10.

Williamson, P.M., Sivasithamparam, K. and Cowling, W.A. (1991) Formation of subcuticular coralloid hyphae by *Phomopsis leptostromiformis* upon latent infection of narrow-leafed lupins. *Plant Disease* 75, 1023–1026.

Williamson, P.M., Highet, A.S., Gams, W., Sivasithamparam, K. and Cowling, W.A. (1994) *Diaporthe toxica* sp. nov., the cause of lupinosis in sheep. *Mycological Research* 98, 1364–1368.

Williamson, P.M., Than, K.A., Sivasithamparam, K., Cowling, W.A. and Edgar, J.A. (1995) Detection of resistance to *Diaporthe toxica* in asymptomatically infected lupin seedlings based on an immunoassay for phomopsin. *Plant Pathology* 44, 95–97.

Wittenberg, H. and Nehring, K. (1965) Untersuchungen uber die Wirkung reiner Lupine-alkaloide auf den tierischen Organism. *Pharmazie* 20, 156–158.

Wood, P.McR. (1986) Epidemiology of *Phomopsis leptostromiformis*. In: *Proceedings of the Fourth International Lupin Conference*. International Lupin Association and Western Australian Department of Agriculture, South Perth, pp. 220–229.

Wood, P.McR. (1989) Infested stubble and seed as inoculum sources of *Diaporthe woodii* infecting *Lupinus angustifolius*. MSc dissertation, The University of Western Australia, Perth, Australia.

Wood, P.McR. and Allen, J.G. (1980) Control of ovine lupinosis: use of a resistant cultivar of *Lupinus albus* – cv. Ultra. *Australian Journal of Experimental Agriculture and Animal Husbandry* 20, 316–318.

Wood, P.McR. and Brown, A.G.P. (1975) Phomopsis – the causal fungus of lupinosis. *Journal of Agriculture, Western Australia* 16, 31–32.

Wood, P.McR. and Petterson, D.S. (1986) *Phomopsis leptostromiformis* infection and phomopsin A content of lupin seed in Western Australia. *Australian Journal of Experimental Agriculture* 26, 583–586.

Wood, P.McR. and Sivasithamparam, K. (1989) *Diaporthe woodii* (anamorph *Phomopsis leptostromiformis*) – a toxigenic fungus infecting cultivated lupins. *Mycopathologia* 105, 79–86.

Wood, P.McR., Brown, A.G.P. and Petterson, D.S. (1973) Production of the lupinosis mycotoxin by *Phomopsis rossiana*. *Australian Journal of Experimental Biology and Medical Science* 51, 557–558.

Wood, P.McR., Petterson, D.S., Hancock, G.R. and Brown, G.A. (1987) Distribution of seed infected with *Phomopsis leptostromiformis* and of phomopsin A within a lupin crop. *Australian Journal of Experimental Agriculture* 27, 77–79.

Zurn, F.A. (1879) Massenerkrankung von Schafen durch den genuss befallener Lupinen. *Vortrage für Thierarzte* 7, 251–277.

15 Marketing and Trade

B. Cox

Grain Pool of Western Australia, 172 St Georges Terrace, Perth, Western Australia 6000

Introduction

Sweet or low alkaloid lupinseed *(Lupinus angustifolius)* has an established international reputation as a high quality, versatile source of protein and energy in the feed markets of Europe, South East Asia and the domestic market within Australia. Sweet lupinseed is a concentrated source of protein and energy, containing a low level of anti-nutritional substances and having excellent handling and storage properties. Sweet lupinseed has long been used as a replacement for other vegetable protein sources in a variety of animal and food applications.

Breeding of sweet lupinseed as a cash crop has largely evolved in Western Australia. Significant advances in production have also occurred in this region, generating the marketing and trade initiatives that were the precursor to the establishment of sweet lupinseed as a highly sought after commodity in the international marketplace. This chapter is primarily focused on that evolution. Information on the marketing and trade of other lupinseed species form the latter part of the chapter.

Sweet Lupinseed

Lupinseed production

Accurate accounts of the most recent world production levels of lupinseed by species and country are difficult to document. Table 15.1 shows the production levels of lupinseed by country in 1991 and 1992 and shows the dominance of Australia in the world's production. A large proportion of the lupinseed production outside Australia is used for purposes other than grain. Alternative uses include forage production and crops as green manure, particularly in the former Soviet Union and Eastern Bloc countries (Ford, 1990).

 Lupins as Crop Plants: Biology, Production and Utilization (eds J.S. Gladstones, C.A. Atkins and J. Hamblin)

Table 15.1. World production of lupinseed expressed as tonnage and in relation to Australian production, expressed as a percentage, in 1991 and 1992 (ACIL, 1994).

Country	1991 (t)	Australian production (%)	1992 (t)	Australian production (%)
Argentina	100	0	100	0
Australia	1,038,000	100	987,000	100
Chile	14,000	1.3	14,000	1.4
Ecuador	1,154	0.1	1,043	0.1
Egypt	5,597	0.5	5,575	0.5
Germany	3,600	0.3	13,000	1.3
Greece	160	0	160	0
Hungary	1,671	0.2	2,000	0.2
Italy	4,300	0.4	3,874	0.4
Morocco	3,200	0.3	3,300	0.3
Peru	3,000	0.3	3,000	0.3
Poland	150,000	14.5	140,000	14.2
Portugal	33	0	33	0
South Africa	7,000	0.7	6,000	0.6
Spain	1,900	0.2	900	0.1
Syria	105	0	107	0
USSR	81,000	7.8	85,000	8.6
Total	1,314,820		1,265,092	

Australian sweet lupin production

With the exception of Australia, the bulk of lupinseed production is consumed locally. Sweet lupinseed has been produced in Australia since the 1960s, and it was in the state of Western Australia where the breeding, development and first release of sweet lupinseed cultivars occurred. As a consequence the bulk of Australian production and exports also originate in Western Australia, where the commercialization of sweet lupinseed, as a seed-producing crop, has predominantly taken place. Over the period 1989/90 to 1994/95 Western Australian lupin production has averaged 815,000 t per annum as against 200,000 t from all other Australian states (Fig. 15.1) (ABARE, 1995).

Lupinseed production in Western Australia commenced in earnest in the early 1970s, but it was not until growers fully recognized the benefits of sweet lupinseed in their cropping regimes and their value as an on-farm feed, and the acceptance of the product in the international marketplace, that production began to increase steadily. The growth in annual production is detailed in Fig. 15.2.

In 1995/96 approximately 75% of the 1,100,000 t of sweet lupinseed produced in Western Australia was exported to the international market, with the balance of production being utilized within the state. Western Australia is the single largest exporter in the world. From small beginnings sweet lupinseed now has an established position in the world market as a valuable protein and energy source and is a highly sought after commodity. This increasing production and its sale has allowed the development of a major new industry and the evolution of its market development follows.

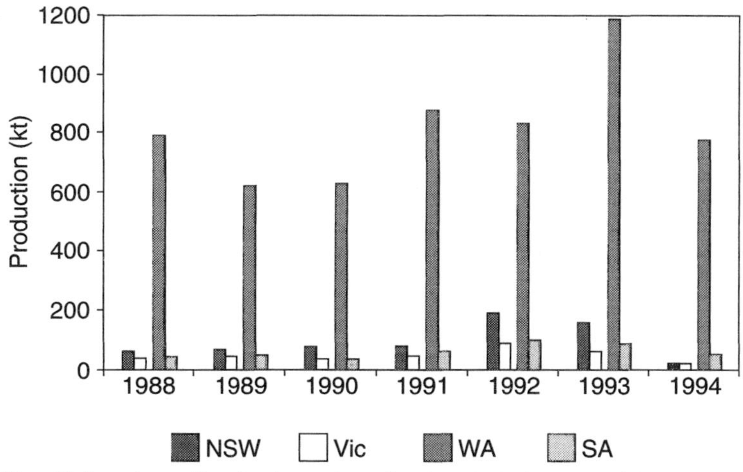

Fig. 15.1. Australian lupinseed production 1989/90–1994/95.

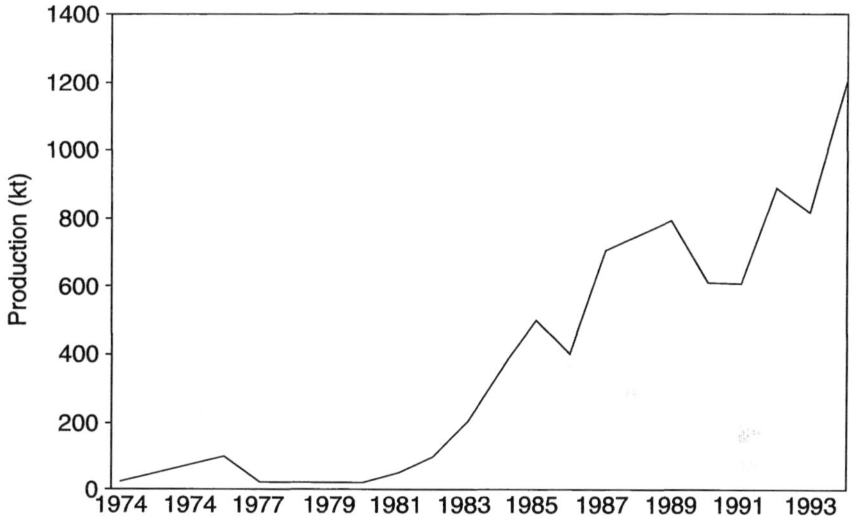

Fig. 15.2. Western Australia lupin production.

Receival, storage and shipment

Most sweet lupinseed produced for export in Western Australia, or other Australian states, is received against strict pre-determined receival standards set by the handling or (statutory) marketing authority; it is stored in inland or port-located storage facilities and held until out-turn is required. In Western Australia sweet lupinseed is generally stored in atmosphere-controlled storage bins, eliminating the need for contact pesticide treatment.

Growers are paid a pre-negotiated cash price in full, when delivering to a domestic outlet, or are advanced around 70% of the expected final price under statutory pooling arrangements in the case of grain delivered for export within Western Australia. Most payments are made within 30 days of harvest.

Shipment is generally effected in bulk cargoes of 5000 to 30,000 t with strict export quarantine procedures in place. There is seldom any stock carry-over from one season to the next.

Marketing arrangements

Sweet lupinseed marketing arrangements vary from state to state within Australia, with free market systems operating in most states. In Western Australia, since 1975, growers selling their seed have been required to deliver their produce to the bulk handling system as a result of state government legislation. Once in the bulk handling facility the seed is then owned by the state-based statutory marketing authority charged with the responsibility of marketing the grain on behalf of growers. The Grain Pool of Western Australia is widely credited with establishing most of the sweet lupinseed markets throughout the world.

Once markets have been established, sales normally takes place well in advance of the crop actually being harvested, with sales often concluded as early as July/August in any given year – some three to four months before the crop is reaped. Sales are concluded when the price negotiations result in satisfactory outcomes for both buyer and seller. They may be through direct negotiation with end-users or via trading companies, brokers or agents.

Sweet lupinseed composition and usage

As with all commodities, to achieve optimal market penetration it is vital for the marketer to understand fully the traits and functionality of the product they intend offering to their chosen markets. With the bulk of Western Australian sweet lupinseed production annually delivered and shipped through a bulk handling and storage facility, representative samples of the commodity are easily obtained for analysis. The Grain Pool of Western Australia used this opportunity to establish a quality database from the very early production stages, enabling the product to be sold with confidence into domestic and international markets.

From early development a valuable quality in sweet lupinseed has been the small incidence of fluctuation in chemical composition when averaged over bulk export shipments. The chemical composition (Table 15.2) remains relatively constant from year to year despite variation experienced in weather conditions and soil characteristics. In turn, this aspect allows for the inclusion of sweet lupinseed in feed or food formulations using a minimum of quality control.

The nutritional value and functionality of sweet lupinseed in animal and food applications are described in detail in other chapters of this book but the major attributes of the commodity are:

- mid-range vegetable protein source;
- suitable for a wide range of feeding conditions and livestock types;
- uniform quality;
- easy to handle and store;
- limited anti-nutritional factors;

Table 15.2. Typical composition of Western Australia sweet lupinseed (*L. angustifolius*).

Chemical compostion	% by weight
Crude protein	32
Crude fat	5
Crude fibre	15
Nitrogen-free extractives	36
Ash	3
Calcium (Ca)	0.22
Phosphorus (P)	0.3
Gross energy value (MJ kg^{-1})	18.3
Moisture	9

- contains very low levels of starch;
- requires little or no processing;
- excellent fermentation substrate (dehulled lupins).

As a result of increased production, the Western Australian Department of Agriculture devoted considerable research effort to examining the practical use of sweet lupinseed in animal application, with particular emphasis on monogastric and ruminant diets, from the early 1970s and through the 1980s. This research established that the grain was a valuable and highly nutritious farm and compound feed source. In turn, this information led to considerable domestic market acceptance and provided the foundations necessary to introduce lupinseed to the international marketplace. In offering lupinseed to the international market the Grain Pool of Western Australia developed an understanding of potential customer information needs and concerns, allowing targeted research programmes to be developed and implemented, thus further building a curriculum vitae for Western Australian sweet lupinseed. Customer feedback allowed marketing and promotion campaigns to target most common customer concerns.

Markets

The domestic market

Prior to the advent of commercial lupinseed production, the Western Australian cropping environment was largely cereal-based, with high protein feed ingredients produced only in minimal quantity. Significant sheep populations relied on cereals as a supplementary feed at times when natural feed was in short supply. Early farm practice and research supported the use of lupinseed as a farm sheep feed. However, once farmers and the local feed industry recognized the value of sweet lupinseed as a protein and energy source, in an environment where there was a historical dearth of protein sources above 15% crude protein, uptake of lupinseed as a feed was rapid. Increased usage led to diversification across a range of animal applications. Today, lupinseed is widely used as a major vegetable protein source in the Australian stockfeed industry with farmers, feedlotters and compound feed manufacturers regularly using lupins, subject to their relative cost against other feed ingredients.

> Approximately 40% of Australian sweet lupinseed production is consumed domestically. Of this percentage a large proportion of the lupinseed is consumed on farm as feed or sold directly to feedlots. A small proportion is supplied to stockfeed manufacturers for incorporation into livestock feeds, including rations for poultry and aquaculture.
>
> ACIL, 1995.

The international environment

The price paid by a stockfeed manufacturer for a particular ingredient depends on its nutritional components and the availability and price of substitutes. Most feed compounders carry out constant computer analysis to reach target feed specifications, both on composition and on price. In these analyses sweet lupinseed is compared with a wide range of feeds that are readily available in the international market place such as soybean meal, rapeseed meal, meat meal, corngluten feed and field peas.

The value of sweet lupinseed in international markets depends on a number of factors, including (but not limited to) the price and availability of domestically produced feed materials, internal and external agricultural policies (e.g. subsidies), current and projected feed consumption patterns and currency movements. Supply and demand of the major competitive products is critical to lupinseed pricing.

In many of the markets where lupinseed is sold its major competitor, in terms of price and functionality, is soybean meal – the meal remaining after oil extraction from whole soybeans. The crude protein level of soybean meal is generally in the range of 43 to 48% as against sweet lupinseed at 32% crude protein in the whole seed. A comparative analysis of the two commodities is shown in Table 15.3.

Table 15.3. Typical composition of sweet lupinseed (*L. angustifolius*) and high protein soybean (*Glycine max*) meal.

Component	Soybean meal[1]	Lupinseed[2]
Ash %	6	3
Dry matter %	90	91
Fat %	1.5	5
Fibre %	7.5	15
Protein %	44	32
Energy for livestock rations		
Cattle (ME) MJ kg^{-1}	11.4	12.7
Pigs (DE) MJ kg^{-1}	14.5	14.4
Poultry (ME) MJ kg^{-1}	10.1	10

[1]Soybean meal is 44% of the weight of whole grain, being the residue following extraction of oil.
[2]Whole grain.

World production of soybean meal has generally exceeded 80 Mt in recent years, and the production of 11 other major world oilmeals typically also totals this quantity (Table 15.4). The USA is a significant exporter, exporting annually around 5.5 Mt; South America (Brazil and Argentina) typically exports 17 Mt per annum (Mielke, 1996). India also is a significant supplier of soybean meal to export markets.

Potential purchasers often compare the crude protein value of sweet lupinseed against the crude protein value of other commodities when assessing its value as a stockfeed. Soybean meal is commonly used by many international stockfeed manufacturers as a benchmark protein source in making price determinations. It is important to note that, while the crude protein level is often a major pricing factor, it is not the sole price determinant. Sweet lupinseed also has the ability to provide a significant contribution to an animal's energy requirements and this factor is integral in the final worth of the commodity to many stockfeed manufacturers. It is for this reason that premium prices have generally been obtained in the markets where sweet lupinseed is used as a ruminant feed. Sweet lupinseed is ideally suited for this use (Chapter 13, this volume).

Soybean meal, a major world commodity, is traded through a number of mediums, including the Chicago Board of Trade Futures Market. Futures markets provide buyers and sellers of commodities with an opportunity to establish prices for future delivery. Individuals and firms use the futures market for this purpose and are able to minimize the risk of adverse price-level changes. A no less vital role is that of a central marketplace where buyers and sellers from virtually every nation in the world continuously interact with one another in the common language of price. Because futures prices serve as a common denominator in a worldwide system of interrelated prices, such a marketplace helps to assure that supply and demand developments, wherever they occur, will be rapidly reflected

Table 15.4. World oilmeal production (million tonnes): the 12 major oilmeals (Mielke, 1996).

Meal	1996/97[f]	1995/96[f]	1994/95	1993/94	1992/93
Soyameal	87.90	85.94	86.65	80.63	76.20
Cotton meal	16.07	15.46	14.49	13.28	14.66
Groundnut meal	6.00	6.00	6.10	5.96	5.63
Sunflower meal	11.00	10.75	10.29	9.11	9.40
Rape meal	17.92	17.83	16.18	15.15	14.30
Sesame meal	0.89	0.88	0.89	0.91	0.89
Corngerm meal	3.20	3.07	2.98	2.80	2.69
Corngluten feed	13.70	13.09	12.76	11.54	11.28
Palmgluten meal	2.66	2.54	2.38	2.26	2.09
Copra meal	1.87	1.81	1.97	1.67	1.78
Linseed meal	1.34	1.31	1.30	1.17	1.11
Fish meal	6.40	6.18	6.66	7.08	6.20
Total	168.95	164.87	162.64	151.55	146.24
Change	+2.5%	+1.4%	+7.3%	+3.6%	+1.0%

[f]Forecast.

in the price of grain, wherever it is offered for sale. Futures markets and future prices thereby help to provide a constant link between local cash prices and world demand (Chicago Board of Trade, 1990).

In essence the soybean meal futures price on any given day reflects buyers' and sellers' views as to the value of soybean meal for specific marketing positions. The strengths and weaknesses of the soybean meal futures prices are affected by a multitude of factors, including (but not restricted to) weather conditions being experienced in the major soybean production areas, the value of other soybean products (i.e. soya oil), the value of cereal products (e.g. corn), international demand for feed ingredients and actual annual production levels of soybeans.

Soybean meal dominates the protein market internationally and the price of sweet lupinseed is often relative to soybean meal. The soybean meal futures trends therefore provide a reasonable guide as to whether prices obtainable for sweet lupinseed are likely to increase, decrease or remain relatively stable. Actual or final prices achieved for sweet lupinseed are governed by the value negotiated between buyer and seller, with this determined not only on the price of competing feed ingredients on that day, but also trading relationships between buyer and seller. Each market tends to have its own idiosyncrasies and customs. For example, the European feed market has, by geographic position, access to a much wider range of feed ingredients than many markets in South East Asia. Sweet lupinseed will often be competing with well over 10 or 15 different feed ingredients at the time it is offered for sale. The European compounders are accustomed to dealing in small parcels and are willing to reformulate rations consistently in order to achieve least-cost formulation. As a result this industry is highly competitive *vis-à-vis* some Asian markets, which often have five or six mainstay ingredients that they are reluctant or unable to change.

Rotterdam is a significant market for sweet lupinseed where feed compounders often purchase the bulk of their soybean meal requirements from South America. The meal is purchased in pellet form with typical protein levels being around 46 to 48%. Figure 15.3 shows the value of Brazilian soybean meal pellets sold on a CIF (cost, insurance, freight) basis over a number of time frames. The figure provides a good snapshot of the volatility of the protein meal market.

Market development – the early stages

Although the long-term potential of lupins as a valuable human food source was recognized by the Australian industry at an early stage, it was foreseen that a

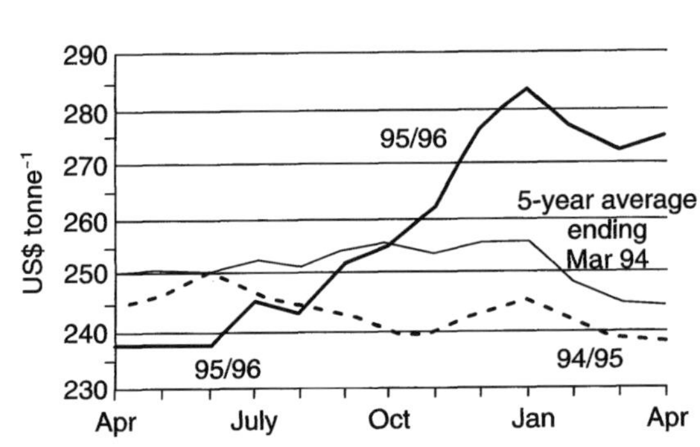

Fig. 15.3. Monthly soya meal prices in Rotterdam (Brazil pellets 48%).

considerable amount of developmental work was required to establish volume sales into this field. Hence, early Western Australian market development strategies were based on opportunities in the animal feed industry.

In 1975, the Western Australian Government, recognizing the potential future production possibilities for sweet lupinseed, appointed the Grain Pool of WA, Perth, Western Australia, as the sole marketing authority. The appointment was formalized through an enactment of state legislation.

With exportable quantities in the early 1970s still at relatively small levels, a single export market was targeted – the compound feed industry in the Netherlands. This market was appealing because of its high level of compound feed manufacture, advanced technology and the willingness of feed nutritionists to investigate the incorporation of sweet lupinseed into a wide range of products. The market sophistication meant that Dutch animal nutritionists quickly recognized the properties of sweet lupinseed as both a protein and an energy source for ruminants and it was soon incorporated into compound feed formulations for cattle.

The Dutch market remained the largest outlet for lupins until the early 1980s, primarily because of the industry's ability to pay an acceptable price as against developing stockfeed markets in other countries. Economies of scale were also achieved in the shipment of sizeable bulk parcels. In addition the limited availability of sweet lupinseed over and above the domestic market requirements precluded the ability of the statutory marketing authority to cater for and reasonably service a multitude of markets.

With expansion of production levels, market diversification became an increasing priority in the early 1980s. The animal feed sector was still considered to be the prime target area and the next stage of the international market development strategy was based around this. However, other marketing considerations were also made, including the following.

- **Livestock type** The composition of the sweet lupinseed with its fibrous yet highly digestible seed coat, comprising some 25% of total seed weight, lends itself to ready incorporation in ruminant diets with little or no processing. The established use of sweet lupinseed in the Western Australian sheep feed industry and the utilization of lupins in the Dutch compound feed market was ready testament to international buyers that lupinseed was a safe and nutritious ruminant feed. Lupinseed was also being used, but to a lesser degree, in monogastric diets within Western Australia and the international marketplace. It was therefore necessary to target countries with large ruminant and/or monogastric animal populations. The Asian markets of Japan and Korea became prime areas for concerted development efforts.
- **Tariff structures** High levels of import duties and/or taxes severely inhibit the profitability of any trading venture as these charges are invariably passed back to the exporter. It is in the producers' best interest to establish markets in countries where tariffs are low.
- **Political considerations** Political stability is also a prime consideration in developing long-term market strategies. An appreciation of the agricultural

policy within a targeted area is needed in order to establish whether the internal market is likely to have the stability for long-term growth.

- **Financial structure** The newly emerging sweet lupinseed industry in Western Australia could not afford to rely on markets where any doubt existed as to financial capabilities. Cash markets were essential.
- **Logistics** The natural market for Western Australian produce is in the Pacific basin, particularly that of the Western Pacific. The Pacific basin is close and provides access to a vast population. It offers an economical freight option for the export of raw commodities between the country of origin and the marketplace. Freight costs are not an inconsequential cost component in determining the viability of a market. The actual cost involved can vary greatly, depending on the mode of transport. The only economical mode of transport is shipment in bulk vessels.
- **Trading partners** The choice of trading partners is a critical issue in market development. Different types of trading partners often need to be selected to suit a particular market and its requirements (Ford, 1990).
- **Economic growth and population** The rate at which economies and population grow can also dictate market opportunities.

In addition to the Dutch compound feed industry, three distinctly different markets that fitted into most of the above parameters were recognized as presenting significant market opportunities for sweet lupinseed. These were Spain/Portugal, Korea and Japan, and they were subsequently targeted and developed through:

- persistent contact with and education of traders/end-users regarding the qualities of lupinseed;
- supply of high quality technical information and support as a result of considerable domestic-based research in conjunction with government and private institutions;
- trialling of sweet lupinseed in the potential country of import;
- competitive price structures for sweet lupinseed *vis-à-vis* other more traditional feed materials;
- ongoing market support;
- a commitment to continually supply markets that were prepared to trial and accept sweet lupinseed provided the market returned acceptable prices.

All of these markets continue to be significant purchasers of sweet lupinseed for incorporation into animal feed products. Average annual purchases by country between 1990 and 1995 are depicted in Table 15.5.

Table 15.5. Average annual purchases of Western Australian sweet lupinseed by country for the years 1990–1995.

Country	Annual average purchase (metric tonnes)
Netherlands	215,000
Spain/Portugal	120,000
Japan	115,000
Korea	105,000

The use of sweet lupinseed in monogastric markets, particularly poultry applications, has been slower to evolve than the utilization in ruminant and swine industries. Development into overseas poultry feed markets was hampered by the limited use of sweet lupinseed by Australian commercial poultry organizations. Over the last decade most Australian poultry organizations have limited the incorporation of sweet lupinseed in broiler diets to less than 10% of the diet and then only after the seed had been dehulled. This limitation was imposed due to a perception that the fibrous seed coat inhibited optimal utilization of the seed in broiler diets, resulting in reduced animal growth and feed conversion. Further, the inclusion of high levels of sweet lupinseed in poultry diets often led to increased faecal moisture, resulting in management problems in intensive broiler operations (Chapter 13, this volume).

With the world poultry markets representing a significant but as yet largely untapped opportunity, it was recognized that a short-term solution was removal of the fibrous seed coat. Suitable equipment was found and the opportunity to establish large-scale dehulling operations was investigated by marketers and domestic and international consumers of sweet lupinseed. The composition of the dehulled product or cotleydon is shown in Table 15.6. It can be seen the crude protein level is around 38% versus 32% crude protein in the whole seed. In this form sweet lupinseed is much more nutritionally comparable to soybean meal in poultry diets. Today, plants dedicated to processing sweet lupinseed exist in several markets, including Korea and Indonesia.

Current Utilization of Sweet Lupinseed by Major Markets

Sweet lupinseed is used in a diverse range of applications throughout the world. An overview of the major markets (primarily of Western Australian origin), typical annual usage and major end-use applications is presented below.

Australia

350,000 t per annum – all livestock types

Sweet lupinseed has been used as a feed since it was first produced within Australia. Initially the most favoured application was its use as a sheep feed. Ongoing research investigating its nutritional value across a range of ruminant and monogastric livestock types resulted in significant expansion of its use as a feed to the point wherein sweet lupinseed is utilized in many intensive and

Table 15.6. Typical composition of whole and dehulled sweet lupinseed.

Chemical composition (% by weight)	Dehulled lupin seed	Whole lupin seed
Crude protein	32	38
Crude fat	5	7
Crude fibre	16	3
Nitrogen-free extractives	36	39
Ash	3	2.5
Calcium	0.21	0.15
Phosphorus	0.29	0.35

non-intensive Australian livestock industries. Australian domestic usage of sweet lupinseed is estimated to be in excess of 350,000 t per annum. Competitive protein products include imported soybean meal, canola meal, field peas and other grain legumes.

Asia

Republic of Korea

105,000 t per annum – all livestock types, aquaculture

A major trading company in Korea established a sweet lupinseed dehulling plant in the late 1980s to enable the seed fractions to be used most effectively over all animal types. The major protein competition in this market includes domestic and imported soybean meal.

Japan

115,000 t per annum – cattle

The whole seed is flaked and incorporated into a balanced diet of partially processed ingredients typically comprising flakes of corn, sorghum and vitamin/mineral supplements. Domestic and imported soybean meals have been partially displaced by sweet lupinseed.

Indonesia

5000–8000 t per annum – human consumption, cattle

The seed is dehulled and the kernel fermented for the production of tempe. The hull material is utilized in cattle diets. Indonesia is the first commercial export market for Western Australian lupins for human consumption.

European Union

The Netherlands, Belgium, Germany

150,000–250,000 t per annum – most compound feed formulations and *dairy rations*

In the Netherlands, Belgium and Germany sweet lupinseed is commonly used in dairy rations. It competes as a protein and energy source with competitive products often including corngluten feed, palm and corn products, rapeseed meal, citrus and beetpulps. The highly sophisticated industry, with access to a broad range of ingredients, creates a very competitive market environment. The industry is generally not committed to any commodity and the major companies constantly trade in and out of materials, depending on price relativities. The market has the depth to absorb large quantities of sweet lupinseed at varying price levels, depending on ration formulation.

Spain, Portugal

120,000 t per annum – all livestock types

Consumption is highest in the ruminant sector, with limited utilization in monogastric diets at low inclusion rates. Chief protein competitors include soybean meal, corn gluten feed, rapeseed and sunflower-seed meals, meat and bonemeal. Sweet lupinseed was initially targeted into limited geographical areas specializing in cattle production in order to extract the highest market price.

Other markets

Countries where trade has been intermittent, are detailed hereunder.

Italy

20,000–50,000 finishing beef cattle

Ready access to inexpensive domestic soybean and sunflower meals preclude wide-scale usage on a cost comparative basis. High freight costs for small parcel shipments is also a significant limiting factor.

Israel

Sales were conducted for several years in the mid 1980s for use in cattle diets. Marketing opportunities tend to be limited because of price competition from cheap alternative protein sources, particularly cottonseed meal. High freight costs from Western Australia often present difficulties.

Taiwan

Significant quantities were sold to feed compounders in the 1980s and now this market is regularly taking small parcels of sweet lupinseed (Ford, 1990).

Other

In addition to the key and regular market outlets of the European Union, Japan and Korea a number of other markets have also been investigated. The Middle East, United Arab Emirates, South Africa, Thailand and the Philippines have feed industries capable of taking a significant portion, if not the bulk, of the Australian sweet lupinseed crop. Although many of these countries have imported sweet lupinseed in the past, they have not been able to compete consistently against the price offered for sweet lupinseed by key traditional markets. Often this is due to the availability of cheaper protein meals from countries in close proximity, restrictions on the allowable import volume of imported protein meals or high tariff levels on sweet lupinseed. These markets continue to evolve and may at some later stage provide valuable opportunities for Australian producers.

A summary of markets is shown in Table 15.7.

Tariff rates

Tariff rates for the importation of feed ingredients play a significant role in a potential customer's interest in importing new commodities. Table 15.8 shows the level of tariffs applying to lupinseed in 1994. All of the major sweet lupinseed markets have low tariff levels.

Utilization of Sweet Lupinseed as a Food

The opportunities for sweet lupinseeds for human consumption, as described in Chapter 12, are diverse. Considerable research is ongoing in this area and the

Table 15.7. Market outlets for Western Australian sweet lupins.

Market		Country
Traditional (current)	Asia	Republic of Korea
		Japan
		Indonesia
		Taiwan
	European Community	The Netherlands
		Belgium
		Germany
		Spain
		Portugal
	Lapsed/spasmodic markets	UK
		Ireland
		Sweden
		Finland
		Israel
Potential or irregular	Middle East	Saudi Arabia
		Yemen
		Jordon
		Syria
	Asia	Singapore
		Malaysia
		Philippines
		People's Republic of China
	Eastern Europe	Former USSR
		Hungary
		Czechoslovakia
		Romania
	Indian Ocean	Mauritius
		Seychelles

most likely long-term application for human consumption is that of a fermentation substrate.

Sweet lupinseed has been used in Indonesia for the manufacture of tempe for more than ten years, albeit at small levels. Future usage will depend on further government approval of the safety of sweet lupinseed as a food in countries outside Australia. Sweet lupinseed has been formally accepted within Australia as being safe for human consumption since 1987. Similar status has been granted in the UK, where lupins are now approved as a 'novel food'. It is likely that further accreditation of sweet lupinseed as a food will be sought in other countries.

Future Production and Varietal Development

With improved farm management and new varieties still under development, the average Australian lupin yield is set to increase over the next decade. The area of lupins will increase rapidly in the wetter southern areas where yield potential

Table 15.8. Tariff rates for Australian sweet lupinseed (ACIL, 1994).

Country	Import tariffs (%)
EU	2.5
China	25
Japan	0
Thailand	30
Korea	3
Saudi Arabia	12
Malaysia	0
Taiwan	0
Philippines	20
Indonesia	10

is high. This will give Western Australia a more geographically balanced production base rather than relying on the traditional northern areas for supply. It would not be unreasonable to suggest that Western Australia could have a million tonnes of lupin seed for export on an annual basis by the year 2000 (ACIL, 1995).

There is reason to believe that the rate of yield increase of most crops has picked up in recent years through plant breeding and from improved management. The pursuit of greater water use efficiency and the adoption of the so-called crop management packages, which involve selecting crops according to soil type and place in the rotation, weed control in the year before cropping, reduced tillage, earlier seeding and more strategic use of nitrogen fertilizer and other chemicals, have been important in shaping this trend (Nelson, 1995).

The future

The Australian cropping environment is such that sweet lupinseed competes for production area with a number of other pulse crops and cereals. Each year producers consider the merits and likely returns of growing lupins against these crops. Two very significant factors that will impact on the future marketing and trade of sweet lupinseed are increasing world population and disposable income, particularly in developing countries.

> In the medium term the world population is expected to grow by over a billion people to a total of more than 6 billion by the year 2005. Almost half of this growth is expected in Asian countries which are also showing high income growth rates. Increased demand for imported feed grains and grain legumes into many of these countries can be expected to meet the increasing demand for livestock production industries as incomes rise.
>
> (ACIL, 1994).

These factors, combined with technological advancement in countries such as China, will provide new market opportunities for lupins in terms of animal feed and food applications. There is likely to be additional potential for market diversification as the Australian and international marketplaces gain a more comprehensive understanding of the attributes of sweet lupinseed. Much of this recognition will be obtained through further research and market development.

Research is currently being conducted into the inclusion of enzymes in prepared poultry feeds, which will decrease faecal moisture and increase energy levels by improving digestibility. Another possible area where sweet lupinseed may find market application is in the aquaculture feed industry. The possibilities for sweet lupinseed incorporation in aquaculture diets is explored in Chapter 13.

Marketing and trade – a world overview

The regional distribution of lupin production on a world basis over the period 1948–1985 is shown in Table 15.9. The regional and country data for lupins on which this table is based serve to highlight fluctuations in world lupinseed production. Interpretation of the data on cultivated areas of production must take into account the fact that in countries such as Poland and the former Soviet Union, a substantial proportion of the lupin grain is for seed stocks for green manure and soil improvement (Watkins Williams, 1986).

World lupin production has fluctuated considerably over the past 40 years with Australia now the dominant sweet lupinseed exporting nation. In Australia the volume of crop and the nature of the marketing systems make monitoring of lupin production by species and marketing and trade activity much easier to monitor and therefore to access statistical information. Australia is the dominant producing and exporting country and this is likely to continue to be the case for the next ten years.

Information on the utilization of lupins by major producing countries is detailed below.

Australia

In addition to the widespread production of *L. angustifolius* in Australia, *L. albus* is also grown but to a much lesser extent. It is estimated that total *L. albus* production in Australia in the 1995–1996 season was around 55,000 t. The commodity can be freely marketed in all Australian states. Approximately half of this tonnage is believed to be destined for human consumption purposes in Egypt and the Middle East, where *L. albus* has long been used as a source of human food. The lupins are predominantly used as a protein source during the holy months of Ramadan – a time of fasting. The export market for *L. albus* for food purposes is currently small, estimated at 20,000 to 25,000 t per annum. Further

Table 15.9. Regional distribution of lupin production 1948–1985 (× 1000 ha).

	1948/1952	1961/1965	1966/1970	1971/1975	1981/1984	1985
USSR	252[1]	607	610	522	310	280[1]
Europe	315	239	185	127	124	148
South Africa	87	185	200	104	20	20
Other African	11	9	5	5	6	7
Latin America	2	4	5	4	10	12
Australia	N/A	2	2	56	242	606
World total	667	1046	1007	818	712	1073

[1]Estimated (Watkins Williams, 1986)

research aimed at investigation of new market opportunities should expand market options in the future. The balance of *L. albus* produced in Australia in the 1995–1996 season was believed to be for livestock feed. There is a possibility of *L. albus* becoming an important crop in world trade if the species becomes well established in either Europe or the USA but this will need further development and the experiences of the development of sweet lupinseed in Australia point to the problems that will be faced.

CIS (formerly USSR)

It is not easy to define the marketing and trade activities associated with lupin production in the former Soviet Union. Many previous authors on the subjects have cited that lupins are used mainly as a green manure in that country.

South America

South American farmers produce a number of lupin species, including *L. angustifolius, L. mutablis, L. luteus* and *L. albus*. It is understood that many of these crops have alkaloid levels in the seed that are much higher than those in Australian sweet lupin cultivars. In Chile, sweet *L. albus* is used for human consumption mainly as an instant coffee substitute and for bread production (Garcia, 1987). Some *L. albus* is also used for animal production. Products from sweet lupins (*L. albus* var. Mutolupa) have been cleared for use as food supplements in milk substitutes for intervention programmes (Aguilera, 1988). An industrial plant has been producing up to 50% of protein requirements for milk substitutes with typical formulations being 68–73% wheat flour, 12% *L. albus* meal, 12% skim milk and 5–6% soybean oil. Other uses are as extrusion – pre-cooked lupin flours and additives in sausages and jams. *L. albus* is also exported from South America as a food to countries such as Egypt, Israel and the Middle East in competition with Australian products. The varieties of *L. albus* produced in South America generally trade at a premium to the Australian product when sold into food markets because of their larger seed size – a trait sought by the market.

Europe

The number of lupin species produced throughout Europe is difficult to track, as are all the lupinseed marketing and trading activities. The lupins are used both as human food and as animal feed, often within the country of production, though there is some internal trade into the European compound feed market.

Conclusion

Australia, as the world's largest exporter of lupins *(L. angustfolius),* is the dominant force in the international marketing and trade of lupins. The bulk of exports have, since the early 1970s, been marketed through a single marketing organization, the Grain Pool of Western Australia. It is through this statutory marketing system that sweet lupinseed has evolved from an unknown product into a highly sought-after food and feed commodity. Grower confidence in marketing opportunities combined with the development and release of new and improved lupin varieties should ensure Australian sweet lupinseed remains a valuable and internationally accepted commodity for many years to come. Export marketing and trade of lupinseed from countries other than Australia is currently limited and likely to remain so in the foreseeable future.

References

ACIL (1994) *A Market Development Strategy for the Australian Lupin Industry. GRDC, Australia*, pp. 1–9.

ACIL (1995) *The Future Production of Grain Crops in Western Australia. GRDC, Australia*, pp. 4–39.

Aguilera, G. (1988) Development of processes and uses of lupins for feed. Vegetable protein utilization in human foods and animal feedstuffs. *American Oil Chemists' Society, USA* 25, 138–142.

Chicago Board of Trade (1990) Future markets and future prices. In: *Understanding Basis – The Economics of Where and When*. Chicago Board of Trade 1, 1–7.

Coffey, R. (1994) International markets and marketing of lupins. In: Dracup, M. and Palta, J. (eds) *First Australian Lupin Technical Symposium*. Agriculture Western Australia, Australia, pp. 235–238.

Cowey, C.B., Mackie, A.M. and Bell, J.G. (1985) *Nutrition and Feeding in Fish*. Academic Press, London, pp. 69–93.

Ford, R. (1994) The international market for lupins. In: von Baer, D. (ed.) *Proceedings of the Sixth International Lupin Conference*. Temuco-Pucon, Chile, pp. 142–153.

Garlinge, J.R., Perry, M.W. and Walton, C.S. (1995) *Pulse Pak*. CLIMA and Agriculture Western Australia, Australia.

Mielke GMBH (1996) *Oil World Annual*. Mielke GmBH, Germany, pp. 19–45.

Morescope (1987) Australian agriculture. In: Cribb, J. (ed.) *The Complete Reference on Rural Industry*. Morescope, Australia, pp. 67–68.

Nelson, P. (1995) The development of the lupin industry in western australia and its role in farming systems. In: *Grain Legume Forum '95: Managing our Future*. Grain Legume Association, Australia, pp. 30–38.

Watkin Williams (1986) The current status of the crop lupins. In: *Proceedings of the Fourth International Lupin Conference*. Agriculture Western Australia, Australia, pp. 1–13.

Index